# Lecture Notes in Computer Science 1557

Edited by G. Goos, J. Hartmanis and J. van Leeuwen

T0188905

**Springer**
*Berlin*
*Heidelberg*
*New York*
*Barcelona*
*Hong Kong*
*London*
*Milan*
*Paris*
*Singapore*
*Tokyo*

Peter Zinterhof  Marian Vajteršic
Andreas Uhl (Eds.)

# Parallel Computation

4th International ACPC Conference
Including Special Tracks on
Parallel Numerics (ParNum'99) and
Parallel Computing in Image Processing,
Video Processing, and Multimedia
Salzburg, Austria, February 16-18, 1999
Proceedings

 Springer

Series Editors

Gerhard Goos, Karlsruhe University, Germany
Juris Hartmanis, Cornell University, NY, USA
Jan van Leeuwen, Utrecht University, The Netherlands

Volume Editors

Peter Zinterhof
Forschungsinstitut für Softwaretechnologie
Hellbrunnerstr. 34, A-5020 Salzburg, Austria
E-mail: Peter.Zinterhof@sbg.ac.at

Marian Vajteršic
Slovac Academy of Sciences, Institute of Mathematics,
Laboratory for Informatics
Dubravska 9, P.O.Box 56, 840 00 Bratislava, Slovakia
E-mail: marian.@iris.savba.sk

Andreas Uhl
Universität Salzburg, RIST++
Hellbrunnerstr. 34, A-5020 Salzburg, Austria
E-mail: uhl@cosy.sbg.ac.at

Cataloging-in-Publication data applied for

Die Deutsche Bibliothek - CIP-Einheitsaufnahme

**Parallel computation** : proceedings / 4th International ACPC Conference Including
Special Tracks on Parallel Numerics (ParNum '99) and Parallel Computing in
Image Processing, Video Processing, and Multimedia, Salzburg, Austria, February
16 - 18, 1999. Peter Zinterhof ... (ed.). - Berlin ; Heidelberg ; New York ;
Barcelona ; Hong Kong ; London ; Milan ; Paris ; Singapore ; Tokyo : Springer,
1999
(Lecture notes in computer science ; Vol. 1557)
ISBN 3-540-65641-3

CR Subject Classification (1998): D.1.3, G.1.0, D.3.2, C.1.2, F.1.2, I.4

ISSN 0302-9743
ISBN 3-540-65641-3 Springer-Verlag Berlin Heidelberg New York

© Springer-Verlag Berlin Heidelberg 1999
Printed in Germany

Typesetting: Camera-ready by author
SPIN: 10692841    06/3142 – 5 4 3 2 1 0    Printed on acid-free paper

# Preface

The Austrian Center for Parallel Computation (ACPC) is a cooperative research organization founded in 1989 to promote research and education in the field of Software for Parallel Computer Systems.

The areas in which the ACPC is active include algorithms, languages, compilers, programming environments, and applications for parallel and high-performance computing systems. The partners of ACPC run research projects in these fields, use a common pool of hardware equipment, and offer a joint curriculum in Parallel Computation for graduate and postgraduate students. Moreover, several national and international workshops and conferences have been organized within the framework of the ACPC.

These proceedings concern the Fourth International Conference of the ACPC (ACPC'99), held on February 16–18 in Salzburg, Austria. This conference is a merge of two established international conference/workshop series devoted to parallel processing: the ACPC conferences which were held previously in Salzburg, Gmunden, and Klagenfurt (all Austria) and the Parallel Numerics (ParNum) workshops which were organized in Smolenice (Slovakia), Sorrento (Italy), Gozd Martuljek (Slovenia), and Zakopane (Poland).

We invited 20 researchers to participate on the program committee. The conference attracted authors from 22 countries around the world who submitted 75 papers, out of which 50 were selected for presentation at the conference. Additionally, a poster session was organized featuring work in progress. Four distinguished researchers presented invited papers with topics related to the two special tracks on Parallel Numerics and Parallel Computing in Image Processing, Video Processing, and Multimedia.

The conference was a result of the dedicated work of a large number of individuals, not all of whom can be mentioned here. In particular, we must acknowledge the reviewers, who formed an international panel and did a competent job in a timely manner. We have to thank also the members of the local organizing committee for their effort in enabling the conference to run smoothly. Finally, we gratefully acknowledge the support of all organizations and institutions sponsoring the conference.

Salzburg, December 1998

Peter Zinterhof
Marian Vajteršic
Andreas Uhl

# Organization

## Program Committee

Chairman:                P. Zinterhof (Univ. of Salzburg, Austria)

Special Tracks Chairmen:    M. Vajteršic (Univ. of Salzburg, Austria and
                                Slovak Academy of Sciences, Slovakia)
                                A. Uhl (Univ. of Salzburg, Austria)

Program Committee Members: P. Brezany (Univ. of Vienna, Austria)
                                L. Böszörményi (Univ. of Klagenfurt, Austria)
                                B. Buchberger (Univ. of Linz, Austria)
                                B. Chapman (University of Southampton, UK)
                                T. Fahringer (Univ. of Vienna, Austria)
                                A. Ferscha (Univ. of Vienna, Austria)
                                S. Grabner (Univ. of Linz, Austria)
                                C. Guerrini (Univ. of Bologna, Italy)
                                G. Haring (Univ. of Vienna, Austria)
                                U. Langer (Univ. of Linz, Austria)
                                L. Montefusco (Univ. of Bologna, Italy)
                                G. di Pietro (IRSIP-CNR Naples, Italy)
                                W. Schreiner (Univ. of Linz, Austria)
                                O. Steinhauser (Univ. of Vienna, Austria)
                                R. Trobec (Jozef Stefan Institute, Slovenia)
                                C. Ueberhuber (Technical Univ. of
                                Vienna, Austria)
                                J. Volkert (Univ. of Linz, Austria)
                                R. Wyrzykowski (Univ. of Czestochowa,
                                Poland)
                                H. Zima (Univ. of Vienna, Austria)
                                W. Zulehner (Univ. of Linz, Austria)

## Organizing Committee

Chairman:                A. Uhl (Univ. of Salzburg, Austria)

Organizing Committee Members: M. Feil, J. Hämmerle,
                                W. Ch. Schmid, M. Traunmüller

# Referees

Each paper was carefully reviewed by at least 2 reviewers and by 2.62 reviewers on average. We thank all reviewers listed below for their valuable work.

| | | |
|---|---|---|
| S. Akramullah | A. Goller | A. Petrosino |
| P. Arbenz | S. Grabner | M. Pic |
| D. Arita | C. Guerrini | E. Pissaloux |
| A. Averbuch | M. Guggisberg | G. Podhájecký |
| K. Balla | J. Hämmerle | I. Podlubný |
| W. Bauer | L. Halada | A. Pommer |
| S. Benkner | S. Hambrusch | P. Purcz |
| A. Bilas | G. Haring | T. Rauber |
| R. Bisseling | Y. He | J. Roman |
| D. Bone | M. Hegland | G. Schaufler |
| R. Borgia | M. Holmström | W. Schreiner |
| L. Böszörményi | T. Hruz | S. Sedukhin |
| P. Brezany | R. Huber | K. Shen |
| A. Broggi | J. Jackson | D. Sparano |
| U. Bueker | D. Jadav | O. Steinhauser |
| H. Bungartz | A. Kiper | W. Stürzlinger |
| B. Chapman | C. Konstantopoulos | K. Tamaru |
| S. Chow | R. Koppler | R. Trobec |
| M. Clint | H. Kosch | P. Tsanakas |
| G.W. Cook | D. Kranzlmüller | C.W. Ueberhuber |
| M. Corvi | G. Larcher | A. Uhl |
| T. Crockett | M. Lea | G. Uytterhoeven |
| P. de Pietro | J. Linhart | M. Vajteršic |
| D. di Serafino | V. Manian | L. van Eycken |
| T. Ertl | S. Manohar | O. Veryovka |
| T. Fahringer | A. Marsh | J. Volkert |
| M. Feil | F. Meunier | I. Vrťo |
| A. Ferscha | L. Montefusco | G. Wesp |
| J. Fridman | A. Murli | R. Wyrzykowski |
| J. Fritts | J. Nabrzyski | D. Xu |
| M. Gabbouj | H. Niederreiter | J. You |
| L. Gatineau | O. Nielsen | E. Zapata |
| B. Gennart | G. Oksa | H. Zima |
| R. Geus | S. Pagnutti | W. Zulehner |
| I. Glendinning | F. Perla | |

# Table of Contents

# Parallel Numerics

## Invited Talks

## Linear Algebra

## Differential Equations and Interpolation

## (Quasi) Monte Carlo Methods

## Numerical Software

## Numerical Applications

# Parallel Computing in Image Processing, Video Processing, and Multimedia

## Invited Talks

# Image Segmentation and Image Understanding

# Motion Estimation and Block Matching

# Video Processing

## Wavelet Techniques

## Satellite Image Processing

# General Aspects of Parallel Computation

## Data Structures

# Posters

# Teraflops Computing:
# A Challenge to Parallel Numerics?

Friedel Hossfeld

John von Neumann Institute for Computing,
Central Institute for Applied Mathematics,
Research Centre Jülich
f.hossfeld@fz-juelich.de

**Abstract.** Following in the wake of the Accelerated Strategic Computing Initiative (ASCI) of the US Department of Energy, in the forthcoming years powerful new supercomputers will be brought into the market by the manufacturers participating in the high-performance computing race. Hence, the large-scale computing facilities in the key research centers and industrial plants world-wide will surpass the teraflops performance barrier, too. The parallel architectures will be further extended to hierarchically clustered parallel computers mainly based on commodity-chip processors and SMP nodes tying together possibly tens of thousands of processing elements. In addition, heterogeneous computing and metacomputing will determine future large-scale computing by interconnecting supercomputers of diverse architectures as giant supercomputer complexes. These developments will challenge not only system reliability, availability and serviceability to novel levels, but also interactivity of concurrent algorithms and, in particular, adaptivity, accuracy and stability of parallel numerical methods.

## 1 Computational Science & Engineering and the ASCI and PACI Impact

Crash simulations are the most intensive supercomputer applications in the automobile industry. In pharmaceutical research and industry, molecular modeling has exploited, and will continue to require, considerable computational capabilities. Aerodynamical flow optimization in car and airplane design is still belonging to the "Grand Challenges" [1]. In recent years, computer simulation has reached even the highest political level, since, in 1996, the United Nations voted to adopt the Comprehensive Test-Ban Treaty banning all nuclear testing for peaceful or military purposes. Banning physical nuclear testing created a need for full-physical modeling and high-confidence computer simulation and, hence, unprecedented steps in supercomputer power, since the Advanced Strategic Computing Initiative (ASCI) of the US Department of Energy (DoE) aims to replace physical nuclear-weapons testing with computer simulations by furthering the simulation technology on a nation-wide basis including the National

P. Zinterhof, M. Vajteršic, A. Uhl (Eds.): ACPC'99, LNCS 1557, pp. 1–12, 1999.

Laboratories in Livermore and Los Alamos and the Sandia Labs, the industry and the universities [2,3]. ASCI is accompanied by the NSF Partnership for Advanced Computational Infrastructure (PACI) centered around NCSA in Urbana-Champaign, Illinois, and SDSC, San Diego, in order to renew the US computing infrastructure by creating a National Technology Grid [4,5]. These initiatives focus much attention and gives terrific technological and scientific impact to an R&D field which developed in parallel with the tremendous increase and ubiquitous distribution of computer capacity over the past five decades: Although born in the 1940s, it has been named "Computational Science" only in the mid-1980s by the Nobel Prize Winner Kenneth Wilson and has been termed recently "Computational Science & Engineering" [6].

Computer simulation has grown and established itself as the third category of scientific methodology. This ever-innovating discipline fundamentally supplements and complements theory and experiment, as the two traditional categories of scientific investigation, in a qualitative and quantitative manner while integrating these into the methodological tripod of science and engineering. Being comparable rather with an experimental discipline, Computational Science and Engineering wastly extends the analytical techniques provided by theory and mathematics; today, in a sense, it is synonymous with investigating complex systems. Its main instrument is the supercomputer; its primary technique is computer simulation. Unsolved complex problems in the areas of climate research and weather forecast, chemical reactions and combustion, biochemistry, biology, environment and ecological as well as sociological systems, order-disorder phenomena in condensed-matter physics, astrophysics and cosmology, quantum chromodynamics, and, in particular, hydrodynamics have been identified as "Grand Challenges".

The various strategic position papers in the 1980s [7,8,9] and the government technology programs in the U.S., in Europe, and in Japan in the early 1990s claimed that the timely provision of supercomputers to science and engineering and the ambitious development of innovative supercomputing hardware and software architectures as well as new algorithms and effective programming tools are an urgent research-strategic response to the grand challenges arising from these huge scientific and technological barriers [10]. Scanning the history since the very birthday of Computational Science and Engineering, which may be dated backs to 1946 when John von Neumann formulated the strategic program in his famous report on the necessity and future of digital computing together with H. H. Goldstine [11], at that time complex systems were primarily involved with flow dynamics. He expected that really efficient high-speed digital computers will "break the stalemate created by the failure of the purely analytical approach to nonlinear problems" and suggested fluid mechanics as a source of problems through which a mathematical penetration into the area of nonlinear partial differential equations could be initiated. John von Neumann envisioned computer output as providing scientists with those heuristic hints needed in all parts of mathematics for genuine progress and to break the deadlock – "the present stalemate" – in fluid mechanics by giving clues to decisive mathematical ideas. In a sense, his arguments sound very young and familiar. As far as fluid mechanics is concerned, in his John von Neumann Lecture at the SIAM National Meeting in 1981 yet Garett Birkhoff came to the conclusion on the development of analytical fluid dynamics that it be unlikely that numerical fluid dynamics would become a truly mathematical science in the near future, although computers might soon rival windtunnels in their capabilities; both, however, would be ever essential for research [12,13,14].

The tripod of science and engineering, thus, has proved to provide scientific research and technology with the stable methodological basis and the instrumental laboratory to effectively approach the solutions of the complex problems which are crucial to the future of science, technology, and society. It will be a crucial factor for the industry in order to meet the requirements of international economic competition especially in the area of high-tech products. Academia in the U.S. was pushing Computational Science and Engineering via a series of important strategic reports and initiatives. Despite the remarkable investments in research centers and universities in building up supercomputing power and skills and also some sporadic efforts in the industry concerning supercomputing in Europe, it took until the 1990s that the U.S. and European as well as national governments started non-military strategic support programs like HPCC, HPCN, and HPSC [15,16,17]. Their goals were also to enhance supercomputing by stimulating the technology transfer from universities and research institutions into industry and by increasing the fraction of the technical community which gets the opportunity to develop the skills required to efficiently access the high-performance computing resources. The ASCI and PACI initiatives will definitely establish computer simulation as a fundamental methodology in science and engineering; and the dedication of the Nobel Prize for Chemistry 1998 to Computational Chemistry will further support its position in the scientific community as well as in industry and politics.

## 2    Responses from Computer Technology

For the first time in computing history, we are able today to build a balanced pyramid of computing power in scientific and technical computation [18]. Whereas local area and wide area network systems with medium speed and bandwidth have been built up almost everywhere in research institutions and universities, and high-speed communication with broadband functionality is promoted in the U.S. on a large scale for scientific as well as commercial applications and also in some european countries strong efforts are made to provide the scientific community with broadband communication services, other european countries are still quite far from having access to broadband communications. There is consensus that the backlash in high-speed communications is a severe barrier to establishing a highly efficient nation-wide infrastructure which is capable to provide supercomputer capacity and functionality to the scientific community on a modern scale with transfer opportunities into the industry [19].

The lack of high-speed communications is certainly an important reason for the retardation of the high-performance computing technology in these countries. In Germany, although still very expensive compared to the US, due to the beneficial activities of the German Research Net (DFN) the Broadband Science Network, B-WiN, is providing communication bandwidths up to 155 megabits per second. For the year 2000, the B-WiN is projected to migrate to gigabit per second bandwidth; the DFN Gigabit Testbeds South and West are pilot projects towards the future G-WiN backbone.

Strategically, for John von Neumann flow-dynamical phenomena have been the primary field where future efforts should have been invested to develop and establish the digital computer – as the "digital windtunnel" – and, thus, by utilizing numerical methods, activate the mathematical penetration of the whole area of partial differential equations. Since then and up to now partial differential equations have been dominating in the advancement of high-speed computers and in the exploitation of their potential. The general solution methodology for such equations leads via discretization of space and time and via linearization into linear algebra and its numerical concepts and algorithms. The response of computer architecture to these early challenges of PDEs have been the vectorcomputers optimizing vector-pipeline processing and creating the effective instruments of vectorization [20].

The exploration of the computing potential of the pipelining principle including programming and compiler techniques, tools, operating system functionality, and shared-memory organization and optimization resulted in the efficient arsenal of knowledge and experience about the strengths and the weaknesses of vectorcomputing. The highest class of vectorcomputers, e.g. the Cray, Fujitsu, and NEC systems, are still dominating many of the supercomputing production environments and the practice of Computational Science and Engineering [21]. Certainly, vectorcomputers will further develop in functionality and performance towards hundreds of gigaflops by exploiting the architectural and technological potential and expanding the "weak" parallelism well beyond hundred processors.

Although today the sustained performance of these systems, e.g. the NEC SX-4, Fujitsu VPP-700 and even CRAY T90, turns out to be competitive with mid-sized massively parallel systems for a wast majority of essential algorithms as well as large applications, the technological progress is tending to replace vectorcomputers as the very workhorses of Computational Science and Engineering by massively parallel computers. But workstations, however powerful they are or will become, cannot replace the potential of parallel computers which are basically built upon this technology of powerful microprocessor chips by tying them together via sophisticated broadbanded interconnection networks in order to support massive parallelism. Massively parallel computers are therefore undoubtedly considered as the – only – remedy to the needs of the demanding applications in the Grand Challenge category and maybe yet unrecognized applications which might soon emerge, for instance, from the expanding multimedia field.

Unfortunately, in the early 1990s the manufacturers of massively parallel systems promised that they would be capable to develop and deliver parallel supercomputers in 1995 which be able to reach the magical "3 T's" (i.e. 1 Teraflops in execution rate, 1 Terabyte in main memory, and 1 Terabyte/s interconnection bandwidth), thus indicating a revolutionary, rather than evolutionary, step of almost three orders of magnitude beyond the then state-of-the-art supercomputer performance. During recent years, nearly thirty companies were offering massively parallel systems and others were planning to enter the market with new products, although many experts predicted that the market will not be able to sustain this many vendors [22]. In the meanwhile, the expected shake-out in the computer industry takes place questioning the health and the future potential of this industry in total. Some went out of the parallel computer business – for quite different reasons – , others became just mergers. The dramatic survival battle in the supercomputer industry is also giving severe damage to the users in the supercomputing arena. Their investments into massively parallel computing may be definitely lost from time to time and the establishment of a new hardware and software platform will require new investments concerning finances and manpower as well as psychological recovery from the frustration caused by unfulfilled soap-bubble promises.

The critical situation of parallel computing has rigorously been analyzed [23] with respect to the possible negative impacts on the future perspectives and the progress of this scientific discipline but also on the support which will be expected and requested from the politicians. The report states that "the history of computing is littered with failed long-term predictions"; it is right in claiming honest answers from the supercomputing arena to some burning questions on the seriosity of predictions concerning the reachability of the goals set in particular in the context of those national research initiatives.

A key issue in massively parallel computing is scalability. Parallelizing "dusty" decks from industry is certainly an important task to do in order to increase the acceptance of parallel computing in commercial environments. However, one cannot expect terrific performance gains in many of these programs from porting such originary sequential, in many cases also organically grown, codes to parallel systems. Therefore, scalability often breaks down when the number of parallel nodes is increased beyond sixty-four, sixteen or even eight. Even the benchmark results on naked algorithmic kernels stress the limiting factor of scalability. There is a big discrepancy between the peak rates of massively parallel systems and the sustained performance which can be reached with algorithmic kernels and, even more significantly, with real application programs and software packages [24]. With kernels, the state of the art of massively parallel computers may still deliver, together with a pretty large variance in performance, not more than 10% of the peak rate as sustained performance. This is disappointing. Since, so far, the microprocessor chips have been developed with a different market goal in mind, it is extremely difficult to exploit the performance hidden in the hardware design of these processors via high-level programming languages and compiler techniques [25]; very often this leads to a loss by a factor of five to ten referred to peak performance of the node [26,27]. But it cannot be accepted as a reasonable software-technological approach to switch back to the very old times of assembler programming to reach reasonable performance levels. Convergence of hardware and compiler design together with the development of valuable programming tools must become the future development strategy.

Another important issue is programming models [28]. While Message Passing – with MPI – is widely and effectively used on distributed memory systems as the only efficiently implemented programming paradigm at present, one can hardly imagine that this programming model will carry all future efforts to introduce massively parallel computing as the overwhelming technology; especially large non-scientific applications will certainly suffer from this obstacle of explicit programming the data communication in message-passing style. Up to now, programming and software technology not only relied on sequential machines, but also on the shared-memory organizational concept. The dominant language in scientific programming has been Fortran with all its strengths and weaknesses. On this language basis, the High Performance Fortran (HPF) seems to be an at least temporary platform to implement parallel applications. Urgent extensions towards more functionality are scheduled for HPF-2. There also seems to arise some renaissance of Cray's Multitasking by the OpenMP programming model [29].

Despite the failure of the first commercially available massively parallel computer system which supported the programming paradigm of the Shared Virtual Memory (SVM), the efforts to explore this programming model should be transferred into the product strategies of manufacturers. From a user's point of view this SVM paradigm [30] seems to carry enough potential to overcome fundamental deficiencies which can be experienced with the Message-Passing paradigm, because SVM hides the physical local memories from the programmer and provides a virtual address space organized in pages which demand-driven move across the parallel processors. The advantage of this paradigm may be especially comfortable for the programmer if dealing with irregular data which are inherent in many scientific and engineering applications like, for instance, in the finite-element methods to treat partial differential equations [31]. Since in many cases data access and communication patterns are unknown prior to the parallel execution, data with efficient domain decomposition or applications with remeshing necessities cannot be realized in advance as is required by the Message-Passing model. However, this needs powerful tools to monitor the progress in the parallelization process and strong support on the hardware level which cannot be seen to be available in the near future due to the present lack of interest in this SVM paradigm on the manufacturers' side. In any case, together with genuinely parallel algorithms, powerful and user-friendly programming tools as well as performance-monitoring capabilities are key issues, too [32].

## 3   System Aspects of Beyond-Teraflops Computing

The experiences with the strengths and weaknesses of the different architectures available for supercomputer applications in Computational Science and Engineering – the technological obstacles for major performance steps in vector-computing, the large variance in performance for algorithms on different parallel machines, and the very low average sustained performance in massively parallel processing relative to the peak rate, the present or even fundamental limitations to the scalability of systems and applications to reach and exploit massive parallelism – quite naturally lead to the concept of heterogeneous computing which requires the coexistence and cooperation of the different computer architectures. In heterogeneous computing [33,34,35], the computational work of – parallel – programs can be split across different computers in order to achieve in total the fastest possible execution, where the individual portions of the work are sent to those computer systems in the heterogeneous ensemble which have been proved to be best for the specific characteristics of the work. This approach could generate results much faster than would be possible on any single system. It also might simplify the programming effort, since program components can be developed using diverse software environments which usually are not available on all machines. Heterogeneous computing is an attractive concept because it takes into account that the individual parallel machines, and vectorcomputers as well, spend much of their time on tasks for which they are unsuited. These effects lead to reduced performance, and also to scalability problems.

On the other hand, it is well known that a user generally invests tremendous efforts in order to extract even small improvements of performance out of an innovative computer system for his specific application, well knowing that the application principally implies a spectrum of heterogeneous requirements which cannot be efficiently satisfied by the single target system he is focussing on just because it is available to him in his specific computing environment. Since the performance of the known supercomputer architectures is a function of the inherent structures of the computations and the data communications involved, it is necessary to discriminate among types of code, algorithms, data, and communications in order to optimize the mapping of tasks onto computer structures.

Researchers in the field of innovative computing believe that there will be no single all-encompassing architecture which will be capable to satisfy heterogeneous requirements with equally optimal performance. Hence, the goal of heterogeneous computing is the efficiency of computation and thereby the effectiveness and cost-effectiveness of both computers and programmers. The price to pay are again grand challenges to respond by hardware and software as well as network designers. Certainly, cumbersome administrative and accounting problems involved in this cross-regional and inter-institutional computing concept will retard the potentiality of getting heterogeneous computing into efficient use. It becomes clear that high-speed networking is the fundamental technical requirement of heterogeneous computing on the way to metacomputing, which brings us back to broadband data communications as the very basis of the "technological pyramid of scientific computing", as discussed earlier.

Despite of taking metacomputing between the supersystems of the National Labs Los Alamos, Livermore, and Sandia into account according to the ASCI-Pathforward plans, the requirements of the ASCI program reach far beyond the available technology and architectures [2,3]. Also, foreseeable trends in developing and implementing supercomputers fall well below the ASCI requirements. Therefore, the supercomputing centers all over the world wait for the ASCI machines to get transformed into market so that they can benefit from the technology jumps in high-end computing achieved within the framework of the ASCI program to harness compute-based modeling, simulation, and virtual prototyping. The ASCI goal is to create the leading-edge computational capabilities. Thus, ASCI requests for near-time performance in the 10-to-30 Teraflops range in the late 1999 to 2001 timeframe, and for future supercomputer developments enabling 100 Teraflops platforms in the 2004 timeframe. The Initiative's applications require a threshold shift of 100 to 1000 times increase in computing capability in order to meet the mission target. The aggregation of new – mainly commodity-based – building blocks for massively parallel supercomputers will challenge significant endeavours of integration and scaling technologies which are not currently driven by commercial markets. Therefore, ASCI is undergoing partnerships with various US manufacturers in order to accelerate the development of the supercomputers required.

As is outlined in the ASCI program, achieving balanced systems at the 10 to 100 Teraflops scale will place stringent requirements on the processor power, the node architecture, the internode interconnect, the I/O systems, and the storage subsystems. Balanced ASCI systems are estimated to scale according to the following approximate ratios:

| | |
|---|---|
| 1 Teraflops | peak performance/ |
| 1 Terabyte | memory size/ |
| 50 Terabyte | disk storage/ |
| 16 Terabyte/s | cache bandwidth/ |
| 3 Terabyte/s | memory bandwidth/ |
| 0.1 Terabyte/s | I/O bandwidth/ |
| 10 Gigabyte/s | disk bandwidth/ |
| 1 Gigabyte/s | archival storage bandwidth/ |
| 10 Petabyte | archival storage. |

The concept includes the following key attributes: multiple high-performance commodity priced compute nodes, which represent regular commercial product lines and not special-purpose designs; hierarchical memory systems, including cache-only memory architectures and distributed shared memory systems with low-latency high-performance memory access; very high performance storage and parallel I/O systems, scalable programming environments and operating systems; a universal programming paradigm; and much more.

## 4 Meet the Needs of CS&E: Requirements Revisited

It is not obvious that, except for the vendors' hardware, the diversity of results of these ASCI developments will be easily and timely available worldwide to the non-ASCI scientific community. Therefore, whenever supercomputer centers outside the ASCI community expect to benefit from these forecast performance steps, significantly enhanced software for distributed operating systems as well as programming environments, tools, and libraries have to be developed and provided in order to enable the users to participate in these technological achievements. Thus, national software initiatives outside ASCI will be urgent to get started now if the emerging technological gap in simulation capabilities and capacities between the ASCI community and the rest of the world shall be kept as narrow as possible. Unfortunately, so far no signs of preparing a response to the ASCI program can be recognized, for instance, in the european countries. There is no doubt that the new level of supercomputing then will put significant pressure also on the numerical methods.

Besides heterogeneous computing and metacomputing, the complex applications in the ASCI program, but also in the other innovative scientific and industrial environments which rely strongly on Computational Science & Engineering and, thus, on supercomputing, require significant upscaling of massively parallel architectures [36]. The development of hierarchical parallel computers with clustered processing elements, e. g. SMP nodes, is on its way. To meet the ASCI performance requirements, the 10-to-30 Teraflops machines in the late 1999 to 2001 timeframe will need the interconnection of more than 10,000 processing elements, or equivalently of the order of 50 to 1000 compute nodes consisting of 256 to, respectively, 8 parallel processing elements.

With this upscaling, the user will have to face severe problems of drastically reduced system stability and reliability [37]. If it is assumed, referring only to hardware failures, that today the mean time between interrupt (MTBI) of such a (SMP) cluster of 128 processing elements, for instance, is around 500 hours and the total system consists of 100 clusters, the overall MTBI will be 500/100 = 5 hours. If a processor-memory module yields a mean time between failure (MTBF) of three years, the MTBF of the whole ensemble of 10,000 modules will end up with 3/10,000 years which corresponds to 2.6 hours. Thus, the effective capability of a teraflops ($10^{12}$ operations per second) computer is limited to an uninterrupted run of these few, say in the average three, hours; hence, a corresponding simulation will involve only about $10^{16}$ operations which is not much measured at the ASCI criteria of 3D real-physics applications.

The situation is getting worse when programming and other software bugs are taken into account, as has been awfully experienced in many large computational projects. Therefore, in order to exploit the performance of such a system, the software – and this means also the numerical as well as non-numerical methods – will have to provide very sophisticated check-point and restart mechanisms. Intelligent hot-swap facilities will be crucial in order to cope with those interrupts and failures from the system side to avoid big capacity losses; new redundancy concepts will have to take care of these technological deficiencies.

Dealing with Computational Science & Engineering problems of ASCI or other Grand Challenge scales, one has to face the fundamental responsibility to verify the results of numerical computer simulations, since these will be used as replacements of experimental explorations as in the ASCI context, but also in other areas like biochemistry, combustion, or crash tests. It can be rigorously shown how easy a numerically illiterate user can be trapped by the error pitfalls of brute force numerical simulations [37,38]. One can also experience the low level of knowledge about numerical analysis and of numerical technical skills of computer scientists who nevertheless enter the field of Computational Science & Engineering. This refers to the issue of education; therefore also university curricula are challenged.

## 5   Support to CS&E Competence "Grids"

But, finally, the new parallel computers seem to become a grand challenge for numerical mathematics itself.As has been already mentioned, governmental support programs are lacking today in order to face the grand challenges of methodological research and development in parallel software and parallel numerical methods and non-numerical algorithms as well; there is an urgent need to start an offensive development program. Its goal will not only be to explore fundamental structures in algorithms and numerical methods, which is undoubtedly necessary, too, if we consider the relatively poor results of just porting sequential methods and programs onto parallel computers. But the support program should also create the capabilities to transfer the results without delay into real and operating implementations on the parallel computer platforms emerging at the horizon.

In order to succeed in these efforts, the vitalization of interdisciplinary competence networks is a prerequisite. In this sense, the Grid as established by the US National Science Foundation within the framework of its PACI program can be a guiding model also for other countries, or even regions like Europe. The interconnection and tight cooperation of Supercomputer Centers would certainly enhance such efforts because these centers are acting as active crystallization kernels and attractors within the national or regional competence structures [39]. Thus, their interconnection and cooperation pattern is well suited to provide a stable skeleton for the growth of an interacting competence network for Computational Science & Engineering.

## References

1. Special Double Issue: *Grand Challenges to Computational Science*, Future Generation Computer Systems 5 (1989), No. 2&3.
2. ASCI: www.llnl.gov/asci/.
3. ASCI-Pathforward: www.llnl.gov/asci-pathforward/.
4. L. Smarr, *Toward the 21st Century*, Comm. ACM 40(1997), No. 11, 28-32. - Ph. L. Smith, *The NSF Partnerships and the Tradition of U.S. Science and Engineering*,

Comm. ACM 40(1997), No. 11, 35-37. - D. A. Reed et al., *Distributed Data and Immersive Collaboration*, Comm. ACM 40(1997), No. 11, 39-48. - R. Stevens et al., *From the I-Way to the National Technology Grid*, Comm. ACM 40(1997), No. 11, 51-60. - K. Kennedy et al., *A Nationwide Parallel Computing Environment*, Comm. ACM 40(1997), No. 11, 63-72. - G. J. McRae, *How Application Domains Define Requirements for the Grid*, Comm. ACM 40(1997), No. 11, 75-83. - J. P. Ostriker and M. L. Norman, *Cosmology of the Early Universe Viewed Through the New Infrastructure*, Comm. ACM 40(1997), No. 11, 85-94.

5. I. Foster and C. Kesselman (eds.), *The Grid: Blueprint for a New Computing Infrastructure*, Morgan Kaufmann Publishers, San Francisco, 1999

6. Rice, J. R.: IEEE Computational Science & Engineering 1(1994), 13

7. Committee on Physical, Mathematical, and Engineering Sciences, Federal Coordinating Council for Science, Engineering, and Technology, Grand Challenges 1993: *High Performance Computing and Communications, The FY 1993 U.S. Research and Development Program*, Office of Science and Technology Policy, Washington, 1992.

8. Board on Mathematical Sciences of the National Research Council (USA), *The David II Report: Renewing U.S. Mathematics - A Plan for the 1990s*, in: Notices of the American Mathematical Society, May/June 1990, 542-546; September 1990, 813-837; October 1990, 984-1004.

9. Commission of the European Communities, *Report of the EEC Working Group on High- Performance Computing* (Chairman: C. Rubbia), February 1991.

10. U. Trottenberg et al., *Situation und Erfordernisse des wissenschaftlichen Höchstleistungsrechnens in Deutschland - Memorandum zur Initiative High Performance Scientific Computing (HPSC)*, Februar 1992; published in: Informatik-Spektrum 15 (1992), H. 4, 218.

11. J. von Neumann and H. H. Goldstine, *On the Principles of Large Scale Computing Machines*, in: Collected Works Vol. V, Pergamon Press, 1963, p. 1-32.

12. G. Birkhoff, *Numerical Fluid Dynamics*, SIAM Review 25 (1983), 1.

13. K. Kuwahara, *Flow Simulation on Supercomputers and its Visualization*, Intern. J. High Speed Computing 4 (1992), No. 1, 49.

14. K. Hwang, *Advanced Computer Architecture: Parallelism, Scalability, Programmability*, McGraw-Hill, 1993.

15. The Congress of the United States, Congressional Budget Office, *Promoting High-Performance Computing and Communications*, Washington, June 1993.

16. High-Performance Computing Applications Requirements Group, *High-Performance Computing and Networking*, Report, European Union, April 1994. - High-Performance Networking Requirements Group, Report, European Union, April 1994.

17. Bundesministerium für Forschung und Technologie, *Initiative zur Förderung des parallelen Höchstleistungsrechnens in Wissenschaft und Wirtschaft*, BMFT, Bonn, Juni 1993.

18. NSF Blue Ribbon Panel on High Performance Computing, *From Desktop To Teraflop: Exploiting the U.S. Lead in High Performance Computing*, Report, NSF-CISE, 19 October 1993.

19. R. Bayer, *Plädoyer für eine Nationale Informations-Infrastruktur*, Informatik-Spektrum 17 (1994), H. 5, 302.

20. F. Hossfeld, *Vector-Supercomputers*, Parallel Computing 7 (1988), 373.

21. J. J. Dongarra, H.-W. Meuer, and E. Strohmaier (eds.), *TOP500 Supercomputer Sites*, 11th Edition, University of Mannheim, RUM 54/98, and University of Tennessee, UT-CS-98-391, June 18, 1998.

22. The Superperformance Computing Service, Palo Alto Management Group, Inc., *Massively Parallel Processing: Computing for the 90s*, SPCS Report 25, Second Edition, Mountain View, California, June 1994.

23. J. R. Gurd, Supercomputing: *Big Bang or Steady State Growth?*, Keynote Address, 7th International Conference on Supercomputing, ICS '93, Tokyo, in: ACM Computer Architecture News 22 (1994), No. 3, 3.

24. D. H. Bailey et al., *NAS Parallel Benchmark Results 3-94; NAS Parallel Benchmarks Update 9-94*, RNR Technical Report RNR-94-006, NASA Ames Research Center, Moffett Field, California, March and September 1994.

25. H. Zima and B. Chapman, *Supercompilers for Parallel and Vector Computers*, ACM Press, 1990.

26. A. Mlynski-Wiese, *Leistungsuntersuchung des iPSC/860 RISC-Knoten-Prozessors: Architekturanalyse und Programmoptimierung*, Zentralinstitut für Angewandte Mathematik, Berichte des Forschungszentrums Jülich Jül-2766, Mai 1993.

27. A. Krumme, *Performance-Analyse des DEC Alpha-Chip: Moderne Compiler-Techniken zur Programmoptimierung*, Zentralinstitut für Angewandte Mathematik, Berichte des Forschungszentrums Jülich Jül-2912, Mai 1994.

28. D. B. Skillicorn and D. Talia, *Models and Languages for Parallel Computation*, ACM Computing Surveys 30(1998), No. 2, 123-169.

29. L. Dagum and R. Menon, *OpenMP: An Industry-Standard API for Shared-Memory Programming*, IEEE CS&E 5(1998), No. 1, 46-55.

30. K. Li, *Shared Virtual Memory on Loosely Coupled Multiprocessors*, PhD Thesis, Yale University, September 1986.

31. R. Berrendorf, M. Gerndt, Z. Lahjomri and Thierry Priol, *A Comparison of Shared Virtual Memory and Message Passing Programming Techniques Based on a Finite Element Application*, in: B. Buchberger and J. Volkert (eds.), Parallel Processing: CONPAR 94 - VAPPVI, Lecture Notes in Computer Science Vol. 854, Springer Verlag, 461-472.

32. W. E. Nagel and A. Arnold, *Performance Visualization of Parallel Programs: The PARvis Environment*, Proceedings 1994 Intel Supercomputer Users Group (ISUG) Conference, 24-31; Technical Report CCSF-47, Concurrent Supercomputing Consortium, Pasadena, California, May 1994.

33. Ch. Pancake, *Area Editor's Message: High-Performance Computing - Lighting the Way*, IEEE Computer 26 (1993), No. 6, 6.

34. R. F. Freund and H. J. Siegel, *Guest Editors' Introduction: Heterogeneous Processing*, IEEE Computer 26 (1993), No. 6, 13.

35. A. A. Khokhar et al., *Heterogeneous Computing: Challenges and Opportunities*, IEEE Computer 26 (1993), No. 6, 18.

36. L. Smarr, *Supercomputers: Directions in Technology, Architecture and Applications*, in: H. - W. Meuer (Hrsg.), *Supercomputer 1998, Anwendungen, Architekturen, Trends*, K. G. Saur, München, 1998, p. 2-16.

37. J. Gustafson, *Computational Verifiability and Feasibility of the ASCI Program*, IEEE CS&E 5(1998), No. 1, 36-45.

38. A. B. Tucker Jr. (ed.), *The Computer Science and Engineering Handbook*, CRC Press, Boca Raton, 1997.

39. F. Hossfeld, *Verbund der Supercomputer-Zentren in Deutschland - Ansichten, Einsichten, Aussichten -*, in: H.-W. Meuer (Hrsg.), *Supercomputer 1998, Anwendungen, Architekturen, Trends*, K. G. Baur, München, p. 160-171.

# Non-standard Parallel Solution Strategies for Distributed Sparse Linear Systems*

Yousef Saad[1] and Maria Sosonkina[2]

[1] Department of Computer Science and Engineering, University of Minnesota,
Minneapolis, MM 55455, USA,
**saad@cs.umn.edu**,
**http://www.cs.umn.edu/ saad**
[2] Department of Computer Science, University of Minnesota, Duluth,
320 Heller Hall, 10 University Drive, Duluth, Minnesota 55812-2496,
**masha@d.umn.edu**.

**Abstract.** A number of techniques are described for solving sparse linear systems on parallel platforms. The general approach used is a domain-decomposition type method in which a processor is assigned a certain number of rows of the linear system to be solved. Strategies that are discussed include non-standard graph partitioners, and a forced load-balance technique for the local iterations. A common practice when partitioning a graph is to seek to minimize the number of cut-edges and to have an equal number of equations per processor. It is shown that partitioners that take into account the values of the matrix entries may be more effective.

## 1  Introduction

Recent years have seen a maturation of parallel processing to a point where the methodology is now beginning to enter many engineering and scientific applications. The innermost part of these applications often requires the solution of large sparse linear systems of equations. The most common architecture used is that of a distributed memory computer, using MPI for message passing. The most natural process for solving Partial Differential Equations and sparse linear systems on distributed memory computers is to employ strategies based on domain decomposition. A typical finite element simulation for example, requires the following steps: (1) The physical mesh is generated, typically on one processor; (2) The mesh is partitioned using a number of publically available tools; (3) The element matrices and right-hand sides are generated in each processor independently; (4) Finally, a solution process, typically based on iterative methods, for the resulting distributed system is undertaken. This course of action seems natural and straightforward. It comes, however, with a few challenging questions.

---

* Work supported by NSF under grant CCR-9618827, and in part by the Minnesota Supercomputer Institute.

P. Zinterhof, M. Vajteršic, A. Uhl (Eds.): ACPC'99, LNCS 1557, pp. 13–27, 1999.

The first of them is related to partitioning. What partitioning approaches will lead to the best overall performance of the solver? Most current partitioners will simply divide up the graph aiming at obtaining about the same number of points in each processor and at reducing the number of edge cuts. A number of heuristics have been developed with this strategy in mind, see e.g., [10, 7, 13]. However, it is easy to imagine that this is far from perfect. First, if load balancing is the main criterion (ignoring communication for a moment) then clearly, the number of points assigned to each processor is not a good measure. One can imagine, for example, a technique based on attempting to equalize the time spent on matrix-vector products in which case, an algorithm that would distribute edges rather than vertices equally would be more appropriate. These two strategies may lead to a similar distribution in many cases, but not always. Another rather complex issue is that the partitioning can affect the quality of the preconditioning in a way that is hard to predict. We may obtain a partitioning that has perfect load balance and a minimal number of cut-edges but which may lead to an unacceptable increase in the number of iterations. This happens almost systematically when the matrix arises from highly discontinuous problems.

This paper illustrates these issues by devising a number of alternative strategies that can be used for partitioning a graph and reducing idle time during an iteration. The intent is not to develop a new general method but rather to show that a few alternative approaches can lead to effective solution methods.

## 2   Graph Partitioning Concepts

The very first task that a programmer faces when solving a problem on a distributed memory parallel computer, be it a dense or a sparse linear system is to decide how to map the data into the processors. We call a *map* of $V$, any set $V_1, V_2, \ldots, V_s$, of subsets of the vertex set $V$, whose union is equal to $V$:

$$V_i \subseteq V , \quad \bigcup_{i=1,s} V_i = V.$$

When all the subsets $V_i$ are not pairwise disjoint, the term partition conflicts with common usage, but we will use the term overlapping partition in this case. The most general way of describing a node-to-processor mapping is to set up a list, containing all the nodes that are mapped to each processor. Thus, in the example shown in Figure 1, the list $\{1, 2, 5, 6\}$ is assigned to Processor 1, the list $\{3, 4\}$ is assigned to Processor 2, the list $\{7, 8, 11, 12\}$ is assigned to Processor 3, and the list $\{9, 10\}$ is assigned to Processor 4. Another representation which is sometimes useful is to set-up an array which lists for each node the processor to which it belongs. This is important when setting up the local data structure in the preprocessing phase of domain decomposition type algorithms [16].

There has been a flurry of activity in finding good partitionings of graphs. Among the most popular techniques are the spectral bisection method [13] and Metis [10]. These methods attempt to provide a partition that will have good load balancing and a small number of edge cuts. A few algorithms to accomplish this have been described in [1, 6, 13, 12, 4, 7, 10].

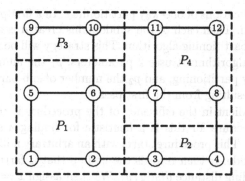

**Fig. 1.** Mapping of a simple $4 \times 3$ mesh to 4 processors.

## 2.1  A Few Graph-Based Partitioners

The Breadth-First-Search (BFS) algorithm is at the basis of many partitioning techniques, including those to be described in this paper. BFS is essentially a level-by-level traversal of a graph. It starts with a given vertex $v_0$ which will constitute the level 0. Then all nodes adjacent to $v_0$ will be visited – and these constitute level 1. In the third step, all non-visited vertices that are adjacent to vertices belonging to level 1, will be visited and they will constitute Level 2, etc.

We note that BFS does need not start with a single node. We can, for example, start with a known level, i.e., a set of nodes forming a string of connected nodes. The one-way partitioning algorithm simply traverse the nodes in the Breadth-First-Search order and assigns nodes to partitions until a given number of assigned nodes is reached.

ALGORITHM 21 *One-way-partitioning using level-sets*

1. *Input: nlev, levels, ip (number of nodes per partition)*
2. *Output: ndom, node-to-processor list.*
3. **Start:** *ndom = 1; siz = 0;*
4. **Loop:** *For lev =1, nlev Do*
5.      *For each j in levels(lev) Do*
6.          *add j to dom(ndom); siz = siz+1;*
7.          *If (siz .ge. ip) then*
8.              *ndom = ndom+1; siz = 0;*
9.          *EndIf*
10.      *EndDo*

In an actual implementation, some care must be exercised to obtain equal sized domains, and to avoid having a last subdomain consisting of the points that are left-over in the very last step. For example, if $n = 101$ and $ip = 10$, we will have 10 subdomains of size 10 each and one of size one. This can be prevented by adjusting the ip parameter as we proceed. Two-way partitioning consists of two applications of one-way partitionings. In a first pass we determine a one-way partition. Then each of the subdomains created is again partitioned using the one-way partitioning algorithm. This strategy will be referred to as the "double-striping" algorithm. It uses 2 parameters: $p_1$ the number of partitions in the first one-way partitioning, and $p_2$ the number of sub-partitions in each of the $p_1$ partitions resulting from the first pass.

A critical ingredient in the efficiency of the procedure is the starting node. A well-known procedure for this is a heuristic for finding a so-called pseudo-peripheral node [5]. This procedure starts with an arbitrary node $x$ and performs a BFS from this node. It then records the node $y$ that is farthest away from $x$ and the corresponding distance $dist(x, y)$. We then assign $x := y$ and perform a new traversal from $x$. We repeat the process and stop when $dist(x, y)$ does not vary between two successive traversals.

## 2.2  Level-Set Expansion Algorithms

As was noted earlier, the Breadth First Search traversal can start from several independent vertices at once instead of one node only. Level-set expansion algorithms consist of building the subdomains by a using a BFS traversal from a number of centers. Thus, these algorithms consist of two phases. The first phase finds 'center' nodes for each partition from which to expand each subdomain. Ideally, these centers are vertices that are far apart from one another, in a graph theory sense. In the second phase these points are used to generate the subdomains using a BFS expansion from them.

Next, we present a version of the algorithm that is sequential with respect to the domains but it is clear that the process is inherently parallel.

ALGORITHM 22 *Level-set-expansion*

| | |
|---|---|
| 1. | **Start:** |
| 2. | *Find an initial set of 'coarse mesh' vertices 3. $v_1, \ldots, v_{ndom}$* |
| 4. | *For $i = 1, 2, \ldots, ndom$ Do label$(v_i) := i$.* |
| 5. | *Define levset $:= \{v_1, \ldots, v_{ndom}\}$ and nodes $= ndom$* |
| 6. | **2. Loop:** *While (nodes $<n$) Do* |
| 7. | *Next_levset $= \phi$* |
| 8. | *For each $v_j$ in levset Do* |
| 9. | *for each neighbor $v_k$ of $v_j$ s.t. label$(v_k) = 0$ Do* |
| 10. | *Next_levset $:= $ Next_levset $\bigcup \{v_k\}$* |
| 11. | *label$(v_k) := $ label$(v_j)$* |
| 12. | *nodes $= $ nodes $+ 1$* |
| 13. | *EndDo* |
| 14. | *EndDo* |
| 15. | *levset $:= $ Next_levset* |
| 16. | *EndWhile* |

The algorithm starts with one node in each processor then expands by adding level sets until all points are labeled. The indicator that is assigned on each node to indicate whether or not the note has already been visited is now a label and the vertices will inherit the labels of their parents in the traversal process. At the end, all nodes having the same label will constitute a subdomain. We must assume here that the initial graph is connected or that there is at least one starting node in each connected component. The parallel version of this algorithm consists of assigning each starting node to a different processor, then expanding the level sets independently. At some point there will be conflicts, i.e., two processors will attempt to 'acquire' the same node which belongs to two level sets originating from two different starting nodes. In such cases, the host program must arbitrate. Our current implementation uses a first-come first served rule, but there are several possible improvements which are not considered here.

The next question is how to find the center points. There are at least three possible options. First, if a coarse mesh is already available from the discretization then the nodes of this mesh can be taken as the centers. If a coarse mesh is not available we can alternatively use points provided from a two-way partitioning algorithm. Second, if the coordinates of the nodes are available then one can easily select the centers from simple geometrical considerations. For example, for a rectangular 2-D mesh, we can choose points that are uniformly distributed in each direction. Other alternatives are required for the cases where only the graph is known and coordinate information is not available.

In general, the two-way partitioning algorithm does not provide as good a splitting of the graph as some of the well-known alternatives such as the Recursive Spectral Bisection technique [13]. It is, however, rather inexpensive to obtain. As a result, we can use this partitioning only to get the centers for the Level-Set Expansion. For example, we can simply take the middle node in the subdomain as a center. The resulting partitioning will be far better than the original two-way partitioning in general. An illustration of the process is given in Figure 2.

## 2.3   Partitioning Strategies Using Shortest Path Methods

Partitioning is a form of reordering and as such it may have an important effect on the quality of partitioning. The purpose of this section is only to illustrate this point and propose, not a solution, but some general guidelines toward developing effective partitioners for iterative solvers.

The key point is to take into account the matrix values when partitioning the problem. This can be done using weights on edges. The weight values are based on the absolute magnitudes of the matrix entries associated with the edges. In particular, if $e_{ij}$ is the edge connecting vertices $i$ and $j$, then the weight $w(e_{ij})$ of this edge is $1/(1 + |a_{ij}| + |a_{ji}|)$, where both matrix entries $a_{ij}$ and $a_{ji}$ are associated with edge $e_{ij}$ under the assumption that the matrix is structurally symmetric. The hypothesis used here is that keeping together those nodes which are strongly coupled would lead to a better preconditioning. Weights have been

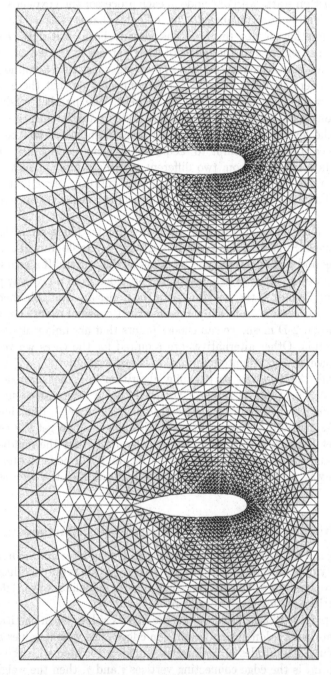

**Fig. 2.** Two-way partitioning (left) and level-set expansion using centers at the "middle node" of resulting subdomains (right).

used in partitioning methods, such as those developed in public domain codes. A drawback of such partitionings, however, is that the decision with respect to the weights is made only locally, that is, the weights are considered only to choose among the possible edge-cut candidates.

To achieve a better quality preconditioning, we would like to "propagate" the weight-related information and group together several matrix entries whose added weights are small. The shortest path algorithm may be employed for this purpose. This algorithm finds the path with the smallest weight from a given node to all the nodes in a graph. Note that this shortest path algorithm starts with a single node. Finding the shortest path results in recording one by one the nodes constituting this path in a proper order, which we will call a shortest-path ordering. Then the nodes are gathered in the subdomains based on this shortest-path ordering rather than on a level-set ordering. At the same time, we would like to keep the edge cut small, i.e., we do not let any shortest path have a large number of (cheap) links. In other words, we would like to preserve the quality of partitions produced by a level-set expansion from the chosen center points. Thus, we introduce a weight coefficient indicating the relative location of a node in the level-set structure produced by BFS starting from a center point. For each edge $e_{ij}$, this coefficient may be computed, for example, as the sum of the levels at which the incident with this edge nodes are located. In general, the partitioning algorithm consists of the following major steps:

1. Find domain centers.
2. Do level-set expansion from these centers.
3. Record the level for each node.
4. Compute the weight coefficients.
5. Multiply weights by the coefficients.
6. Do shortest path ordering of the nodes.
7. Collect nodes into subdomains.

The full shortest-path algorithm is quite expensive: A standard implementation costs $\mathcal{O}((|V| + |E|) \log |V|)$. Thus, some heuristic approximations were used instead of the complete shortest path algorithm. A test has been performed on the RAEFSKY3 matrix [2]. This matrix is of size 21,200 and has 1,488,768 of nonzeros. The Schur-LU preconditioner [17] has been applied to solve the corresponding linear system using FGMRES(20) [15] on 16 processors. A brief description of the Schur-LU preconditioner is given in Section 3. Table 1 shows a comparison between using a standard level-set expansion versus a method using a full shortest-path ordering in the expansion. The times are in seconds on the Paragon parallel computer. Note that these timings measure the solution phase only. The major time gains occur in the preconditioning phase — 142.04 vs. 97.84 seconds — which is not surprising since the preconditioning application, being expensive, benefits the most from improving the quality of the local matrices. The importance of using the knowledge of matrix values in partitioning can be demonstrated in another example discussed next.

|              | Time   | Iterations |
|--------------|--------|-----------|
| Standard     | 175.27 | 161       |
| Shortest-Path| 122.95 | 106       |

**Table 1.** Impact of two different partitioning strategies on the performance of the Schur-LU preconditioner.

## 2.4   Partitioning for Problems with Discontinuities

Consider an elliptic partial differential equation of the form

$$-\frac{\partial}{\partial x}\left(a\frac{\partial u}{\partial x}\right) - \frac{\partial}{\partial y}\left(b\frac{\partial u}{\partial y}\right) = h \qquad (1)$$

on rectangular regions with general mixed-type boundary conditions. In the test problems, the regions are the square $\Omega = (0,1)^2$, or the cube $\Omega = (0,1)^3$; the Dirichlet condition $u = 0$ is always used on the boundary. Only the discretized matrix is of importance, since the right-hand side will be created artificially. Therefore, the right-hand side $h$ is not relevant here. The mesh is 96×96 grid points *per processor*, meaning that the problem size is scaled with the number of processors: the more processor used, the larger the problem solved. For example, on 4 processors, the problem size is $96 \times 96 \times 4 = 36,864$; on 16 processors, the problem size is $96 \times 96 \times 16 = 147,456$.

In the region $0.25 < x, y < 0.75$, the coefficient $a = 100$ (Figure 3) and $a = 1$ elsewhere, while the function $b$ is constant and equal to 1. We can partition the

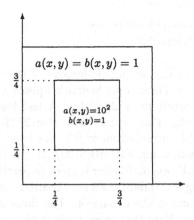

**Fig. 3.** A test problem with discontinuous coefficient.

resulting mesh without consideration given to the coefficients or we can partition it by trying to ensure that the discontinuity lines will not cross subdomains.

Having coefficients $a = 100$ and $b = 1$ in the region creates the discontinuity in only one direction, which already shows the advantages of treating this region separately and allows a greater freedom in partitioning it separately. We use a two-way level-set partitioning ("double-stripe") as described in Section 2.1. This consists of taking a pseudo-peripheral node [5], then doing a breadth-first traversal from the peripheral node and keeping as many levels in the traversal as needed to have about $n/p_1$ nodes in each processor. The process is repeated to partition each of the resulting subgraphs into $p_2$ subpartitions. The result is a partition into $p = p_1 \times p_2$ subgraphs, where $p$ must be multiple of 4 in our example.

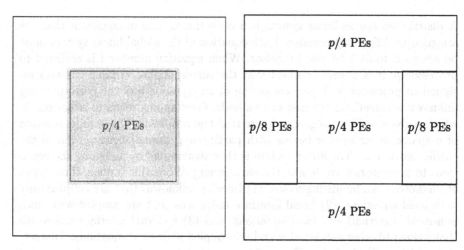

**Fig. 4.** A macro-partitioning of the problem in Figure 3.

Table 2 shows the solution times using this algorithm (DoubleStripe) and its modification (DoubleStripe_m) in which the partitioning is done in two phases. First, the double-stripe algorithm is applied to the area outside the region ("low coefficient $a$" area) only to partition it into $3p/4$ processors (Figure 4—left). Second, the rest of the processors ($p/4$) is assigned to the "high coefficient $a$" area inside the region using the same algorithm.

One might argue that using a general purpose partitioning for this problem is not adequate since we can do the partitioning "By hand", i.e., provide a rectangular partitioning for this regular mesh. In Table 2, ByHand represents such a partitioning. To consider separately the "low coefficient $a$" region, we proceed in the same way as with double-stripe partitioning and assign the region to the $p/4$ processors as shown in Figure 4(right). The results for the two-phase partitioning "By hand" are labeled ByHand_m in Table 2.

|  | Time | | | Iterations | | |
|---|---|---|---|---|---|---|
| PEs | 16 | 24 | 48 | 16 | 24 | 48 |
| DoubleStripe | 9.66 | 14.34 | 51.29 | 81 | 132 | 341 |
| DoubleStripe_m | 5.58 | 6.33 | 6.17 | 50 | 61 | 58 |
| ByHand | 6.39 | 12.73 | 9.98 | 52 | 120 | 87 |
| ByHand_m | 6.3 | 7.34 | 7.25 | 53 | 62 | 59 |

**Table 2.** Solution times (Paragon seconds), numbers of outer iterations for the Schur-LU preconditioner with four different partitioning strategies.

## 3   Solution of Distributed Sparse Systems

A distributed sparse linear system is a collection of sets of equations that are assigned to different processors. Each equation of the global linear system must be assigned to at least one processor. When equation number $i$ is assigned to processor $p$, it is always assumed that the corresponding variable $i$ is also assigned to processor $p$. A pair consisting of an equation and the corresponding unknown is sometimes referred to as a node. Overlapping refers to situations in which a given node is assigned to more that one processor. The parallel solution of a sparse linear system begins with partitioning the adjacency graph of the coefficient matrix. The linear system is then distributed by assigning the equations to processors according to the partitioning. When this is done, three types of unknowns can be distinguished: (1) Interior unknowns that are coupled only with local equations; (2) Local interface unknowns that are coupled with both non-local (external) and local equations; and (3) External interface unknowns that belong to other subdomains and are coupled with local equations. This setting which is illustrated in Figure 5, is common to most packages for parallel iterative solution methods [14, 16, 8, 19, 19, 3, 18, 9].

The matrix assigned to a certain processor is split into two parts: the *local* matrix $A_i$, which acts on the local variables and an *interface matrix* $X_i$, which acts on the external variables. Accordingly, the local equations can be written as follows:

$$A_i x_i + X_i y_{i,ext} = b_i. \tag{2}$$

where $x_i$ represents the vector of local unknowns, $y_{i,ext}$ are the external interface variables, and $b_i$ is the local part of the right-hand side vector. It is common to reorder the local equations in such a way that the interface points are listed last after the interior points. This ordering leads to an improved interprocessor communication and to reduced local indirect addressing during matrix-vector multiplication. Thus, the local variables form a local vector of unknowns $x_i$ which is split into two parts: the subvector $u_i$ of internal vector components followed by the subvector $y_i$ of local interface vector components. The right-hand side $b_i$ is conformally split into the subvectors $f_i$ and $g_i$. When the block is partitioned according to this splitting, the local equations (2) can be written as follows:

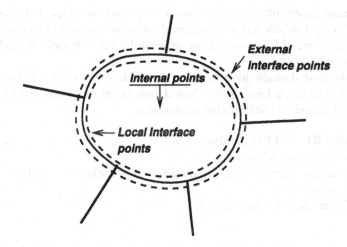

**Fig. 5.** A local view of a distributed sparse matrix.

$$\begin{pmatrix} B_i & F_i \\ E_i & C_i \end{pmatrix} \begin{pmatrix} u_i \\ y_i \end{pmatrix} + \begin{pmatrix} 0 \\ \sum_{j \in N_i} E_{ij} y_j \end{pmatrix} = \begin{pmatrix} f_i \\ g_i \end{pmatrix}. \tag{3}$$

Here, $N_i$ is the set of indices for subdomains that are neighbors to the subdomain $i$. The term $E_{ij} y_j$ is a part of the product $X_i y_{i,ext}$ which reflects the contribution to the local equation from the neighboring subdomain $j$. The result of the multiplication by $X_i$ affects only the local interface unknowns, which is indicated by a zero in the top part of the second term of the left-hand side of (3).

### 3.1 Distributed Krylov Subspace Solvers

Implementation issues when developing Preconditioned Krylov subspace algorithms for solving distributed sparse systems have been discussed elsewhere, see e.g., [16, 8, 19, 3, 15]. Once the data structure associated with the local data has been built in each processor, then the operations required for implementing a Krylov subspace method are (1) vector operations such as SAXPY and dot products, (2) matrix vector products, and (3) preconditioning operations. The SAXPY operations are entirely local, since vectors are all partitioned conformally. The dot products require a global sum of partial inner products. This reduction operation requires communication but if the domains are large enough, the related overhead is usually small.

To multiply a given vector $x$ by the global matrix, we only need to multiply the local matrix by the local part of $x$ to obtain a vector $z$ then get the external interface variables, multiply them by the $X$ matrix and add the result to $z$. Thus, there are three costs to be added: the cost of the local matvec, then the cost of the exchange of interface data and finally the cost of the second matvec.

The most important operation when solving distributed sparse linear systems is undoubtedly the preconditioning operation. A discussion of preconditioners is beyond the scope of this paper. Here we only discuss two such methods for the sake of completeness.

The simplest domain-decomposition preconditioner is the additive Schwarz procedure, which is a form of the block Jacobi iteration, where the blocks refer to matrices associated with entire subdomains.

ALGORITHM **31** *Additive Schwarz*
1. *Obtain external data* $y_{i,ext}$
2. *Compute (update) local residual* $r_i = (b - Ax)_i = b_i - A_i x_i - X_i y_{i,ext}$
3. *Solve* $A_i \delta_i = r_i$
4. *Update solution* $x_i = x_i + \delta_i$

The systems which arise in line 3, are solved by either a standard (sequential) ILUT preconditioner [15] combined with GMRES or the application of one ILU preconditioning operation. Of particular interest in this context are the overlapping additive Schwarz methods. In the domain decomposition literature [18] it is known that overlapping is a good strategy to reduce the number of steps.

Another preconditioning method described in [17] is the Schur-LU preconditioner. The main step in this technique is to solve approximately the Schur complement system, i.e., the (global) system which involves the interface variables $y$. This system is obtained by eliminating the variable $u_i$ from equation (3), using the first equation. This global system in $y$ can be solved approximately by a form of block Jacobi preconditioner. An ILUT factorization for the matrix $A_i$ yields as a by-product an ILUT factorization for the local Schur complement. This is used to precondition the global Schur system. Once the approximation to the $y$ variable is obtained, the $u$ variables are extracted. The step of obtaining the approximate $u_i, y_i$ pair in this manner from a right-hand side constitutes one step of the Schur-LU preconditioner. For further details, see [17].

## 3.2   Reducing Idle Time in Preconditioning Operations

For the local preconditioning strategies, such as Additive Schwarz, the amount of work each processor accomplishes in the preconditioning application is different and depends on the properties of the local submatrices. Since the properties of the local submatrices may vary greatly, the times of the preconditioning phase may also differ substantially leading to a load imbalance among processors. Thus when the processor synchronizations take place (in the orthogonalization phase of FGMRES and during the matrix-vector product computation), the processors with a small preconditioning workload must wait for the rest of the processors. One way to avoid this idling is to force all the processors to spend the same time in the preconditioning application in each outer iteration. The rationale is that it is better to spend the time that would otherwise be wasted, to perform more iterations in the "faster" processors. A better accuracy may be achieved in these processors which would eventually propagate to others, resulting in a reduction of the number of iterations. The time may be fixed, for example, based on the time required by the processor with the largest workload to apply a preconditioning step.

There are several approaches to control the "fixed-time" condition for an Additive Schwarz application which uses an iterative process (say, preconditioned GMRES). One of these approaches is to change the number of inner iterations at a certain iteration of FGMRES for each processor based on some criterion comparing the time of the previous preconditioning step. Thus the processors with small workloads will proceed for more iterations and compute a more accurate preconditioning vector. As a result, the number of the outer iterations may be reduced.

In testing this approach, the following iteration adjustment has been determined experimentally and applied after each preconditioning step in processor $i$, $(i = 1, \ldots, p)$:

$$\text{if } (\Delta_j^i > n_{j-1}^i/3) \quad n_j^i = n_{j-1}^i + \Delta_j^i,$$

where $n_j^i$ is the number of the inner iterations in the $j$th iteration of FGMRES; $\Delta_j^i$ is the number of iterations that processor $i$ can fit into the time to be wasted in idling otherwise at the (next) $j$th outer iteration of FGMRES. Specifically,

$$\Delta_j^i = \frac{(T_{\max} - T^i)N^i}{T^i},$$

where $T_{\max}$ is the maximum time among all the processors and $T^i$ the time for processor $i$ to perform preconditioning operations during $j - 1$ previous outer iterations, $N^i$ is the total number of preconditioning operations performed by processor $i$ so far. The number of inner iterations $n_j^i$ can be updated provided that the limit $n_{\lim}$ on the number of inner iterations is not reached. Figure 6 compares the time and iteration results for the standard (Jacobi in Figure 6) and the "fixed-time" (Jacobi_ft in Figure 6) Additive Schwatz preconditionings accelerated with FGMRES(20). The experiments have been performed on the IBM SP with the problem AF23560 from the Harwell-Boeing collection [2]. This problem has 23,560 unknowns and 484,256 nonzeros in the matrix. The preconditioning phase consists of ILUT-preconditioned GMRES(20) with the following parameters: lfil=25, $n_0^i = 5$, $n_{\lim} = 20$, and the relative accuracy of $10^{-2}$. Note that this accuracy is increased to the accuracy of FGMRES ($10^{-8}$) whenever $n^i$ is increased for the "fixed-time" algorithm. Figure 6 indicates that Jacobi_ft exhibits a better overall performance on a wide range of processor numbers, thus showing an advantage of forced load balancing. Earlier testings of this approach with a slightly different criterion, reported in [11] for small processor numbers, are in agreement with our numerical experiments.

## 4    Conclusion

A few strategies have been described for enhancing the performance of preconditioned Krylov subspace methods for solving distributed sparse linear systems.

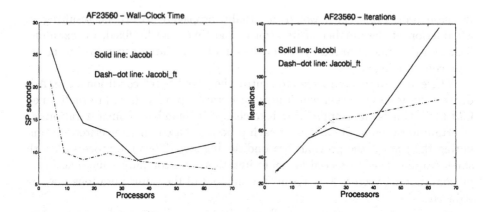

**Fig. 6.** Solution times and iterations for the standard and "fixed-time" Additive Schwatz preconditionings.

The techniques described show that much can be gained from using partitioners that take into account the values of the matrix entries. In addition, the standard stopping criteria used when iterating locally in a processor may lead to wasted idle time, either because of load imbalance or because some subdomain systems are much easier to solve than others. We found that some improvements can be made by simply forcing all processors to iterate about the same amount of time. Such non-standard heuristics may be necessary because the impact of a given partitioning on the overall preconditioner is difficult to predict or analyze.

# References

[1] X. C. Cai and Y. Saad. Overlapping domain decomposition algorithms for general sparse matrices. *Numerical Linear Algebra with Applications*, 3:221–237, 1996.

[2] I. S. Duff, R. G. Grimes, and J. G. Lewis. Sparse matrix test problems. *ACM Transactions on Mathematical Software*, 15:1–14, 1989.

[3] V. Eijkhout and T. Chan. ParPre a parallel preconditioners package, reference manual for version 2.0.17. Technical Report CAM Report 97-24, UCLA, 1997.

[4] C. Farhat and M. Lesoinne. Mesh partitioning algorithms for the parallel solution of partial differential equations. *Applied Numerical Mathematics*, 12, 1993.

[5] J. A. George and J. W. Liu. *Computer Solution of Large Sparse Positive Definite Systems*. Prentice-Hall, Englewood Cliffs, NJ, 1981.

[6] T. Goehring and Y. Saad. Heuristic algorithms for automatic graph partitioning. Technical Report UMSI 94-29, University of Minnesota Supercomputer Institute, Minneapolis, MN, February 1994.

[7] B. Hendrickson and R. Leland. An improved spectral graph partitioning algorithm for mapping parallel computations. Technical Report SAND92-1460, UC-405, Sandia National Laboratories, Albuquerque, NM, 1992.

[8] Scott A. Hutchinson, John N. Shadid, and R. S. Tuminaro. Aztec user's guide. version 1.0. Technical Report SAND95-1559, Sandia National Laboratories, Albuquerque, NM, 1995.

[9] M. T. Jones and P. E. Plassmann. BlockSolve95 users manual: Scalable library software for the solution of sparse linear systems. Technical Report ANL-95/48, Argonne National Lab., Argonne, IL., 1995.

[10] G. Karypis. *Graph Partitioning and its Applications to Scientific Computing*. PhD thesis, Department of Computer Science, University of Minnesota, Minneapolis, MN, 1996.

[11] S. Kuznetsov, G. C. Lo, and Y. Saad. Parallel solution of general sparse linear systems. Technical Report UMSI 97/98, Minnesota Supercomputer Institute, University of Minnesota, Minneapolis, MN, 1997.

[12] J. W. H. Liu. A graph partitioning algorithm by node separators. *ACM Transactions on Mathematical Software*, 15:198–219, 1989.

[13] A. Pothen, H. D. Simon, and K. P. Liou. Partitioning sparse matrices with eigenvectors of graphs. *SIAM Journal on Matrix Analysis and Applications*, 11:430–452, 1990.

[14] Y. Saad. Parallel sparse matrix library (P_SPARSLIB): The iterative solvers module. In *Advances in Numerical Methods for Large Sparse Sets of Linear Equations, Number 10, Matrix Analysis and Parallel Computing, PCG 94*, pages 263–276, Keio University, Yokohama, Japan, 1994.

[15] Y. Saad. *Iterative Methods for Sparse Linear Systems*. PWS publishing, New York, 1996.

[16] Y. Saad and A. Malevsky. PSPARSLIB: A portable library of distributed memory sparse iterative solvers. In V. E. Malyshkin et al., editor, *Proceedings of Parallel Computing Technologies (PaCT-95), 3-rd international conference, St. Petersburg, Russia, Sept. 1995*, 1995.

[17] Y. Saad and M. Sosonkina. Distributed Schur complement techniques for general sparse linear systems. Technical Report UMSI 97/159, Minnesota Supercomputer Institute, University of Minnesota, Minneapolis, MN, 1997. Submitted, Revised.

[18] B. Smith, P. Bjørstad, and W. Gropp. *Domain decomposition: Parallel multilevel methods for elliptic partial differential equations*. Cambridge University Press, New-York, NY, 1996.

[19] B. Smith, W. D. Gropp, and L. C. McInnes. PETSc 2.0 user's manual. Technical Report ANL-95/11, Argonne National Laboratory, Argonne, IL, July 1995.

# Optimal Tridiagonal Solvers on Mesh Interconnection Networks

Eunice E. Santos

Department of Electrical Engineering and Computer Science,
Lehigh University, Bethlehem, PA 18015, USA.

**Abstract.** We consider the problem of designing optimal and efficient algorithms for solving tridiagonal linear systems on a mesh interconnection network. We derive precise upper and lower bounds for these solvers using odd-even cyclic reduction. We present various important lower bounds on execution time for solving these systems including general lower bounds which are independent of initial data assignment, lower bounds based on classifications of initial data assignments which classify assignments via the proportion of initial data assigned amongst processors, and lower bounds for commonly-used data layouts for tridiagonal solvers. Finally, algorithms are provided which have running times not only within a small constant factor of the lower bounds provided but which are within a small constant additive term of the lower bounds.

## 1 Introduction

In this paper, we consider the problem of designing algorithms for solving tridiagonal linear systems. Such algorithms are referred to as tridiagonal solvers. A method for solving these systems in which there is particular interest by designers is the well-known odd-even cyclic reduction method which is a direct method. Due to this interest, we chose to focus our attention to odd-even cyclic reduction. Thus, our results will be applicable to tridiagonal solvers designed on a mesh utilizing cyclic reduction. Clearly, much research has been spent exploring this problem, most deal with designing and analyzing algorithms that solve these systems on specific types of interconnection networks [1,5,7,8] such as hypercube or butterfly. However, very little has been done on determining lower bounds for solving tridiagonal linear systems [4,6] on any type of interconnection network or on specific general parallel models [10].

The main objective of this paper is to present precise upper and lower bounds on the running time for solving tridiagonal systems which utilize odd-even cyclic reduction on meshes. Our decision to work with the mesh topology is based on the simple fact that meshes are a very common and frequently used interconnection network. The results obtained in this paper will provide not only a means for measuring efficiency of existing algorithms but also provide a means of determining what kinds of data layouts and communication patterns are needed to achieve optimal or near-optimal running times.

P. Zinterhof, M. Vajteršic, A. Uhl (Eds.): ACPC'99, LNCS 1557, pp. 28–37, 1999.

Some of the interesting results we shall show include the following: The skewness in the proportion of data will significantly effect running time. Using common data layouts and straightforward communication patterns do not result in significantly higher complexities than assuming that all processors have access to all data items regardless of communication pattern. In fact, in most cases, common data layouts and straightforward communication patterns can be used to obtain optimal running times. The threshold for processor utilization is $\Omega(N^{\frac{2}{3}})$.

The paper is divided as follows. Section 2 contains a description of the mesh topology. In Section 3 we discuss the odd-even cyclic reduction method for solving tridiagonal systems. We then derive various important lower bounds on execution. First, we derive general lower bounds for solving tridiagonal systems, i.e. the bounds hold regardless of data assignment. We follow this by deriving lower bounds which rely on categorizing data layouts via the proportion of data assigned amongst processors. Furthermore, we describe commonly-used data layouts designers utilize for this problem. We follow by deriving lower bounds utilizing these specific layouts. Lastly, running times for algorithms which are not only within a small constant factor of the lower bounds derived but are within a constant additive term (dependent on $P$) are provided. For brevity, proofs and algorithms are not provided in this paper. Many of the proofs and algorithms can be found in [9]. Section 4 gives the conclusion and summary of results.

## 2   2-Dimensional Mesh Interconnection Network

A *mesh network* is a parallel model on $P$ processors in which processors are grouped as two types: border processors and interior processors. Each interior processor is linked with exactly four neighbors. Each but for four border processors have three neighbors. And the remaining four border processors have exactly two neighbors. More precisely:

- Denote processors by some $p_{i,j}$ where $1 \leq i, j \leq \sqrt{P}$
  - For $1 < i, j < \sqrt{P}$ the four neighbors of $p_{i,j}$ are $p_{i+1,j}$, $p_{i-1,j}$, $p_{i,j+1}$, and $p_{i,j-1}$
  - For $i = 1$ and $1 < j < \sqrt{P}$ the three neighbors of $p_{i,j}$ are $p_{i+1,j}$, $p_{i,j+1}$, and $p_{i,j-1}$
  - For $i = \sqrt{P}$ and $1 < j < \sqrt{P}$ the three neighbors of $p_{i,j}$ are $p_{i-1,j}$, $p_{i,j+1}$, and $p_{i,j-1}$
  - For $j = 1$ and $1 < i < \sqrt{P}$ the three neighbors of $p_{i,j}$ are $p_{i+1,j}$, $p_{i-1,j}$, and $p_{i,j+1}$
  - For $j = \sqrt{P}$ and $1 < i < \sqrt{P}$ the three neighbors of $p_{i,j}$ are $p_{i+1,j}$, $p_{i-1,j}$, and $p_{i,j-1}$
  - For $i = 1$ and $j = 1$ the two neighbors of $p_{i,j}$ are $p_{i+1,j}$, and $p_{i,j+1}$
  - For $i = 1$ and $j = \sqrt{P}$ the two neighbors of $p_{i,j}$ are $p_{i+1,j}$, and $p_{i,j-1}$
  - For $i = \sqrt{P}$ and $j = 1$ the two neighbors of $p_{i,j}$ are $p_{i-1,j}$, and $p_{i,j+1}$
  - For $i = \sqrt{P}$ and $j = \sqrt{P}$ the two neighbors of $p_{i,j}$ are $p_{i-1,j}$, and $p_{i,j-1}$

Communication between neighbor processors require exactly 1 time step. By this, we mean that if a processor transmits a message to its neighbor processor *at the beginning* of time $x$, it's neighbor processor will receive the message *at the beginning* of time $x + 1$. Moreover, we assume that processors can be receiving[transmitting] a message while computing a local operation.

## 3    Odd-Even Cyclic Reduction Method

**The Problem:**    Given $Mx = b$ solve for $x$, where

- $M$ is a tridiagonal $N \times N$ matrix,
- $b = (b_j)$ is a vector of size $N$, and
- $x = (x_j)$ a vector of size $N$.

We assume for the discussion that $1 < P = 2^{2k} \leq N$. An algorithm is simply a set of arithmetic operations such that each processor is assigned a sequential list of these operations. An initial assignment of data to the processors is called a data layout. A list of message transmissions and receptions between processors is called a communication pattern. These three components (algorithm, data layout, communication pattern) are needed in order to determine running time.

Odd-even cyclic reduction [2,3,7] is a recursive method for solving tridiagonal systems of size $N = 2^n - 1$. This method is divided into two parts: reduction and back substitution.

The first step of reduction is to remove each odd-indexed $x_i$ and create a tridiagonal system of size $2^{n-1} - 1$. We then do the same to this new system and continue on in the same manner until we are left with a system of size 1. This requires $n$ phases. We refer to the tridiagonal matrix of phase $j$ as $M^j$ and the vector as $b^j$. The original $M$ and $b$ are denoted $M^0$ and $b^0$. The three non-zero items of each row $i$ in $M^j$ are denoted $l_i^j, m_i^j, r_i^j$ (left, middle, right). Below are the list of operations needed to determine the items of row $i$ in matrix $M^j$.

$$e_i^j = -\frac{l_i^{j-1}}{m_{i-2^{j-1}}^{j-1}}, \qquad f_i^j = -\frac{r_i^{j-1}}{m_{i+2^{j-1}}^{j-1}},$$

$$l_i^j = e_i^j l_{i-2^{j-1}}^{j-1}, \qquad r_i^j = f_i^j r_{i+2^{j-1}}^{j-1},$$

$$m_i^j = m_i^{j-1} + e_i^j r_{i-2^{j-1}}^{j-1} + f_i^j l_{i+2^{j-1}}^{j-1},$$

$$b_i^j = b_i^{j-1} + e_i^j b_{i-2^{j-1}}^{j-1} + f_i^j b_{i+2^{j-1}}^{j-1}$$

Clearly each system is dependent on the previous systems. Figure 1 shows the dependency between all the $M^j$'s. The nodes in level $j$ of the figure represent the rows of $M^j$. A directed path from node $a$ in level $k$ to node $b$ in level $j$ implies

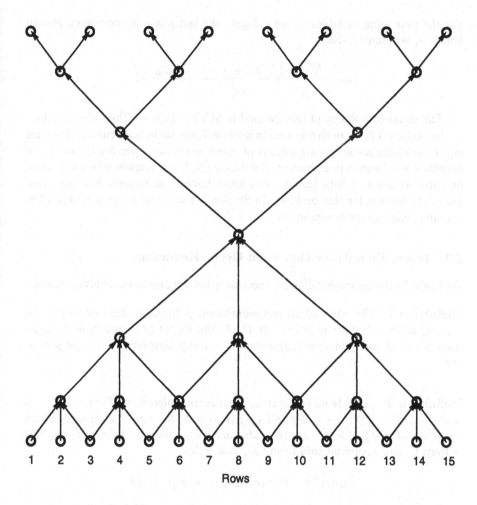

**Fig. 1.** *Row dependency for odd-even cyclic reduction for $N = 15$*

that row $a$ in $M^k$ is needed in order to determine the entries for row $b$ in $M^j$. In this paper, we assume that if an algorithm employs odd-even cyclic reduction, we assume that a processor computed items of whole rows of a matrix (i.e. the three non-zero data items) and the appropriate item in the vector.

The back substitution phase is initiated after the system of one equation has been determined. We recursively determine the values of the $x_i$'s. The first operation is

$$x_{2^{n-1}} = \frac{b_{2^{n-1}}^{n-1}}{m_{2^{n-1}}^{n-1}}.$$

For the remaining variables $x_i$, let $j$ denote the last phase in the reduction step before $x_i$ is removed then

$$x_i = \frac{b_i^{j-1} - l_i^{j-1}x_{i-2^{j-1}} - r_i^{j-1}x_{i+2^{j-1}}}{m_i^{j-1}}.$$

The serial complexity of this method is $S(N) = 17N - 12\log(N+1) + 11$.

In Section 3.1, we derive several important lower bounds on running time for odd-even cyclic reduction algorithms on mesh networks. Specifically, we derive general lower bounds independent of data layout, lower bounds which are based on categorization of data layouts, and lower bounds on running time for common data layouts for this problem. In Section 3.2 we present optimal algorithm running times on mesh topologies.

## 3.1   Lower Bounds for Odd-Even Cyclic Reduction

We begin by listing some definitions necessary for the discussion of lower bounds.

**Definition 1.** *The class of all communication patterns is denoted by $\mathcal{C}$. The class of all data layouts is denoted by $\mathcal{D}$. A data layout $D$ is said to be a single-item layout if each non-zero matrix-item is initially assigned to a unique processor.*

**Definition 2.** *Let $A$ be an odd-even cyclic reduction algorithm. For $i = 1, 2, \cdots n$ and $j = 1, 2, \cdots 2^{n-i+1} - 1$, define $T_{A,D,C}(i,j)$ to be the minimum time at which the items of row $j$ of level $i$ are computed using algorithm $A$ and communication pattern $C$ and assuming data layout $D$, and define*

$$T_{A,D,C}(i) = min_{1 \le j \le 2^{n-i+1}-1}T_{A,D,C}(i,j).$$

In other words, $T_{A,D,C}(i)$ is the minimum time at which the items of any row in matrix $M^{i-1}$ is computed using odd-even cyclic reduction algorithm $A$ and communication pattern $C$ and assuming data layout $D$.

It follows that for all odd-even cyclic algorithms $A$, data layout $D$, communication pattern $C$, and any $i < n$,

$$T_{A,D,C}(i+1) > T_{A,D,C}(i).$$

**Definition 3.** *Let $A$ be an odd-even cyclic reduction algorithm and let $\mathcal{O}$ denote the class consisting of all odd-even cyclic reduction algorithms. For all $i \le n$,*

$$T_{A,D}(i) = min_{C \in \mathcal{C}}T_{A,D,C}(i),$$

$$T_A(i) = min_{D \in \mathcal{D}}T_{A,D}(i) \text{ and}$$

$$T_{\mathcal{O}}(i) = min_{A \in \mathcal{O}}T_A(i).$$

In other words:

- $T_{A,D}(i)$ is the minimum time needed to compute the items of any row in matrix $M^{i-1}$ using algorithm $A$ and data layout $D$ regardless of communication pattern,
- $T_A(i)$ is the minimum time needed to compute the items of any row in matrix $M^{i-1}$ using algorithm $A$ regardless of data layout and communication pattern, and
- $T_{\mathcal{O}}(i)$ is the minimum time needed to compute the items of any row in matrix $M^{i-1}$ by any algorithm in the class $\mathcal{O}$ regardless of data layout and communication pattern.

In the following sections we shall provide lower bounds on $T_{A,D}(n)$ for algorithms $A \in \mathcal{O}$ and certain types of data layouts $D$. The lower bounds hold regardless of the choice of communication pattern.

### 3.1.1 A General Lower Bound for Odd-Even Cyclic Reduction

In this subsection we assume the data layout is the one in which each processor has a copy of every non-zero entry of $M^0$ and $\mathbf{b}^0$. We denote this layout by $\bar{D}$. Since $\bar{D}$ is the most favorable data layout,

$$T_{\mathcal{O}}(i) = min_{A \in \mathcal{O}} T_{A,\bar{D}}(i).$$

**Theorem 1.** *Let $A$ be an odd-even cyclic reduction algorithm. The following is a lower bound for $A$ regardless of data layout and communication pattern:*

$$\begin{cases} max(\frac{S(N)}{P}, 17n - 16, \sqrt{P}) & \text{if } P \leq [S(N)^{\frac{2}{3}}] \\ \quad = \Omega(\frac{N}{P} + \log N + \sqrt{P}) \\ max([S(N)]^{\frac{1}{3}}, 17n - 16) & \text{otherwise} \\ \quad = \Omega(N^{\frac{1}{3}} + \log N) \end{cases}$$

In Section 3.2 we will provide optimal algorithms, i.e. the running times are within a small constant additive term (dependent on $P$) of the lower bounds. Therefore, we note that our results show that when $P$ is sufficiently large, i.e. $P = \Omega(N^{\frac{2}{3}})$, using more than $O(N^{\frac{2}{3}})$ processors will not lead to any substantial improvements in running time if utilizing data layout $\bar{D}$.

### 3.1.2 Lower Bounds for $\mathcal{O}$ on $\frac{c}{P}$-Data Layouts

Many algorithms designed for solving tridiagonal linear systems assume that the data layout is single-item and that each processor is assigned roughly $\frac{1}{P}^{th}$ of the rows of $M^0$ where $P$ is the number of processors available. However, in order to determine whether skewness of proportion of data has an effect on running time and if so, exactly how much, we therefore classify data layouts by initial assignment proportions to each processor. In other words, in this section, we consider single-item data layouts in which each processor is assigned at most a fraction $\frac{c}{P}$ of the rows of $M$ where $1 \leq c \leq P(\frac{N-P-1}{N})$.

**Definition 4.** *Consider c where* $1 \leq c < P(\frac{N-P-1}{N})$. *A data layout D on P processors is said to be a $\frac{c}{P}$-data layout if*

- *D is single-item,*
- *no processor is assigned more than a fraction $\frac{c}{P}$ of the rows of $M^0$,*
- *at least one processor is assigned exactly a fraction $\frac{c}{P}$ of the rows of $M^0$, and*
- *each processor is assigned at least one row of $M^0$.*

*Denote the class of $\frac{c}{P}$ data layouts by $\mathcal{D}(\frac{c}{P})$.*

**Theorem 2.** *If $D \in \mathcal{D}(\frac{c}{P})$, then for any $A \in \mathcal{O}$,*

$$T_{A,D}(n) \geq max(\frac{S(N)}{P}, 17n - 16, \sqrt{P}, \frac{Nc}{P})$$

$$= \Omega(\frac{Nc}{P} + \log N + \sqrt{P}).$$

Analyzing the result given in the above theorem, we see that $\frac{1}{P}$-data layouts will result in the best lower bounds for this class of data layouts. I.e. :

**Corollary 1.** *If $D \in \mathcal{D}(\frac{1}{P})$, then for any $A \in \mathcal{O}$,*

$$T_{A,D}(n) \geq max(\frac{S(N)}{P}, 17n - 16, \sqrt{P})$$

$$= \Omega(\frac{N}{P} + \log N + \sqrt{P}).$$

Comparing results against the general lower bound, we see that for sufficiently large $N$ (i.e. $N >> P$, or more precisely $P \leq [S(N)]^{\frac{4}{3}}$), the lower bound for $\frac{1}{P}$-data layouts is precisely equal to the general lower bound. The complexity of any algorithm $A \in \mathcal{O}$ using a $\frac{1}{P}$-data layout is $\Omega(\log N + \frac{N}{P} + \sqrt{P})$. In Section 3.2, we present algorithms that are within an additive term of the bounds.

### 3.1.3 Lower Bounds for $\mathcal{O}$ on Common Data Layouts

We now derive lower bounds on the running time for odd-even cyclic reduction algorithms using specific data layouts commonly used by algorithm designers, namely blocked data layouts. Definition is given below.

**Definition 5.** *A single-item data layout on $(1 \leq)P(\leq N)$ processors $p_1, \cdots p_P$ is blocked if for all $i \leq P$, $p_i$ is assigned the nonzero items in rows $(i-1)\frac{N}{P}+1$ to $i\frac{N}{P}$ of $M^0$ and $b^0$. We denote this layout by $D_B$.*

The following definitions are needed for the discussion of lower bounds in this section.

**Definition 6.** *A row $j$ of level $i$ is said to be an original row of some processor $p$ if the items of rows $j - 2^{i-1} + 1$ to $j + 2^{i-1} - 1$ of level 0 are originally assigned to $p$. If a row is not an original of $p$ it is said to be a non-original row of $p$.*

**Definition 7.** *Two rows $j$ and $j + 1$ of level $i$ are referred to as* neighbors.

**Theorem 3.** *Let $A$ be an algorithm in $\mathcal{O}$, i.e. $A$ is an odd-even cyclic reduction algorithm. The following is a lower bound assuming the blocked data layout:*

$$max(\frac{S(N)}{P}, 17n - 16, 2\sqrt{P})$$

$$= \Omega(\frac{N}{P} + \log N + \sqrt{P})$$

In Section 3.2 we provide an algorithm using blocked data layout whose running time is not only asymptotically equal, i.e. $O(\log N + \frac{N}{P} + \sqrt{P})$ but is also within a small additive term of the lower bound.

Comparing the complexity of blocked layout with the lower bounds for $\frac{1}{P}$-data layouts, we see that the complexity of blocked layout is almost exactly equal to that of the lower bounds for $\frac{1}{P}$-data layouts. Furthermore, comparing the complexity of blocked layout with the general lower bound (which we show is achievable up to a small additive constant term in Section 3.2), we see that for sufficiently large $N$ (i.e. $P \leq [S(N)]^{\frac{2}{3}}$) the complexity of blocked layout is not only within a constant factor of the general lower bound but, again, actually almost equal to the bound. Therefore we can use the much more realistic blocked data layout rather than $\bar{D}$ and still achieve the lower bounds (up to an constant additive term).

## 3.2 Algorithms and Communication Patterns

We have designed algorithms and communication patterns where when used with the appropriate data layouts have running times matching the lower bounds presented, i.e. the running times differ from the lower bounds by at most a small constant factor. Below is a table of running times.

We must map the mesh processors to the processors in the definition of the data layouts defined. Consider processor $p_{i,j}$, this processor would be equivalent to

- $p_{(i-1)\sqrt{P}+j}$ if $i$ is odd
- $p_{(i-1)\sqrt{P}+\sqrt{P}-j+1}$ if $i$ is even

For brevity, algorithms and communication schedules have been omitted. Full algorithm and communication schedules can be found in [9].

| Running Time: | Data Layout |
|---|---|
| $max(\frac{S(N)}{P}, 17n - 16) + 4\sqrt{P}$ | Blocked Layout |
| $= O(\frac{N}{P} + \log N + \sqrt{P})$ | |
| $max(\frac{3}{2}[S(N)]^{\frac{1}{3}}, 17n - 16)$ | $\bar{D}$ – Best Layout |
| $= O([\frac{N}{P}]^{\frac{1}{3}} + \log N)$ | when $P > [S(N)]^{\frac{2}{3}}$ |
| $S(N) = O(N)$ | Serial Algorithm |

We note the following:

$$max(\frac{3}{2}[S(N)]^{\frac{1}{3}}, 17n - 16) < max([S(N)]^{\frac{1}{3}}, 17n - 16) + \sqrt{P}.$$

## 4    Conclusion

In this paper we tackled the problem of designing optimal tridiagonal solvers on mesh interconnection networks using odd-even cyclic reduction. We were able to derive precise lower bounds on the execution time. Moreover we were able to provide algorithms whose running times differ by only a constant additive term (dependent on $P$) of these lower bounds.

Specifically, we proved that the complexity for solving tridiagonal linear systems regardless of data layout is

$$max(\frac{S(N)}{P}, 17(\log N + 1) - 16) + \Theta(\sqrt{P})$$

for $P = O(N^{\frac{2}{3}})$ and

$$max(S(N)^{\frac{1}{3}}, 17(\log N + 1) - 16) + \Theta([S(N)]^{\frac{1}{3}}$$

$$= max(S(N)^{\frac{1}{3}}, 17(\log N + 1) - 16) + O(\sqrt{P})$$

for $P = \Omega(N^{\frac{2}{3}})$. This shows that utilizing more than $\Omega(N^{\frac{2}{3}})$ processors will not result in any substantial run-time improvements. In fact, utilizing more processors may result in much slower run-times. When we added the realistic assumption that the data layouts are single-item and the number of data items assigned to a processor is bounded, we derived lower bounds for classes of data layouts. We showed that for these classes, the best types of layouts in general are those in which each processor is assigned an equal number of rows of $M^0$. Therefore, in other words, the skewness of proportion of data will significantly affect performance. Thus, the proportion of data must be equally assigned amongst processors if optimal or near-optimal run-time is to be achieved at all. Comparing the lower bounds for the $\frac{1}{P}$-layouts with the general lower bounds with this class of layouts, we see that restricting the proportion of data items assigned to a processor to $\frac{N}{P}$ does not result in a significantly higher complexity than assuming all processors have all the data items for sufficiently large $N$ (i.e. $P \leq [S(N)]^{\frac{2}{3}}$).

Lastly, we show that there are algorithms, data layouts, and communication patterns whose running times are within a constant factor of the lower bounds provided. This provides us with the $\Theta$-bounds stated above. To achieve the general lower bound, i.e. the complexity for these methods regardless of data layout, we used $\bar{D}$ the best data layout, i.e. the data layout in which every processor is assigned all the data items for $P \leq [S(N)]^{\frac{2}{3}}$. For the $\frac{1}{P}$-data layout lower bound, we used blocked data layout. Clearly blocked data layout is more realistic than $\bar{D}$ and is easy to assign across processors. Also, since for sufficiently large $N$ this lower bound is not only asymptotic to the general lower bound but within a small constant additive term of the bound, this makes the algorithm and communication pattern provided for blocked data layout practical and efficient.

Since all of the optimal algorithms discussed were variants of standard algorithms using straightforward communication patterns, this shows that it is futile to search for sophisticated techniques, and complicated communication patterns to significantly improve the running times of algorithms on mesh interconnection networks using odd-even cyclic reduction.

# 5  Acknowledgments

Research supported in part by an NSF CAREER Grant.

# References

1. C. Amodio and N. Mastronardi. A parallel version of the cyclic reduction algorithm on a hypercube. *Parallel Computing*, 19, 1993.
2. D. Heller. A survey of parallel algorithms in numerical linear algebra. *SIAM J. Numer. Anal.*, 29(4), 1987.
3. A. W. Hockney and C. R. Jesshope. *Parallel Computers*. Adam-Hilger, 1981.
4. S. L. Johnsson. Solving tridiagonal systems on ensemble architectures. *SIAM J. Sci. Stat. Comput.*, 8, 1987.
5. S. P. Kumar. Solving tridiagonal systems on the butterfly parallel computer. *International J. Supercomputer Applications*, 3, 1989.
6. S. Lakshmivarahan and S. D. Dhall. A Lower Bound on the Communication Complexity for Solving Linear Tridiagonal Systems on Cube Architectures. In *Hypercubes 1987*, 1987.
7. S. Lakshmivarahan and S. D. Dhall. *Analysis and Design of Parallel Algorithms : Arithmetic and Matrix Problems*. McGraw-Hill, 1990.
8. F. T. Leighton. *Introduction to Parallel Algorithms and Architectures: Arrays-Trees-Hypercubes*. Morgan Kaufmann, 1992.
9. E. E. Santos. Designing Optimal Tridiagonal Solvers on Mesh Interconnection Networks Technical Report PDPL-98-003, Parallel & Distributed Processing Laboratory, Lehigh University, 1998.
10. E. E. Santos. Optimal Parallel Algorithms for Solving Tridiagonal Linear Systems. In *Springer-Verlag Lecture Notes in Computer Science #1300*, 1997.

# Parallel Pivots LU Algorithm on the Cray T3E*

Rafael Asenjo and Emilio L. Zapata

Computer Architecture Department
University of Málaga, Spain,
{asenjo,ezapata}@ac.uma.es,
http://www.ac.uma.es

**Abstract.** Solving large nonsymmetric sparse linear systems on distributed memory multiprocessors is an active research area. We present a loop-level parallelized generic LU algorithm which comprises analyse-factorize and solve stages. To further exploit matrix sparsity and parallelism, the analyse step looks for a set of compatible pivots. Sparse techniques are applied until the reduced submatrix reaches a threshold density. At this point, a switch to dense routines takes place in both analyse-factorize and solve stages. The SPMD code follows a sparse cyclic distribution to map the system matrix onto a $P \times Q$ processor mesh. Experimental results show a good behavior of our sequential algorithm compared with a standard generic solver: the MA48 routine. Additionally, a parallel version on the Cray T3E exhibits high performance in terms of speed-up and efficiency.

## 1 Introduction

The kernel of many computer-assisted scientific applications is to solve large sparse linear systems. Furthermore, this problem presents a good case study and is a representative computational code for many other irregular problems.

We say that a matrix is sparse if it is advantageous to exploit its null elements with the development of a sparse version of an algorithm, instead of a dense one. However, if the matrix suffers from fill-in it will be worthwhile to combine sparse and dense approaches. That way, our parallel nonsymmetric sparse system solver algorithm follows sparse processing techniques until the reduced submatrix reaches a certain threshold density. At this point, we switch to a parallel dense LU factorization code which uses BLAS as much as possible.

For the sake of brevity, a sparse LU factorization problem survey and a more detailed description of our algorithm are presented in [2]. In this paper we briefly comment the proposed algorithm, but the main section focus on the experimental results, validation of the sequential code, and presentation of the parallel performance. Finally, the related work and conclusions section close the paper.

* The work described in this paper was supported by the Ministry of Education and Science (CICYT) of Spain under project TIC96-1125-C03, by the European Union under contract BRITE-EURAM III BE95-1564, by the Human Capital and Mobility programme of the European Union under project ERB4050P1921660, and by the Training and Research on Advanced Computing Systems (TRACS) at the Edinburgh Parallel Computing Centre (EPCC)

P. Zinterhof, M. Vajteršic, A. Uhl (Eds.): ACPC'99, LNCS 1557, pp. 38–47, 1999.
© Springer-Verlag Berlin Heidelberg 1999

## 2 Algorithm Outline and Input Parameters

Summarizing, the main characteristics of the proposed code, called **SpLU**, are:

- The algorithm is right-looking. The code, written in C, is SPMD and is designed for a distributed memory multiprocessor. There is a portable version thanks to the MPI message passing interface, and a more optimized one for the Cray T3D or T3E using the SHMEM library.
- Data distribution follows the two dimensional sparse cyclic scheme (scatter), mapping the system matrix $A$ $(n \times n)$ onto a $P \times Q$ processor mesh.
- The data structure used to store local matrices is a semi-ordered two-dimensional doubly linked list. Entries are linked in an ordered way by rows and by columns in any order.
- We exploit both inherent parallelism in the updating loops and the parallelism we achieve by selecting m compatible pivots. Analyse and factorize stages are joined into a single analyse-factorize one.
- We use a threshold-based heuristic to ensure numerical stability, and the Markowitz criterion (min row in min column) to preserve sparsity. The number of columns in which we search for compatible pivots change according to matrix density in an adaptive way. Explicit full pivoting reduces the unbalancing problems [3, 13].
- When matrix density reaches certain threshold, we switch to a parallel dense LU factorization code.
- The solve stage is also a parallel phase, with sparse and dense sub-stages for the forward and backward substitution.

In the parallel sparse factorization, the sequential outermost loop, k, nests three parallel stages: look for a set of m compatible pivots, called PivotSet; parallel rows and columns permutations; and reduced submatrix update. For each iteration, k=k+m.

Apart from the data structure for matrix $A$, we will need two nonzero count vectors, $R$ and $C$ to implement the Markowitz strategy. The value $R_i^{(k)}$ $(C_j^{(k)})$ represents the number of entries in row i (column j) of active matrix at iteration k. For a selected pivot $A_{ij}^{(k)}$, the maximum number of new entries that can be created may be $M_{ij}^{(k)} = (R_i^{(k)} - 1)(C_j^{(k)} - 1)$, where $M_{ij}^{(k)}$ is the Markowitz count at iteration k for the mentioned pivot.

Therefore, to preserve sparsity, selected pivots should have a minimum Markowitz count and a maximum absolute value to ensure stability. On the other hand this search is prohibitive, since one needs to visit the whole active matrix. To keep the search for compatible pivots effective and simple, we will only search in the ncol columns with the least $C_j^{(k)}$ in each column of the processor mesh. In these columns we will select candidates to be pivots to those with minimum $R_i^{(k)}$ (min row in min column technique) and complying with the following equation to ensure numerical stability: $|A_{ij}^{(k)}| \geq u \cdot \max_l |A_{lj}^{(k)}|$. This input parameter u, $0 < u \leq 1$, will prioritize stability when u $\to$ 1, or sparsity when u $\to$ 0.

To control sparsity, an additional input parameter a will be used to reject pivots with an unacceptable Markowitz count. In particular, candidates with $M_{ij} > a \cdot M_{i_0,j_0}$ will be rejected, where $M_{i_0,j_0}$ is the minimum Markowitz count of the candidates.

Vectors $C^{(k)}$ or $R^{(k)}$ are also used to decide the switch iteration to a dense factorization code. Active submatrix density, dens, is calculated at each k iteration as dens$= (\sum_{i=k}^{n-1} R^{(k)})/(n - k)$. When dens>maxdens, where maxdens is an input parameter, and $n-k$ is big enough to compensate the data structure change, the switch to a dense code takes place. In addition, when active matrix density increases, looking for parallel pivots tends to be unproductive. For this reason, initially, ncol contains the number of columns per processor in which the search for candidates is performed, but this ncol parameter will change dynamically during factorization, automatically adapting to the density.

# 3   Experimental Results

This section aims to analyse the sequential and parallel behavior of our SpLU algorithm, when changing the input matrix characteristics and some of the input parameters. Experimental results will be conducted on a Cray T3E with 16 DEC 21164 (Alpha EV-5) processors at 300 MHz with a peak performance of 600 Mflops per processor. We have selected some heterogeneous unsymmetric sparse matrices from Harwell-Boeing [9] and the University of Florida [6] matrix collection. As a message-passing interface, SHMEM routines have been used since they are supported by the CRAY T3E supercomputer. The sequential version of the program is obtained by simplifying the parallel code, removing all redundant or never executed sentences when $P = 1$ and $Q = 1$.

## 3.1   Fill-In and Stability

Two input parameters can be tuned to control stability and fill-in: u and a. A study of u parameter incidence is presented in table 1. We can see the variation of the average size of diagonal blocks $\overline{m}$, the number of sparse LU iterations, fill-in and factorization errors, for different u values. For the sake of brevity, we present these results for the LNS3937 matrix. Other matrices show the same behavior, but the LNS3937 is the worst conditioned and the u effect can be better appreciated. In the experiment we fixed a= 4 and ncol= 16.

| Values for u | 0.9 | 0.5 | 0.1 | 0.05 | 0.01 | 0.001 |
|---|---|---|---|---|---|---|
| $\overline{m}$ | 5.84 | 6.17 | 6.85 | 6.90 | 7.28 | 8.48 |
| Sparse iterations | 493 | 475 | 429 | 420 | 410 | 349 |
| Fill-in | 283772 | 250748 | 241163 | 222655 | 216017 | 216196 |
| Error | 2.32E-2 | 1.05E-2 | 9.56E-3 | 2.27E-2 | 2.57E-2 | 4.02E-1 |

**Table 1.** The influence of the u parameter on LNS3937

In table 1 we can see that the smaller u is, the bigger is the average size of PivotSet, allowing us to exploit more parallelism. The same effect can be appreciated in the next row: increasing $\overline{m}$ makes the number of outermost loop iterations decrease, thus reducing both sequential and parallel execution time. Additionally, fill-in is reduced when u is diminished, since there are more candidates to choose from with a smaller Markowitz count. On the other hand, the factorization error increases when reducing u, which leads to the necessity of choosing a trade-off value. Furthermore, we have observed that the more density is achieved on factors $L$ and $U$, the bigger is the factorization error, as the number of floating point operations increases. For this reason, for u= 0.1 we get the minimum error. These experiments corroborate that the trade-off u$\approx$ 0.1 [8, 7] leads to good results in many situations. In any case, the best selection of u is problem dependent, so we may need to test some u values to find the best one.

The algorithm behavior as a function of the a input parameter is shown in table 2 for the same matrix LNS3937, with u= 0.1 and ncol= 16.

| Values for a | 10 | 8 | 6 | 4 | 2 | 1 |
|---|---|---|---|---|---|---|
| $\overline{m}$ | 8.38 | 7.91 | 7.69 | 6.85 | 4.27 | 1.84 |
| Sparse Iterations | 361 | 377 | 384 | 429 | 687 | 1611 |
| Fill-in | 254176 | 251622 | 245374 | 241163 | 238923 | 220624 |

**Table 2.** The influence of the a parameter on LNS3937

The greater a is, the bigger will be the average size of the compatible pivots set, and the less the number of sparse iterations. At the same time, if we do not limit the Markowitz count, we can select pivots which will bring about more fill-in. To keep a high $\overline{m}$ without provoking an excessive fill-in, the trade-off value for a will be around 4 (also selected by other authors [4, 15]).

As ncol value is dynamically adjusted during program execution, the initial value is not specially significant. In any case, we found an appropriate initial value ncol=16.

Searching for a parallel pivots set is worthwhile even in the sequential code, as we can see in figure 1 (a), where we study the execution time versus the value of maxncol. In this experiment we have fixed ncol=maxncol, cancelling the adaptive function to update ncol. For the more sparse matrix in our set (SHERMAN5) and the second most dense one (SHERMAN2), we present in this figure the execution time normalized by the time when maxncol=1.

When factorizing sparse matrices, we see that it is interesting to search for big sets of compatible pivots. Regarding SHERMAN5, the sequential time when fixing ncol=16 is over 45% less than the one we get when ncol=1. The variable $\overline{m}$ reaches the value 26.7 with ncol=36, although execution time is worse due to wastage of time looking for compatible pivots. However, when factorizing more dense matrices, such as SHERMAN2, setting ncol to a large value is unproductive. For example, when ncol=48, we search for a large set of candidates which later turned out to be incompatible due to high matrix density.

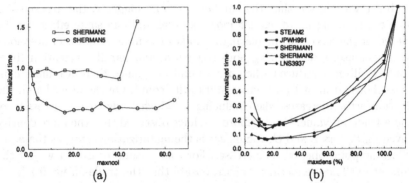

**Fig. 1.** Fixed `ncol` (a) and `maxdens` (b) influence on execution time

Finally, we have also studied execution time dependence with the threshold `maxdens` which decides the switch to a dense code. In figure 1 (b) we show the relation between `maxdens` and execution time for some matrices. Execution times are normalized by the worst case (when there is no switch to a dense factorization code, identified in the figure by `maxdens`=110). We see that the switch leads to a significant execution time saving. Minimum execution times are obtained when `maxdens`≈15%.

## 3.2   Comparison with the MA48 Routine

Before studying the parallel algorithm performance, it is important to check that the sequential version is good enough. The MA48 routine [10] is one of the generic sparse system solver more widely used. The good performance of this routine is mainly due to its left-looking organization which leads to low data traffic with memory and a subsequent good exploitation of the cache. However, this left-looking organization leads to low efficiencies in a loop level parallelized version of the MA48. On the other hand, SpLU exhibits more loop level parallelism due to the *right-looking* and parallel pivots organization, but should be comparable to MA48 performances and execution time if we want to get a competitive generic code.

In table 3 we present a comparison for the more significant characteristics of both algorithms: execution time, factorization error and fill-in. Common parameters are set equally: u=0.1 and `maxdens`=15%.

We can see how the execution time ratio (MA48 time divided by SpLU time) is greater than one, for five matrices (ratios in boldface). In these cases SpLU is faster than MA48, reaching a 4.26 factor for the EX10HS matrix. However, for the remaining 12 matrices MA48 is faster than SpLU, although the ratio do not decrease below 0.5, except for WANG1, WANG2, and LHR04C matrices. For the latter, LHR04C, factorization time in SpLU is clearly the worst, but this is in exchange for a numerical error around 20 times better.

For 12 of the 17 matrices the factorization error in SpLU is better than in the MA48 routine. For the remaining 5 matrices, there is never more than an

| | Time | | Error | Fill-in |
|---|---|---|---|---|
| Matrix | SpLU–MA48 | ratio | SpLU–MA48 | SpLU–MA48 |
| STEAM2 | 1.10–.61 | (0.55) | **.29E-13**–.13E-11 | **79552**–110466 |
| JPWH991 | 1.05–.88 | (0.84) | **.18E-14**–.82E-13 | **89810**–101892 |
| SHERMAN1 | .37–.19 | (0.51) | .22E-12–.16E-12 | 43320–43171 |
| SHERMAN2 | 6.87–16.7 | **(2.43)** | **.69E-6**–.15E-5 | **325706**–656307 |
| EX10 | 7.49–24.51 | **(3.27)** | **.23E-6**–.31E-6 | **283378**–296270 |
| ORANI678 | 9.23–6.48 | (0.70) | .15E-12–.74E-13 | **406568**–439280 |
| EX10HS | 10.24–43.71 | **(4.26)** | **.72E-7**–.20E-6 | **321031**–336832 |
| CAVITY10 | 51.78–25.94 | (0.50) | **.16E-9**–.36E-9 | 1139121–1087769 |
| WANG1 | 46.99–21.18 | (0.45) | .26E-12–.97E-13 | 1124807–808989 |
| WANG2 | 49.82–21.02 | (0.42) | **.47E-13**–.52E-13 | 1178085–808989 |
| UTM3060 | 42.30–26.13 | (0.62) | **.16E-8**–.58E-8 | **1066896**–1073933 |
| GARON1 | 69.73–35.07 | (0.50) | **.17E-8**–.21E-8 | 1431657–1257874 |
| EX14 | 131.75–206.63 | **(1.56)** | **.19E+1**–.93E+1 | **2293851**–2658661 |
| SHERMAN5 | 8.69–11.02 | **(1.26)** | **.17E-12**–.59E-12 | **363186**–519855 |
| LNS3937 | 34.1–25.8 | (0.75) | .95E-2–.13E-2 | 1078221–1002494 |
| LHR04C | 101.43–14.05 | (0.13) | **.89E-5**–.16E-3 | 1988258–870784 |
| CAVITY16 | 193.46–109.86 | (0.56) | .85E-9–.49E-9 | 2683852–2581086 |

**Table 3.** SpLU and MA48 comparison

order of magnitude of difference. With regard to fill-in, $L$ and $U$ matrices are sparser on 9 occasions if they are computed by SpLU code.

In spite of the high optimization of the MA48 code, we believe that it can be improved by the SpLU in some cases due to the analyse stage. Even when the MA48 analyse stage is also based on Markowitz and threshold strategies, the fact that this analyse stage takes place before factorizing has its own drawbacks: permutation vectors are selected in advance, but during the factorize stage, the numerical partial pivoting is also allowed and this may undo the analyse decisions to some extent.

SpLU shows a joined analyse-factorize stage where for each iteration a proper set of compatible pivots are selected over the candidates in the active matrix. In many cases this enables a better pivot selection during factorization, yielding better numerical precision. In exchange, the analyse fragment of code is more expensive than the corresponding one in the MA48, due to it searching for a single pivot instead of many which are mutually compatible.

## 3.3  Parallel Performance

In this section we will compare the parallel algorithm execution time over a $P \times Q$ processor mesh with the sequential version, executed over a single Alpha processor. To make times comparable for both versions, input parameters will be equally fixed. As we saw in subsection 3.1 it seems appropriate to set u=0.1 and a=4. As for the initial local ncol, it will be set to $16/Q$, to make the initial maximum number of compatible pivots independent of the mesh size.

Parallel version exhibits the same fill-in and factorization error as sequential version, as u, a, and maxdens, do not affect the parallel version in a different way to the sequential one.

Table 4 presents the speed-up we get when factorizing the 14 biggest matrices in our set. The last three columns in this table show dimension, $n$, initial density, $\rho_0$, and the final one, $\rho_n$. Figure 2 shows speed-up and efficiency when factorizing the 9 computationally more expensive matrices for four mesh sizes.

| Matriz | Speed-up | | | | Density | | |
|---|---|---|---|---|---|---|---|
|  | 2 | 4 | 8 | 16 | n | $\rho_0$ | $\rho_n$ |
| SHERMAN2 | 1.85 | 3.62 | 5.98 | 9.82 | 1080 | 1.98% | 27.92% |
| EX10 | 1.74 | 2.99 | 4.25 | 4.96 | 2410 | 0.94% | 4.87% |
| ORANI678 | 1.77 | 3.02 | 4.96 | 6.67 | 2529 | 1.41% | 6.35% |
| EX10HS | 1.74 | 3.65 | 5.39 | 5.63 | 2548 | 0.88% | 4.94% |
| CAVITY10 | 1.94 | 3.72 | 5.59 | 8.77 | 2597 | 1.13% | 16.88% |
| WANG1 | 2.16 | 3.76 | 7.01 | 10.44 | 2903 | 0.22% | 13.94% |
| WANG2 | 2.06 | 4.15 | 6.60 | 12.45 | 2903 | 0.22% | 13.97% |
| UTM3060 | 1.88 | 3.41 | 5.87 | 10.07 | 3060 | 0.45% | 11.39% |
| GARON1 | 2.32 | 3.76 | 7.18 | 12.02 | 3175 | 0.88% | 14.20% |
| EX14 | 2.18 | 4.04 | 7.22 | 13.17 | 3251 | 0.63% | 21.70% |
| SHERMAN5 | 1.63 | 3.00 | 4.28 | 5.66 | 3312 | 0.19% | 3.31% |
| LNS3937 | 1.93 | 3.69 | 6.33 | 11.01 | 3937 | 0.16% | 6.95% |
| LHR04C | 2.02 | 3.93 | 7.04 | 11.79 | 4101 | 0.49% | 11.82% |
| CAVITY16 | 1.99 | 3.85 | 7.48 | 14.11 | 4562 | 0.66% | 12.89% |

**Table 4.** Speed-up for different mesh sizes

We see that speed-up monotonically increases with the number of processors. When changing from 8 to 16 processors, EX10 and EX10HS exhibit a less notable increment of speed-up due to the low computational load presented by these matrices. In these cases, communications dominate local computations and messages comprise a small number of data, so latency prevails over communication bandwidth. We should take into account the high ratio between the power of Alpha 21164-300Mhz and the 500Mbytes/s peak bandwidth and 0.5 to 2 $\mu s$ latency for the shmem-put communication routine.

It is noteworthy that, contrary to dense LU factorization, the computational load depends not only on the matrix dimension but also on the initial or (even more) final density. This way, EX10, EX10HS, and SHERMAN5 are the only matrices with $\rho_n$ <5% and with lowest speed-up on 16 processors.

Therefore, better speed-ups on 16 processors are reached for matrices with high n and high $\rho_n$. The best speed-up is presented for CAVITY16. For some of the bigger matrices, such as WANG1, WANG2, and EX14, parallel factorization exhibits super-lineal speed-up even for four processors.

Regarding the solve stage, we did not reach speed-up when using more than 4 processors. The reason is the low computational load presented by these matrices

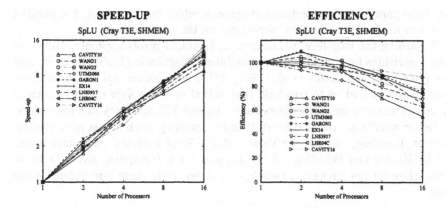

**Fig. 2.** SpLU speed-up and efficiency

for this step. The time expended on the solve stage never exceeded 0.3 seconds for any of the 17 matrices (the more expensive is CAVITY16 expending 0.25 seconds on this stage). Additionally, our solve stage comprises a sparse and a dense part, and the last one can only exploit unidimensional parallelism. Therefore, redistribution for the dense submatrix implies some overhead. In any case, the parallel solve stage is worthwhile as it permits us to solve matrices which do not fit on an single processor. Moreover, the parallel solve stage avoids collecting the whole matrix in one processor. As future work we will try to reduce this communication overhead by using a block cyclic distribution.

## 4 Related Work

In this section we summarize recent works, organizing them according to the level of parallelism. Regarding task level parallelism, Gallivan, Marsolf, and Wijshoff (1996) [12] carry out a matrix reordering to a bordered block triangular form. In the same direction, Zlatev et al. (1995) [19] have developed the tool PARASPAR, to solve the linear system on shared memory multiprocessors. By using a better reordering stage, LORA-P[5] code, the Y12M3 program achieves speed-ups between 3.0 and 4.7 on 8 processors of the Alliant FX/80 [18].

Apart from reordering, another source of task parallelism is multifrontal or supernode methods. As these methods were traditionally applied to symmetric matrices, parallel cholesky multifrontal codes were quickly developed. Gupta, Karypis, and Kumar (1995) [14] were the authors who reported probably the best performance for the sparse cholesky.

A significant more difficult problem appears when matrices are not symmetric. Here, the supernode tree is the tool to exploit task level parallelism. A parallel version of the SuperLU is presented by Li et al. [16], achieving on 8 processors shared memory machines, and for 21 unsymmetric sparse matrices, an average speed-up of less than 4. Better results can be achieved on distributed memory machines as recently shown by Fu, Jiao, and Yang (1998) [11]. However, they have parallelized the factorize stage only, which can be executed in parallel thanks to fill-in overestimation carried out on the analyse stage.

Regarding the loop level parallelism and parallel pivots approach, there are some algorithms for shared memory machines: Alaghband (1995) [1], Davis and Yew (1990) [5], and Zlatev et al. (1995) [19]. The experimental results of these previous works lead us to conclude that actual system solver implementations for shared memory multiprocessors hardly exceed 50% efficiency in 8 processors.

Better results are reached for distributed memory machines as presented by Stappen, Bisseling, and van der Vorst (1993) [17] for a square Transputer mesh, and by Koster and Bisseling (1994) [15] also for a transputer mesh. In these codes, they do not present a switch to a dense code stage nor parallel solve stage.

## 5   Conclusions

This work presents a complete tool, SpLU, to solve large nonsymmetric linear systems on shared memory multiprocessors. SpLU code comprises analyse-factorize and solve stages. Both of them were split into sparse and dense steps to avoid applying sparse techniques when fill-in turns the problem into a dense one. The algorithm follows a generic approach exploiting loop-level parallelism and takes advantage of matrix sparsity due to parallel pivoting selection. We have compared sequential SpLU with another generic sequential nonsymmetric sparse solver: the high optimized MA48 routine. Our SpLU code leads in many cases to fewer numerical errors and fill-in than MA48 does. On the other hand, MA48 is usually slightly faster than the sequential SpLU, but to the best of our knowledge MA48 can not be parallelized efficiently. Therefore, since SpLU exhibits a high degree of parallelism, speed-up computed as MA48 sequential time divided by parallel SpLU execution time is still competitive.

As far as we know, there is no published work for the whole parallel nonsymmetric sparse system solver on current distributed memory machines, including sparse analyse-factorize stage, switch to dense LU factorization stage, and forward and backward substitution.

On the other hand, SpLU could be improved mainly in two areas. The first is further reducing communication overhead by using a block cyclic distribution instead of a cyclic one. The second one is directed at reducing data movements and to make entries insertion easier using an unordered linked list both by rows and columns. This two points joined with better care of cache exploiting would result in higher performances.

## Acknowledgements

We gratefully thank Iain Duff and all the members in the parallel algorithm team at CERFACS, Toulouse (France), for their kind help and collaboration. We also thank the CIEMAT (Centro de Investigaciones Energéticas, Medioambientales y Tecnológicas), Spain, for giving us access to the Cray T3E multiprocessor.

# References

[1] G. Alaghband. Parallel sparse matrix solution and performance. *Parallel Computing*, 21(9):1407–1430, 1995.

[2] R. Asenjo and E.L. Zapata. Parallel pivots lu algorithm on the Cray T3E. Technical Report UMA-DAC-99/01, Dept. of Computer Architecture, University of Mlaga, Spain, http://www.ac.uma.es/, 1998.

[3] E. Chu and A. George. Gaussian elimination with partial pivoting and load balancing on a multiprocessor. *Parallel Comput.*, 5:65–74, 1987.

[4] T. A. Davis. *A parallel algorithm for sparse unsymmetric LU factorization*. PhD thesis, Center for Supercomputing Research and Development, Univ. of Illinois, Urbana, IL, September 1989.

[5] T. A. Davis and P. C. Yew. A nondeterministic parallel algorithm for general unsymmetric sparse LU factorization. *SIAM J. Matrix Anal. Appl.*, 11:383–402, 1990.

[6] Tim Davis. Sparse matrix collection. At URL http://www.cise.ufl.edu/ davis/.

[7] J.J. Dongarra, I.S. Duff, D.C. Sorensen, and H.A. van der Vorst. *Solving Linear Systems on Vector and Shared Memory Computers*. Society for Industrial and Applied Mathematics, 1991.

[8] I.S. Duff, A.M. Erisman, and J.K. Reid. *Direct Methods for Sparse Matrices*. Oxford University Press, Oxford, U.K., 1986.

[9] I.S. Duff, R.G. Grimes, and J.G. Lewis. User's guide for the Harwell-Boeing sparse matrix collection (Release I). Technical report, CERFACS, Toulouse, France, 1992.

[10] I.S. Duff and J.K. Reid. The design of MA48: A code for the direct solution of sparse unsymmetric linear systems of equations. *ACM Trans. Math. Softw.*, 22(2):187–226, June 1996.

[11] C. Fu, X. Jiao, and T. Yang. Efficient sparse lu factorization with partial pivoting on distributed memory architectures. *IEEE Transaction on Parallel and Distributed Systems*, 9(2):109–125, February 1998.

[12] K. Gallivan, B. Marsolf, and H.A.G. Wijshoff. Solving large nonsymmetric sparse linear systems using MCSPARSE. *Parallel Computing*, 22(10):1291–1333, 1996.

[13] G. A. Geist and C. H. Romine. LU factorization algorithm on distributed-memory multiprocessor architecture. *SIAM J. Sci. Statist. Comput.*, 9:639–649, 1989.

[14] A. Gupta, G. Karypis, and V. Kumar. Highly scalable parallel algorithms for sparse matrix factorization. *IEEE Transactions on Parallel and Distributed Systems*, 8(5), 1995. Available at URL: http://www.cs.umn.edu/~kumar.

[15] J. Koster and R.H. Bisseling. An improved algorithm for parallel sparse LU decomposition on a distributed memory multiprocessor. In J.G. Lewis, editor, *Fifth SIAM Conference on Applied Linear Algebra*, pages 397–401, 1994.

[16] X. Li. *Sparse Gaussian Elimination on High Performance Computers*. PhD thesis, CS, UC Berkeley, 1996.

[17] A. F. van der Stappen, R. H. Bisseling, and J. G. G. van de Vorst. Parallel sparse LU decomposition on a mesh network of transputers. *SIAM J. Matrix Anal. Appl.*, 14(3):853–879, July 1993.

[18] A.C.N. van Duin, P.C. Hansen, T. Ostromsky, H.A.G. Wijshoff, and Z. Zlatev. Improving the numerical stability and the performance of a parallel sparse solver. *Computers Math. Applic.*, 30:81–96, 1995.

[19] Z. Zlatev, J. Waśniewski, P.C. Hansen, and T. Ostromsky. PARASPAR: a package for the solution of large linear algebraic equations on parallel computers with shared memory. Technical Report 95-10, Tech. Univ. Denmark, Lyngby, 1995.

# Experiments with Parallel One–Sided and Two–Sided Algorithms for SVD [*]

Martin Bečka[1], Sophie Robert[2], and Marian Vajteršic[3,1]

[1] Institute for Informatics, Slovak Academy of Sciences, Bratislava, Slovakia
[2] LORIA, Vandoeuvre-Lès-Nancy Cedex, France
[3] Institute for Mathematics, University of Salzburg, Austria

**Abstract.** A paper reports on testing parallel SVD algorithms for matrices arising from selected scientific and industrial applications. The codes for the SVD are based respectively on the one–sided and the two–sided Jacobi approach. The matrices come from solving problems of the diffraction process in the crystallography, the diffusion equation in the reactor physics and from the aircraft industry. A parallelization of each of these approaches is described. Results from computational experiments performed on the Paragon machine with 56 processors are presented and discussed.

## 1 Introduction

This paper brings results of parallel computational experiments for a computation of SVD (Singular Value Decomposition) of dense matrices. Let $A$ be a rectangular complex matrix of order $m \times n$, the problem of SVD is to get $A = U\Sigma V^H$, where $U$ and $V$ are unitary matrices of orders $m$ and $n$ respectively and $\Sigma$ is an $m \times n$ diagonal matrix. The diagonal elements of $\Sigma$ are the singular values of $A$ and the columns of $U$ and $V$ are the left and right singular vectors, respectively. ($H$ denotes the Hermitian transpose.) There are two parallel algorithms considered. Their mathematical background is respectively the one–sided [4], [7] and two–sided Jacobi method [2], [3].

The matrices selected come from real application problems. The Tolosa matrices arise by solving problems in the aircraft industry and is belonging to the Harwell-Boeing collection. By solving the diffraction process in the crystallography, thin and large matrices arise whose SVD decompositions are required in a huge amount repeatedly. The boundary problem for the two–group diffusion equation delivers square matrices with a rather regular occupancy of non-zero elements with a Toeplitz–like structure.

The paper is organized as follows. Section 2 is devoted to a presentation of the background information of the test matrices. The kernel of the work brings Section 3 where besides the parallelization approaches also results of parallel computational experiments are presented. Comparison and some conclusions are in the final Section 4.

---

[*] The research described in this paper has been supported by the EC project STABLE CP96-0237.

P. Zinterhof, M. Vajteršic, A. Uhl (Eds.): ACPC'99, LNCS 1557, pp. 48–57, 1999.
© Springer-Verlag Berlin Heidelberg 1999

## 2   Test Matrices

### 2.1   Aircraft–Industry Matrices

A technical discipline which these matrices arise from, is aeroelasticity. Their working acronym is Tolosa due to the city of Toulouse, where the Aerospatiale Aircraft Division and CERFACS are located. They were provided by the Aerospatiale Aircraft Division, as an industrial problem [1]. The Tolosa matrices are involved in the Harwell–Boeing collection.

The goal was to compute some eigenvalues of large Tolosa matrices. These eigenvalues are directly related to the flutter frequencies: the damping is represented by the ratio (real part)/(imaginary part) and the frequency is expressed in a form of the ratio (imaginary part)/$(2\,\pi)$.

These matrices have an order $n$ greater than or equal to 90 ($n$ is always a multiple of 5). The Tolosa matrices have a $5 \times 5$ block–structure

$$A = \begin{pmatrix} 0 & I & 0 & 0 & 0 \\ Y_1 & Y_2 & Y_3 & Y_4 & Y_5 \\ 0 & I & l_1 I & 0 & 0 \\ 0 & I & 0 & l_2 I & 0 \\ 0 & I & 0 & 0 & l_3 I \end{pmatrix} ,$$

where each block has the size $m \times m$ ($m = n/5$). $0$ ($I$) represents the null (identity) matrix of order $m$. $Y_1, Y_2, Y_3, Y_4$ and $Y_5$ are five dense blocks of dimensions $m \times m$. $l_1, l_2$ and $l_3$ are three given scalars. The blocks $Y_1, Y_2, Y_3, Y_4$ and $Y_5$ have the structure

$$Y_1 = \begin{pmatrix} \frac{X_1}{\quad} & \\ x_1 & \\ & \ddots \\ & \quad x_{m-18} \end{pmatrix} , \qquad Y_2 = \begin{pmatrix} \frac{X_2}{\quad} & \\ y_1 & \\ & \ddots \\ & \quad y_{m-18} \end{pmatrix} ,$$

where $x_i = -\omega_i^2$ and $y_i = -2\alpha_i\omega_i$, $\alpha_i$ and $\omega_i$ are given scalars ($1 \le i \le m - 18$). The blocks $X_1$ and $X_2$ are the same as the blocks $Y_1$ and $Y_2$ for $n = 90$. For $k = 3, 4, 5$:

$$Y_k = \begin{pmatrix} \frac{X_k}{\quad} & \\ z_1 & \\ & \ddots \\ & \quad z_{m-18} \end{pmatrix} ,$$

where $z_i = 0$ ($1 \le i \le m - 18$). The blocks $X_k$ are the same as the blocks $Y_k$ for $n = 90$.

### 2.2   Diffraction–Process Matrices

Four matrices (va–w1.asc, va–vsqr.asc, va–wvar.asc, va–wyo.asc) have been chosen for testing from this application. They arise in investigation of the diffraction process in the crystallography by a newly developed approach at the Institute of Anorganic Chemistry of the Slovak Academy of Sciences in Bratislava.

In this method, an application of different weights is performed to structures obtained by X–ray produced data. Mathematical method behind is based on weighted m–linear least squares. In order to get quantities which characterize the diffraction, SVD of a large amount of matrices has to be computed [8]. To process one of them on a Pentium-equipped PC takes about 1.5 min. By serial processing, the computational demand of hours is needed for analysis of the behaviour of the process for a reasonable set of weighting parameters.

The matrices are dense and no particular regularities in their structure can be observed. Their shape is thin and large ($13 \times 3327$) and could be seen as extreme in the context of some parallelization strategies. From the nature of the process, their singular values are steeply ordered in the magnitude, whereby the first of them is possessing more than 90% of the total spectrum.

### 2.3    Neutron–Diffusion Matrices

These matrices are related to solving the steady-state two-group neutron diffusion equation for the WWER-440 type nuclear power reactors [6]. The reactor core is divided into $N_z$ horizontal layers the top one with thickness of 10 cm and the rest with equal thicknesses. Thus each fuel assembly is divided into regular hexagonal prisms, called meshes or nodes. The region of solution includes only the meshes occupied by fuel. Since the control assemblies are movable up and down and consist of an absorber part and fuel follower, some of the meshes in the core may be occupied by absorber and they are excluded from the region of solution. A mesh can be either fully occupied by fuel or fully occupied by absorber.

The matrices coming out from the discretization of the equation are square with a relatively large amount of zeros, but when their order is not too large they can be considered as dense and hence, suitable for our tests. They are possessing a Toeplitz–like structure and their working notation is $BG_A$, $BG_B$ and $BG_C$ respectively.

## 3    Parallel Implementation and Test Results

### 3.1    One–Sided Jacobi Method

**Parallelization.** An implementation of this algorithm in Fortran has been done on the Paragon machine of IRISA. The communication library used is NX. The implementation consists in using a linear network of $p$ processors. Each processor receives $\lfloor \frac{n}{2p} \rfloor$ or $\lceil \frac{n}{2p} \rceil$ couples of column vectors. They orthogonalization proceeds according to the Brent and Luk scheme which is depicted on Fig. 1. To overlap communications and calculations, the computations begin by the column vectors which need to be communicated to a neighbour.

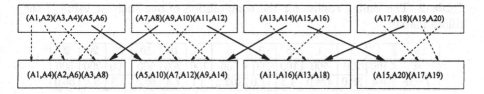

**Fig. 1.** The communications between two steps of a sweep for a matrix of order 20 distributed on a linear network of 4 processors

**Tests with the Aircraft–Industry Matrices.** The tests with the Tolosa matrices were performed in the context of spectral portrait which needs the computation of the smallest singular values of $A - zI$ for all $z$ of a grid from the complex plane. For this problem, a new approacht has been developed.

The aim was to integrate data parallelism coming from the independence between the computation of smallest singular values for two different $z$. Because of this independence, the grid chosen can be shared to blocks of processors. These blocks compute in parallel the singular values of $A - zI$ for their $z$. By this way the use of this method of Jacobi can be reasonable only when the matrix is of small order to allow us to distribute it to only one processor.

The timing results *Time* (in sec.) obtained on 50 processors are given in Table 1 for the Tolosa matrix of the size 100, 200 and 500 respectively. The grid of $z$ considered is of the size $50 \times 50$.

**Table 1.** The timing results for the computation of the spectral portrait of the Tolosa matrix

| $p$ | 50 | 50 | 50 |
|---|---|---|---|
| $n$ | 100 | 200 | 500 |
| *Time* | 545 | 1972 | 21104 |

Table 2 shows also timings (in sec.) for the computation of a singular value for different number of processors from 1 to 20. In particular, it shows that the gain of a parallelization of a one-sided Jacobi algorithm is not enough to win when we choose to distribute the computations of the orthogonalizations on more than a processor when the matrix is of order less than 500.

**Tests with the Diffraction–Process Matrices.** The computations were performed for four matrices of order $3327 \times 13$. Since the number of column vectors is odd and the Brent and Luk ordering is formulated for even orders, a null vector column has been added to each matrix. Since the parallelism is applied across the columns the maximum of processors employed is 7. (This corresponds

52      Martin Bečka, Sophie Robert, and Marian Vajteršic

**Table 2.** The timing results for the computation of one singular value of the Tolosa matrix

| $p$ | 1 | 1 | 1 | 5 | 5 | 5 | 10 | 10 | 10 | 20 | 20 | 20 |
|------|-----|-----|-------|-----|------|-------|-----|-----|------|-----|-----|------|
| $n$ | 100 | 200 | 500 | 100 | 200 | 500 | 100 | 200 | 500 | 100 | 200 | 500 |
| $Time$ | 9.4 | 38.0 | 424.0 | 3.3 | 11.0 | 102.0 | 2.1 | 7.8 | 60.0 | 1.9 | 5.8 | 36.0 |

to the distribution of two matrix columns to one processor, as prescribed by the Brent and Luk ordering.)

The timings $Time$ (in sec.), the performance $V$ (in Mflops) and the error $e$ are shown in Table 3 where $MAT$ is the name of the matrix issued by the crystallography application. The error $e$ is the relative value related to the smallest singular value $\tilde{\sigma}_{min}$ (computed by the Lapack function) and our estimation $\sigma_{min}$:

$$e = |\frac{\sigma_{min} - \tilde{\sigma}_{min}}{\tilde{\sigma}_{min}}|. \tag{1}$$

The value chosen to stop the iteration is $10^{-10}$.

**Table 3.** The timing results, the performance and the error for the diffraction–process matrices

| $MAT$ | va-w1.asc | va-wsqr.asc | va-wvar.asc | va-wyo.asc |
|-------|-----------|-------------|-------------|------------|
| $p$ | 7 | 7 | 7 | 7 |
| $Time$ | 0.9 | 1.0 | 0.9 | 1.1 |
| $V$ | 61 | 63 | 61 | 64 |
| $e$ | $10^{-11}$ | $10^{-14}$ | $10^{-12}$ | $10^{-12}$ |

**Tests with the Neutron–Diffusion Matrices.** The tests have been performed with matrices $BG_A$ and $BG_C$ (abbreviated as A and C respectively) which are square of order 634. We were interested at the smallest singular values. In this case, these values are very close for both matrices. As observed from the timing results $Time$ (in sec.) and the performance $V$ (in Mflops) obtained on the different number of processors $p$, these are also very close (Table 4). The criterion to stop the iterative process is the same as above. The error, according to the criterion (1) is of the order $10^{-14}$.

**3.2   Two–Sided Jacobi Method**

**Parallelization.** For parallel execution of the algorithm it is reasonable to distribute $A$ column-wise (row-wise) among the processors in such a way, that two block columns (rows) of $A$ belong to one processor. We use the column-wise

**Table 4.** The timing results and the performance for the neutron–diffusion matrices.

| $MAT$ | A or C | A or C | A or C | A or C | A or C | A or C |
|-------|--------|--------|--------|--------|--------|--------|
| $p$   | 5      | 10     | 20     | 30     | 40     | 56     |
| $Time$ | 410   | 213    | 114    | 83     | 64     | 51     |
| $V$   | 82     | 158    | 295    | 410    | 525    | 660    |

method. Let $p$ be the number of processors. Then the slicing with a parameter $l = 2*p$ produces the submatrices $A_{ij}$ of the size $(m/l + w1) \times (n/l + w2)$, where

$$w1 = \begin{cases} 1 & i < m \bmod l \\ 0 & \text{otherwise} \end{cases} \quad \text{and} \quad w2 = \begin{cases} 1 & i < n \bmod l \\ 0 & \text{otherwise.} \end{cases}$$

A crucial problem of the parallelization represents the ordering scheme for SVD computations on submatrices

$$S_{ij} = \begin{pmatrix} A_{ii} & A_{ij} \\ A_{ji} & A_{jj} \end{pmatrix}.$$

One outer iteration (sweep) of the algorithm consists of solving SVD for every pair $(i,j)$ $(i < j,\ i,j = 0,1,\dots,l-1)$ until

$$F(A) = \sqrt{\sum_{i,j=1, i \neq j}^{l} \|A_{ij}\|^2} < \epsilon,$$

for a given accuracy $\epsilon$ ($F(A)$ denotes the Frobenius norm of $A$ and $\|.\|$ is a proper matrix norm). A subproblem is solved when the condition

$$F(S_{ij}) = \sqrt{\|A_{ij}\|^2 + \|A_{ji}\|^2} \geq \delta$$

is satisfied for a given $\delta > 0$.

The subproblems are solved in an order that can be characterized by the ordering $(i_0, j_0), \dots, (i_t, j_t)$, $t = l*(l-1)/2 - 1$. Let us arrange the sequence of all pairs $(i,j)$, $i < j$, $i,j = 0,1,\dots,l-1$, in a $(l-1) \times p$ table. This table implies an initial distribution (the first row) and interchanges of block columns between processors. The tables suitable for our purposes should satisfy the following conditions :

- in any row it holds : $((a,b) \neq (c,d)) \Rightarrow (a \neq c \ \& \ b \neq d)$
- in any column every two consecutive couples differ only in one element.

A number of orderings, that satisfy this restriction is proposed in [3], Fig. 2 shows one example of them. Moreover, they are suitable for such architectures as rings and hypercubes because the interchanges take place between neighbouring processors. We have chosen $CO(0)$ ordering [3].

$$\begin{array}{cccc}
P(0) & P(1) & P(2) & P(3) \\
(0,1) & (2,3) & (4,5) & (6,7) \\
(0,2) & (4,3) & (6,5) & (1,7) \\
(0,7) & (4,2) & (6,3) & (1,5) \\
(0,4) & (6,2) & (1,3) & (7,5) \\
(0,5) & (6,4) & (1,2) & (7,3) \\
(0,6) & (1,4) & (7,2) & (5,3) \\
(0,3) & (1,6) & (7,4) & (5,2) \\
(0,1) & (7,6) & (5,4) & (3,2)
\end{array}$$

**Fig. 2.** The $CO(0)$ ordering for p=4

For $m = n$, the complexity of the parallel algorithm is [9]

$$TIME_{par} = 2\,p\,niter\left(c_1\,(n/p)^3 + 2\,c_2\,n^3/p^2 + startup\,\log p + 3/2\,c_3\,n^2/p\right)$$

$$= 2\,niter\left(\frac{c_1 + 2\,c_2\,p}{p^2}\,n^3 + 3/2\,c_3\,n^2 + startup \cdot p\,\log p\right) \qquad (2)$$

$$= 2\,niter\left(startup \cdot p\,\log p + \frac{2c_2 n^3}{p} + \frac{c_1 n^3}{p^2} + 3/2\,c_3\,n^2\right). \qquad (3)$$

Here, $c_1$ is a constant for the embedded SVD computation on subproblems, $c_2$ is a constant for the matrix multiplication and $niter$ is the number of global iterations which is a function of $A, m, n, p$.

In our model of communication, the time to send a message of *datasize* units is expressed by the formula

$$startup + c_3\,datasize$$

where *startup* represents the latency - the amount of time every message requires as overhead, and $c_3$ represents the amount of time to send one unit.

When comparing (2) with the serial complexity

$$TIME_{seq} = 2\,p^2\,niter\left(c_1\,(n/p)^3 + 2\,c_2\,n^3/p^2\right) = \frac{2\,niter\,(c_1 + 2\,c_2\,p)}{p}\,n^3, \qquad (4)$$

the parallel behaviour of the algorithm is quite nice. For the computational part we obtain an optimal speed–up $p$, but there is an additional time for the communication $(3/2\,c_3\,n^2 + startup \cdot p\,\log p)$. When the size of $n$ is doubled then the increase in the computational part of the algorithm is 8-fold and the amount of transmited data is 4-fold (see (2)). On the contrary, when the size of $p$ is doubled, the time spent by the SVD computation and the matrix multiplications decreases respectively 4– and 2 times, but the increase in the number of startups is slightly more than 2–fold (see (3)).

**Implementation of the Two–Sided Algorithm.** The above parallel algorithm was implemented on the Paragon machine [5] at IRISA Rennes. For programming of the communication steps, the MPI (Message Passing Interface) has been used. For the matrix multiplication occurring in our algorithm, the procedure *zgemm* of BLAS has been applied.

We have performed experiments with two procedures for the computation of SVD in the inner loop of the algorithm where the subproblem solving is realized. These are the one–sided Jacobi method and *zgesvd* function from LAPACK. After the tests we have decided to use the LAPACK routine for further computations.

For the termination of the global iterative process we have set either

$$F(A) \geq \epsilon \text{ AND } iter \leq threshold \tag{5}$$

(*threshold* denotes a prescribed maximum number of sweeps and *iter* is the actual iteration number) or from experiments there follows a better choice of the criterion which is expressed (for a properly chosen accuracy *precision*$_1$) by

$$F(A) \geq \epsilon \text{ AND } \|\sigma_{min}^{iter} - \sigma_{min}^{iter-1}\| \geq precision_1. \tag{6}$$

**Tests with the Aircraft–Industry Matrices.** Table 5 contains the correct reference values for Tolosa matrices. The results of computational experiments with these matrices is summarized in Table 6 (*niter* denotes the number of sweeps). The convergence criterion is given as a conjunction of (5) and (6).

**Table 5.** Reference values of $\tilde{\sigma}_{min}$ for the Tolosa matrix

| $n$ | $\tilde{\sigma}_{min}$ |
|-----|------------------------|
| 100 | 0.99862149106250 |
| 200 | 0.99862149106281 |
| 500 | 0.99862149106237 |

**Tests with the Neutron–Diffusion Matrices.** We have performed tests for the neutron–diffusion matrices BG$_A$ and BG$_B$ which are of the order 634 and 1226 respectively. The results are collected in Table 7 and Table 8.

The relative error $e$ given in Table 7 is computed at the same manner as given in (1). The reference value $\tilde{\sigma}_{min}$ is equal to 0.98814467236727.

## 4   Conclusions

Two fast parallel Jacobi algorithms have been tested by computing the SVD for special matrices of three properly chosen scientific and industrial applications.

**Table 6.** Timing results for the Tolosa matrix with $precision = 10^{-5}$, $treshold = 10$ and $precision1 = 10^{-5}$

| $n$ | $p$ | $\sigma_{min}$ | $Time$ | $niter$ |
|-----|-----|----------------|--------|---------|
| 100 | 5 | 0.9986248872035146 | 7.71 | 8 |
| 100 | 10 | 0.9985115620289451 | 7.15 | 7 |
| 100 | 20 | 0.9985427240975232 | 10.94 | 8 |
| 100 | 30 | 0.9986185494253728 | 16.95 | 8 |
| 100 | 40 | 0.9986217795919820 | 21.85 | 7 |
| 100 | 50 | 0.9986213362495867 | 25.91 | 7 |
| 200 | 5 | 0.9986216328458206 | 9.37 | 4 |
| 200 | 10 | 0.9981468664059168 | 12.83 | 7 |
| 200 | 20 | 0.9983950119547799 | 12.59 | 6 |
| 200 | 30 | 0.9982546792048150 | 15.87 | 6 |
| 200 | 40 | 0.9946569039266127 | 21.70 | 6 |
| 200 | 50 | 0.9974083279045680 | 28.51 | 6 |
| 500 | 5 | 0.9986215498497516 | 67.61 | 3 |
| 500 | 10 | 0.9986216330843882 | 26.46 | 3 |
| 500 | 20 | 0.9841441847838350 | 43.52 | 6 |
| 500 | 30 | 0.9984056526099007 | 42.06 | 8 |
| 500 | 40 | 0.9861434709368520 | 36.59 | 7 |
| 500 | 50 | 0.9982855574596819 | 27.23 | 5 |

**Table 7.** Timing results for the $BG_A$ matrix with $precision = 10^{-5}$, $treshold = 10$ and $precision1 = 10^{-5}$

| $n$ | $p$ | $\sigma_{min}$ | $Time$ | $niter$ | $e$ |
|-----|-----|----------------|--------|---------|-----|
| 634 | 10 | 0.9881446742207031 | 323.39 | 6 | $10^{-8}$ |
| 634 | 20 | 0.9881446728032011 | 122.86 | 6 | $10^{-8}$ |
| 634 | 30 | 0.9881446859165189 | 80.71 | 5 | $10^{-7}$ |
| 634 | 40 | 0.9881446725499764 | 77.31 | 6 | $10^{-9}$ |
| 634 | 50 | 0.9881446729495019 | 65.87 | 5 | $10^{-9}$ |

**Table 8.** Timing results for the $BG_B$ matrix with $precision = 10^{-5}$, $treshold = 10$ and $precision1 = 10^{-5}$

| $n$ | $p$ | $\sigma_{min}$ | $Time$ | $niter$ |
|-----|-----|----------------|--------|---------|
| 1226 | 10 | 1.000086456879128 | 2300.69 | 7 |
| 1226 | 20 | 1.000086500491703 | 682.77 | 6 |
| 1226 | 30 | 1.000086496939367 | 463.98 | 6 |
| 1226 | 40 | 1.000086457148098 | 460.19 | 6 |
| 1226 | 50 | 1.000086457871167 | 373.87 | 6 |

The codes are written in Fortran and have been run on the distributed–memory parallel system Intel Paragon of IRISA in Rennes.

From the timing comparisons of the one– and two–sided Jacobi–based codes it follows that a preference concerning the time consumption for the solution has to be given to the one–sided approach. (See e.g. the timings in Table 2 and Table 4 and those in Table 6 and Table 7 respectively.)

From the tests performed with the one-sided Jacobi algorithm it is apparent that this is well adapted to rectangular matrices where $m \ll n$ because it encompasses block–column operations (for $m \gg n$ the transpose can be considered). However, for the matrices from the diffraction–process application the maximum number of processors is restricted to only 7 because of their extremely thin shape. A possible improvement could be the insertion of the parallelism onto the level of the column orthogonalizations and to use a rectangular grid of processors instead of the linear network of processors as it was the case in our implementation. The accuracy constraints have seem not to cause any problems when running this code with the test matrices.

The higher timings for the two–sided approach are not surprising because this requires significantly more operations than its one–sided counterpart. The two–sided Jacobi code was not reasonably applicable to the thin diffraction–process matrices because there arises a multiplication by square matrices of the order which is equal to the larger block–matrix size. This multiplication is to be done in each iteration which makes the whole process unproportionally expensive. By this method also some convergence problems have been observed. Since from a certain point the increase of the number of iterations did not lead to an improvement of the accurracy to a prescribed value a problem–dependent criterion with a fixed number of iterations was used to stop the iterations.

# References

1. Godet-Thobie, S.: Eigenvalues of large highly nonnormal matrices. Ph.D. thesis, Paris IX Dauphine University, CERFACS thesis report TH/PA/93/06, 1992.
2. Golub, G.H., Van Loan C.H.: Matrix Computations. The Johns Hopkins University Press, 1989.
3. Guerrini, C., Vajteršic, M.: Optimal parallel ordering scheme for solving SVD on a hypercube multiprocessor. Int.J.Mini & Microcomputers, 16, 1994, 49-56.
4. Luk, F.T., Park, K.: On parallel Jacobi orderings. SIAM J. Sci. Stat. Comput., 10, 1989, 18–26.
5. Paragon System User's Guide. Intel Corporation, 1995.
6. Petkov, P.T.: SPPS-1.6—A 3D Diffusion Code for Neutronics Calculations of the VVER-440 Reactors. Proc. of the Fourth Symposium of AER, Sozopol, 10–15 Oct., 1994.
7. Rijk, P.M.: A one-sided Jacobi algorithm for computing the singular value decomposition on a vector computer. SIAM J. Sci. Statist. Comput., 10, 1989, 359–371.
8. Smrčok, L.: Solving the diffraction process in crystallography by the weighted regression. (Personal communication), 1997.
9. Vajteršic, M., Robert, S., Bečka, M. and Dimov, I.: Testing the parallel SVD codes for dense matrices. Deliverable WP 1.1 of the INCO–COPERNICUS–96 0237 project, 1997.

# Combined Systolic Array for Matrix Portrait Computation

Gabriel Okša

Institute for Informatics, Slovak Academy of Sciences,
Bratislava, Slovak Republic

**Abstract.** Given the matrix $A \in C^{n \times n}$ and scalars $\lambda_1, \lambda_2, \ldots, \lambda_m \in C$, our task is to design a systolic implementation of the matrix portrait computation – i.e., the singular value decomposition of matrices $A - \lambda_k I$, $k = 1, 2, \ldots, m$. We propose the triangular-rectangular and hexagonal systolic subarrays for the recursive QR updating of matrices $A - \lambda_k I$, and another triangular subarray for the singular value decomposition of the R-factor. Let $m$, $n$ and $r$ be the number of various $\lambda$s, the matrix order and the number of repeated loops in the SVD algorithm, respectively. Due to the large amount of overlap between subarrays, the time complexity of our solution is $O(3mn)$ whereas the straightforward systolic implementation requires $O(\lceil \frac{7}{2} mn \rceil + 4rm)$ times steps. The number of PEs and delays is $O(\lceil c\,n^2 \rceil)$, where $c = \frac{37}{8}$ for our solution and $c = \frac{5}{8}$ for the straightforward solution.

## 1 Problem Formulation

Let $A \in C^{n \times n}$ be a complex matrix of order $n$ and $\lambda_1, \lambda_2, \ldots, \lambda_m \in C$ a given set of complex numbers. Our task is to design a systolic algorithm for the computation of singular value decomposition (SVD) of the sequence of matrices $B_k = A - \lambda_k I$, $k = 1, 2, \ldots, m$ where $I \in \mathcal{R}^{n \times n}$ is the identity matrix. The set of minimal singular values $\sigma_{min}^{(k)}$, $k = 1, 2, \ldots, m$ defines the portrait of matrix $A$. The matrix portrait is needed, for example, in the analysis of the stability of dynamic systems.

We restrict ourselves to the real problem $A \in \mathcal{R}^{n \times n}$, $\lambda_1, \lambda_2, \ldots, \lambda_m \in \mathcal{R}$. The case of complex numbers $\lambda_i$ has no impact on the structure of designed systolic subarrays, only the complex arithmetic should be used instead of real one.

The numerically stable method for computing the SVD of matrix $B_k$ consists of its QR decomposition $B_k = Q_k R_k$ and of the reduction of $R_k$ to the diagonal form. Because $Q_k$ is orthogonal, the matrices $B_k$ and $R_k$ have the same singular values and they appear on the diagonal after reduction of $R_k$.

Both basic steps – QR decomposition and reduction of R-factor to the diagonal form – have well-known systolic algorithms. QR decomposition of square matrix of order $n$ can be computed on the triangular array of $O(n^2)$ processing elements (PEs) in time $O(n)$ by means of one-sided Givens rotations [3].

P. Zinterhof, M. Vajteršic, A. Uhl (Eds.): ACPC'99, LNCS 1557, pp. 58–67, 1999.
© Springer-Verlag Berlin Heidelberg 1999

The reduction of R-factor to its diagonal form requires another triangular array of $O(n^2)$ PEs that implements one-sided or two-sided Givens rotations (Kogbetliantz's method) [2], [5]. Givens rotations are organized in sweeps during which the "norm" of the off-diagonal elements of R-factor is successively reduced. Because the computation of SVD is inherently an iterative procedure, the number of sweeps depends on the matrix under consideration. But experience shows that – for given $n$ – a constant number of sweeps is required for the sufficient convergence of R-factor to its diagonal form. Hence, the time needed for the SVD triangular array is $O(n)$.

We begin with the description of QR decomposition by means of one-sided Givens rotations. Our main result is the algorithmical design of *recursive* QR decomposition of matrices $B_k$ for various $\lambda_k$. "Recursive" means that we utilize the maximum amount of information when proceeding from $k$th to $k + 1$th recursion step – i.e., we use the triangular factor $R_k$ in the computation of $R_{k+1}$. Once obtained, the R-factor is then reduced to the diagonal form using two-sided Givens rotations with permutations [2], [8].

The combination of algorithms is mirrored in the combination of systolic subarrays, which implement the matrix portrait computation. We show that the recursive QR decomposition can be realized on two subarrays: triangular-rectangular one that computes the QR decomposition, and hexagonal one that serves for the addition of certain matrices. Both arrays co-operate closely and their work can be overlapped. This is true also for the third array of triangular shape that implements the reduction of R-factor to the diagonal form. The smallest element on the main diagonal is then used for the construction of matrix portrait.

## 2   Basic Algorithms

### 2.1   Recursive QR Update

Let $B_1 = A - \lambda_1 I$ be the first matrix for which the portrait has to be computed. As mentioned in the first section, we compute its QR decomposition $B_1 = \tilde{Q}_1^t \tilde{Q}_1 (A - \lambda_1 I) = Q_1 R_1$ where $R_1 = \tilde{Q}_1 (A - \lambda_1 I)$ is upper triangular and the orthogonal matrix $\tilde{Q}_1$ incorporates the product of elementary Givens rotations that are needed to nullify the lower triangular part of $B_1$. The numerically stable algorithm for the computation of Givens rotation can be found in [4].

Suppose that the QR decomposition of $B_1 = A - \lambda_1 I$ was computed and we are interested in the QR decomposition of $B_2 = A - \lambda_2 I$. Then

$$B_2 = A - \lambda_2 I = (A - \lambda_1 I) + (\lambda_1 - \lambda_2)I$$
$$= Q_1 R_1 + (\lambda_1 - \lambda_2)Q_1 Q_1^t$$
$$= Q_1 [R_1 + (\lambda_1 - \lambda_2)Q_1^t]$$
$$= \underbrace{Q_1 \tilde{Q}_2^t}_{Q_2} \underbrace{\tilde{Q}_2 [R_1 + (\lambda_1 - \lambda_2)Q_1^t]}_{R_2}.$$

The last row in the above array of equations displays yet another product of elementary Givens rotations – the othogonal matrix $\tilde{Q}_2$, so that $B_2 = Q_2 R_2$, where $Q_2 = Q_1 \tilde{Q}_2^t$ is again orthogonal (the product of two orthogonal matrices) and $R_2 = \tilde{Q}_2[R_1 + (\lambda_1 - \lambda_2)Q_1^t]$ is upper triangular.

It is now clear how to proceed with computations when passing from $B_k$ to $B_{k+1}$, i.e., from $\lambda_k$ to $\lambda_{k+1}$. The computations are summarized in the algorithm A1.

- **Algorithm A1 -- recursive QR updating**

  1. Input: $A \in \mathcal{R}^{n \times n}$, $\lambda_i \in \mathcal{R}$, $i = 1, 2, \ldots, m$.
  2. Initialization:

$$B_1 = A - \lambda_1 I, \tag{1}$$
$$R_1 = \tilde{Q}_1 B_1, \tag{2}$$
$$Q_1^t = \tilde{Q}_1 I. \tag{3}$$

  3. Recursive computations:
     for $k = 1, 2, \ldots, m - 1$:

$$C_k = R_k + (\lambda_k - \lambda_{k+1})Q_k^t, \tag{4}$$
$$R_{k+1} = \tilde{Q}_{k+1} C_k, \tag{5}$$
$$Q_{k+1}^t = \tilde{Q}_{k+1} Q_k^t. \tag{6}$$

  **end**

- **End of algorithm A1**

The upper triangular matrix $R_{k+1}$ is the triangular factor in the QR decomposition of matrix $B_{k+1} = A - \lambda_{k+1} I$ and has to be reduced to the diagonal form if we wish to compute the SVD of $B_{k+1}$ (this reduction is *not* part of the algorithm A1).

Note that in the step $k + 1$ the matrices $R_k$ (the triangular factor) and $Q_k^t$ (*not* $Q_k$) from step $k$ are needed. Both computations of $R_{k+1}$ and $Q_{k+1}^t$ require the *same* transformation matrix $\tilde{Q}_{k+1}$ applied on the *rows* of $C_k$ and $Q_k^t$, respectively.

## 2.2   Reduction of R-Factor to Diagonal Form

After obtaining the upper triangular factor $R_k$ in the $k$th recursion step we need to reduce it to the diagonal form. The basic idea of an inherently iterative algorithm consists of the nullifying of non-diagonal elements of the matrix $R_k$ while preserving its upper triangular form.

Formally, the SVD steps can be defined as the series of transformations described by the algorithm A2 (cf. [2,8]):

- **Algorithm A2 -- SVD steps**

  **for** $k = 1, 2, \ldots, m$:
  1. Input: upper triangular matrix $R_k$.
  2. Computation:
      **for** $i = 1, 2, \ldots, n - 1$:

      $$R_k \Leftarrow G(i, i+1, \phi_k^i)^t \, R_k \, G(i, i+1, \psi_k^i). \tag{7}$$

      **end**
  **end**

- **End of algorithm A2**

The matrices $G(i, i+1, \phi_k^i)$ and $G(i, i+1, \psi_k^i)$ represent plane rotations (the $i$th rotation in the $k$th recursion) through angles $\phi_k^i$ and $\psi_k^i$ in the $(i, i+1)$-plane. The rotation angles should be chosen so that the $(i, i+1)$ element in $R_k$ is zeroed, while $R_k$ remains upper triangular. Each iteration therefore solves a $2 \times 2$ SVD subproblem on the main diagonal of matrix $R_k$.

The inner loop in the algorithm A2 represents zeroing of only certain non-diagonal matrix elements $(i, i+1)$, $i = 1, 2, \ldots, n-1$, i.e., of those lying on the first upper diagonal of $R_k$. However, to ensure the convergence of the Jacobi method we need to nullify *all* non-diagonal elements of $R_k$ by solving a $2 \times 2$ SVD subproblem on the main diagonal. This can be achieved by the pipelining of $n$ sequences of two-sided Givens rotations in (7) *including* the permutation $P_{i|i+1}$ which swaps the $i$th and $i + 1$th row (column) of the matrix $R_k$ (cf. [2,7]) and moves "fresh" elements on the first upper diagonal where they can be nullified. Then equation (7) becomes

$$R_k \Leftarrow P_{i|i+1} \, G(i, i+1, \phi_k^i)^t \, R_k \, G(i, i+1, \psi_k^i) \, P_{i|i+1}. \tag{8}$$

Note that the row (column) permutation can be incorporated into the appropriate elementary Givens rotation so that the permutation does not represent any computational overhead as compared with an ordinary Givens rotation.

It can be shown that $n$ repeated sequences of the inner loop in the algorithm A2 together with permutations in (8) are equivalent to a so called *pipelined double sweep* for a row-by-row cyclic ordering of elements of the matrix $R_k$ [9]. Based on the experience and the quadratic convergence of the Jacobi-like SVD algorithm (cf. [2,7]) we can restrict the number of sequences to $r$, $1 \le r \le n$.

For the parallel implementation of the algorithm A2 on a systolic array it is important to note that the $2 \times 2$ SVD subproblems on the main diagonal may not overlap. Hence, in the first turn, we can compute in parallel the subproblems with *odd* pivot indeces (i.e., $i = 1, 3, 5, \ldots, n-1$ - we assume $n$ even) which nullify the elements $(1, 2), (3, 4), \ldots, (n - 1, n)$, and then in parallel the subproblems with *even* pivot indeces, which nullify the elements $(2, 3), (4, 5), \ldots, (n-2, n-1)$. These alternative sub-loops of algorithm A2 are repeated $r$-times, $1 \le r \le n$. The splitting of pivot indeces in two groups (so called *odd-even ordering*, cf. [8]) is equivalent (in the sence of convergence to the diagonal matrix) to the cyclic row-by-row ordering (cf. [6]).

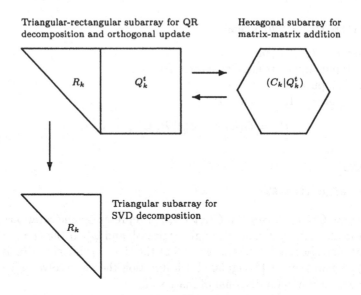

Triangular-rectangular subarray for QR decomposition and orthogonal update

Hexagonal subarray for matrix-matrix addition

Triangular subarray for SVD decomposition

**Fig. 1.** Systolic subarrays for the matrix portrait computation.

# 3 Systolic Implementation

## 3.1 Systolic Subarrays for Recursive QR Decomposition

In this subsection we design the systolic implementation of the algorithm **A1**. At the beginning, the matrix $B_1$ in (1) is computed by the host computer and augmented by the identity matrix $I$ into the form $(B_1|I) \in \mathcal{R}^{n \times 2n}$. This rectangular composed matrix is sent by host into the composed triangular-rectangular systolic subarray with $n^2 + n(n-1)/2$ PEs (Fig. 1). The composed matrix enters the systolic subarray row-oriented and in the skewed manner – the adjacent elements in the row are read by PEs one time step apart. The triangular part of the subarray serves for the computations in (2) of the triangular factor $R_1$. The elementary Givens rotations are produced by diagonal PEs and are pipelined to the right where they are used for the modification of stationary and moving rows of matrix $B_1$. When pipelined through the "border" between the triangular and rectangular parts, the Givens rotations begin to modify the original identity matrix $I$. Hence, the rectangular part serves for the computations in (3) and its PEs store the elements of $Q_1^t$.

Next, we need to form the matrix $C_1$ according to (4). Instead of forming it by the host, we propose another systolic subarray of hexagonal shape (Fig. 2). The subarray contains $2n^2 - n$ PEs and $n^2 - n$ delay elements (see below the discussion of the timing function). The input of the hexagonal subarray consists of the row-oriented matrices $R_1$, $Q_1^t$ that come from the triangular-rectangular subarray, and of the scalar $\lambda_1 - \lambda_2$ that is stored in the special memory cell and is read by the central line of PEs that perform the addition of two matrices according

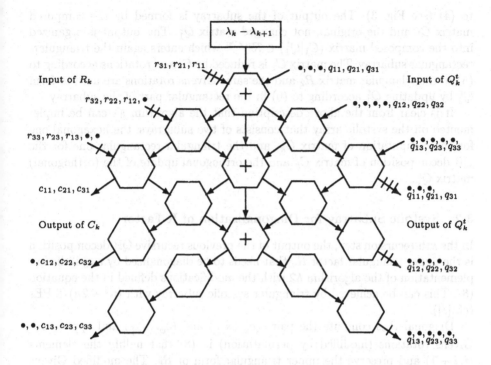

**Fig. 2.** Hexagonal subarray for the matrix-matrix summation.

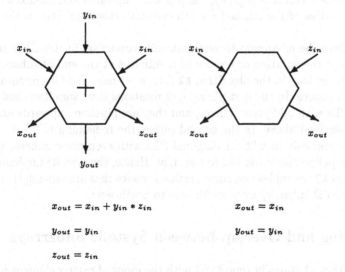

$$x_{out} = x_{in} + y_{in} * z_{in}$$

$$y_{out} = y_{in}$$

$$z_{out} = z_{in}$$

$$x_{out} = x_{in}$$

$$y_{out} = y_{in}$$

**Fig. 3.** Internal structure and function of PEs in the hexagonal subarray.

to (4) (see Fig. 3). The output of the subarray is formed by the computed matrix $C_1$ and the original, not changed matrix $Q_1^t$. This output is organized into the composed matrix $(C_1 | Q_1^t) \in \mathcal{R}^{n \times 2n}$ which enters again the triangular-rectangular subarray. The matrix $C_1$ is reduced by Givens rotations according to (5) to the triangular matrix $R_2$ and the same Givens rotations are used to form $Q_2^t$ by updating $Q_1^t$ according to (6) in the rectangular part of the subarray.

It is clear from the above description that the algorithm A1 can be implemented on the systolic array that consists of two subarrays: the hexagonal one for the computation of matrix $C_k$, and the triangular-rectangular one for the QR decomposition of matrix $C_k$ and the orthogonal update of the (orthogonal) matrix $Q_k^t$.

## 3.2   Systolic Subarray for Diagonalization of R-Factor

In the $k$th recursion step, the output of the previous recursive QR decomposition is the upper triangular factor $R_k$ that needs to be diagonalized by a systolic implementation of the algorithm A2 with the modifications defined in the equation (8). This can be achieved by triangular systolic subarray with $(n^2 + 2n)/8$ PEs (cf. [8]).

Diagonal PEs compute the pairs $(c_{\phi_k^i}, s_{\phi_k^i})$ and $(c_{\psi_k^i}, s_{\psi_k^i})$ and apply both Givens rotations (modified by permutation) in (8) that nullify the elements $(i, i+1)$ and preserve the upper triangular form of $R_k$. The modified Givens rotation $(c_{\phi_k^i}, s_{\phi_k^i})$ is then pipelined to the right and subsequently modifies all $2 \times 2$ data submatrices stored in PEs on the horizontal line. These submatrices consist of appropriate elements of the $i$th and $i + 1$th row of $R_k$. Similarly, the modified Givens rotation $(c_{\psi_k^i}, s_{\psi_k^i})$ is pipelined upwards and modifies all $2 \times 2$ data submatrices of the $i$th and $i + 1$th columns, stored in PEs on the vertical line.

To achieve the maximum throughput and parallelism in this SVD subarray, we can adopt the splitting of indices as mentioned at the end of subsection 2.2. Then one inner loop in the algorithm A2 (i.e., $n - 1$ modified Givens rotations) is split in two phases. In the first phase, $n/2$ rotations start simultaneously in $n/2$ diagonal PEs with odd pivot indices, and they are pipelined upwards and to the right as described above. In the second phase, the remaining $n/2 - 1$ rotations start simultaneously in $n/2 - 1$ diagonal PEs with even pivot indices, and they are again pipelined upwards and to the right. Hence, the systolic implementation of algorithm A2 resembles two computational waves that are moving through the triangular SVD subarray from south-east to north-west.

## 4   Timing and Overlap between Systolic Subarrays

The algorithm A1 starts in time $t = 1$ with the input of matrix element $b_{11}$ of the composed matrix $(B_1 | I)$ into the PE $p_{11}$ of the triangular-rectangular systolic subarray. In time step $t = n$, the last element $b_{n1}$ from the first column of $(B_1 | I)$ is read by PE $p_{11}$, the last Givens rotation is generated and performed in $p_{11}$ and the element $r_{11}$ is computed. Other elements of the first row of $R_1$ are computed in PEs $p_{1j}$ in subsequent time steps $t = n + j$, $j = 1, 2, \ldots, n - 1$.

In time step $t = 3n - 1$, the last rotation of the first row of composed matrix $(B_1|I)$ is computed and the first row of $(R_1|Q_1^t)$ is ready for the computation of the first row of $C_1$ in the hexagonal subarray according to (4). However, to achieve the maximum throughput, we should not wait until the *whole* first row is ready; instead, we utilize the elements $R_1$ and $Q_1^t$ *immediately* after their computation.

Note, that the appropriate elements of the first row of $R_1$ and $Q_1^t$ that are needed for the computation of $C_1$ in the hexagonal subarray, are computed in the triangular-rectangular subarray exactly $n$ time steps apart. Hence, we need to slow down the elements of $R_1$ on the input of hexagonal subarray. The number of delay elements depends on the column-number of matrix $(R_1|Q_1^t)$: to schedule the computation of $i$th row of $C_1$ in the hexagonal subarray correctly, we need $n - i + 1$ delay elements for the input of $i$th column of matrix $R_1$ (see Fig. 2). On the other side, to achieve the subsequent output of elements of matrix $Q_1^t$ we need $n - i + 1$ delay elements on the output line for the $i$th column of matrix $Q_1^t$. This modification enables the row-wise skewed output of the matrix $(C_1|Q_1^t)$ that matches perfectly the needed row-wise skewed input into the triangular-rectangular subarray for the next recursion step.

The elements of $R_1$ and $Q_1^t$ enter the hexagonal subarray in the skewed row-wise manner – exactly as they are produced in the triangular-rectangular subarray. Because of maximum degree of pipelining, the following timings can be easily verified:

- $t = n$: the element $r_{11}$ of $R_1$ is ready in the triangular-rectangular subarray.
- $t = n + 1$: input of $r_{11}$ into the hexagonal subarray that begins the computation of $C_1$.
- $t = 3n - 1$: the last element $q_{1n}^t$ in the first row of triangular-rectangular subarray is computed. First row of PEs is ready to start the new QR decomposition.
- $t = 3n + 1$: input of the new element $c_{11}$ into the triangular-rectangular subarray, and the start of QR decomposition of the new matrix $C_1$ together with the update of $Q_1^t$.
- $t = 4n - 2$: the last rotation in the triangular-rectangular subarray is performed and the element $q_{nn}^t$ is computed.
- $t = 6n - 2$: output of the last element $q_{nn}^t$ from the hexagonal subarray.

Hence, it takes $4n - 2$ time steps for the triangular-rectangular array to complete the QR decomposition and orthogonal update of the composed matrix $(B_1|I)$, and $5n - 2$ time steps for the hexagonal array to compute the sum $C_1 = R_1 + (\lambda_1 - \lambda_2)Q_1^t$.

It is clear that the same analysis with the same relative timings is valid for the transition from $(C_k|Q_k^t)$ (hexagonal subarray) to the $(R_{k+1}|Q_{k+1}^t)$ (triangular-rectangular subarray), i.e., from $k$th to $(k + 1)$th recursion step. The important consequence of this analysis is the threefold *overlap* of both subarrays:

- Overlap in the operation of triangular-rectangular and hexagonal subarrays in the same recursion for $3n - 2$ time steps.
- Overlap in two consecutive recursions in the triangular-rectangular array for $n - 2$ time steps.
- Overlap in two consecutive recursions in the hexagonal array for $2n - 2$ time steps.

The situation is a bit different when analysing the possible overlap between the recursive QR decomposition and the operation of SVD subarray. As mentioned in the subsection 3.2, the SVD subarray alternatively produces two computational waves from "odd" and "even" diagonal PEs, where all PEs in one group start their operation simultaneously. This means that the triangular SVD subarray can start its operation when the *whole* matrix $R_k$ is known and stored in the SVD subarray. Therefore, we have to measure the performance of the SVD subarray relatively to the triangular-rectangular subarray and our analysis can be restricted to the time steps when only matrix $R_k$ alone is computed (the matrix $Q_k^t$ is not needed in the SVD subarray).

It is straightforward to show that the triangular systolic subarray with $n(n + 1)/2$ PEs computes R-factor in $3n - 2$ time steps. Therefore, $R_1$ is computed in $t_1^e = 3n-2$ and first SVD steps can start in $\tau_1^b = t_1^e + 1 = 3n - 1$. Due to the overlap between triangular-rectangular and hexagonal subarrays, the computation of $R_k$ starts in $t_k^b = 3(k-1)n + 1$, ends in $t_k^e = 3(k-1)n + 1 + 3n - 3 = 3kn - 2$, so that the SVD steps in $k$th recursion can start in time $\tau_k^b = 3kn - 1$.

How many time steps do we need for the SVD subarray in each recursion? This depends on the number $r$ of repeated inner loops in the algorithm **A2** as already discussed in subsection 2.2. One loop ($r = 1$) consists of two computational waves ("odd" and "even" ones). Each wave travels $n/2$ time steps through the SVD subarray and the gap between them is one time step. Hence, one inner loop takes $n/2 + 2$ time steps. But $r$ loops can be pipelined and the end of SVD steps is defined by the last "even" computational wave that ends its movement through the subarray $4r + n/2 - 2$ time steps from the beginning of SVD in each recursion. Then the (approximate) SVD of $R_k$ in the $k$th recursion will be ready in $\tau_k^e = 3kn - 1 + 4r + n/2 - 2 = 3kn + n/2 + 4r - 3$, so that our systolic implementation of matrix portrait computation finishes with the last value $\lambda = \lambda_m$ in time $\tau_m^e = 3mn + n/2 + 4r - 3$.

It is interesting to note that the whole triangular-rectangular subarray ends the computation of $(R_k|Q_k^t)$ (i.e., *including* the orthogonal update) in time step $3kn + n - 2$. Hence, for $r < 1 + n/8$, the SVD subarray finishes its operation in each recursion *before* the orthogonal update of $Q_k^t$ is ready, so that there is a total overlap between the SVD and triangular-hexagonal subarrays. When $r$ increases, the amount of overlap decreases. It should be stressed, however, that the length of one synchronized time step for the computation in all systolic subarrays is defined by the length of the most complex operation – i.e., by the computation and application of two-sided modified Givens rotations in the diagonal PEs of the triangular SVD subarray.

# 5 Conclusion

We have designed systolic subarrays for the matrix portrait computation. Our main result is the algorithmical as well as systolic design of the recursive QR decomposition of matrices $B_k = A - \lambda_k I$, $k = 1, 2, \ldots, m$. The recursive QR decomposition can be realized on triangular-rectangular and hexagonal systolic subarrays. For the (approximate) SVD decomposition of the upper triangular matrices we have used the triangular SVD subarray. Our solution needs $O(n^2)$ PEs and delays, and the whole computation takes $3mn + \lceil \frac{n}{2} \rceil + 4r - 3$ time steps (valid also for $n$ odd) where $n$ is the matrix order, $m$ is the number of various $\lambda$s and $r$ is the number of repeated loops in the (iterative) SVD algorithm A2. The subarrays exhibit a large amount of overlap in their operation.

Our approach can be compared with the straightforward method, when the matrix portrait computation is implemented on two triangular systolic subarrays without any overlap of their operation. The first subarray computes the QR decomposition of matrix $B_k$ (where each $B_k$ is formed by the host), and the second one computes the SVD steps. The time complexity of the straightforward approach is then $m(3n - 2 + \lceil \frac{n}{2} \rceil + 4r - 3)$, which is significantly worse than the time complexity of our solution. The number of PEs and delays is $O(\lceil cn^2 \rceil)$, where $c = \frac{37}{8}$ for our solution and $c = \frac{5}{8}$ for the straightforward solution.

# References

1. Brent, R.P., Luk, F.T.: The solution of singular-value and symmetric eigenvalue problems on multiprocessor arrays. SIAM J. Sci. Statist. Comput. **6** (1985) 69-84
2. Charlier, J.P., Vanbegin, M., Van Dooren, P.: On efficient implementations of Kogbetliantz's algorithm for computing the singular value decompositions. Numer. Math. **52** (1988) 279-300
3. Gentleman, W.M., Kung, H.T.: Matrix triangularization by systolic arrays. Proc. SPIE **298** (1983) 19-26
4. Golub, G., Van Loan, C.: Matrix computations. 2nd edn. North Oxford Academic, John Hopkins U.P. (1989)
5. Luk, F.T.: A triangular processor array for computing the singular value decomposition. Linear Algebra Appl. (Special Issue on Parallel Computation) **77** (1986) 259-273
6. Luk, F.T., Park, H.: A proof of convergence for two parallel Jacobi SVD algorithms. IEEE Trans. Computers **38** (1989) 806-811
7. Moonen, M., Van Dooren, P., Vandewalle, J.: A singular value decomposition updating algorithm for subspace tracking. SIAM J. Matrix Anal. Appl. **13** (1992) 1015-1038
8. Moonen, M., Van Dooren, P., Vandewalle, J.: A systolic array for SVD updating. SIAM J. Matrix Anal. Appl. **14** (1993) 353-371
9. Stewart, G.W.: A Jacobi-like algorithm for computing the Schur decomposition of a nonhermitian matrix. SIAM J. Sci. Statist. Comput. **6** (1985) 853-863

# A Class of Explicit Two-Step Runge-Kutta Methods with Enlarged Stability Regions for Parallel Computers

Helmut Podhaisky and Rüdiger Weiner

Universität Halle-Wittenberg, Institut für Numerische Mathematik
Theodor-Lieser-Straße 5, D-06120 Halle, Germany
{podhaisky,weiner}@mathematik.uni-halle.de

**Abstract.** In this paper we study a class of explicit pseudo two-step Runge-Kutta (EPTRK) methods for first-order ODEs for parallel computers. We investigate linear stability and derive methods with enlarged stability regions. In numerical experiments on a shared memory computer we compare a parallel variable step size EPTRK implementation with the efficient sequential Runge-Kutta method **dopri5**.

**Key words:** Runge-Kutta methods, parallelism, two-step methods, stability
**AMS(MOS) subject classification (1991):** 65M12, 65M20

## 1 Introduction

For the numerical solution of systems of first-order ordinary differential equations (ODEs)

$$y' = f(t, y), \quad y(t_0) = y_0, \quad y, f \in \mathbb{R}^n, \tag{1}$$

Fei [3], Cong [1] and Cong et al. [2] have recently investigated a class of explicit pseudo two-step Runge-Kutta methods (EPTRK methods). EPTRK methods compute an approximation $u_m \approx y(t_m)$ by the $s$-stage scheme

$$U_{m+1} = \mathbb{1} \otimes u_m + h_m(A \otimes I)F(t_{m-1}\mathbb{1} + h_{m-1}c, U_m), \tag{2a}$$

$$u_{m+1} = u_m + h_m(b^T \otimes I)F(t_m\mathbb{1} + h_m c, U_{m+1}), \tag{2b}$$

with $s$ (external) stage approximations $U_{m+1} \in \mathbb{R}^{n \times s}$, parameters $A = (a_{ij}) \in \mathbb{R}^{s \times s}$, $c, b \in \mathbb{R}^s$, $\mathbb{1} := (1, \ldots, 1)^T$ and step size $h_m$. $F$ denotes the straightforward extension of $f$ to $\mathbb{R}^{n \times s}$. Here, $\otimes$ denotes the Kronecker tensor product, e.g. see [4].

The EPTRK methods are well suited for parallelization with $s$ processors, especially in a shared memory environment. They do not have a sequential bottleneck and need only one effective function evaluation per step. Notice that the parallelism is already in the method. Consequently, using an EPTRK method

P. Zinterhof, M. Vajteršic, A. Uhl (Eds.): ACPC'99, LNCS 1557, pp. 68–77, 1999.
© Springer-Verlag Berlin Heidelberg 1999

does not require any additional effort to split the ODE system. The ODE system can be solved by a black-box (library) call yielding coarse grain parallelism.

Numerical tests of a special class of EPTRK methods of high order have shown the efficiency for problems, where the step size is determined by accuracy only [2]. However, this class suffers from very small stability regions. Therefore, in this paper we construct EPTRK methods of order 3 and 4 with enlarged stability regions.

## 2   Order Conditions and Convergence

In this section we consider order conditions for EPTRK methods. Assume exact values $U_m = y(t_{m-1}\mathbb{1} + h_{m-1}c)$ and $u_m = y(t_m)$. We perform one EPTRK step with step size $h_m$ and compare the numerical solutions $U_{m+1}$ and $u_{m+1}$ with the exact values $y(t_m\mathbb{1} + h_m c)$ and $y(t_{m+1})$. Let $\alpha := h_m/h_{m-1}$ be the step size ratio. Comparing the Taylor expansions of numerical and analytical solutions yields, analogously to implicit Runge-Kutta Methods, the simplifying conditions

$$C(q): \quad \sum_{j=1}^{s} a_{ij}(c_j - 1)^{\nu-1} = \alpha^{\nu-1}\frac{c_i^\nu}{\nu}, \qquad i = 1,\ldots,s, \quad \nu = 1,\ldots,q \quad (3a)$$

$$B(p): \quad \sum_{i=1}^{s} b_i c_i^{j-1} = \frac{1}{j}, \qquad j = 1,\ldots,p \tag{3b}$$

for stage order $q$ and step order $p$. From $C(p-1)$ and $B(p)$ follows consistency order $p$, i.e. $\|u_{m+1} - y(t_{m+1})\| = \mathcal{O}(h^{p+1})$, and convergence order $p$, as stated in the following theorem.

**Theorem 1 (Convergence for EPTRK methods).** *An EPTRK method with order $p$ and stage order $q$, i.e. with $B(p)$ and $C(q)$, converges with order $p^* := min(p, q+1)$ if the local errors of starting values $U_1$ and $u_1$ for the first EPTRK step are of order $p^*$.*

*Proof.* The theorem can be proved by standard techniques. We use $C(q)$ and $B(p)$ to get a recursion for the global error. Using this recursion we can bound the global error. For details see [7]. □

## 3   Stability of EPTRK Methods

We investigate the stability of the EPTRK methods by means of the linear test equation

$$y' = \lambda y, \quad \mathrm{Re}\lambda \le 0 \tag{4}$$

and constant step size $h$. The EPTRK solution of (4) fulfills the recursion

$$(U_{m+1}, u_{m+1})^T = M(z)(U_m, u_m)^T, \quad z = h\lambda , \tag{5}$$

with the amplification matrix

$$M(z) = \begin{pmatrix} zA & \mathbb{1} \\ z^2 b^T A & 1+z \end{pmatrix} . \tag{6}$$

We define the stability region $S^*$ of an EPTRK method by

$$S^* := \{z : \varrho(M(z)) \le 1\}$$

and have $\varrho(M(z)) < 1$ if $z$ is an inner point of $S^*$ and hence $(U_m, u_m) \to 0$ for $m \to \infty$. We compute $\varrho(M(z))$ by computing the zeros of the stability polynomial $\varrho(x,z) := \det(xI - M(z))$. From (6), it can be seen that EPTRK methods are always zero-stable. In $z = 0$ the matrix $M(z)$ has the single eigenvalue $x = 1$ and the $s$-fold eigenvalue zero. Furthermore, by calculating the derivative by the implicit function theorem, we can see that the absolute value of the maximal eigenvalue $x$ is less than 1 in a whole region locally left to the origin and hence an EPTRK method is stable there. To construct methods with large stability regions we calculate the coefficients of the stability polynomial in the following theorem.

**Theorem 2.** *A consistent $s$ stage EPTRK method has the stability polynomial*

$$\varrho(x,z) = \sum_{k=0}^{s} x^{s-k} \left[ \left(p_k - \sum_{i=0}^{k-1} p_{k-i-1} b^T A^i \mathbb{1}\right)x - p_k \right] z^k \tag{7}$$

*with parameters $p_k$ defined by*

$$\det(\lambda I - A) = \sum_{i=0}^{s} p_i \lambda^{s-i}, \quad p_0 = 1 . \tag{8}$$

*Proof.* Starting with the definition

$$\varrho(x,z) = \det(xI - M(z)) = \det \begin{pmatrix} xI - zA & -\mathbb{1} \\ -z^2 b^T A & x - 1 - z \end{pmatrix}$$

we get

$$\varrho(x,z) = \det \left\{ \begin{pmatrix} xI - zA & -\mathbb{1} \\ -z^2 b^T A & x-1-z \end{pmatrix} \begin{pmatrix} (xI - zA)^{-1} & (xI - zA)^{-1}\mathbb{1} \\ 0 & 1 \end{pmatrix} \right\}$$

$$\times \frac{1}{\det(xI - zA)^{-1}}$$

$$= [x - 1 - z - z^2 b^T A (xI - zA)^{-1} \mathbb{1}] \det(xI - zA) .$$

Using (8) and a Neumann series for the inverse matrix yields

$$\varrho(x,z) = \left[ (x - 1 - z) - z^2 b^T A \sum_{i=0}^{\infty} z^i \left(\frac{1}{x}\right)^{i+1} A^i \mathbb{1} \right] \left[ \sum_{i=0}^{s} p_i z^i x^{s-i} \right] . \tag{9}$$

The polynomial $\varrho(x,z)$ has degree $s$ in $z$ and degree $s+1$ in $x$ and therefore almost all terms in (9) vanish. This proves the theorem.      □

Since for many applications with (nearly) real eigenvalues only stability is crucial along the real axis we optimize the boundary $r_{max}$ of the real stability interval defined as $r_{max} = \sup\{x : (-x, 0) \subset S^*\}$. We determine analytically optimal values $r_{max}$ for one-stage EPTRK methods and for two-stage EPTRK methods of order $p = 2$ in Theorem 3. For EPTRK methods with more than 2 stages we have found stability polynomials with large values $r_{max}$ by a numerical search.

**Theorem 3.** *The optimal value for $r_{max}$ of a consistent $s$-stage EPTRK method of order $p$ is given by*

*(i) $r_{max} = 4$, if $s = p = 1$ , and*
*(ii) $r_{max} = 2$, if $s = p = 2$ .*

*Proof.* We prove only (ii). The result (i) can be obtained similarly. According to Theorem 2, we have

$$\varrho(x, z) = x^3 + (-1 + p_1 z - z)x^2 + (-p_1 z + p_2 z^2 - p_1 z^2 - b^T A \mathbb{1} z^2)x - p_2 z^2.$$

Order $p = 2$ implies $b^T A \mathbb{1} = b^T c = 1/2$. Assume $r_{max} \geq 2$. Then all zeros $w_1, w_2$ and $w_3$ of

$$\varrho(x, -2) = x^3 + (-2p_1 + 1)x^2 + (4p_2 - 2 - 2p_1)x - 4p_2 \qquad (10)$$

lie in the unit circle. Define

$$g = w_1 + w_2 + w_3 + w_1 w_2 + w_1 w_3 + w_2 w_3 - w_1 w_2 w_3$$
$$= w_1 + (1 + w_1)[w_2 + w_3] + (1 - w_1)w_2 w_3$$

By Vieta's Theorem and (10) we have $g = -3$. Let $w_1$ be a real zero. Assume $|w_1| < 1$. Then we get

$$g > w_1 + (1 + w_1)(-2) + (1 - w_1)(-1) = -3 ,$$

in contradiction to $g = -3$. Therefore, $|w_1| = 1$. A simple calculation shows now that $\varrho(x, -2)$ has the zeros $-1, -1$ and $1$. A study of the unique polynomial with these zeros shows $r_{max} = 2$. $\qquad\square$

By using Theorem 2 we can determine stability polynomials of EPTRK methods with more than 2 stages and large stability regions. In Sect. 4 we will construct appropriate methods corresponding to these polynomials. The properties for these EPTRK methods are given in Tab. 1. Here, $p_{emb}$ denotes the order of an embedded method used for step size selection and $\#procs$ the number of processors for parallelization. Note, that the parameters $b^T A^i \mathbb{1}$, $i = 0, \ldots, p-1$ are already determined by $C(p-1)$ and $B(p)$, but we can choose $p_k$, $k = 1, \ldots, s$ and $b^T A^i \mathbb{1}$, $i = p \ldots, s$ arbitrarily to optimize the stability regions. Figure 1 shows these stability regions. For comparison we have also drawn the stability region of the 5-stage 5-order collocation EPTRK method cong5 [2] and the stability region $S^*_{\text{dopri}}$ of the well known explicit Runge-Kutta method dopri5 [4] and the smaller scaled stability region $S^{\text{scal}}_{\text{dopri}} := \{z/6 : z \in S^*_{\text{dopri}}\}$ which takes into account the six effective sequential functions evaluations per step. Numerical experiments [2] show that cong5 performs well for non-stiff problems. But for mildly stiff problems dopri5 is superior due to better stability properties. The new methods p33, p43 and p44 of this paper have a larger stability region than dopri5 (scaled). Therefore we can expect that these methods are more efficient than dopri5 if stability is important and the function evaluations are not too cheap.

**Table 1.** EPTRK methods

| name | stages | conditions | order $p^*$ | emb. order $p_{\text{emb}}$ | #proc |
|------|--------|------------|-------------|------------------------------|-------|
| p33 | 3 | $C(2), B(3)$ | 3 | 2 | 3 |
| p43 | 4 | $C(3), B(3)$ | 3 | 4 | 4 |
| p44 | 4 | $C(4), B(4)$ | 4 | 3 | 4 |
| cong5 | 5 | $C(5), C(5)$ | 5 | 3 | 5 |

## 4  Constructing the Methods from the Polynomials

The 5-stage 5-order method cong5 [2] fulfills $C(5)$. Here, for constructing the $s$-stage EPTRK methods p33, p43 and p44 with better stability properties, we require $C(s-1)$ only. First consider fixed step sizes. We start by choosing the vector $c$ of distinct nodes and a stability polynomial calculated in Sect. 3. We have $C(s-1)$ and the polynomial, i.e.

$$AV = P, \qquad V_{ij} = (c_i - 1)^{j-1}, \quad P_{ij} = c_i^j/i, \quad P, V \in \mathbb{R}^{s \times s-1}, \quad (11a)$$

$$\det(\lambda I - A) = \sum_{i=0}^{s} p_i \lambda^{s-i}, \quad p_0 = 1 \tag{11b}$$

hold. We calculate $A$ from (11): the solution of (11a) has $s$ free parameters which can be determined by substituting this solution into (11b). Now, $A$ is known and we get the weights $b$ directly from the polynomial coefficients $b^T A^i \mathbb{1}$, $i = 0, \ldots, s-1$.

For variable step sizes, $C(s-1)$ becomes

$$A(\alpha)V = PD(\alpha), \quad \text{with} \quad D(\alpha) = \text{diag}(1, \alpha, \ldots, \alpha^{s-2}), \tag{12}$$

and with (11a) we have

$$(A(\alpha) - A)V = P(D(\alpha) - I) . \tag{13}$$

We calculate a row minimal 2-norm solution $A(\alpha) - A$ of this underdetermined system (13) in order to preserve the stability properties for $A(1) = A$.

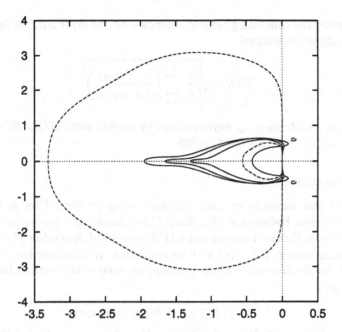

**Fig. 1.** Stability regions for the EPTRK methods drawn with solid lines and for dopri5 with dashed lines. Regions from biggest to smallest: dopri5, p43, p33, p44, dopri5^scal, cong5

# 5   Numerical Experiments

## Implementation

We have implemented the EPTRK methods in FORTRAN 77. The code for the collocation method cong5 is available at:

   http://www.mathematik.uni-halle.de/institute/numerik/software.

The numerical tests below were performed on a shared memory HP/Convex-Exemplar X-Class computer. Parallelism is expressed by means of the compiler-directives for loop-parallelization. For step size selection we computed an approximation of the local error

$$err = \sqrt{\frac{1}{n}\sum_{i=1}^{n}\left(\frac{u_{m+1,i} - \tilde{u}_{m+1,i}}{\text{ATOL} + \text{RTOL}|u_{m+1,i}|}\right)^2},$$

and the new step size

$$h_{m+1} = \begin{cases} h_m & \text{for } \alpha \in [0.9, 1.1] \\ \alpha h_m & \text{else} \end{cases}, \quad \text{with}$$

$$\alpha = \min\left\{3, \max\left\{0.3, 0.8 \cdot err^{-1/\tilde{p}}\right\}\right\}, \quad \tilde{p} := \min(p^*, p_{\text{emb}}) + 1 .$$

We compare the computing time in seconds with the error $ERR$ in the endpoint of the integration interval,

$$ERR = \sqrt{\frac{1}{n} \sum_{i=1}^{n} \left( \frac{u_i - u_{\text{ref},i}}{1 + |u_{\text{ref},i}|} \right)^2} .$$

The reference solutions $u_{\text{ref}}$ were obtained by dop853 with ATOL=RTOL=$10^{-14}$. The version of dopri5 was 03/14/93.

## Test Problems

As a first test example we take the seven body problem **Plei** in two spatial dimensions from Hairer et al. [4]. This problem models the gravity forces between seven stars leading to a second order ODE system of dimension 14, respectively a first order system $y' = f(y)$ with 28 equations. To illustrate how the speedup depends on the dimension of the problem we enlarge this system by a factor $d$ and we get

$$Y' = F(Y), \quad Y(0) = Y_0, \quad \text{with} \tag{14}$$

$$Y = (1 \otimes y), \quad F(Y) = (1 \otimes f(y)), \quad Y_0 = 1 \otimes y_0, \quad \text{and} \quad 1 \in \mathbb{R}^d . \tag{15}$$

We discuss the speedup for different values $d = 1, \ldots, 1000$.

A well known test problem is the **Brusselator**, see [5] and [6], a two-dimensional diffusion-reaction equation

$$u_t = B + u^2 v - (A+1)u + \alpha \left( \frac{\partial^2 u}{\partial x^2} + \frac{\partial^2 u}{\partial y^2} \right) ,$$

$$v_t = A - u^2 v + \alpha \left( \frac{\partial^2 v}{\partial x^2} + \frac{\partial^2 v}{\partial y^2} \right)$$

with $A = 3$, $B = 1$ and initial values

$$u(0, x, y) = 0.5 + y, \qquad v(0, x, y) = 1 + 5x$$

and Neumann boundary conditions

$$\frac{\partial u}{\partial n} = 0 , \quad \frac{\partial v}{\partial n} = 0 \quad \text{auf} \quad \partial\Omega, \quad \Omega = [0,1] \times [0,1], \quad t \in [0,1] .$$

Semidiscretization by the method of lines with second order central differences in $x$ and $y$ on a $100 \times 100$ uniform grid leads to a first order system of 20000 ordinary differential equations. For small values $\alpha$ this system is non-stiff. We consider $\alpha = 2 \cdot 10^{-4}$ and $\alpha = 5 \cdot 10^{-3}$.

The second test problem is the two dimensional diffusion-convection equation **Difkon** from [8]:

$$u_t = \alpha \Delta u + \beta(-y u_x + x u_y) + g(t, x, y), \quad \text{with}$$

$$\alpha = 10^{-3}, \quad \beta = 1, \quad \Omega = [0,1]^2, \quad t \in [0,1] ,$$

with function $g$ and initial values and Dirichlet boundary conditions chosen according to the exact solution $u(t, x, y) = \sin(\pi x) \sin(\pi y) \left(1 + 4xy \exp(-2\alpha \pi^2 t)\right)$. We use second order differencing on a $69 \times 69$ grid to get a semidiscrete system of ODEs. Due to the convection term the Jacobian of this system is nonsymmetric and has non-real eigenvalues. We obtain a modification of this problem with even larger imaginary parts if we set $\beta = 10$.

## Results and Discussion

Figure 2 shows the speedups for the $s$-stage EPTRK methods on $s$ processors over the same method on one processor. Although the function evaluations are not very expensive we get speedups between 3 and 4 with five processors (cong5). Of course, increasing the dimension of an ODE systems yields better speedups as demonstrated with example Plei. Figure 2 depends strongly on the parallel computer.

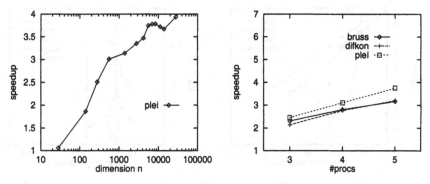

**Fig. 2.** The speedup for cong5 for different dimensions of the ODE system Plei (left figure) and for EPTRK methods with $s = \#procs$ (right figure: Plei with $n = 7000$)

For the celestial mechanic example Plei (see Fig. 3) we get a speedup of about 4 of cong5 over dopri5 with 5 processors. For low tolerances the parallel p44 is also competitive with dopri5. For the non-stiff Brusselator (see Fig. 4, left), for crude tolerances p33 and for more stringent tolerances cong5 are most efficient. For the mildly stiff Brusselator with increased $\alpha$ (right figure) the stability enlarged EPTRK methods p33 and p43 become the most efficient ones, except for very high accuracy. Consider problem Difkon (see Fig. 5) with $\beta = 1$ (left). The best method for low tolerances is p33 and for high tolerances dopri5. If we set $\beta = 10$ (right figure) the imaginary parts of the eigenvalues of the Jacobian become larger and the stability of the method near the imaginary axis becomes crucial. Accordingly p44 and p33 perform well.

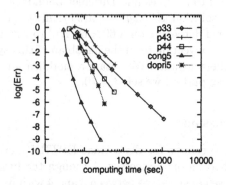

**Fig. 3.** Test problem Plei with dimension $n = 7000$

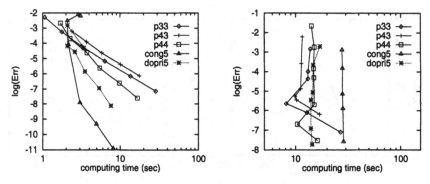

**Fig. 4.** Test problem Brusselator, left $\alpha = 4 \cdot 10^{-4}$, right $\alpha = 5 \cdot 10^{-3}$

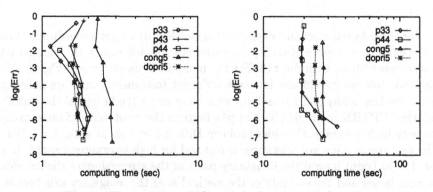

**Fig. 5.** Test problem Difkon, left $\beta = 1$, right $\beta = 10$

# 6  Conclusions

The class of explicit two-step Runge-Kutta (EPTRK) methods is an interesting subject for solving ODE systems of large dimension and/or with expensive $f$-evaluations in a parallel environment. For problems where accuracy is more important than stability the high order collocation method cong5 [2] performs better than the sequential Runge-Kutta method dopri5. On the other hand, for mildly stiff problems and not too stringent tolerances, where stability is important, EPTRK methods p33, p43 and p43 with large stability regions are a good choice.

# References

[1] N.H. Cong, *A general family of pseudo two-step Runge-Kutta methods*, submitted for publication, 1997.

[2] N.H. Cong, H. Podhaisky, and R. Weiner, *Numerical experiments with some explicit pseudo two-step RK methods on a shared memory computer*, Computers and Mathematics with Applications **36** (1998), no. 2, 107–116.

[3] J.G. Fei, *A class of parallel explicit Runge-Kutta formulas*, Chinese Journal of Numerical Mathematics and Applications **16** (1994), 23–36.

[4] E. Hairer, S.P. Nørsett, and G. Wanner, *Solving Ordinary Differential Equations I*, Springer, 1993.

[5] E. Hairer and G. Wanner, *Solving Ordinary Differential Equations II*, Springer, 1996.

[6] M. Hochbruck, C. Lubich, and H. Selhofer, *Exponential integrators for large systems of differential equation*, SIAM J. Sci. Comput. **19** (1997), no. 5, 1552–1574.

[7] Helmut Podhaisky, *Parallele explizite Zweischritt Runge-Kutta-Methoden*, Diplom-Thesis, Universität Halle, 1998.

[8] B.A. Schmitt and R. Weiner, *Matrix-free W-methods using a multiple Arnoldi iteration*, Appl. Numer. Math. **18** (1995), 307–320.

# A Parallel Strongly Implicit Algorithm for Solving of Diffusion Equations

Ladislav Halada[1] and Mária Lucká[2]

[1] Slovak Technical University,
Nám.slobody, 812 31 Bratislava, Slovak Republic,
halada@dekan.sjf.stuba.sk,
[2] Institute of Control Theory and Robotics, SAS,
Dúbravská cesta 9, 842 37 Bratislava, Slovak Republic,
geofluck@savba.sk

**Abstract.** We present a parallel algorithm for the solution of partial differential equations representing a 3-D diffusion process of the underground water by a finite difference method. The algorithm belongs to a class of the incomplete $LU$ factorization methods, where corresponding system of linear algebraic equations is solved by a quasi $LU$ decomposition in every time step. A code realizing the algorithm was written in Fortran 90 programming language using the MPI message passing interface system and was tested on a SGI Origin 2000 parallel computer.

## 1 Introduction

Modeling of the water diffusion process is an important problem in hydrology. For the solution of corresponding multidimensional differential equations finite difference methods are often used. The corresponding system of linear algebraic equations in this case is represented by a symmetric, positive definite matrix of the order $10^6 - 10^9$. The coefficients of the matrix are time dependent, and thus the corresponding system must be solved separately for each time step. This is very time consuming so that parallel methods are of great importance. For the solution of large linear algebraic systems the mostly used methods are: relaxation methods ([1]), ADI methods (Alternating Direction Iteration) and SIP (Strongly Implicit Procedure) methods and during the last decade the preconditioned conjugate gradient methods ([2], [3], [4], [5]). The effectiveness of the methods mentioned above is different. The serious problematic aspect of the parallel computation is the triangular solver used in the ADI ([6], [7]) and SIP ([8], [9]) methods. Compared to the above-mentioned methods the advantages of SIP methods are in fewer number of iterations for required accuracy and lower computational cost for each time step. In this paper a parallel method for one-time step solution of a differential equation related to the 3-D diffusion process of the underground water in hydrology is described.

The paper is organized as follows: In Section 2 the theoretical background of the method proposed is given. In Section 3 a parallel algorithm for one-time step solution of a SIP method is described, and in Section 4 numerical results on parallel Cray Origin 2000 computer are presented.

P. Zinterhof, M. Vajteršic, A. Uhl (Eds.): ACPC'99, LNCS 1557, pp. 78–84, 1999.
© Springer-Verlag Berlin Heidelberg 1999

## 2    Formulation of the Problem

Consider the following partial differential equation

$$\frac{\partial}{\partial x}(K_x \frac{\partial H}{\partial x}) + \frac{\partial}{\partial y}(K_y \frac{\partial H}{\partial y}) + \frac{\partial}{\partial z}(K_z \frac{\partial H}{\partial z}) = S \frac{\partial H}{\partial t} + f(x, y, z, t), \qquad (1)$$

which describes the ground water flow in an anisotropic and heterogeneous aquifer with flow off, or an infiltration process, where $K_x, K_y, K_z$ are coefficients related to hydraulic conductivity in x, y, z - directions , $H := H(x, y, z, t)$ is the piezometric level to be found, $f(x, y, z, t)$ is the function characterizing water infiltration in a given point, and $S := S(x, y, z)$ denotes the piezometric level of an elastic reservoir. The boundary conditions of equation (1) are given by the initial values of $H(x, y, z, t)$ at time $t = t_0$ in a given point.

The numerical solution of this equation will be performed by a finite difference method using space and time discretization. Consider a rectangular 3-D space and a grid created by $I, J, K$ grid points in the corresponding $x$, $y$ and $z$-directions of the coordinate rectangular system. To solve (1), let us consider a finite difference method using 3-point difference formulas in the $x$, $y$ and $z$-directions, respectively, so that we obtain the following system of linear algebraic equations

$$F_j x_{j-IJ} + D_j x_{j-I} + B_j x_{j-1} + A_j x_j + C_j x_{j+1} +$$
$$+ H_j x_{j+I} + Z_j x_{j+IJ} = q_j,$$

for $j := 1, 2, ..., IJK$, where $IJK := I \times J \times K$, and $F_j, D_j, B_j, A_j, C_j, H_j, Z_j$ and $q_j$ are scalars.

This system of equations can be expressed in the matrix form

$$Mx = q, \qquad (2)$$

where $M$ is a symmetric, sparse, regularly structured matrix containing at most seven non-zero coefficients in each row. The structure of the sparse matrix $M$ is evident, but it is worth to remark that each diagonal except the main one has several elements equal to zero, which are regularly distanced from each other. They allow to split the matrix and calculate it in parallel.

The main idea of the SIP method ([10]) is : Find a matrix $N$, such that $M + N$ can be decomposed into two matrices $L$ and $U$, where $M + N$, $L$ and $U$ meet the following conditions:

- $M + N$ is "close" to $M$,
- $L$ is lower triangular form while $U$ is upper triangular form and all entries along the main diagonal $U$ are equal to unity, and
- both $L$ and $U$ have just four non-zero diagonals.

The corresponding system of linear algebraic equations has the form

$$(M + N)x = (M + N)x - (Mx - q) \qquad (3)$$

and a suitable iterative algorithm for solving (3) is

$$(M + N)x^{(i+1)} = (M + N)x^{(i)} + (q - Mx^{(i)}), \qquad (4)$$
$$(M + N)(x^{(i+1)} - x^{(i)}) = q - Mx^{(i)}, \qquad (5)$$
$$LUh^{(i+1)} = r^{(i)}, \qquad (6)$$

where $h^{(i+1)} := x^{(i+1)} - x^{(i)}$ and $r^{(i)} := q - Mx^{(i)}$ is the residual. The last equation can be solved as follows

$$Ly^{(i+1)} = r^{(i)}, \quad Uh^{(i+1)} = y^{(i+1)}. \qquad (7)$$

Below we present a suitable choice for the non-zero coefficients of $L$ and $U$, respectively, that enables an efficient parallel implementation on a distributed memory parallel computer system.

## 3   A Parallel Implementation of the Algorithm

Let $f_j, d_j, b_j, a_j$ be four non-zero coefficients of $L$ and $1, c_j, h_j, z_j$ be four non-zero coefficients of $U$ related to each row, in the positions of the non-zero entries of the original matrix $M$. It can be easily verified that $LU$ has elements on six non-zero diagonals which are not present in $M$. The influence of these coefficients can be eliminated by using an interpolation procedure to obtain a system $(M+N)x \doteq q$, where $LU = M + N$. For details see [10].

Let the non-zero coefficients be chosen as follows :

$$f_j := \frac{F_j}{(1 + c_{j-IJ} - h_{j-IJ})}, \qquad (8)$$

$$d_j := \frac{D_j}{(1 + c_{j-I} + z_{j-I})}, \qquad (9)$$

$$b_j := \frac{B_j}{(1 + h_{j-1} + z_{j-1})}, \qquad (10)$$

$$a_j := A_j - f_j z_{j-IJ} - d_j h_{j-I} - b_j c_{j-1} \qquad (11)$$
$$+ f_j c_{j-IJ} + f_j h_{j-IJ} + d_j c_{j-I} \qquad (12)$$
$$+ b_j h_{j-1} + d_j z_{j-I} + b_j z_{j-1}, \qquad (13)$$

$$c_j := \frac{(C_j - f_j c_{j-IJ} - d_j c_{j-I})}{a_j}, \qquad (14)$$

$$h_j := \frac{(H_j - f_j h_{j-IJ} - b_j h_{j-1})}{a_j}, \qquad (15)$$

$$z_j := \frac{(Z_j - d_j z_{j-I} - b_j z_{j-1})}{a_j}, \qquad (16)$$

for $j := 1, 2, ..., IJK$.

We will now address the problem of distributing the work to evaluate (8)-(16) in parallel on $P$ processors denoted by $p := 0, 1, ..., P-1$. Let $W := K/P$, where

$W$ is an integer and $K$ is the number of the grid points in the $z$-direction. Furthermore, assume that all nonzero elements of the matrix $M$ are represented by into seven vectors $A := (A_j), B := (B_j), C := (C_j), D := (D_j), H := (H_j), F := (F_j), Z := (Z_j)$, for $j := 1, 2, ..., IJK$. Consider the same for the non-zero coefficients of $L$ and $U$, respectively, $a := (a_j), b := (b_j), c := (c_j), d := (d_j), h := (h_j), f := (f_j), z := (z_j)$, for $j := 1, 2, ..., IJK$. Assume block-cyclic distribution of all vectors onto P processors, that to each processor has an associated block of elements,

$$j \in ([P \times (k-1) + p] \times IJ + 1, [P \times (k-1) + p + 1] \times IJ), \qquad (17)$$

for $p := 0, 1, ..., P - 1$, $k := 1, 2, ..., W$.

Suppose that $ST$ is an integer, which is a multiplier of $I$ and a divisor of $IJ$, where $I$ and $J$ are the number of grid points in the $x$, $y$ - directions, respectively. The structure of the matrix $M$ and the positions of the non-zero elements on subdiagonals allows the following parallel calculation of the equations (8) - (16) that comprise the evaluation of the coefficients of $L$, $U$, respectively:

```
do  k := 1, W
   do  parallel  for  p := 0, P - 1
      do  ij := 1, IJ, ST
      kij := (k - 1) × P × IJ
         do  j := ij + kij, ij + kij + ST - 1
            calculate  (8) - (16)
            if  ((p ≠ (P - 1)) or ((p = (P - 1)) and (k < W)))
               send  ST elements of c, h, z to (p + 1)modP
            end  if
            if  ((p ≠ 0) or ((p = 0) and (k = 1)))
               receive  ST elements of c, h, z
            end  if
         end  do  j
      end  do  ij
   end  do  p
end  do  k
```

Having calculated the coefficients of $L$ and $U$, the final solution is obtained by solving (7).

## 4  Experimental Results

The numerical experiments were conducted on a Cray Origin 2000 parallel computer with "distributed shared memory" and 128 processors at Parallab, University of Bergen [11]. Beside being used in a shared-memory mode the system can be also used as a traditional message passing computer. For our purposes we have used it as a message passing computer. The algorithm was written in Fortran 90 language using the MPI system, which allows a high degree of portability across different machines. The experiments have confirmed that by increasing the number of processors the time complexity was decreased. There are two basic features of the machine which complicate the performance evaluation: The first is that the computer is a "shared-everything-system", i.e. the operating system tries to share CPUs, memory, intercommunication and I/O among all running processes. So a running job depends heavily on other jobs running at the same time. The greatest difference between Origin 2000 and other distributed memory machines like the Paragon, SP2, T3E, Meiko CS2, etc. is that if on these systems a job gets a node, then it gets all its resources and the CPU time can be measured easily. This is not possible on the Origin 2000 computer, because the system distinguish only processes and data. The system assigns CPU's and memory for processes and data dynamically, what can influence the runtime.

| | NUMBER OF PROCESSORS | | | |
|---|---|---|---|---|
| *ST* | 2 | 4 | 8 | 16 |
| 50 | 38.93 | 18.45 | 9.67 | 5.14 |
| 100 | 29.06 | 14.68 | 6.99 | 3.05 |
| 250 | 24.10 | 12.06 | 6.02 | 3.37 |
| 400 | 22.94 | 11.13 | 5.31 | 2.76 |
| 1000 | 21.46 | 10.58 | 4.93 | 2.69 |
| 2000 | 24.26 | 12.17 | 5.69 | 3.03 |
| 4000 | 21.54 | 10.46 | 5.84 | 3.92 |

**Table 1:** Time in seconds for I = 200, J= 100, K = 16

The results of the numerical experiments shown in Tab.1 correspond to $IJK$ = 320000, and the computation was performed in double precision. The code was tested for different values of $ST$. The best results were achieved for $ST = 1000$. The length of messages depends on the value of $ST$ and on the length of the MPI representation of the data. The delay between the first and the $(p+1)$-st processor when calculating the elements of the matrices $L$ and $U$ is equal to $T_{ST} \times p$, where $T_{ST} = ST \times T_{fpo} + (\alpha + 3 \times ST \times \beta)$. Here $\alpha$ is the startup time, $\beta$ is the time required for sending one 8-byte word, and the $T_{fpo}$ is the time of calculating all elements in one row. This value is influenced also by the placement of the elements in the memory during the computation.

Finally, a sequential code written in Fortran 90 ran on the same computer approximately 49.90 seconds.

# 5  Conclusion

We have presented a parallel algorithm for the solution of partial differential equations representing a 3-D diffusion process of the underground water by a finite difference method. The used method is based on a quasi-$LU$ factorization applied in every time step. We have written a code implementing the algorithm on a SGI Origin 2000 parallel computer and the time measurement results are enclosed. The experimental results published in [13] demonstrate the suitability of the SIP method for some well structured matrices.

Our future work will concern the application of the existing decomposition of the matrix as an incomplete Cholesky preconditioning for a modified variant of the preconditioned conjugate-gradient method.

# 6  Acknowledgement

This work was supported by Grant No.2/5131/98, VEGA, Slovak Republic. The second author is indebted for the opportunity of using the parallel computer Cray Origin 2000 at Parallab, at the University of Bergen, during her stay at Parallab.

# References

1. Hofhaus, J., van de Velde, E.F.: Alternating-direction line-relaxation methods on multicomputers. SIAM J. Sci. Comput., Vol 27, No. 2, March 1996 454–478
2. Ortega, J.M.: Introduction to Parallel and Vector Solution of Linear Systems, Plenum Press, New York and London, (1988)
3. Benzi, M., Tuma, M.: A sparse approximate inverse preconditioner for nonsymmetric linear systems. SIAM Journal of Scientific Computing 19, (1998) 968–994
4. Chow, E., Saad, Y.: Approximate inverse preconditioners via sparse-sparse iterations. SIAM Journal of Scientific Computing 19, (1998) 995–1023
5. Gould, N.I.M., Scott, J.A.: Sparse approximate-inverse preconditioners using norm-minimization techniques. SIAM Journal of Scientific Computing 19, (1998) 605–625
6. Bruno, J., Cappello, P.: Implementing the 3-D alternating direction method on the hypercube. Journal of Parallel and Distributed Computing 23, (1994) 411–417
7. Tsompanopoulou, P., Vavalis, E.: ADI methods for cubic spline discretizations of elliptic PDEs. SIAM Journal of Scientific Computing 19, (1998) 341–363
8. Walters, R.W., Dwoyer, D.L., Hassan, H.A.: A strongly implicit procedure for the compressible Navier-Stokes equations. AIAA Journal, Vol. 24, No. 1, (1985) 6–12
9. Alden, J.A., Booth J., Compton, R.G., Dryfe, R.A.W., Sanders, G.H.W.: Diffusional mass transport to microband electrodes of practical geometries: A simulation study using the strongly implicit procedure. Journal of Electroanalytical Chemistry 389, (1995) 45–54
10. Stone, H.L.: Iterative solution of implicit approximations of multidimensional partial differential equations. SIAM J. Numer. Anal., Vol. 5, No. 3, (1968) 530–558

11. Technical Overview of the Origin Family,
    http://www.sgi.com/Products/hardware/servers/technology/
12. Snir, M., Otto ,S., Huss-Lederman, S., Walker, D., Dongarra, J.: MPI: The Complete Reference, The MIT Press, Cambridge, Massachusetts, London, England (1996)
13. Hluchy, L., Godlevsky A., Halada, L., Dobrucky, M., Tran, D.V.: Ground water flow modeling in distributed environment. Proc. of DAPSY'98, Budapest, (Sept. 27-29, 1998) 155–161.

# A Parallel Algorithm for Lagrange Interpolation on $k$-ary $n$-Cubes

Hamid Sarbazi-Azad[1], Lewis M. Mackenzie[1], Mohamed Ould-Khaoua[2]

[1] Department of Computing Science, University of Glasgow, Glasgow G12 8QQ, U.K.
e-mail: {has,lewis}@dcs.gla.ac.uk
[2] Department of Computer Science, University of Strathclyde, Glasgow G1 1XH, U.K.
e-mail: mohamed@cs.strath.ac.uk

**Abstract.** Most current multicomputers employ $k$-ary $n$-cube networks for low-latency and high-bandwidth inter-processor communication. This paper introduces a parallel algorithm for computing an $N=k^n$ point Lagrange interpolation on these networks. The algorithm consists of three phases: initialisation, main and final. While there is no computation in the initialisation phase, the main phase is composed of $\lceil k^n/2 \rceil$ steps, each consisting of four multiplications and four subtractions, and an additional step including one division and one multiplication. The final phase is carried out in $n \times \lceil k/2 \rceil$ steps, each using one addition.

## 1 Introduction

Lagrange interpolation for a given set of points $(x_0, y_0)$, $(x_2, y_2)$, ... , $(x_{N-1}, y_{N-1})$ and value $x$ is carried out via the formula

$$f(x) = \sum_{i=0}^{N-1} L_i(x) y_i \qquad (1)$$

where the term $L_i(x)$ is called a Lagrange polynomial, and is given by:

$$L_i(x) = \frac{(x-x_0)(x-x_1)\cdots(x-x_{i-1})(x-x_{i+1})\cdots(x-x_{N-1})}{(x_i-x_0)(x_i-x_1)\cdots(x_i-x_{i-1})(x_i-x_{i+1})\cdots(x_i-x_{N-1})} \qquad (2)$$

When the number of points, $N$, is very large, long computation time and large storage capacity may be required to carry out the above computation. To overcome this, several authors have recently proposed parallel implementations for Lagrange interpolation. For instance, Goertzel [11] has introduced a parallel algorithm suitable for a tree topology with $N$ processors, augmented with ring connections. The algorithm requires $\frac{N}{2} + O(\log N)$ steps, each composed of two subtractions and four multiplications. Capello, et al. [7] have described another algorithm using $2N-1$ steps on $N/2$ processors where each step, after implementation, requires two subtractions,

P. Zinterhof, M. Vajteršic, A. Uhl (Eds.): ACPC'99, LNCS 1557, pp. 85-95, 1999.
© Springer-Verlag Berlin Heidelberg 1999

two multiplications and one division. More recently, a parallel algorithm-architecture has been discussed in [14] which uses a $\sqrt{\frac{N}{2}} \times \sqrt{\frac{N}{2}}$ node mesh with toroidal wrap-around connections, consisting of $\frac{N}{4} + O(\sqrt{\frac{N}{2}})$ steps, each with 12 multiplications and subtractions.

K-ary $n$-cubes have been the most popular multicomputer networks due to their desirable properties, such as ease of implementation, recursive structures, and ability to exploit communication locality to reduce message latency. The hypercube and torus are the most common instances of $k$-ary $n$-cubes. The former has been used in early multicomputers like the Cosmic Cube [16] and iPSC/2 [12] while the latter has become popular in recent systems, such as the J-machine [8], iWarp [5] and CRAY T3D [13]. The topological and performance properties of the $k$-ary $n$-cubes have been widely studied in the literature ( see [1],[2],[4],[6],[9] and [10] for typical studies).

The $k$-ary $n$-cube, where $k$ is referred to as *radix* and $n$ as *dimension*, has $N=k^n$ nodes, arranged in $n$ dimensions, with $k$ nodes per dimension. Each node, $P$, can be identified by an $n$-digit radix $k$ address $(a_1,a_2,...,a_n)$ or $a_1,a_2,...,a_n$. The $i^{th}$ digit of the address, $a_i$, represents the node's position in the $i^{th}$ dimension. Nodes $P_{a_1,a_2,...,a_n}$ and $P_{b_1,b_2,...,b_n}$ are connected if and only if there exists $i$, $(1 \le i \le n)$, such that $a_i =(b_i \pm 1)$ $mod$ $k$ and $a_j = b_j$ for $1 \le j \le n ; i \ne j$. Thus, each node has $2n$ bi-directional links connected to $2n$ neighbours (2 neighbours in each dimension).

Fig.1 shows various examples of $k$-ary $n$-cubes topologies. The $k$-ary 1-cube is the well-known ring, while the $k$-ary 2-cube is best known as the torus; a variation of the mesh with added wrap-around connection. Note when $k$=2, the network collapses to the hypercube topology.

This paper proposes a parallel algorithm for computing an $N=k^n$ point Lagrange interpolation on a $k$-ary $n$-cube. Since $k$-ary $n$-cubes are the underlying topology of a number of practical parallel machines, it is possible to implement our algorithm on these available systems, eliminating the need of building any special-purpose hardware. This algorithm relies at some stages during computation, as we shall see below, on broadcast communication. The algorithm needs a Hamiltonian ring topology which should be embedded in the host $k$-ary $n$-cube network to ensure efficient broadcast operations. The rest of the paper is organised as follows. Section 2 introduces the parallel algorithm, and discusses its main structure. Section 3 draws some conclusions from this study.

## 2    The Parallel Algorithm

The proposed algorithm consists of three phases: *initialisation, main* and *final* phase. Firstly, the set of points to be interpolated are allocated to the nodes, one point for each node. Then, the Lagrange polynomials, $L_i(x)$ $(0 \le i < N)$, are computed. Finally the sum of the terms $L_i(x) \times y_i$ $(0 \le i < N)$, is calculated to obtain the final result, $y= f(x)$, according to equation (1).

Before describing these phases in more details, this section introduces some of the notations which will be used in the development of the algorithm. To compute $y=f(x)$ on a $k$-ary $n$-cube given the points $(x_0,y_0)$, $(x_1,y_1)$, ... , $(x_{N-1},y_{N-1})$ and value $x$, where $N=k^n$, let each processing node have four registers, $R_1$, $R_2$, $R_3$ and $R_4$. Registers $R_1$ and $R_2$, are used to compute terms required for evaluating Lagrange polynomial $L_i(x)$ and registers $R_3$ and $R_4$ are two buffers to implement an efficient all-to-all broadcast between nodes. Let $P_{a_1,a_2,...,a_n}$ indicate the node located at the $a_i{}^{th}$ position in the $i^{th}$ dimension and $P_{a_1,a_2,...,a_n}(R_m)$ denote the content of register $R_m (m=1,2,3,4)$ of that node. Furthermore, let $P_{a_1,a_2,...,a_n}^{(t)}(R_m)$ indicate the content of register $R_m$ of $P_{a_1,a_2,...,a_n}$ after step $t$, where a step may involve a set of communication and computational operations. Symbol '$\Leftarrow$' denotes a communication operation between two adjacent nodes.

## 2.1   The Initialisation Phase

In this phase, the values $x$, $x_{i_1+i_2\times k+i_3\times k^2+...+i_n\times k^{(n-1)}}$ and $y_{i_1+i_2\times k+i_3\times k^2+...+i_n\times k^{(n-1)}}$ are first given to the processor $P_{i_1,i_2,...,i_n}$, for $i_m = 0,1, ... , k-1$; $m=1,2,..., n$. The registers $R_1$- $R_4$ of each processor are then set to their initial values by the following instruction sequence:

$$P_{i_1,i_2,...,i_n}^{(0)}(R_1) \leftarrow 1;$$

$$P_{i_1,i_2,...,i_n}^{(0)}(R_2) \leftarrow 1;$$

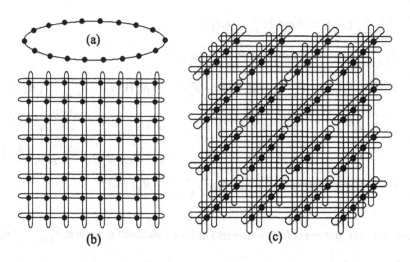

**Fig. 1.** Three examples of *k*-ary *n*-cube networks. (a) An 18-ary 1-cube or 18-node ring, (b) An 8-ary 2-cube or an 8x8 torus, and (c) a 4-ary 3-cube.

$$P_{i_1,i_2,\dots,i_n}^{(0)}(R_3) \leftarrow x_{i_1+i_2\times k+i_3\times k^2+\dots+i_n\times k^{(n-1)}};$$

$$P_{i_1,i_2,\dots,i_n}^{(0)}(R_4) \leftarrow x_{i_1+i_2\times k+i_3\times k^2+\dots+i_n\times k^{(n-1)}};$$

The initialisation phase can be implemented in a pipeline fashion, requiring only communication steps and no computation.

## 2.2   The Main Phase

Examining the communication pattern inherent in the interpolation algorithm reveals that a given node needs at some point to broadcast its x-value $x_{i_1+i_2\times k+i_3\times k^2+\dots+i_n\times k^{(n-1)}}$ to all the other nodes in the network. This broadcast operation is best performed according to the ring topology. To support efficient communication, the algorithm needs, in the main phase, a *Hamiltonian* ring topology that is embedded in the host *k*-ary *n*-cube; a Hamiltonian ring (or cycle) is a ring embedded in the host network including all its nodes [2]. To achieve this, we shall use the functions *Next* and *Previous* applied to processor $P_{i_1,i_2,\dots,i_n}$ to denote, respectively, the next and previous nodes of $P_{i_1,i_2,\dots,i_n}$ in the embedded ring. Fig.2 shows these two functions.

Fig.3 shows a 3-ary 4-cube in which a Hamiltonian ring is embedded so that for every node of the embedded ring, the next and previous nodes in the ring can be determined using functions *Next* and *Previous* (the formal proof of the Hamiltonian ring embedding method used here appears in [15]).

During this phase terms $L_m(x)$ and partial products $L_m(x)\times y_m$ for $m=0,1,\dots,$ $N$-1; are computed. To do this, firstly, all the processors perform the following instruction sequence simultaneously, for $t=0, 1,\dots, [N/2]$-1:

$$P_{i_1,i_2,\dots,i_n}^{(t+1)}(R_3) \Leftarrow Next\ P_{i_1,i_2,\dots,i_n}^{(t)}(R_3);$$

$$P_{i_1,i_2,\dots,i_n}^{(t+1)}(R_4) \Leftarrow \mathrm{Pr}\,evious\,P_{i_1,i_2,\dots,i_n}^{(t)}(R_4);$$

$$P_{i_1,i_2,\dots,i_n}^{(t+1)}(R_1) \leftarrow P_{i_1,i_2,\dots,i_n}^{(t)}(R_1)\times(x-P_{i_1,i_2,\dots,i_n}^{(t+1)}(R_3))\times(x-P_{i_1,i_2,\dots,i_n}^{(t+1)}(R_4));$$

$$P_{i_1,i_2,\dots,i_n}^{(t+1)}(R_2) \leftarrow P_{i_1,i_2,\dots,i_n}^{(t)}(R_2)\times(x_{i_1+i_2\times k+i_3\times k^2+\dots+i_n\times k^{(n-1)}}-P_{i_1,i_2,\dots,i_n}^{(t+1)}(R_3))\times$$

$$(x_{i_1+i_2\times k+i_3\times k^2+\dots+i_n\times k^{(n-1)}}-P_{i_1,i_2,\dots,i_n}^{(t+1)}(R_4));$$

if ($N$ *is even*)and($t=[\frac{N}{2}]-2$) then $P_{i_1,i_2,\dots,i_n}^{(t+1)}(R_3)\leftarrow 1;$

where the last statement ("if" statement) has been included to prevent $P_{i_1,i_2,\dots,i_n}(R_1)$ and $P_{i_1,i_2,\dots,i_n}(R_2)$ from multiplying by t he same factor twice when $N$ is an even number. The data path used in these $\lfloor N/2 \rfloor$ steps is the embedded Hamiltonian ring on which the values $x_0, x_1, ..., x_{N-1}$ rotate using an all to all broadcasting method described in [3]. These values rotate through the embedded Hamiltonian cycle enabling each node to obtain a copy. Each step consists of one data communication (note that the first two communication instructions can be realised in parallel because of bi-directional links between nodes), four subtractions and four multiplications. The two subtractions, in the third instruction, could be removed by adding two extra registers

```
Function Next P_{i_1,i_2,...,i_n} ;
{ If (i_1+i_2+...+i_n+1)mod k=0 then
    If (i_2+...+i_n+1)mod k=0 then
        ...
        ...
        ...
            If (i_{n-1}+i_n+1)mod k=0 then Return P_{i_1,i_2,...,i_n+1 mod k}

            Else Return P_{i_1,i_2,...,i_{n-1}+1 mod k,i_n} ;
        ...
        ...
        ...
    Else Return P_{i_1,i_2+1 mod k,i_3,...,i_n} ;

  Else Return P_{i_1+1 mod k,i_2,...,i_n} ;
};

Function Previous P_{i_1,i_2,...,i_n} ;
{ If (i_1+i_2+...+i_n)mod k=0 then
    If (i_2+...+i_n)mod k=0 then
        ...
        ...
        ...
            If (i_{n-1}+i_n)mod k=0 then Return P_{i_1,i_2,...,i_n-1 mod k}

            Else Return P_{i_1,i_2,...,i_{n-1}-1 mod k,i_n} ;
        ...
        ...
        ...
    Else Return P_{i_1,i_2-1 mod k,i_3,...,i_n} ;

  Else Return P_{i_1-1 mod k,i_2,...,i_n} ;
};
```

**Fig. 2.** Functions *Next* and *Previous*

to communicate terms $x - x_{i_1 + i_2 \times k + i_3 \times k^2 + ... + i_n \times k^{(n-1)}}$ through the ring as well as $x_{i_1 + i_2 \times k + i_3 \times k^2 + ... + i_n \times k^{(n-1)}}$ because the terms $x - x_{i_1 + i_2 \times k + i_3 \times k^2 + ... + i_n \times k^{(n-1)}}$ are computed over and over again by all processors during this phase. It adds just one subtraction in the initialisation phase. Using temporary registers in each node, it is also possible to overlap the communication time in the first two instructions and the execution time of the next three instructions. Fig.2 shows the data paths used in this phase for an 81-point interpolation, established with applying either *Next* or *Previous* function. At this point, we have

$$P_{i_1, i_2, ..., i_n}(R_1) = (x - x_0) \times (x - x_1) \times \cdots \times (x - x_{i_1 + i_2 \times k + i_3 \times k^2 + ... + i_n \times k^{(n-1)} - 1}) \times$$
$$(x - x_{i_1 + i_2 \times k + i_3 \times k^2 + ... + i_n \times k^{(n-1)} + 1}) \times \cdots \times (x - x_{N-1})$$

and

$$P_{i_1, i_2, ..., i_n}(R_2) = (x_{i_1 + i_2 \times k + i_3 \times k^2 + ... + i_n \times k^{(n-1)}} - x_0) \times (x_{i_1 + i_2 \times k + i_3 \times k^2 + ... + i_n \times k^{(n-1)}} - x_1)$$
$$\times \cdots \times (x_{i_1 + i_2 \times k + i_3 \times k^2 + ... + i_n \times k^{(n-1)}} - x_{i_1 + i_2 \times k + i_3 \times k^2 + ... + i_n \times k^{(n-1)} - 1}) \times$$
$$(x_{i_1 + i_2 \times k + i_3 \times k^2 + ... + i_n \times k^{(n-1)}} - x_{i_1 + i_2 \times k + i_3 \times k^2 + ... + i_n \times k^{(n-1)} + 1}) \times \cdots \times (x - x_{N-1}).$$

As the last step in this phase, all the processors execute the following instruction

$$P_{i_1, i_2, ..., i_n}^{([\frac{N}{2}]+1)}(R_1) \leftarrow \frac{P_{i_1, i_2, ..., i_n}^{([\frac{N}{2}])}(R_1)}{P_{i_1, i_2, ..., i_n}^{([\frac{N}{2}])}(R_2)} \times y_{i_1 + i_2 \times k + i_3 \times k^2 + ... + i_n \times k^{(n-1)}};$$

Therefore, at the end of this phase, $P_{i_1, i_2, ..., i_n}(R_1)$ contains $L_{i_1 + i_2 \times k + i_3 \times k^2 + ... + i_n \times k^{(n-1)}} \times y_{i_1 + i_2 \times k + i_3 \times k^2 + ... + i_n \times k^{(n-1)}}$ for $(0 \le i_j \le k - 1; 1 \le j \le n)$. In the main phase, each processor performs $N/2$ data communications(because of bi-directional links), $2N+1$ multiplications, $2N$ subtractions and one division.

## 2.3 The Final Phase

In this phase, the contents of register $R_l$ in all the nodes are added together to obtain the final result. This phase consists of $n$ sub-phases, each including $\lceil k/2 \rceil$ additions. We start from the $n^{th}$ dimension, adding the content of all registers $R_l$ in the $k$ nodes in that dimension. Dividing the dimension into two pieces, one includes the processors 0 to $k/2$ and another the nodes $\frac{k}{2}+1$ to $k$-1, and doing the addition process in these two parts in parallel , the sum of the registers $R_l$ of the nodes in dimension $n$ will be in the register $R_l$ of node 0 of that dimension after $\lceil k/2 \rceil$ steps, each consisting of one data communication and one addition. After each sub-phase, the dimension of problem is

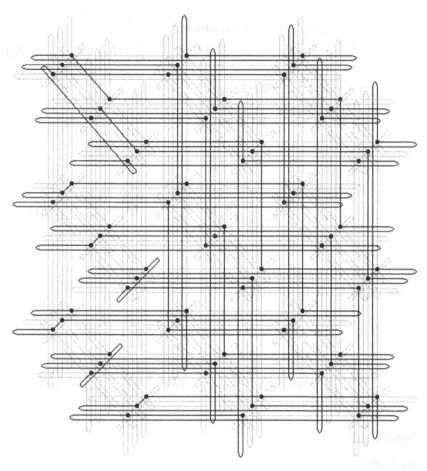

**Fig. 3.** Hamiltonian ring embedding in a 3-ary 4-cube employed for an 81-point interpolation

reduced by one until it reaches to the first dimension and after that sub-phase ($n^{th}$ sub-phase), the final result will be in register $R_l$ of node $P_{0,0,...,0}$. Therefore this phase can be realised in $n \times \lceil k/2 \rceil$ steps. The following parallel code must be executed in this phase:

**_Sub-phase 1:_**

*In parallel*
$$\{$$
$$\text{for } i'_n = \lfloor k/2 \rfloor, \lfloor k/2 \rfloor - 1, \ldots, 1$$
$$\text{for } i'_{n-1} = 0, 1, \ldots, k-1 \text{ in parallel}$$
$$\text{for } i'_{n-2} = 0, 1, \ldots, k-1 \text{ in parallel}$$
$$\cdots$$
$$\cdots$$

...

for $i'_l = 0,1,\ldots,k\text{-}1$ in parallel

$$P_{i'_1,i'_2,\ldots,i'_n-1}(R_1) \Leftarrow P_{i'_1,i'_2,\ldots,i'_n-1}(R_1) + P_{i'_1,i'_2,\ldots,i'_n}(R_1);$$

for $i_n = \lfloor k/2 \rfloor + 1, \lfloor k/2 \rfloor + 2,\ldots,k\text{-}2$
  for $i_{n\text{-}1} = 0,1,\ldots,k\text{-}1$ in parallel
    for $i_{n\text{-}2} = 0,1,\ldots,k\text{-}1$ in parallel

    ...

    ...

    ...

    for $i_l = 0,1,\ldots,k\text{-}1$ in parallel

$$P_{i_1,i_2,\ldots,i_n+1}(R_1) \Leftarrow P_{i_1,i_2,\ldots,i_n+1}(R_1) + P_{i_1,i_2,\ldots,i_n}(R_1);$$

};

for $i_{n\text{-}1} = 0,1,\ldots,k\text{-}1$ in parallel
  for $i_{n\text{-}2} = 0,1,\ldots,k\text{-}1$ in parallel

  ...

  ...

  ...

  for $i_l = 0,1,\ldots,k\text{-}1$ in parallel

$$P_{i_1,i_2,\ldots,i_{n-1}-1,0}(R_1) \Leftarrow P_{i_1,i_2,\ldots,i_{n-1}-1,0}(R_1) + P_{i_1,i_2,\ldots,i_{n-1}-1,k-1}(R_1);$$

### Sub-phase 2:

*In parallel*
    {
        for $i'_{n\text{-}1} = \lfloor k/2 \rfloor, \lfloor k/2 \rfloor - 1,\ldots,1$
          for $i'_{n\text{-}2} = 0,1,\ldots,k\text{-}1$ in parallel
            for $i'_{n\text{-}3} = 0,1,\ldots,k\text{-}1$ in parallel

            ...

            ...

            ...

          for $i'_l = 0,1,\ldots,k\text{-}1$ in parallel

$$P_{i'_1,i'_2,\ldots,i'_{n-1}-1,0}(R_1) \Leftarrow P_{i'_1,i'_2,\ldots,i'_{n-1}-1,0}(R_1) + P_{i'_1,i'_2,\ldots,i'_{n-1},0}(R_1);$$

        for $i_{n\text{-}1} = \lfloor k/2 \rfloor + 1, \lfloor k/2 \rfloor + 2,\ldots,k\text{-}2$
          for $i_{n\text{-}2} = 0,1,\ldots,k\text{-}1$ in parallel
            for $i_{n\text{-}3} = 0,1,\ldots,k\text{-}1$ in parallel

            ...

            ...

$$\cdots$$
$$\text{for } i_l = 0, 1, \ldots, k\text{-}1 \text{ in parallel}$$
$$P_{i_1,i_2,\ldots,i_{n-1}+1,0}(R_1) \Leftarrow P_{i_1,i_2,\ldots,i_{n-1}+1,0}(R_1) + P_{i_1,i_2,\ldots,i_{n-1},0}(R_1);$$
$$\};$$

for $i_{n-2} = 0, 1, \ldots, k\text{-}1$ in parallel
   for $i_{n-3} = 0, 1, \ldots, k\text{-}1$ in parallel
$$\cdots$$
$$\cdots$$
$$\cdots$$
   for $i_l = 0, 1, \ldots, k\text{-}1$ in parallel
$$P_{i_1,i_2,\ldots,i_{n-2}-1,0,0}(R_1) \Leftarrow P_{i_1,i_2,\ldots,i_{n-2}-1,0,0}(R_1) + P_{i_1,i_2,\ldots,i_{n-2}-1,k-1,0}(R_1);$$

$$\cdots$$
$$\cdots$$
$$\cdots$$

## Sub-phase n-1:

*In parallel*
   {
      for $i'_2 = \lfloor k/2 \rfloor, \lfloor k/2 \rfloor\text{-}1, \ldots, 1$
         for $i'_l = 0, 1, \ldots, k\text{-}1$ in parallel
$$P_{i'_1,i'_2-1,0,\ldots,0}(R_1) \Leftarrow P_{i'_1,i'_2-1,0,\ldots,0}(R_1) + P_{i'_1,i'_2,0,\ldots,0}(R_1);$$

      for $i_2 = \lfloor k/2 \rfloor + 1, \lfloor k/2 \rfloor + 2, \ldots, k\text{-}2$
         for $i_l = 0, 1, \ldots, k\text{-}1$ in parallel
$$P_{i_1,i_2+1,0,\ldots,0}(R_1) \Leftarrow P_{i_1,i_2+1,0,\ldots,0}(R_1) + P_{i_1,i_2,0,\ldots,0}(R_1);$$
$$\};$$

for $i_l = 0, 1, \ldots, k\text{-}1$ in parallel
$$P_{i_1,0,\ldots,0}(R_1) \Leftarrow P_{i_1,0,\ldots,0}(R_1) + P_{i_1,k-1,0,\ldots,0}(R_1);$$

## Sub-phase n:

*In parallel*
   {
      for $i'_1 = \lfloor k/2 \rfloor, \lfloor k/2 \rfloor\text{-}1, \ldots, 1$
$$P_{i'_1-1,0,\ldots,0}(R_1) \Leftarrow P_{i'_1-1,0,\ldots,0}(R_1) + P_{i'_1,0,\ldots,0}(R_1);$$

$$\text{for } i_2 = \lfloor k/2 \rfloor + 1, \lfloor k/2 \rfloor + 2, \ldots, k-2$$
$$P_{i_1+1,0,\ldots,0}(R_1) \Leftarrow P_{i_1+1,0,\ldots,0}(R_1) + P_{i_1,0,\ldots,0}(R_1);$$
$$\};$$
$$P_{0,0,\ldots,0}(R_1) \Leftarrow P_{0,0,\ldots,0}(R_1) + P_{k-1,0,\ldots,0}(R_1);$$

Fig.4 shows the data paths used in this phase for an 81-point interpolation. As can be seen in this figure, since the employed network is an 3-ary 4-cube, the final phase is carried out in 4 sub-phases.

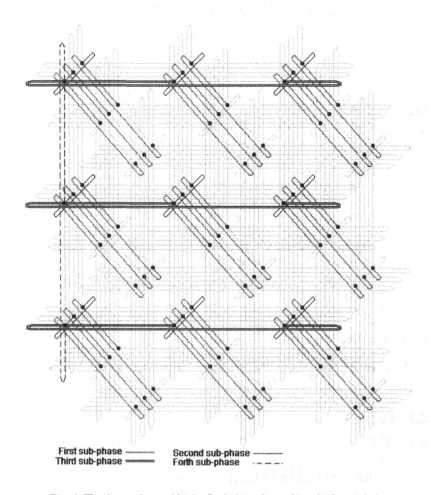

First sub-phase ————    Second sub-phase ————
Third sub-phase ══════    Forth sub-phase  · — · — ·

**Fig. 4.** The data paths used in the final phase for an 81-point interpolation

## 3.  Conclusion

$K$-ary $n$-cubes have been widely used in practice due to their desirable properties, such as ease of implementation and ability to exploit communication locality. This paper has proposed a parallel algorithm for Lagrange interpolation on $k$-ary $n$-cubes. The algorithm computes an $N(=k^n)$-point interpolation, requiring $2N+1$ multiplications, $N$ subtractions ( if two extra registers are used in each node; otherwise $2N$ subtractions will be required), $n \times \lceil k/2 \rceil$ additions and one division, without taking in to account any parallelisation in the internal architecture of the processors.

## References

1.      Ashir, Y., Stewart, I.A.: On embedding cycles in $k$-ary $n$-cubes. Parall. Process. Lett. 7 (1997) 49-55.
2.      Ashir, Y., Stewart, I.A.: Fault tolerant embedding of Hamiltonian circuits in $k$-ary $n$-cubes. Technical report, Dept of Math. Computer Sci., Univ. of Leicester, Leicester, U.K (1997).
3.      Bertsekas, D., Tsitsiklis, J.: Parallel and Distributed computation: Numerical methods. Prentice-Hall (1989).
4.      Bettayeb, S.: On the $k$-ary hypercube. Theoret. Comp. Sci. 140 (1995) 333-339.
5.      Borkar, S., et al.: iWarp: an integrated solution to high speed parallel computing. Proc. Supercomputing'88, IEEE Computer Society Press (1988) 330-339.
6.      Bose, B., Broeg, B., Kwon, Y., Ashir, Y. : Lee distance and topological properties of $k$-ary $n$-cubes. IEEE Trans. Computers 44 (1995) 1021-1030.
7.      Capello, B., Gallopoulos, Koc : Systolic computation of interpolation polynomials. Parallel Computing Journal 45 (1990) 95-117.
8.      Dally, W.J., et al.: The J-machine: a fine-grain concurrent computer. Information Processing'89, Elsevier Science Publisher (1989) 114-1153.
9.      Dally,W.J.: Performance analysis of $k$-ary $n$-cube interconnection networks. IEEE Trans. Computers 39 (1990) 775-785.
10.     Day, K., Al-Ayyoub, A.: Fault diameter of $k$-ary $n$-cube networks. IEEE Trans. Parallel & Distributed Systems 8 (1997) 903-907.
11.     Goertzel, B.: Lagrange interpolation on a tree of processors with ring connections. Journal Parallel & Distributed Computing 22 (1994) 321-323.
12.     Nugent, S.F.: The iPSC/2 direct-connect communication technology. Proc. Conf. Hypercube Concurrent Computers & Applications Vol. 1 (1988) 51-60.
13.     Oed, W.: The Cray research massively parallel processing system: Cray T3D. Cray Research Inc.Technical Report (1993).
14.     Sarbazi-Azad, H., Mackenzie, L.M., Ould-khaoua, M. : A parallel algorithm-architecture for Lagrange interpolation. Proc. Parallel & Distributed Processing: Techniques & Applications (PDPTA'98) Las Vegas (1998) 1455-1458.
15.     Sarbazi-Azad, H., Mackenzie, L.M., Ould-khaoua, M. :  A novel method for Hamiltonian ring embedding in $k$-ary $n$-cube networks. Tech. Report, Dept of Computing Sci., Glasgow University (1998).
16.     Seitz, C.L.: The Cosmic Cube. CACM 28 (1985) 22-33.

# Parallel Quasi-Monte Carlo Integration Using (t,s)-Sequences

Wolfgang Ch. Schmid[1] and Andreas Uhl[2]

[1] Department of Mathematics
University of Salzburg, AUSTRIA
[2] RIST++ & Department of Computer Science and System Analysis
University of Salzburg, AUSTRIA
{wolfgang.schmid,andreas.uhl}@sbg.ac.at

**Abstract.** Currently, the most effective constructions of low-discrepancy point sets and sequences are based on the theory of $(t, m, s)$-nets and $(t, s)$-sequences. In this work we discuss parallelization techniques for quasi-Monte Carlo integration using $(t, s)$-sequences. We show that leapfrog parallelization may be very dangerous whereas block-based parallelization turns out to be robust.

## 1  Introduction

Currently, the most effective constructions of low-discrepancy point sets and sequences, which are of great importance for quasi-Monte Carlo methods in multidimensional numerical integration, are based on the concept of $(t, m, s)$-nets and $(t, s)$-sequences. A detailed theory was developed in Niederreiter [10].

High dimensional numerical integration problems may require a significant amount of computations. Therefore, substantial effort has been invested into finding techniques for performing these computations on all kinds of parallel architectures (see [8] for an exhaustive overview). In order to keep the communication amount within a parallel system to a minimum, each processing element (PE) requires its own source of integration nodes. Therefore, our aim is to investigate techniques for using separately initialized and disjoint portions of a given point set on single PEs.

In practice, it is usually not possible to determine a priori the number of integration nodes $N$ necessary to meet a given error requirement. Therefore, it is of great importance that $N$ may be increased without loosing previously calculated function values. Additionally, unbalanced load within a parallel system makes it extremely difficult to predict the amount of integration nodes required on a single PE. For these reasons we restrict the discussion to infinite $(t, s)$-sequences.

We may choose between two possible approaches of generating separately initialized and disjoint substreams of a $(t, s)$-sequence:

- **Blocking**: disjoint contiguous blocks of the original sequence are used on the PEs. This is achieved by simply using a different start point on each PE (e.g. employing $p$ PEs, $PE_i$ generates the points $x_i, x_{i+1}, x_{i+2}, \ldots$ where $i$ is the first index of the block specific for each PE).

P. Zinterhof, M. Vajteršic, A. Uhl (Eds.): ACPC'99, LNCS 1557, pp. 96–106, 1999.
© Springer-Verlag Berlin Heidelberg 1999

– **Leaping**: interleaved substreams of the original sequence are used on the PEs. Each PE skips those points handled by other PEs (*leap-frogging*) (e.g. employing $p$ PEs, $PE_i$ generates the points $x_i, x_{i+p}, x_{i+2p}, \ldots$ with $i = 0, \ldots, p-1$).

These techniques have their corresponding counterparts in the field of pseudo random number generation where it is also desirable to obtain separately initialized and disjoint portions of the output stream of a pseudo random number generator for parallel applications. In this context, both techniques may lead to dangerous side-effects (see e.g. [4] for blocking and [5] for leaping).

Only a little amount of work has been done using $(t, s)$-sequences for parallel numerical integration. Bromley [3] describes a leap-frog parallelization technique to break up the so-called Sobol' sequence into interleaved subsets. In this work we extend Bromley's technique to all types of binary digital $(t, s)$-sequences, and we will investigate the effects which occur when using blocking and leaping parallelization, respectively. Most important, we will demonstrate that leaping parallelization may lead to dramatic defects in the results whereas blocking behaves very stable.

## 2 Numerical Integration by Means of Digital $(t, s)$-Sequences

We consider quasi-Monte Carlo methods for multidimensional numerical integration, with the half-open $s$-dimensional unit cube $I^s = [0, 1)^s$ for $s \geq 2$ as a normalized integration domain. These methods are based on the integration rule

$$\int F(\mathbf{u}) d\mathbf{u} \approx \frac{1}{N} \sum_{n=0}^{N-1} F(\mathbf{x}_n)$$

with deterministic nodes $\mathbf{x}_0, \mathbf{x}_1, \ldots, \mathbf{x}_{N-1}$ which are customarily taken from $I^s$.

One basic error estimate for the integration error is given by the Koksma-Hlawka inequality, where for the rather general class of functions $F$ of bounded variation $V(F)$ in the sense of Hardy and Krause we have for any $\mathbf{x}_0, \ldots, \mathbf{x}_{N-1} \in I^s$ :

$$\left| \int F(\mathbf{u}) d\mathbf{u} - \frac{1}{N} \sum_{n=0}^{N-1} F(\mathbf{x}_n) \right| \leq V(F) D_N^*(\mathbf{x}_0, \ldots, \mathbf{x}_{N-1}) .$$

We recall that for a point set $P$ consisting of $\mathbf{x}_0, \mathbf{x}_1, \ldots, \mathbf{x}_{N-1} \in I^s$, its *star discrepancy* $D_N^*(P)$ is defined by

$$D_N^*(P) = \sup_J \left| \frac{A(J; P)}{N} - \text{Vol}(J) \right|,$$

where the supremum is extended over all intervals $J$ of the form $J = \Pi_{i=1}^s [0, u_i)$ with $0 < u_i \leq 1$ for $1 \leq i \leq s$. Here, for an arbitrary subinterval $J$ of $I^s$,

$A(J; P)$ is the number of $0 \leq n \leq N - 1$ with $\mathbf{x}_n \in J$ and $\mathrm{Vol}(J)$ denotes the $s$-dimensional volume of $J$.

So to guarantee small integration errors, the nodes should form a low-discrepancy point set, i.e., a point set with small star discrepancy (see Niederreiter [12] for a survey of quasi-Monte Carlo methods).

The concepts of $(t, m, s)$-nets and of $(t, s)$-sequences in a base $b$ provide low-discrepancy point sets of $b^m$ points, respectively infinite sequences, in $I^s$, $s \geq 1$, which are extremely well distributed if the quality parameters $t \in \mathbb{N}_0$ are "small". We follow Niederreiter [12] in our basic notation and terminology.

**Definition 1** *Let $b \geq 2$, $s \geq 1$, and $0 \leq t \leq m$ be integers. Then a point set consisting of $b^m$ points of $I^s$ forms a $(t, m, s)$-net in base $b$ if every subinterval $J = \prod_{i=1}^{s} [a_i b^{-d_i}, (a_i + 1)b^{-d_i})$ of $I^s$ with integers $d_i \geq 0$ and $0 \leq a_i < b^{d_i}$ for $1 \leq i \leq s$ and of volume $b^{t-m}$ contains exactly $b^t$ points of the point set.*

Until now all construction methods which are relevant for applications in quasi-Monte Carlo methods are digital methods over finite fields $\mathbb{F}_q$, in most cases over $\mathbb{F}_2$.

**Definition 2** *Let $q$ be a prime-power and $s \geq 1$ be an integer. Let $C^{(1)}, \dots, C^{(s)}$ be $(\infty \times \infty)$–matrices over $\mathbb{F}_q$. For $0 \leq n$ let $n = \sum_{k=0}^{\infty} a_k q^k$ be the $q$-adic representation of $n$ in base $q$. Consider the digits $a_0, a_1, \dots$ as elements of $\mathbb{F}_q$. Let*

$$(y_1^{(i)}(n), \dots, y_\infty^{(i)}(n))^T := C^{(i)} \cdot (a_0, \dots, a_\infty)^T \quad for\ i = 1, \dots, s .$$

*The sequence*

$$\mathbf{x}_n := \left( \sum_{k=1}^{\infty} \frac{y_k^{(1)}(n)}{q^k}, \dots, \sum_{k=1}^{\infty} \frac{y_k^{(s)}(n)}{q^k} \right) \in I^s \quad for\ n = 0, 1, \dots$$

*is called a digital $(t, s)$-sequence constructed over $\mathbb{F}_q$ if, for all integers $k \geq 0$ and $m > t$, the point set consisting of the $\mathbf{x}_n$ with $kq^m \leq n < (k + 1)q^m$ is a $(t, m, s)$-net in base $q$.*

For a more general definition of digital $(t, s)$-sequences and digital $(t, m, s)$-nets (over arbitrary finite commutative rings) see for example Niederreiter [12] or Larcher, Niederreiter, and Schmid [9].

**Remark:** The construction of the points of digital nets and sequences and therefore the quality of their distribution only depend on the matrices $C^{(1)}, \dots, C^{(s)}$. So, the crucial point for a concrete implementation is the construction of the matrices. Examples for the construction of "good" matrices (and also properties providing a good distribution) can be found for example in Niederreiter [12].

# 3   Generating Substreams of Digital $(t, s)$-Sequences

In the following we only consider digital sequences over the binary field $\mathbb{F}_2$. In this case we may use the following notation for the definition of the sequence (see Definition 2): $\mathbf{y}_n^{(i)} := (y_1^{(i)}(n), \ldots, y_\infty^{(i)}(n))^T = \mathbf{c}_0^{(i)} a_0 \oplus \mathbf{c}_1^{(i)} a_1 \oplus \mathbf{c}_2^{(i)} a_2 \oplus \cdots$, where $\oplus$ denotes a binary exclusive-or operation and $\mathbf{c}_j^{(i)}$ the $j$-th column vector of the matrix $C^{(i)}$, $j \geq 0$.

Since there is a bijection between the vector $\mathbf{y}_n^{(i)}$ and the $i$-th coordinate of the $n$-th point $\mathbf{x}_n^{(i)}$ of the sequence, we frequently will use the vector notation without further mention.

Due to a suggestion of Antonov and Saleev [1] one can rewrite the sequence and calculate $\mathbf{y}_n^{(i)}$ from $\mathbf{y}_n^{(i)} = \mathbf{c}_0^{(i)} g_0 \oplus \mathbf{c}_1^{(i)} g_1 \oplus \cdots = \mathbf{c}_{l(n-1)}^{(i)} \oplus \mathbf{y}_{n-1}^{(i)}$, where $g_0, g_1, \ldots$ are the binary digits of the Gray-code representation of $n$ and $l(n)$ is the position of the least-significant-zero bit in the binary representation of $n$. (For details see for example [1,2]; for $k = 1, 2, \ldots$ the use of the Gray-code representation only shuffles each segment of length $2^k$, but does not affect the property of being a digital $(t, s)$-sequence.)

**Remark:** The shuffled sequence is provided by matrices $D^{(i)} := C^{(i)} \cdot Z$ for $i = 1, \ldots, s$, where $Z$ is a regular $(\infty \times \infty)$-matrix over $\mathbb{F}_2$ with 1 as elements of the main diagonal and the diagonal below, and 0 elsewhere.

In our experiments we will confine ourselves to the so-called Niederreiter sequences (defined in Niederreiter [11] for arbitrary base $b$; see also Bratley, Fox, and Niederreiter [2]). These sequences are widely used in many applications of quasi-Monte Carlo methods, such as physics, technical sciences, financial mathematics, and many more.

## 3.1   Blocking of $(t, s)$-Sequences

The starting point $\mathbf{x}_n$ is calculated by the usual Gray-code notation ($\mathbf{y}_n^{(i)} = \mathbf{c}_0^{(i)} g_0 \oplus \mathbf{c}_1^{(i)} g_1 \oplus \cdots$), the further points are given by the recursion relation ($\mathbf{y}_k^{(i)} = \mathbf{c}_{l(k-1)}^{(i)} \oplus \mathbf{y}_{k-1}^{(i)}$).

Note that if the number of points is of the form $2^m$, $m \in \mathbb{N}$, and the index of the starting point is a multiple of the number of points, say $n2^m$, then the chosen point set $\mathbf{x}_{n2^m}, \mathbf{x}_{n2^m+1}, \ldots, \mathbf{x}_{(n+1)2^m-1}$ forms a $(t, m, s)$-net by definition.

Starting points with rather small indices different from 0 already were considered for example by Fox [6], Bratley et al. [2], or Radovic et al. [14].

For a parallel execution, different blocks (i.e. different starting points) are assigned to different PEs. In order to guarantee non-overlapping blocks, the starting points should be sufficiently far apart. Note that even if it happens by chance that an identical point set is used in parallel and sequential execution the values of the integration error will be different for almost all samplesizes $\leq N$ (except for $= N$) since the ordering of the integration nodes is extremely shuffled. In general, the point sets used in parallel and sequential execution are different.

## 3.2   Leaping of $(t, s)$-Sequence

Bromley [3] describes a leap-frog parallelization technique to break up the Sobol'
sequence into interleaved subsets. In our experiments we only use leaps which
are an integral power of 2, say $2^L$. Then, in the terminology of Bromley, we have
the new recursion relation for leaped substreams

$$\mathbf{y}^{(i)}_{k2^L+o} = \mathbf{c}^{(i,L)}_{l(k2^L-1)} \oplus \mathbf{y}^{(i)}_{(k-1)2^L+o} \, , \quad o = 0, 1, \dots, 2^L-1; \ k = 1, 2, \dots \, ,$$

where

$$\mathbf{c}^{(i,L)}_r = \mathbf{c}^{(i)}_{l(2^r-1)} \oplus \mathbf{c}^{(i)}_{l(2^r-2)} \oplus \cdots \oplus \mathbf{c}^{(i)}_{l(2^r-2^L)} \, , \quad r \geq L$$

are the column vectors of the new $(\infty \times \infty)$-matrices $C^{(i,L)}$, $i = 1, \dots, s$, such
that each point of the new sequence is obtained by skipping $2^L - 1$ points of
the original sequence (for $r < L$ we can set $\mathbf{c}^{(i,L)}_r = 0$ since these vectors are
never touched in the recursion relation). There are $2^L$ different streams (the
same number as the leap factor) which are uniquely defined by their offset $o$.
The first point of each stream is calculated by the usual Gray-code notation:
$\mathbf{y}^{(i)}_o = \mathbf{c}^{(i)}_0 g_0 \oplus \mathbf{c}^{(i)}_1 g_1 \oplus \cdots$, $o = 0, \dots, 2^L - 1$.

For the details of this technique we refer to Bromley [3]. Kocis and Whiten
[7] already have mentioned a leaped Sobol' sequence, but without any numerical
experiments.

For a parallel execution, we assign each of the $2^L$ different streams (assuming
leap factor $2^L$) to a different PE. Note that in this case we use identical point
sets in parallel and sequential execution, even the ordering of the points may be
completely preserved. A different strategy is to use a leap factor larger than the
number of PEs which results in the use of different point sets used in parallel
and sequential execution (a reason for doing this might be the identification of a
specific leap factor leading to high quality streams as suggested for Halton and
Sobol sequences [7]).

# 4   Parallel Numerical Integration Experiments

## 4.1   Experimental Settings

The selection of the test function $F$ is a very crucial point in any attempt of
rating low-discrepancy point sets with quasi-Monte Carlo integration. Among
others, we consider the following test functions:

$$F1(x_1, x_2, \dots, x_s) = \prod_{i=1}^{s} (-2.4\sqrt{7}(x_i - 0.5) + 8\sqrt{7}(x_i - 0.5)^3) \, ,$$

$$F2(x_1, x_2, \dots, x_s) = \prod_{i=1}^{s} \left( \frac{|4x_i - 2| + i^2}{1 + i^2} \right) - 1 \, ,$$

$$F3(x_1, x_2, \dots, x_s) = \prod_{i=1}^{s} \left( x_i^{20} - \frac{1}{21} + 1 \right) - 1 \, ,$$

$$F4(x_1, x_2, \ldots, x_s) = \sqrt{\frac{45}{4s}} \left( \sum_{i=1}^{s} x_i^2 - \frac{s}{3} \right) .$$

For all these functions, $\int_{I_s} F(\mathbf{x})dx = 0$ which permits numerically stable calculation of the value $\frac{1}{N} \sum_{i=0}^{N-1} F(\mathbf{x}_i)$ up to sample sizes $N \approx 2^{30}$ and dimensions $s \approx 300$. All these functions have been used before in the context of rating point sets for quasi-Monte Carlo or Monte Carlo integration (F1 and F4 in [7], F2 in [14], and a variant of F3 in [5]).

The error of an integral at a particular value of $N$ varies in an apparently random manner within the accuracy range as $N$ changes [7]. Therefore, we use the maximal error in the range $(N_{j-1}, N_j]$ as our error-criterion ($N_j$ are the exponentially increasing measurement positions depicted in the plots). The maximal samplesize is set to $N = 2^{24}$, all figures show results employing Niederreiter sequences. We restrict the splitting parameters to powers of two (e.g. leap 3 denotes a leap factor $2^3$, whereas start 3 denotes a block with start point $2^3$).

In parallel (quasi)-Monte Carlo integration there is a trade-off between communication minimization and integration efficiency. The partial results calculated on the single PEs need to be combined at certain stages in order to perform an error estimation to determine whether the entire computation may terminate or has to continue (to achieve higher accuracy). For example, in quasi-Monte Carlo integration this error estimation may be done using randomized rules (see e.g. [13], [8, p. 166, 174]). Such a procedure requires communication or access to shared variables. Therefore, it should be kept to a minimum on the one hand. On the other hand, performing the calculations independently for too long may result in wasted computational effort since the desired accuracy could have been reached employing fewer integration nodes. We denote the number of integration nodes $\mathbf{x}_i$ evaluated independently on a single PE before computing the overall integration error as *syncstep* and fix its value to 500 (which is too small for practical applications but is already sufficient for demonstrating severe defects). We denote the $j^{th}$ set of integration nodes $\{\mathbf{x}_{j \cdot syncstep}, \ldots, \mathbf{x}_{(j+1) \cdot syncstep-1}\}$ produced on PE$_i$ as $\psi_i^j$, $j = 0, 1, 2, \ldots$ and $i = 0, 1, \ldots, \#PEs - 1$.

All computations have been carried out on a SGI POWERChallenge GR using a data parallel programming model employing the language PowerC.

## 4.2  Experimental Results

The possible effects of sequential leaping are displayed in Figure 1. We notice severe worsening of the integration results for all leaps $\neq 0$ for test function F4 (Figure 1.b), significant quality decrease concerning F2 is observed only for leaps $\geq 3$ (Figure 1.a). The results for F3 are similar to those of F4, whereas F1 does not exhibit any sensitivity against leaped sequences for leaps $\leq 4$.

Figure 2 reports on results for F4 employing parallel leaping on 2 PEs (therefore, leap 1 should result in the same error as compared to sequential execution).

Synchronized execution means that the partial results of the single PEs are combined only if $|j - k| \leq 1$ (with $j, k$ belonging to the most recently generated

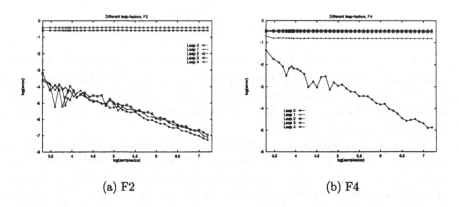

(a) F2                                    (b) F4

**Fig. 1.** Comparing integration results using differently leaped Niederreiter sequences.

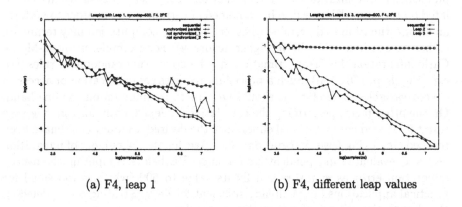

(a) F4, leap 1                        (b) F4, different leap values

**Fig. 2.** Parallel integration with leaping on 2 PEs.

set of integration nodes $\psi_0^j$ and $\psi_1^k$). Note that $|j - k| = 0$ is required to obtain the sequential result (this situation may be typical in a synchronized execution on a SIMD machine, but even that needs to be assured explicitly). Not synchronized execution does not put any restriction onto $|j - k|$ which means that PEs are allowed to "overtake" each other (which obviously happens and is due to scheduling mechanisms). In the following plots execution is synchronized if not stated differently.

We observe that using leap 1 even with synchronized execution (which already decreases computational efficiency) the integration precision already decreases slightly. A severe decrease takes place with not synchronized execution for samplesize $N > 10^6$ (see Figure 2.a). Parallel leaping with leap 2 decreases the precision slightly whereas we again observe catastrophic results for leap 3 (see Figure 2.b). Similar results are obtained for F3, whereas for F1 and F2 no problems arise.

The effects employing parallel leaping are even more severe when using a higher number of PEs. Figure 3 reports on results for F2 and F4 employing parallel leaping on 16 PEs (which means that leap 4 should result in the same error as compared to sequential execution). Synchronized execution means in that case that the absolute difference between any two exponents of $\psi_i^j$ needs to be $\leq 1$.

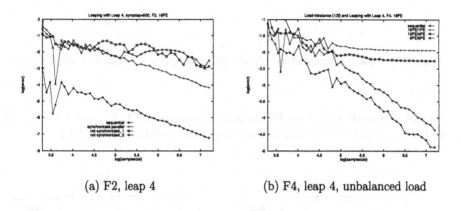

(a) F2, leap 4                    (b) F4, leap 4, unbalanced load

**Fig. 3.** Parallel integration with leaping on 16 PEs.

Even in synchronized execution mode we now notice a severe degradation of the results for F2, even more pronounced for samplesize $N > 10^5$ in not synchronized mode (see Figure 3.a). Very bad results are as well obtained using leaps 5 and 6, the results for F3 and F4 are almost identical in all cases. Figure 3.b reports on results using leap 4 for F4 if certain PEs show load level 2 instead of 1 (such an effect may be due to other users or system processes and causes these PEs to produce less integration nodes (i.e. half of the amount the others do)). The effects of this scenario are already dramatic in the case of one PE showing load 2 and even worse for the other cases considered (see Figure 3.b).

Now we switch to the case of blocking. Figure 4.a shows only hardly noticeable decrease in integration precision for F2 using different start points whereas Figure 4.b even displays significant precision gain for F3. For F4 we obtain a slight gain as well whereas F1 again does not show any sensitivity against different start points.

As one might expect, the properties of sequential blocking are propagated to parallel blocking (as it is the case for leaping). Employing 2 PEs, start 23 denotes the use of the blocks starting at start point $0 \cdot 2^{23}$ and $1 \cdot 2^{23}$ on the PEs, which results in the sequential integration node set for $N = 2^{24}$ in shuffled ordering. We observe a slight precision gain for F3 in synchronized and not synchronized

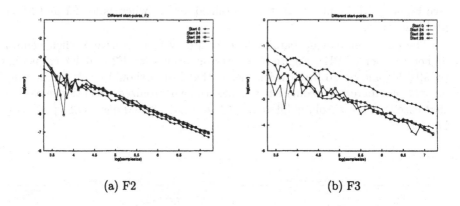

(a) F2                              (b) F3

**Fig. 4.** Comparing integration results using blocks with different start points.

parallel execution mode (see Figure 5.a). Blocking for F1, F2, and F4 does not lead to any significant effects on the integration precision with 2 PEs.

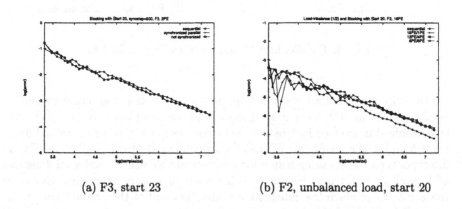

(a) F3, start 23              (b) F2, unbalanced load, start 20

**Fig. 5.** Parallel integration with blocking on 2 and 16 PEs.

The worst result obtained when employing parallel blocking is shown in Figure 5.b. The unbalanced load scenario as described for leaping leads to a moderate reduction of integration accuracy for F2. Similar results are obtained for F4, whereas F3 again shows moderate improvement and the results concerning F1 are not affected at all.

# 5  Conclusion

In this work we investigate possible effects of blocking and leaping $(t, s)$-sequences for parallel quasi-Monte Carlo integration. The properties of blocked and leaped substreams of these sequences are propagated to their use in parallel environments. Therefore, leaping may lead to dramatic degradation of the integration results whereas blocking turns out to be very robust. However, leaping offers the only possibility of obtaining an identical result as compared to sequential execution. To achieve this, a significant amount of caution has to be invested into proper synchronization of the parallel execution.

## Acknowledgements

The first author was partially supported by the FWF (Austrian Science Fund) project P12441-MAT. We thank Andreas Pommer for his help in identifying and resolving a major software problem on the POWERChallenge.

## References

1. I. A. Antonov and V. M. Saleev. An economic method of computing LP$_\tau$-sequences. *U.S.S.R. Comput. Maths. Math. Phys.*, **19**:256–259, 1979.
2. P. Bratley, B. L. Fox, and H. Niederreiter. Algorithm 738: Programs to generate Niederreiter's low-discrepancy sequences. *ACM Trans. Math. Software*, **20**:494–495, 1994.
3. B. C. Bromley. Quasirandom number generators for parallel Monte Carlo algorithms. *J. Parallel and Distributed Computing*, **38**:101–104, 1996.
4. A. DeMatteis and S. Pagnutti. Parallelization of random number generators and long-range correlations. *Numer. Math.*, **53**:595–608, 1988.
5. K. Entacher, A. Uhl, and S. Wegenkittl. Linear congruential generators for parallel Monte-Carlo: the Leap-Frog case. *Monte Carlo Methods and Appl.*, **4**:1–16, 1998.
6. B. L. Fox. Algorithm 647: Implementation and relative efficiency of quasirandom sequence generator. *ACM Trans. Math. Software*, **12**:362–376, 1986.
7. L. Kocis and W. J. Whiten. Computational investigations of low-discrepancy sequences. *ACM Trans. Math. Software*, **23**:266–294, 1997.
8. A. R. Krommer and C. W. Überhuber. *Numerical Integration on Advanced Computer Systems*, volume 848 of *Lecture Notes in Computer Science*. Springer, Berlin, 1994.
9. G. Larcher, H. Niederreiter, and W. Ch. Schmid. Digital nets and sequences constructed over finite rings and their application to quasi-Monte Carlo integration. *Monatsh. Math.*, **121**:231–253, 1996.
10. H. Niederreiter. Point sets and sequences with small discrepancy. *Monatsh. Math.*, **104**:273–337, 1987.
11. H. Niederreiter. Low-discrepancy and low-dispersion sequences. *J. Number Theory*, **30**:51–70, 1988.
12. H. Niederreiter. *Random Number Generation and Quasi-Monte Carlo Methods*. Number **63** in CBMS–NSF Series in Applied Mathematics. SIAM, Philadelphia, 1992.

13. A. B. Owen. Randomly permuted $(t, m, s)$-nets and $(t, s)$-sequences. In H. Niederreiter and P. J.-S. Shiue, editors, *Monte Carlo and Quasi-Monte Carlo Methods in Scientific Computing*, volume 106 of *Lecture Notes in Statistics*, pages 299–317. Springer, New York, 1995.
14. I. Radović, I. M. Sobol, and R. F. Tichy. Quasi-Monte Carlo methods for numerical integration: Comparison of different low discrepancy sequences. *Monte Carlo Methods and Appl.*, **2**:1–14, 1996.

# Parallel Random Number Generation: Long-Range Correlations Among Multiple Processors

Karl Entacher[1], Andreas Uhl[2], and Stefan Wegenkittl[1]

[1] Department of Mathematics
University of Salzburg, AUSTRIA
[2] RIST++ & Department of Computer Science and System Analysis
University of Salzburg, AUSTRIA
{karl.entacher,andreas.uhl,stefan.wegenkittl}@sbg.ac.at

**Abstract.** We use an empirical study based on simple Monte Carlo integrations to exhibit the well known long-range correlations between linear congruential random numbers. In contrast to former studies, our long-range correlation test is carried out to assess more than only two parallel streams. In addition we perform our test also with explicit inversive generators which from the theoretical point of view have to be stable against long-range correlations.

## 1 Introduction

The well known concept of long-range correlations between consecutive blocks of random numbers is due to a series of papers written by DeMatteis, Pagnutti, Eichenauer-Herrman and Grothe, e.g. see [5, 6, 7, 8, 9, 13].

Consecutive blocks of random numbers provide an easy method to get parallel streams of random numbers for parallel and distributed simulation [1, 3, 6] and such blocks are supported by various simulation languages [21, 23]. Additional papers which already suggest an assessment of correlations between parallel streams obtained from consecutive blocks are [11, 30], see also [27] for a recent article related to the latter concepts.

Consider a full-period sequence of a common random number generator $\mathbf{x} = (x_n)_{n=0}^{\rho-1}$. Using different seeds of such generators one can easily assign consecutive blocks

$$\omega_k := (x_{k \cdot l + n})_{n=0}^{l-1}, \quad k \geq 0. \tag{1}$$

of proper length $l$ to the $k$-th processor.

DeMatteis et al. studied pairs of random numbers of the form $\sigma_i := (x_i, x_{i+l})$, $i \geq 0$, hence correlations between pairs of processors, mainly for linear congruential generators ("LCGs"). The latter authors observed that for large block lengths $l$ the numbers between such blocks exhibit strong correlations ("long-range correlations") i.e. the pairs $\sigma_i$ show "strong" linear structures. They derived critical distances for several examples of well known generators.

P. Zinterhof, M. Vajteršic, A. Uhl (Eds.): ACPC'99, LNCS 1557, pp. 107–116, 1999.
© Springer-Verlag Berlin Heidelberg 1999

In this paper we use an empirical test based on Monte Carlo integration to analyze correlations between a larger number $s \geq 2$ of consecutive blocks, i.e. we assess vectors of the form

$$\sigma_i := (x_i, x_{i+l}, x_{i+2\cdot l}, \dots, x_{i+(s-1)\cdot l}), \quad i \geq 0, \quad s \geq 2. \tag{2}$$

For our linear congruential generators we use a previous calculated spectral test which analyzes the lattice structure behind the vectors (2). With the latter test we easily predict strong correlations down to *very* small block lengths $l$. With the empirical test we verify the presence of these correlations, which in an impressive way demonstrates the danger of using consecutive blocks of linear generators in parallel settings, and, secondly, shows the power of theoretical predictions by the spectral test.

In addition we performed our empirical study also for explicit inversive generators [12] which should be uncorrelated from the theoretical point of view whenever consecutive blocks are used in parallel settings, see [29, p. 15].

The definitions and basic properties of linear congruential and explicit inversive congruential random number generators are given in [12, 17, 28, 29]. We will denote the linear congruential generator with underlying recurrence $y_{n+1} \equiv ay_n + b \pmod{m}$ with seed $y_0$ by $LCG(m, a, b, y_0)$, $a, b, y_0 \in \mathbf{Z}_m$, and the explicit inversive congruential generator with prime modulus $p$, parameters $a$ and $b$ and seed $n_0$ by $EICG(p, a, b, n_0)$. Explicit inversive random numbers are generated by $y_n = \overline{a(n + n_0) + b}$, $n \geq 0$, where $\overline{z} \equiv z^{p-2} \pmod{p}$, $z \in \mathbf{Z}_p$. Note that $\overline{z}$ is the multiplicative inverse of $z$ in $\mathbf{Z}_p$ if $z \neq 0$. Implementations of these generators can be found at [20].

## 2   Long-Range Correlation and the Spectral Test

The spectral test (geometric version) measures the coarseness of $s$-dimensional lattices $L := \{\sum_{i=1}^{s} k_i \cdot \mathbf{b}_i : k_i \in \mathbf{Z}, 1 \leq i \leq s\}$, with lattice basis $\mathbf{b}_1, \dots, \mathbf{b}_s \in \mathbf{R}^s$. Overlapping tuples $\mathbf{x}_i := (x_i, x_{i+1}, \dots, x_{i+s-1})$, generated from linear congruential random numbers $x_i$, $i \geq 1$, and also from other linear generators, form grid structures which can be described as the intersection of certain lattices $L$ with the $s$-dimensional unit cube $[0, 1[^s$, see [14, 17, 24, 28]. The spectral test thus measures the quality of the linear congruential generator by determining the maximum distance $d_s$ between adjacent hyper-planes, taken over all families of parallel hyper-planes which contain all points $\mathbf{x}_i$. Hence $d_s$ gives the maximal size of empty slices (containing no points $\mathbf{x}_i$) within $[0, 1[^s$, the smaller $d_s$ the more uniform is the sample space of points.

Widely used is also a normalized spectral test $S_s := d_s^*/d_s$, $2 \leq s \leq 8$, for which $0 \leq S_s \leq 1$. The latter figure of merit should be viewed as a correlation measure (values near 1 imply a "good" lattice structure, whereas values near 0 exhibit strong correlations within the generated sequence). The constants $d_s^*$ are absolute lower bounds on $d_s$, see [22, p. 105] and [17, Sect. 7.7].

The calculation of the spectral test is realized using the *dual lattice* of $L$, since the maximal distance of adjacent hyper-planes $d_s$ is equal to one over the length of the shortest vector of the dual lattice [10].

Historically this test is due to Coveyou and MacPherson [4] who used multivariate Fourier analysis to study the quality of LCGs. An efficient implementation of the spectral test for arbitrary multiple recursive generators is given in [26]. Derived from the original form [4] of the spectral test, Hellekalek [19] proposed the general notion of weighted spectral test, which is not limited to random number generators with lattice structures.

For our purposes we use the spectral test to study correlations between consecutive blocks from linear congruential random numbers, i.e. we assess the lattices which contain vectors (2) produced by such generators, see [14] for details.

We consider for example[1] the well-known Cray-system generator $\texttt{Ranf} = LCG(2^{48}, 44485709377909, 0, 1)$ [1, 2, 3, 6, 8, 9, 11, 17] and the generator behind the ANSI C $\texttt{drand48()}$ function, BSD version, $\texttt{drand48} = LCG(2^{48}, 25214903917, 11, 0)$. Table 1 shows normalized spectral test analyses of consecutive blocks obtained from $\texttt{Ranf}$ and $\texttt{drand48}$. The table contains results of $S_s$ for block lengths $l = 2^j$, $9 \leq j \leq 17$ and dimensions $2 \leq s \leq 5$. From the spectral test results we observe that for both generators, and similarly for other power-of-two LCGs consecutive blocks with lengths $l$ where $\gcd(l, 2^{48})$ equals a power-of-two should not be used in parallel.

| Ranf | | | | | | | | |
|---|---|---|---|---|---|---|---|---|
| $s$ \| $j = 9$ | 10 | 11 | 12 | 13 | 14 | 15 | 16 | 17 |
| 2 \| 0.51 | 0.76 | 0.58 | 0.72 | 0.98 | 0.24 | 0.48 | 0.94 | 0.55 |
| 3 \| 0.79 | 0.83 | 0.46 | 0.69 | 0.55 | 0.83 | 0.22 | **0.054** | **0.014** |
| 4 \| 0.55 | 0.85 | 0.17 | **0.021** | **0.0026** | **0.0013** | **0.0013** | **0.0013** | **0.0013** |
| 5 \| **0.046** | **0.012** | **0.012** | **0.012** | **0.0044** | **0.0044** | **0.0044** | **0.0044** | **0.0044** |

| drand48 | | | | | | | | |
|---|---|---|---|---|---|---|---|---|
| $s$ \| $j = 9$ | 10 | 11 | 12 | 13 | 14 | 15 | 16 | 17 |
| 2 \| 0.4 | 0.62 | 0.57 | 0.35 | 0.52 | 0.65 | 0.87 | 0.65 | 0.67 |
| 3 \| 0.77 | 0.58 | 0.44 | 0.45 | 0.53 | 0.62 | 0.55 | 0.14 | **0.034** |
| 4 \| 0.62 | 0.81 | 0.47 | **0.059** | **0.0073** | **0.00092** | **0.00092** | **0.000 92** | **0.00092** |
| 5 \| 0.14 | **0.0088** | **0.0088** | **0.0088** | **0.0088** | **0.0033** | **0.0033** | **0.0033** | **0.0033** |

**Table 1.** Correlation analysis of consecutive blocks obtained from $\texttt{Ranf}$ and $\texttt{drand48}$.

In the sections below we will analyze the impact of these bad spectral test results with an empirical study. The latter is carried out in order to confirm that generators rejected by the spectral test fail also in simple practical applications. Note that the spectral test assesses the lattice structure of point sets consisting of the entire period of the linear congruential generators whereas the set of

---

[1] Further empirical results from a Monte Carlo simulation using $\texttt{Ranf}$ and $\texttt{drand48}$ are contained in [2].

vectors (2) used in our empirical test is of significantly smaller magnitude since the sample size has to be smaller than the block size.

# 3  Empirical Testing by Monte Carlo Integration

## 3.1  Experimental Settings

Theoretical analysis of a random number generators has always to be complemented by empirical testing. Here, the generator is treated as a black box. A sample of PRNs from the generator is used to compute a realization of a test statistic whose distribution is known under a null-hypothesis $H_0$, which states that the PRNs are independent random variables distributed uniformly on $[0, 1[$. Monte Carlo Integration is a particular worth-while example of an empirical test since it is at the core of many simulation techniques and is easy to implement and to verify. By varying the test function it is possible to imitate properties of several real-world models.

To be more specific, we shall consider the problem of numerical integration in dimension $s \geq 1$ (see e.g. [28]). Denote by $N$ a sample size and choose a test function $f : I^s \to \mathbf{R}$, where $I = [0, 1[$ is the half open unit interval. Put

$$\epsilon(f, \psi, N) := \frac{1}{N} \sum_{i=0}^{N-1} f(\psi_i) - \int_{I^s} f(\xi) d\xi$$

the difference between sum and integral, arising from Monte Carlo Integration with the sequence $\psi = (\psi_i)$, $i = 0, 1, \ldots, N - 1$, $\psi_i \in I^s$ of integration nodes. Under the assumption that $\psi$ is a sequence of independent random variables distributed uniformly on $I^s$ and that $f \in L^1$, i.e. the integral exists and is finite, the strong law of numbers guarantees almost sure convergence of the error to zero as $n$ increases to infinity. Here we shall use the error $\epsilon$ as criterion for a comparison of different sequences $\psi$ (our test setup is similar to [16, 15, 18]).

For the purpose of illustrating the impact of long-range correlations on the quality of several random number generators, we have chosen the following polynomial as test function $f$. For any integer $r \geq 1$ put $g^r : [0, 1] \to \mathbf{R}$, $g^r(x) := x^r - 1/(r + 1)$, and let $f(x_1, x_2, \ldots, x_s) = \prod_{i=1}^{s} g^r(x_i)$. For this function, $\int_{I^s} f(\xi) d\xi = 0$. The calculations are numerically stable in all considered dimensions.

In order to relate the absolute error $\epsilon$ to the sample size and to the properties of the test function, we first calculate 32 independent samples (sample errors) $\epsilon_k$, $1 \leq k \leq 32$, where each sample depends on $N$ integration nodes $(\psi_i^k)$, $0 \leq i < N$. We compute the mean error $\hat{\epsilon} = \frac{1}{32} \sum_{k=1}^{32} \epsilon_k$ and the sample variance $\hat{\sigma}^2 = \frac{1}{31} \sum_{k=1}^{32} (\epsilon_k - \hat{\epsilon})^2$. An estimator for a 99% confidence interval for the true integral $\int_{I^s} f(\xi) d\xi$ is given by $C = (\hat{\epsilon} - 0.485\hat{\sigma}, \hat{\epsilon} + 0.485\hat{\sigma})$, where 0.485 is $1/\sqrt{32}$ times the 0.005 quantile of the Student-$T$-distribution with 31 degrees of freedom.

In order to investigate long-range correlations using block length $l$ among $s$ processors, we use Monte Carlo integration in dimension $s$.

The following procedure for the construction of the integration nodes $\psi_i^k$ assures that no PRN will be used twice during the simulation. We assume that the blocklength $l$ divides the sample size $N$. This condition will be fulfilled in all our examples. First, generate $s \cdot N$ PRNs $x_n \in [0,1[$, $0 \le n < s \cdot N$. For $i = \rho \cdot l + j$ with $0 \le i < N$, $0 \le j < l$, put

$$\psi_i^1 = (x_{(\rho s)\cdot l+j}, x_{(\rho s+1)\cdot l+j}, \ldots, x_{(\rho s+s-1)\cdot l+j}).$$

Integration with these nodes $\psi_i^1$ yields the first sample $\epsilon_1$. Subsequently, we generate the next $s \cdot N$ PRNs, construct according nodes $\psi_i^2$ and calculate $\epsilon_2$. This is repeated 32 times to get the 32 independent samples $\epsilon_1, \ldots, \epsilon_{32}$.

Note that in this way we construct an integration node $\psi_i^k$ (a $s$-tuple for integration in dimension $s$) by using the corresponding PRNs out of $s$ disjoint blocks (corresponding to streams on $s$ different processors) separated by the block length $l$. By doing this, we may measure the impact of the correlation among the $s$ different streams on the integration error.

If $(x_i)$ is a sequence of independent random variables distributed uniformly on $[0,1[$, then the integration nodes $(\psi_i^k)$ are independent and distributed uniformly on $I^s$. We thus can reject a node set (i.e. a specific way of using PRNs) at the 99% level of significance if the true value (zero) of the integral is not contained in the confidence interval $C$.

## 3.2 Experimental Results

In this section we report some experimental results obtained in the above Monte Carlo integration and relate the spectacular failings of certain random number generators using specific values for dimension $s$ (number of processors or streams) and block length $l$ to their corresponding scores in the spectral test.

In the graphics, the ticks on the $x$ axis denote the dual logarithm of the sample size $N$. The dotted line reports the value of $\hat{\epsilon}$, whereas the shaded area corresponds to the estimated 99% confidence interval $C$. The horizontal line labeled with 0 represents the true value of the integral. Each figure shows dimensions (i.e. processors or streams) $s \in \{2, \ldots, 5\}$. The upper two graphics show (left to right) dimensions 2 and 3, the lower graphics show dimensions 4 and 5. The parameter $r$ is indicated in the figure captions. From the numerous results with $r$ varying in $\{3, \ldots, 20\}$ we give some representative examples where $r \in \{3, 7\}$. A more detailed analysis of the impact of the parameter $r$ on integration results can be found in [16].

We investigate the following random number generators:

- Ranf $= LCG(2^{48}, 44485709377909, 0, 1)$ a CRAY system generator with period length $\rho = 2^{46}$ [1, 2, 3, 6, 8, 9, 11, 17].
- drand48 $= LCG(2^{48}, 25214903917, 11, 0)$ the ANSI C generator ($\rho = 2^{48}$)[2].

- primeLCG = $LCG(2^{48} - 59, 49235258628958, 0, 1)$ from the recent tables of L'Ecuyer ($\rho = 2^{48} - 60$) [25].
- EICG = $EICG(2^{48} - 59, 3141592653, 0, 0)$ ($\rho = 2^{48} - 59$).

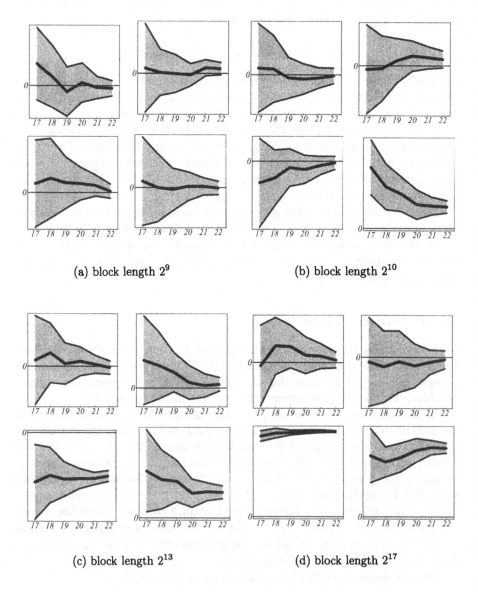

(a) block length $2^9$          (b) block length $2^{10}$

(c) block length $2^{13}$          (d) block length $2^{17}$

**Fig. 1.** Effects of critical distances using different block lengths with Ranf. Test parameter $r = 3$.

In approximate accordance to the predictions by the spectral test (compare Table 1) we observe a failing of **Ranf** in our test in dimension 5 (correlations among 5 processors) for block lengths $l \geq 2^{10}$ (see Figure 1.b). Additionally, correlations among 4 processors are detected for block lengths $l \geq 2^{13}$ (see Figure 1.c). Note the surprising fact that empirical evidence is exhibited at the next block length considered as compared to the failings in the spectral test which we consider to be due to the fact that the lattice structure does not show up for the (relatively small) sample sizes used in our computations.

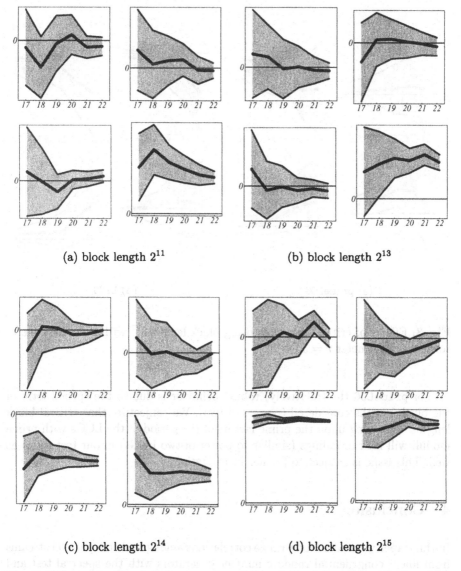

(a) block length $2^{11}$          (b) block length $2^{13}$

(c) block length $2^{14}$          (d) block length $2^{15}$

**Fig. 2.** Effects of critical distances using different block lengths with **drand48**. Test parameter $r = 7$.

Similar behavior as for **Ranf** is exhibited for **drand48** (see Figure 2) Again we notice a correspondence to the spectral test given in Table 1. Due to the larger period ($2^{48}$) our tests start to fail for block length $l \geq 2^{11}$.

For **primeLCG** and for **EICG** we could not detect any systematic correlations among power-of-two blocks (see e.g. Figure 3 for the largest block length considered). Concerning **EICG** this result is to be expected since the latter type of generator has been theoretically proven to be stable with respect to splitting [29].

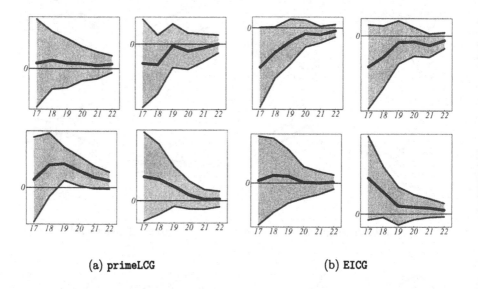

(a) **primeLCG**                              (b) **EICG**

**Fig. 3.** Effects of critical distances using block length $2^{17}$ with different generators. Test parameter $r = 3$.

For **primeLCG** the (relatively) robust behavior is due to the special form of the block lengths considered (powers of two). We conjecture that critical block lengths depending upon the prime factors of the period of the LCGs with prime moduli will lead to failings (similar to power-of-two LCGs) in our test setup as well. This issue is subject to further investigations.

## 4   Conclusion

In this paper we studied long-range correlations among multiple parallel streams from linear congruential random number generators with the spectral test and verified the results obtained by the latter test using an empirical study based on Monte Carlo integration. In addition we performed our empirical study also with explicit inversive congruential generators which from the theoretical point of view have to be stable against long-range correlations.

Our results demonstrate (i) the close coherence of the assessment of linear congruential random numbers by means of the spectral test and their behavior in applications, (ii) the danger of using consecutive blocks of linear congruential generators in parallel settings and (iii) empirical stability of explicit inversive congruential generators in parallel settings.

The well-known defect [22] of low-order bits obtained from power-of-two LCGs also shows up in empirical behaviour of parallel streams with power-of-two lags in our tests, so that we do not recommend the use of these popular but dangerous generators (see also [24]).

**Acknowledgments:**

The first and the third author are supported by the Austrian Science Fund (FWF), project P11143-MAT led by P. Hellekalek.

# References

[1] S.L. Anderson. Random number generators on vector supercomputers and other advanced architectures. *SIAM Rev.*, **32**:221–251, 1990.

[2] P. Coddington. Analysis of Random Number Generators Using Monte Carlo Simulation. *Int. J. Mod. Phys.*, C **5**:547, 1994.

[3] P. Coddington. Random Number Generators for Parallel Computers. NHSE Review, Second Issue, Northeast Parallel Architectures Center, 1996. Available at: http://nhse.cs.rice.edu/NHSEreview/RNG/.

[4] R.R. Coveyou and R.D. MacPherson. Fourier analysis of uniform random number generators. *J. Assoc. Comput. Mach.*, **14**:100–119, 1967.

[5] A. DeMatteis, J. Eichenauer-Herrmann, and H. Grothe. Computation of critical distances within multiplicative congruential pseudorandom number sequences. *J. Comp. Appl. Math.*, **39**:49–55, 1992.

[6] A. DeMatteis and S. Pagnutti. Parallelization of random number generators and long-range correlations. *Numer. Math.*, **53**:595–608, 1988.

[7] A. DeMatteis and S. Pagnutti. A class of parallel random number generators. *Parallel Comput.*, **13**:193–198, 1990.

[8] A. DeMatteis and S. Pagnutti. Critical distances in pseudorandom sequences generated with composite moduli. *Intern. J. Computer Math.*, **43**:189–196, 1992.

[9] A. DeMatteis and S. Pagnutti. Controlling correlations in parallel Monte Carlo. *Parallel Comput.*, **21**:73–84, 1995.

[10] U. Dieter. How to calculate shortest vectors in a lattice. *Math. Comp.*, **29**:827–833, 1975.

[11] M.J. Durst. Using linear congruential generators for parallel random number generation. In E.A. MacNair, K.J. Musselman, and P. Heidelberger, editors, *Proceedings of the 1989 Winter Simulation Conference*, pages 462–466, 1989.

[12] J. Eichenauer-Herrmann. Statistical independence of a new class of inversive congruential pseudorandom numbers. *Math. Comp.*, **60**:375–384, 1993.

[13] J. Eichenauer-Herrmann and H. Grothe. A remark on long-range correlations in multiplicative congruential pseudo random number generators. *Numer. Math.*, **56**:609–611, 1989.

[14] K. Entacher. Parallel Streams of Linear Random Numbers in the Spectral Test. Preprint, Department of Mathematics, University of Salzburg, Austria, submitted for publication, 1998.

[15] K. Entacher, A. Uhl, and S. Wegenkittl. Linear and Inversive Pseudorandom Numbers for Parallel and Distributed Simulation. In *Twelfth Workshop on Parallel and Distributed Simultation* PADS'98, May 26th - 29th, pages 90–97, Banff, Alberta, Canada, 1998. IEEE Computer Society, Los Alamitos, California.

[16] K. Entacher, A. Uhl, and S. Wegenkittl. Linear Congruential Generators for Parallel Monte-Carlo: the Leap-Frog Case. *Monte Carlo Methods and Appl.*, 4(1):1–16, 1998.

[17] G.S. Fishman. *Monte Carlo: Concepts, Algorithms, and Applications*, volume 1 of *Springer Series in Operations Research*. Springer, New York, 1996.

[18] P. Hellekalek. Don't Trust Parallel Monte Carlo. In *Twelfth Workshop on Parallel and Distributed Simultation* PADS'98, May 26th - 29th, pages 82–89, Banff, Alberta, Canada, 1998. IEEE Computer Society, Los Alamitos, California.

[19] P. Hellekalek. On Correlation Analysis of Pseudorandom Numbers. In H. Niederreiter, P. Hellekalek, G. Larcher, and P. Zinterhof, editors, *Monte Carlo and Quasi-Monte Carlo Methods 1996*, volume 127 of *Lecture Notes in Statistics*, pages 251–265. Springer, 1998.

[20] P. Hellekalek, K. Entacher, H. Leeb, O. Lendl, and S. Wegenkittl. The PLAB www-server. http://random.mat.sbg.ac.at, 1995. Also accessible via ftp.

[21] W.D. Kelton, R.P. Sadowski, and D.A. Sadowski. *Simulation with Arena*. WCB/McGraw-Hill, 1998.

[22] D.E. Knuth. *The Art of Computer Programming*, volume 2: Seminumerical Algorithms. Addison-Wesley, Reading, MA, 2nd edition, 1981.

[23] A.M. Law and W.D. Kelton. *Simulation Modeling and Analysis*. McGraw-Hill, New York, 2 edition, 1991.

[24] P. L'Ecuyer. Random Number Generation, 1998. Chapter 4 of *Handbook of Simulation*, Jerry Banks Ed., Wiley.

[25] P. L'Ecuyer. Tables of Linear Congruential Generators of Different Sizes and Good Lattice Structure. *Mathematics of Computation*, 68(225):249–260, 1999.

[26] P. L'Ecuyer and R. Couture. An Implementation of the Lattice and Spectral Tests for Multiple Recursive Linear Random Number Generators. *INFORMS Journal on Computing.*, 9(2):206–217, 1997.

[27] M. Mascagni. Parallel linear congruential generators with prime moduli. *Parallel Computing*, 24(5–6):923–936, 1998.

[28] H. Niederreiter. *Random Number Generation and Quasi-Monte Carlo Methods*. SIAM, Philadelphia, 1992.

[29] H. Niederreiter. New developments in uniform pseudorandom number and vector generation. In H. Niederreiter and P. Jau-Shyong Shiue, editors, *Monte Carlo and Quasi-Monte Carlo Methods in Scientific Computing*, volume 106 of *Lecture Notes in Statistics*. Springer, 1995.

[30] O.E. Percus and M.H. Kalos. Random Number Generators for MIMD Parallel Processors. *Journal of Parallel and Distributed Computing*, 6:477–497, 1989.

# A Monte-Carlo Method with Inherent Parallelism for Numerical Solving Partial Differential Equations with Boundary Conditions

Erika Hausenblas

Department of Mathematics
University of Salzburg
AUSTRIA
erika.hausenblas@sbg.ac.at

**Abstract.** One can construct the representation of solutions to various problems for parabolic differential equation with boundary conditions in the framework of classical stochastic analysis. This fact yields Monte Carlo methods for solving PDE's numerically. Instead of solving a PDE by common numeric techniques, one can simulate a stochastic system which allows for a simple parallelism with linear speed up.

A number of numerical schemes exists for solving stochastic differential equations, i.e. the Euler scheme. If reflection is concerned, most methods have some shortcomings. In this article an efficient algorithm for simulating a reflected stochastic differential equation is developed. Results of numerical experiments are referenced revealing significant reduction in computational time gained by parallelization.

## 1 General Setting

Probabilistic methods are an old approach, recently recovered, for solving partial differential equations (PDEs). The underlying idea for solving differential equations is to search for stochastic systems, for example diffusion processes or particle systems, which can be described by stochastic differential equations (SDEs). When the number of trajectories or particles tends to infinity, the behavior of the system is determined by the strong law of large numbers. The stochastic part vanishes or, more exactly, can be neglected and the limit is determined by a certain partial differential equation. Thus, such a PDE can be solved by simulating the corresponding stochastic system until a fixed time $T$ has elapsed and sampling a certain functional for computing the expectation of the functional.

Now, what is the advantage of stochastic methods over other methods, particularly with regard to the finite element method, a method which works rather well? Initially the probabilistic methods may not seem useful when a finite difference method, a finite element method, or finite volume method is numerical stable. In contrast to finite elements, where the solution of one element depends on prior solutions, the trajectories or particles can be computed almost independently. Therefore, on a single machine, common methods are better. However,

P. Zinterhof, M. Vajteršic, A. Uhl (Eds.): ACPC'99, LNCS 1557, pp. 117–126, 1999.

for parallel architectures, probabilistic methods are much more efficient. These methods are suitable for SIMD architectures, especially for loosely coupled systems. Communication time can mostly be neglected such that the speedup is proportional to the number of processes. A further advantage is that the problem can be divided into smaller parts, which allows for load balancing. Thus SDEs, resp. stochastic analysis becomes an increasingly important issue in the numerics of PDE's. A large amount of literature was published in the recent years on this topic (See, e.g., [1, 2, 9, 18, 16]).

For PDE's without boundary conditions, i.e. diffusion without boundary, the problem is well investigated. The simplest method is the Euler scheme, where we can rely on many results (See e.g. [20, 13]). For solving parabolic PDE's with mixed boundary conditions we need a quantity which measures the time the process remains at the infinitesimal neighborhood of the boundary. This measure turns out to be the local time of the boundary. Further, the existence of a solution turns out to be related to the Skorohod problem, or the question of existence of reflected diffusion. But there exists a few algorithms which are difficult to implement (See e.g. [6, 17, 14]). In [11] we suggested a new approach which is easy to implement and includes local time.

In this article we report on the numerical experiments and on the implementation of the algorithm on a workstation cluster. First we give a sketch of the theoretical background and motivate the algorithm. The third chapter describes the implementation and parallelization carried out on a FDDI connected workstation cluster of six DEC AXP 3000/400 using the hots-node model. Lastly we present the numerical results and indicate some problems which arise using Monte-Carlo methods on parallel architectures.

## 2    A Sketch of the Theoretical Background

The PDE considered in this article is the following:

$$Lf = \frac{\partial}{\partial t}f \quad \text{on} \quad \bar{D} \tag{1}$$

$$\frac{\partial}{\partial n}f = \gamma(x) \quad \text{on} \quad \partial D, \tag{2}$$

where $D$ is a smooth domain in $\mathbb{R}^d$, $L$ a strictly elliptic differential operator and $f : \mathbb{R}^d \times \mathbb{R}^+ \to \mathbb{R}$ is twice continuously differentiable. Let $(\Omega, \mathcal{F}, \{\mathcal{F}_t\})$ be a filtered probability space where the standard $d$-dimensional Brownian motion $B_t$ is defined. The probabilistic approach to the problem is to identify the operator $L$ by an infinitesimal generator of a Markov-process and to rewrite the equation (1) as a SDE of Skorohod type, i.e. $X_t$ is solution of the stochastic system

$$dX_t = \sigma(X_t)dB_t + b(X_t)dt + n\gamma(x)L_t, \tag{3}$$

where $x \in \bar{D}$ for all $t \geq 0$, a.s. and $L_t$ is a bounded variation process with

$$L_t = \int_0^t 1_{X_s \in \partial D} dL_s.$$

$L_t$ is called local time. $\sigma : \mathbb{R}^d \to \mathbb{R}^r \times \mathbb{R}^d$, $b : \mathbb{R}^d \to \mathbb{R}^d$ are supposed to be Lipschitz continuous, $\in \mathcal{C}^\infty(\mathbb{R}^d)$ and satisfy the linear growth condition. The solution of system (1) coincides with the function $f(x) = \mathbb{E}^x[X_t]$, where the semimartingal $X_t$ solves system (3). For more details we refer for examples to [19, 12, 15].

**Splitting into excursion:** The idea of approximating the local time is to approximate the underlying Poisson point process, where the local time corresponds to the parameter and the real time or the time the process $X_t$ passes through corresponds to the sum over the time intervals, where $X_t$ is removed from the boundary (See e.g. [3, 4]).

$L_t$ grows exactly at those $t$ for which $X_t$ is in $\partial D$, i.e. $supp(dL_t) = \{t \mid X_t \in \partial D\}$. Thus, we can split a path $X_t(\omega)$ into pieces by cutting at the time points $\{X_t(\omega) \in \partial D\}$. Each piece is a process starting somewhere at the boundary and being killed upon reaching the boundary again. Such a piece is called an excursion. $\tau = \inf_{t>0}\{X_t \in \partial D\}$ is the lifetime of such an excursion (See figure 1 for illustration).

The idea is to now use local time to parametrize the excursion. To take into account the case in which $X_t$ spends a positive real time at the boundary, we create a set of empty paths $\Theta$, isomorphic to $\partial D$. Thus, to every strong Markov process corresponds an excursion process, i.e., a set $\mathcal{U} \cup \Theta$ and a point process of Markov type $e = (e(s), s > 0)$ defined by

$$e : [0, \infty) \to \mathcal{U} \cup \Theta$$

$$e_t(s) = \begin{cases} X_{L_{s^-}^{-1}+t} & \text{if } L_s^{-1} - L_{s^-}^{-1} > 0 \\ \delta_t^d \text{ with } d = \lim_{r \to s, r < s} e_\tau(r) & \text{otherwise.} \end{cases}$$

It follows that $s \mapsto e(s) \notin \Theta$ for only countable many $s$. Let $\tau(e) = \inf_{t>0}\{e_t \in \partial D\}$ if $e \in \mathcal{U}$ and $\tau(e) = 0$ if $e \in \Theta$. The parameter $s$ turns out to be the local time of the boundary. If we sum up the lengths of the excursion, i.e.,

$$T_s = \sum_{u \leq s} \tau(e(u)),$$

the result is the time the process passed through: $L_s^{-1} = T_s$ or $L_{T_s} = s$. It follows that $T_s$ is a subordinator of $s$ with associated Lévy measure $\eta^d(\tau > s)$.

**Sticking together the excursion:** Now, we can reverse this construction by starting with a set of excursions equipped with an excursion law and the corresponding point process and linking together the excursions to get a Markov process. By doing this, our time parameter passes through the local time. The real time of the process is given by the subordinator $T_s$. Let $T_s^- = \lim_{u \to s, u < s} T_s$ be the left limit point of $T_s$. Let $t > 0$ and $s = L_t$. We define $X_t$ by:

$$X_t = \begin{cases} e_{t-T_s^-}(s) \text{ if } T_s^- < T_s \\ \lim_{r \to s, r < s} e_\tau(r) \text{ otherwise.,} \end{cases}$$

**Approximation of the underlying Poisson process.** For smooth $D$ the Lévy measure has infinite mass at zero. This means that the number of points in a finite time interval $[0, s]$ where $X_t$ goes on an excursion with lifetime larger than $\epsilon$, tends to infinity as $\epsilon$ tends to zero, i.e.

$$\sum_{\substack{\tau(e(r))>\epsilon \\ 0\leq r\leq s}} 1 \to \infty \quad \text{as} \quad \epsilon \searrow 0.$$

But, in the case of Brownian motion, for example, it holds almost surely.

$$\sqrt{\frac{\pi}{2\epsilon}} \sum_{\substack{\tau(e(r))<\epsilon \\ 0\leq r\leq s}} \tau(e(r)) \to s \quad \text{as} \quad \epsilon \searrow 0.$$

The time interval between two excursions with lifetime $\tau > \epsilon$ are exponentially distributed with holding parameter tending to infinity as $\epsilon \searrow 0$. Therefore, to give an explicit construction, we fix $\epsilon > 0$ and approximate $X_t$ by only considering excursions with lifetime $\tau > \epsilon$. Let $e$ be an excursion with lifetime larger than $\epsilon$, i.e. $\tau(e) > \epsilon$. Since we have cancelled the excursion with lifetime smaller than $\epsilon$, we know that $\tau(e)$ is at least $\epsilon$ and therefore, to preserve the Markov property of the limit process for $\epsilon \to 0$, we have to start the excursion in the algorithm at point $e_\epsilon$. This has the effect of jumping into $\bar{D}$ and leads to an entrance law $\eta^d$ which coincides with the excursion law, a $\sigma$-finite measure which is strong Markov for the transition semigroup of $\bar{X}_t$ defined by $\bar{X}_t = X_t$ if $t < \tau$ and $= \delta$ otherwise. After jumping in, $e_t$ behaves like $X_t$ until the lifetime

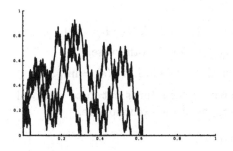 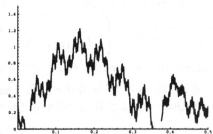

**Fig. 1.** Set of excursions          **Fig. 2.** Point process with entrance law

$\tau(e)$ where it is killed (for illustration see figure 2). The approximated time $T_s^\epsilon$ of $T_s$ satisfies

$$T_s^\epsilon = \sum_{\substack{\tau(e(r))>\epsilon \\ r\leq s}} (\tau(e(r)) + \epsilon) + \mathbb{E}\left[ \sum_{\substack{\tau(r)\leq\epsilon \\ r\leq s}} \tau(e(r)) \right].$$

**Approximation of the entrance law.** Since the entrance law of an arbitrary diffusion is not explicitly given, we approximate it by a Brownian motion frozen at time $\tau$. The entrance law of a d-dimensional Brownian motion can be found in Burzdy [4]. The time change formula leads to the entrance law of a Brownian motion with a constant diffusion coefficient $\sigma$. The drift term can be included as an application of the Cameron-Martin-Girsanov formula and the optional sampling theorem. Thus, if $\hat{X}_t$ is a Brownian motion with constant diffusion $\sigma$ and drift $b$, a short argument shows that the entrance law satisfies

$$\eta_\epsilon^{\sigma B - bt}(x) = \frac{\langle \sigma^{-1}(x - \epsilon b), \sigma^{-1} n \rangle}{\sqrt{\pi^d \epsilon^{d+2}}} \; exp(-\frac{\|\sigma^{-1}(x - \epsilon b)\|_2^2}{2\epsilon}) \tag{4}$$

with mass $\|\eta_\epsilon^{\sigma B - bt}\| = \sqrt{\frac{1}{\epsilon}} \sqrt{\left(\frac{2}{\pi}\right)^d} det(\sigma^{-1}) \int_{\sqrt{\epsilon}\sigma^{-1}b}^{\infty} e^{-\langle z,z\rangle/2} dz$ which we approximated by

$$\sqrt{\frac{1}{\epsilon}} det(\sigma^{-1}) \left(1 + C_1 \sqrt{\epsilon} + C_2 \epsilon + C_3 \sqrt{\epsilon}^3 + \cdots \right), \tag{5}$$

where $h_i = -(\sigma^{-1}b)_i$, $1 \leq i \leq r$ and $C_i$ are the coefficients: $C_1 = \sum_{i=1}^{d} h_i$, $C_2 = \sum_{i \neq j}^{d} h_i h_j$, $C_3 = \sum_{i \neq j \neq k}^{d} h_i h_j h_k - \sum_{i=1}^{d} h_i^3, \cdots$ .

**Illustration of the algorithm.** Roughly spoken, the algorithm simply simulates the point process at which the excursions with lifetime smaller than $\epsilon$ are cancelled. The excursions are approximated by the Euler scheme and the entrance law by the entrance law of a frozen Brownian motion as described above. The Monte-Carlo error, or the error of weak convergence, can be estimated by

$$\mathbb{E}^x[f(X_T)] - \mathbb{E}^x[f(\hat{X}_T^{\epsilon,n})] \leq C(T) \left(\epsilon + \sqrt{\frac{1}{n}}\right) + C(T) \max(\frac{1}{n}, \epsilon)$$

$$\mathbb{E}^x[L_T] - \mathbb{E}^x[\hat{L}_T^{\epsilon,n}] \leq C(T)\epsilon^{\frac{3}{2}}.$$

For illustration we present the sequence of figures 3-8.

# 3    Implementation and Parallelization Results

This section deals with our computer experiments and their parallelization. First, we describe some parallelization aspects, especially the problems arising by random number generation. Second, we determine the complexity of the operator. The third point concerns the implementation, and afterwards we solve a reflected Brownian motion where the distribution of the local time is known. Finally, we compare the numerical solution with the exact one to arrive at the degree of approximation.

## 3.1    Parallelization Aspects

For parallelization we choose the host-node model, implemented in $C^{++}$ using the PVM library. The PVM system handles message routing, data conversion for incompatible architectures and other tasks that are necessary for operation in a heterogeneous network environment. User programs written in, e.g. $C^{++}$, access PVM by passing calls to PVM library functions such as process initiation or message passing. For more details see [8]. Since the algorithm allows small partitioning, load balancing is applied.

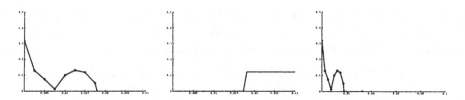

**Fig. 3.** We start with an ar- **Fig. 4.** The increment of **Fig. 5.** For preserving the bitrary point $x \in \bar{D}$ and ap- the local time is simulated Markov property, $X_\epsilon$ to proximate the excursion by by an exponential random jump into $\bar{D}$. The density of the Euler scheme until it hits variable with holding pa- the random variable is given the boundary. rameter $\|\hat{\eta}_\epsilon\|$. by the entrance law $\eta_\epsilon$.

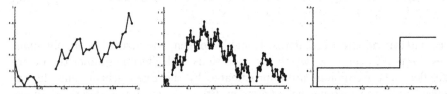

**Fig. 6.** After jumping in we **Fig. 7.** Thus, we get a re- **Fig. 8.** Each time $X_t^n$ hits approximated the excursion flected diffusion. the boundary the local time by the Euler scheme. is simulated by an exponentially distributed random variable.

**Random Number Generation on Parallel Architectures:** For our parallel streams of random numbers we used lagged subsequences with prime-step-size. The random number generator is a $C^{++}$-system generator. This approach should be safe against correlations between parallel streams. For problems concerning parallel random number generation we refer to [5, 7].

## 3.2 Complexity

Table 1 lists the complexity for one step of the Euler scheme and starting one excursion. Column (a) represents the operation of one step of the Euler scheme, while column (b) represents the operation of starting an excursion.

| Operation | | (a) | (b) | Operation | (a) | (b) |
|---|---|---|---|---|---|---|
| Addition | (+) | 2d | (2r)d | exponential function $exp(\cdot)$ | 1 | 0 |
| Subtraction | (-) | 2d+1 | 3d | logarithm $\qquad\qquad\log(\cdot)$ | 0 | 2 |
| Comparison | (¡) | 1 | 2 | Square root $\qquad\qquad\sqrt{\cdot}$ | 0 | 1 |
| Multiplication | (*) | d(r+3) | (r+2)d+2 | generating of uniform r.v. | 1 | 2 |
| Division | (÷) | 1 | 1 | generating of a gaussian r.v. | 1 | 0 |
| evaluating $\sigma, b$ | | 4 | 2 | evaluating $\sigma(\cdot)$ | 1 | 0 |

**Table 1.** Complexity of the algorithm

$\frac{[Tn]}{n}+1$ is an upper bound of the number of steps in the Euler scheme. Because of the random nature of Monte-Carlo simulations, we cannot determine the exact number of excursions, and we can only consider its expectation. Therefore the expectation of the complexity is proportional to

$$\frac{[Tn]}{n} \; comp(\text{Euler step}) + \frac{1}{\sqrt{2\epsilon}} \; comp(\text{starting one excursion})$$

### 3.3  Random Number Generation

We must generate three types of random variables:

1. Gaussian random variables
2. random variables distributed according to the entrance law
3. exponentially distributed random variables.

The Gaussian random variables are generated by summing and renormalizing seven uniform random numbers. The Box Muller method (See e.g. [13, p. 13]) leads to distortions. The random variables distributed according to the entrance law  are generated in analogy to its construction of the entrance. First, we must generate a random variable $x_1$ according to the entrance law of a one-dimensional Brownian motion. Since we have $z/\epsilon \exp(z^2/2\epsilon) = d/dz \exp(z^2/2\epsilon)$, the inverse transform method (See e.g. [13, p. 12]) can be applied. Next we must generate a (d-1)-dimensional Gaussian random variable $(x_2, x_3, \ldots, x_d)$. Assuming $n_1, n_2, \ldots, n_d$ is a orthonormal base in $\mathbb{R}^d$ where $n_1 = n$ and $T$ the corresponding matrix, the arising vector $x \in \mathbb{R}^d$ must be transformed by $x \mapsto Tx \mapsto \sigma Tx + \epsilon b$ where $\sigma = \sigma(d)$ and $b = b(d)$. In order to generate an exponential random variable we use the inverse transform method.

## 4  Numerical Results

Since the distribution of the local time is only known in a few cases we use the reflected Brownain motion to test the algorithm. This corresponds to

$$\frac{1}{2}\Delta f = \frac{\partial}{\partial t} f \text{ on } \bar{D}$$

$$\frac{\partial f}{\partial n} = 1 \text{ on } \partial D,$$

where $D = \{x_1 > 0\} \subset \mathbb{R}^d$.

To compute the local time we must wait until $B_t^x$ hits the boundary, and then we apply the Skorohod embedding theorem:

$$\mathbb{E}^x[L_T] = \int_0^T \underbrace{\frac{x}{\sqrt{2\pi}} t^{-\frac{3}{2}} e^{-\frac{x}{2t}}}_{\substack{\text{hitting the boundary} \\ \text{at time } t}} \underbrace{\sqrt{\frac{2(T-t)}{\pi}}}_{\mathbb{E}^0[L_{T-t}]} dt$$

As an exact value we have $\mathbb{E}^1[L_5] = 0.9986$.

## 4.1   Local Time Computed by the Algorithm

The accuracy of the algorithm depends on the sample size, the discretization step $\frac{1}{n} = h$ and the parameter $\epsilon$. The influence of the sample size is given by the weak law of large numbers, but our main interest is the ratio between $\epsilon$ and $n$. We used a sample size of 50.

Figure 9 shows the local time as a function of the step size $h$, where the range of $h$ is $3^{-i}$, $i = 2, \ldots 14$. $\epsilon$ is varying from 0.1 to 0.0005 (See table 2). The line denotes the theoretical value. Figure 10 shows the corresponding variance of the sample, figure 11 shows the corresponding point $X_T^{n,\epsilon}$.

| $\epsilon =$ | 0.1 | 0.05 | 0.01 | 0.005 | 0.001 | 0.0005 | 0.0001 | 0.00005 |
|---|---|---|---|---|---|---|---|---|
| sign | • | o | + | - | * | % | # | ' |

**Table 2.** Symbols used in figures 9 - 11.

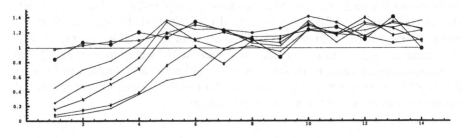

**Fig. 9.** Convergence behavior of the local time

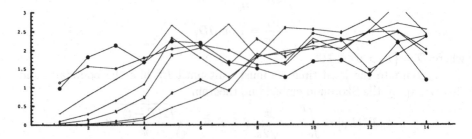

**Fig. 10.** Corresponding variance of the sample

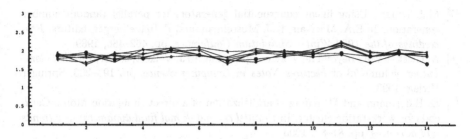

**Fig. 11.** Convergence behavior of $X_t^n$

If $\epsilon \ll h$, the Euler scheme is too clumsy with regard to the entrance law, i.e. the error induced by the difference $\tau^n - \tau$ is too large. For example if $X^n$ hits the boundary at each step of the simulation, the expectation of the local time is only 0.00225. Therefore, the local time in this case is too small.

## 5  Conclusion

First experiments lead to good results, but the work is in the initial stage. This approach can be easily extended to schemes of higher order, i.e. the Milshtein scheme. One difficulty with Monte-Carlo methods is reproducing the numerical experiments. The random nature of the Monte Carlo method and the dynamic load balancing of the processors, i.e. the slight differences between processor speed, makes the reproduction of results difficult (for further reading, see e.g. [10]). However, this problem can be solved by statistical tests.

**Acknowledgment:** I would like to thank Peter Zinterhof, the head of the RIST++, for enabling me to do this research.

## References

[1] P. Altevogt and A. Linke. An algorithm for dynamic load balancing of synchronous Monte Carlo simulations on multiprocessor systems. *Comput. Phys. Commun.*, 79(3), pp. 373–380, 1994.

[2] V.C. Bhavsar and J.R. Isaac. Design and analysis of parallel Monte Carlo algorithms. *SIAM J. Sci. Stat. Comput.*, 8, pp. 73–95, 1987.

[3] R. Blumenthal. *Excursions of Markov Processes*. Birkhäuser Berlin, 1992.

[4] K. Burdzy. *Multidimensional Brownian excursions and potential theory*, volume 164 of *Pitman research notes in mathamtics series*. Longman Scientific and Technical, 1987.

[5] P. Coddington. Random Number Generators for Parallel Computers. NHSE Review, Second Issue, Northeast Parallel Architectures Center, 1996. Available at: http://nhse.cs.rice.edu/NHSEreview/RNG/.

[6] C. Costantini, B. Pacchierotti, and F. Sartoretto. Numerical Approximation for Functionals of Reflecting Diffusion Processes. (submitted) May 1995.

[7] M.J. Durst. Using linear congruential generators for parallel random number generation. In E.A. MacNair, K.J. Musselman, and P. Heidelberger, editors, *Proceedings of the 1989 Winter Simulation Conference*, pp. 462–466, 1989.

[8] A. Geist. PVM3 beyond network computation. In J. Volkert, editor, *Parallel Computing*, volume 73 of *Lectures Notes in Computer Sience*, pp. 194–203. Springer Verlag, 1993.

[9] D. Hafermann and M. Kuntz. Parallelization of a direct simulation Monte Carlo code for workstation cluster. In *Parallel computational fluid dynamics: new trends and advances*, pp. 81–88, 1995.

[10] R. Hanxleden and L. Scott. Correctness and determinism of parallel Monte Carlo processes. *Parallel Comput.*, 18(2), pp. 373–380, 1994.

[11] E. Hausenblas. Numerical Approximation for Reflecting Diffusion Processes using Ito-Excursion and Local Time. To appear in *Osaka Journal of Mathematics*.

[12] N. Ikeda and S. Watanabe. *Stochastic Differential Equations and Diffusion Processes*. North-Holland, Amsterdam, 1981.

[13] P. E. Kloeden and E. Platen. *Numerical Solution of Stochastic Differential Equations*, volume 23 of *Application of Mathematics*. Springer Verlag, 1992.

[14] Slominski L. On the existence, uniqueness and stability of solutions of multidimensional SDE's with reflecting boundary. *Annales de l'Institut Henri Poincaré, Probabilités et Statistiques*, 29(2), pp. 163–198, 1993.

[15] P.L. Lions and A.S. Sznitzman. Stochastic Differential Equations with Reflecting Boundary Conditions. *Communications on Pure and Applied Mathematics*, 37, pp. 511–537, 1984.

[16] M. Mastrangelo, V. Mastrangelo, D. Gassilloud, and F. Simon. Stochastic modelling of diffusion equations on a parallel machine. *Comput. Phys. Commun.*, 76(2-3), pp. 159–183, 1993.

[17] J. Menaldi. Stochastic variational inequality for reflected diffusion. *Indiana University Journal of Mathematics*, 32(5), pp. 733–743, 1983.

[18] W.P. Petersen. Numerical simulation of Ito stochastic differential equations on supercomputers. *Random media, IMA Vol. Math. Appl.*, 7, pp. 215–225, 1987.

[19] A.V. Skorohood. Stochastic Equation for Diffusion Processes in a Bounded Region. *Theory of Probability and its Applications*, 6(2), pp. 264–274, 1975.

[20] D. Talay. Elements of Probabilistic Numerical Methods for Partial Differential Equations. In *Probabilistic Models for Nonlinear Partial Differential Equations*, volume 1627 of *Lectures Notes in Mathematics*, pp. 148–196. D. Talay, L.Tubaro, 1996.

# Blocking Techniques in Numerical Software*

Wilfried N. Gansterer[1], Dieter F. Kvasnicka[2], and Christoph W. Ueberhuber[1]

[1] Institute for Applied and Numerical Mathematics,
University of Technology, Vienna
ganst@aurora.tuwien.ac.at, christof@uranus.tuwien.ac.at
[2] Institute for Physical and Theoretical Chemistry,
University of Technology, Vienna
dieter@titania.tuwien.ac.at

**Abstract.** When developing high performance algorithms blocking is a standard procedure to increase the locality of reference. Conflicting factors which influence the choice of blocking parameters are described in this paper. These factors include cache size, load balancing, memory overhead, algorithmic issues, and others. Optimal block sizes can be determined with respect to each of these factors. The resulting block sizes are independent of each other and can be implemented in several levels of blocking within a program. A tridiagonalization algorithm serves as an example to illustrate various blocking techniques.

## 1 Introduction

In modern computer systems the gap between raw computing power and memory throughput is increasing tremendously. This leads to the effect that for many algorithms the processors are not running at full speed since they have to wait for data from memory.

One solution to this problem is to insert one or several levels of fast cache memory between main memory and the processor. These caches hold a subset of the data stored in main memory. If the processor accesses a data item from memory and this data item is resident in cache the memory access time is much shorter. The *locality of reference* is high if a high fraction of the memory accesses can be served from the high levels of the memory hierarchy, which consist (in top down order) of registers, one to three levels of caches, local memory, distributed memory, and secondary storage.

Many algorithms can be restructured to reuse data by rearranging the order of the operations performed. To exploit the higher speed of cache memory linear algebra algorithms are recast as block algorithms which operate on submatrices (blocks) of the original matrix. Partitioning a matrix into disjoint submatrices is referred to as *blocking* (see Section 2).

---

* This work was supported by the Austrian Science Fund (*Österreichischer Fonds zur Förderung der wissenschaftlichen Forschung*).

P. Zinterhof, M. Vajteršic, A. Uhl (Eds.): ACPC'99, LNCS 1557, pp. 127–139, 1999.
© Springer-Verlag Berlin Heidelberg 1999

Section 3 summarizes conflicting aspects which sometimes favor large blocks and under other circumstances favor small blocks. Section 5 describes a tridiagonalization algorithm, which serves as an example for the various blocking techniques discussed in this paper. Subsequent sections go into detail with these blocking methods.

## 2  Blocked Memory Access

The key to high-performance computing is to reuse data which have been transferred to a faster level of the memory hierarchy as often as possible before they are replaced by new data and restored at a slower level (see Lam et al. [29], Schreiber, Dongarra [32], Laderman et al. [28], Pan [31], Carr, Lehoucq [8]).

In order to transform a matrix algorithm in such a way that the highest levels of the memory hierarchy are exploited to a high degree, the algorithm is reformulated as a block algorithm which operates on submatrices (blocks) of the original matrix. (The concept of blocking as well as hierarchical blocking of matrix operations is explained in detail in Ueberhuber [34].)

On the one hand, the submatrices have to be chosen small enough such that for each operation of the block algorithm, all blocks needed as arguments can be stored in the respective level of the memory hierarchy. On the other hand, the blocks should be large enough so that the number of operations executed on elements of those blocks is as large as possible.

Other factors that influence the optimal block sizes are cache line sizes and page sizes. If a cache line is already transferred to a high level of the memory hierarchy, it should be used completely before fetching the next cache line. As a result of the complexity of all these factors, the optimal blocking strategy can often only be determined by way of experiment (see, for example, Anderson et al. [1]).

## 3  Block Size

At first sight the size of submatrices to be worked on should be chosen so that they fit into a certain level of the memory hierarchy. To exploit more than one memory level, several blocking levels, i. e., *hierarchical blocking*, can be applied (see Ueberhuber [34], Haunschmid, Kvasnicka [21]). However, as will be shown in this paper, blocking techniques are also influenced by aspects which were not mentioned in Section 2.

**Large Block Sizes.** The main reasons for choosing large block sizes are (i) the exploitation of all cache levels (Section 4), (ii) the reduction of the number of messages in data parallel implementations (Section 10), and (iii) the reduction of algorithmic complexity (obtained, for instance, using the Strassen algorithm or particular complex matrix multiplication algorithms; Sections 8 and 9).

**Small Block Sizes.** Reasons for choosing small block sizes are (i) the reduction of the number of Level 2 (matrix-vector) operations in relation to Level 3 (matrix-matrix) operations (Section 6), (ii) load balancing purposes in data parallel implementations (Section 10), (iii) algorithmic requirements, which can sometimes produce significant overhead when using large block sizes (Section 7), (iv) storage requirements (Section 11), and (v) memory throughput for complex floating-point numbers (Section 9).

All these aspects are investigated in more detail in the sections indicated above.

## 4  Basic Linear Algebra Subprograms

To account for the great variety of architectures available and in order to combine efficiency with portability, a set of matrix routines has been defined: the Basic Linear Algebra Subprograms (BLAS; see Lawson et al. [30], Dongarra et al. [11, 10]), which are optimized for each new computer system by hardware or compiler vendors. The determination of the optimal parameters may also be done automatically (see Bilmes et al. [2, 3], Whaley, Dongarra [36], Kagstrom et al. [25]). These hardware-optimized BLAS routines usually run at the highest level of floating-point performance achieveable on a certain computer.

Table 1 illustrates the reason for the advantage of Level 3 (matrix-matrix) BLAS by a comparison of the ratios of floating-point operations to data movement for three closely related operations from the Level 1, 2, and 3 BLAS (see Dongarra et al. [12]).

**Table 1.** Ratio of floating-point operations to data movement.

| BLAS | Routine | Memory Accesses | Flops | Flops per Memory Access |
|---|---|---|---|---|
| Level 1 | **daxpy** | $3n$ | $2n$ | $2/3$ |
| Level 2 | **dgemv** | $n^2$ | $2n^2$ | $2$ |
| Level 3 | **dgemm** | $4n^2$ | $2n^3$ | $n/2$ |

Numerical algorithms should exploit Level 3 BLAS as much as possible in order to achieve satisfactory floating-point performance on a wide variety of computers. This is precisely what can be achieved with block algorithms, where Level 3 BLAS routines are applied to submatrices. These routines yield higher performance for larger matrices (see Ueberhuber [34]).

## 5  Tridiagonalization

One particular algorithm whose performance is strongly determined by an appropriate way of blocking is the tridiagonalization algorithm. This algorithm is used in eigensolvers to transform a symmetric input matrix into a similar tridiagonal matrix (see Golub, Van Loan [19] and Anderson et al. [1]).

Unblocked transformation of a symmetric $n \times n$ matrix $A$ into a tridiagonal matrix $T = Q^T A Q$ requires the product $Q = H_{n-2} H_{n-3} \ldots H_1$ of Householder reflectors $H_i = I - \tau u_i u_i^T$. The vector $u_i$ is chosen to eliminate the $i$th column of $A$ except for the diagonal and subdiagonal elements.

The matrix $Q$ is not explicitly accumulated, but each Householder reflector is applied to the matrix $A$ as soon as it becomes available: $A^{(i)} = H_i^T A^{(i-1)} H_i$, with $A^{(0)} \equiv A$ and $A^{(n-2)} \equiv T$. The computation of $T$ can be achieved by a series of rank-two updates to the matrix $A$ (see Gansterer et al. [16]).

## 6   Blocked Tridiagonalization in One Sweep

LAPACK provides a blocked *one-sweep* tridiagonalization algorithm (see Anderson et al. [1] and Dongarra et al. [13, 14]). Like in most linear algebra routines, this blocked tridiagonalization routine computes a matrix panel from a block of columns and then applies this panel to the rest of the matrix in a rank-$2k$ update. The computation of this panel cannot be implemented using efficient Level 3 (matrix-matrix) operations, but has to be done vector-wise, based on Level 2 (matrix-vector) operations (see Dongarra et al. [13, 14]).

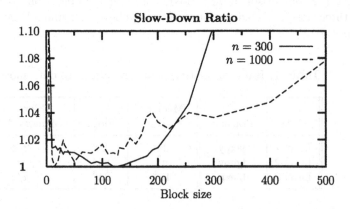

**Fig. 1.** Slow-down effect of suboptimal choice of the block size in the blocked one-sweep tridiagonalization. The minimum (optimum) execution time is scaled to 1 for each matrix size ($n = 300$ and $n = 1000$).

The optimum blocking parameter is heavily influenced by the amount of work which has to be done using Level 2 (matrix-vector) operations. The influence of the blocking parameter is illustrated in Fig. 1, which shows the slow-down effect for a suboptimal choice of the block size in the LAPACK routine LAPACK/dsytrd on one MIPS R8000 processor of an SGI Power Challenge. Note, the optimum block size decreases with increasing matrix size $n$, so that the current panel fits into the largest cache. This is, however, a minor effect: for a $300 \times 300$ matrix block sizes from 6 to 300 lead to a performance variation which is within a 10% range near the optimum floating-point performance. Execution times of this algorithm (and of the two-sweep algorithm presented in Chapter 7) are shown in Fig. 2.

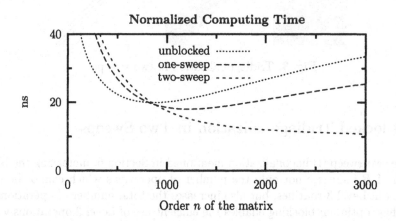

**Fig. 2.** Comparison of normalized execution times $(T/n^3)$ of the unblocked, the one-sweep and the two-sweep tridiagonalization on one MIPS R8000 processor of an SGI Power Challenge.

The one-sweep tridiagonalization routine LAPACK/dsytrd is able to utilize Level 3 BLAS routines for about one half of the floating-point operations. The default blocking parameter for the tridiagonalization in LAPACK is set to 1, which was the best choice for many architectures at the time LAPACK was released. Nowadays, the blocking parameters can be adapted to any given architecture during the LAPACK installation process. However, most installations do not take advantage of this feature.

The Level 3 operations of the one-sweep algorithms occur as rank-$2k$ updates. The remaining half of the floating-point operations are performed as matrix-vector products, i.e., Level 2 operations. These matrix-vector products cannot be avoided in a one-sweep tridiagonalization (see Haunschmid, Kvasnicka [21] and Gansterer et al. [16]). Thus, one-sweep tridiagonalization of large matrices on cache-based computers may spend more than 80% of the overall execution time in the 50% of floating-point operations required by these matrix-vector products (see Haunschmid, Kvasnicka [21]).

To enable the utilization of Level 3 BLAS for all accesses to the full matrix, the algorithm has to be modified. A *two-sweep* algorithm has to be used (see Fig. 3, Gansterer, Kvasnicka [16], and Kvasnicka et al. [27]): A first sweep reduces the full matrix $A$ to a band matrix $B$ using Level 3 BLAS, and a second sweep reduces $B$ to a tridiagonal matrix $T$. If the band width of $B$ is chosen reasonably small, most computational effort is spent in the first sweep, where fast Level 3 BLAS routines can be applied.

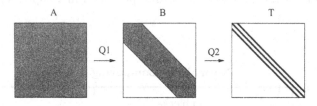

**Fig. 3.** Tridiagonalization in two sweeps.

# 7    Blocked Tridiagonalization in Two Sweeps

In the two-sweep tridiagonalization mentioned in Section 6, increasing the band width of $B$ increases not only the number of operations which cannot be performed in Level 3 routines, but also increases the total number of operations.

Thus, optimum blocking enables the efficient use of Level 3 operations while the total number of operations is not increased excessively. The influence of the choice of the blocking parameter is shown in Fig. 4 for a two-sweep tridiagonalization using the routines SBR/dsyrdb and LAPACK/dsbtrd on one processor of an SGI Power Challenge.

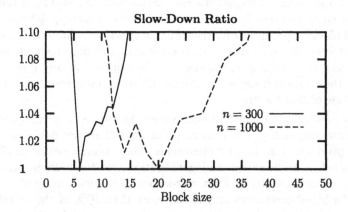

**Fig. 4.** Slow-down effect of suboptimal choice of blocking parameters in the blocked tridiagonalization. The minimum (optimum) execution time is scaled to 1 for each matrix size.

Note, the range of useful blocking parameters is narrower than for the one-sweep algorithm (see Fig. 1). The optimum block size increases with increasing

$n$, because the overhead is of lower complexity than $O(n^3)$. Computing times required by the three tridiagonalization algorithms discussed in this paper can be found in Fig. 2.

The SBR toolbox (see Bischof et al. [4, 5, 6]) is based on QR factorization to reduce a dense symmetric matrix to a band matrix. This algorithm (see Algorithm 1) reveals that, in contrast to the LAPACK blocking, no access to the rest of the matrix $A$ has to be made to accumulate the update matrices. There are two accesses to the matrix $A$ in the matrix update. Both of them are blocked with block size $\bar{p}$.

An implementation of a two-sweep reduction is included in the SBR toolbox (see Bischof et al. [4, 5, 6]). A complete algorithm for the symmetric eigenproblem is described in Gansterer, Kvasnicka [16, 17] and Kvasnicka et al. [27].

---

**Algorithm 1**

> $n$ is the order of the matrix $A$
> $\bar{p}$ is the block size
> $\bar{N} = (n-2)/\bar{p}$
> $A^{(0)} = A$
> **do** $k = 1, \bar{N}$
> > $\bar{s} = (k-1)\bar{p}+1$
> > $\bar{r} = k\bar{p}$
> > – QR factorization with $Q^{(k)} = I - W^{(k)}(Y^{(k)})^T$
> > $(W^{(k)}, Y^{(k)}) = \text{QR}(A^{(k-1)}(\bar{s}:\bar{r}))$
> > – matrix update
> > $A^{(k)} = (I - W^{(k)}(Y^{(k)})^T)^T A^{(k-1)}(I - W^{(k)}(Y^{(k)})^T)$
> **end do**
> $B = A^{(\bar{N})}$

---

## 8  Strassen's Algorithm

Using Strassen's algorithm for the multiplication of matrices of arbitrary order the complexity of this Level 3 operation is less than $28n^{\log_2 7}$ arithmetic operations (see Strassen [33]).

It is not easy to implement the Strassen algorithm efficiently. Thus, for a long time the Strassen algorithm was regarded to be of theoretical interest only. However, Douglas et al. [15] published efficient implementations for use on both single-processor and parallel computers. Some software libraries (e. g., IBM's ESSL) already comprise implementations of the Strassen algorithm.

On most computers the break-even point $n_{min}$, i. e., the minimum matrix order for which a Strassen implementation takes less time than the corresponding Level 3 BLAS routine, is between 32 and 256. For larger matrices a significant complexity and run-time reduction is achieved (see Fig. 5 and Ueberhuber [34]).

Thus, to exploit the advantages of efficient implementations of the Strassen algorithm in blocked tridiagonalization the block size used in matrix-matrix multiplications should be chosen as large as possible.

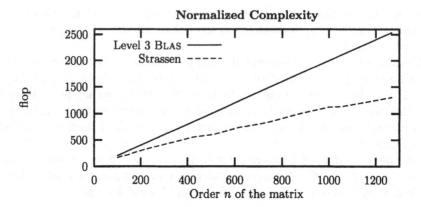

**Fig. 5.** Number *flop* of floating-point operations *per element* of the resulting matrix required by the Strassen algorithm and the Level 3 BLAS implementation based on standard matrix-matrix multiplication.

It has to be mentioned that the Strassen algorithm is less stable than the conventional method for multiplying two matrices. Similar to the modified error bounds for the multiplication of two complex matrices (cf. Section 9), the best bound known for the rounding error in Strassen's algorithm is a *normwise* and not an elementwise bound (Higham [23]). It turned out that in most practical cases the computed solution is of acceptable accuracy.

## 9   Multiplication of Complex Matrices

Blocked matrix multiplication algorithms are characterized by a high fraction of Level 3 operations. This can be utilized to reduce the overall complexity of the multiplication of complex matrices (see Winograd [37], Higham [24]).  The multiply-add operation, $a = c \times d + e$, in its complex form

$$(a_r + a_i i) = (c_r + c_i i) \times (d_r + d_i i) + (e_r + e_i i),$$

can be written using four real multiplications and four additions

$$a_r = c_r \times d_r - c_i \times d_i + e_r, \qquad a_i = c_r \times d_i - c_i \times d_r + e_i$$

and in a form using only three real multiplications and seven additions

$$X_1 = c_r \times d_r, \qquad X_2 = c_i \times d_i, \qquad X_3 = (c_r + c_i) \times (d_r + d_i)$$

$$a_r = X_1 - X_2 + e_r, \qquad a_i = X_3 - X_1 - X_2 + e_i.$$

On modern computer architectures, the latter method has the major drawback of a larger number of add instructions. However, for blocked matrix algorithms, a new side of this method is revealed: A multiply-add operation on subblocks requiring only three block-multiplies instead of four enables a great complexity reduction. A matrix multiply requires (roughly) $n^3$ multiply-add instructions. A matrix add requires $n^2$ add instructions[1]. So reducing the effort for the complex matrix-matrix multiply from $4n^3$ to $3n^3 + O(n^2)$ operations saves asymptotically 25 % of the required work.

**Block Sizes.** The aforementioned technique yields increasing complexity reduction with growing matrix size. For blocked algorithms matrix multiplications are applied to submatrices. So algorithms having large block sizes are to be preferred.

Nonetheless, if performance on a specific computer system is limited by memory throughput rather than processor performance, block sizes must not be too large. A multiply-add operation on complex data involves four multiplications and four additions on real data. The data transfer for complex data is only twice as much as for real data. So memory bound algorithms may have twice the floating-point performance when operating on complex data instead of real data.

If the block size is too large to store all required data in the first level cache the speed-up factor of two for memory bound complex algorithms is lost and only a speed-up factor of 4/3, resulting from complexity reduction, can be obtained. The block size has to be decreased in order to approach the full potential speedup of $2 \times 4/3$.

It has to be mentioned that the multiplication technique of this section is numerically less stable than conventional complex matrix-matrix multiplication (see Higham [24]). Nevertheless, when *normwise* instead of componentwise rounding errors are considered the resulting bounds are comparable to that of conventional complex matrix-matrix multiplication.

# 10   Blocking for Parallel Computing

In parallel computing the choice of an appropriate data distribution heavily depends on the data flow features of the involved algorithms. For dense matrix computations, SCALAPACK routines (see Blackford et al. [7]), PLAPACK routines (see van de Geijn [35]), and HPF programs (see High Performance Fortran Forum [22]) achieve their best floating-point performance if a two-dimensional block-cyclic data layout scheme is used.

---

[1] Addition operations require only add instructions and no multiplies, but since the time required for one add is the same as the time needed for one multiply-add on most of the modern processors the respective numbers of instructions are directly comparable. A different situation occurs when the algorithm is memory bound. In this case the add as well as the multiply instructions are executed faster than the fused multiply-add instructions because they require one memory transfer less (only two reads and one write instead of three reads and one write).

The two main issues in choosing a data layout for dense matrix computations are (i) load balancing, i.e., splitting the work reasonably well among the processors, and (ii) extensive use of Level 3 BLAS for local computations on all processors, to avoid excessive movement of data to and from memory.

Another issue is that on parallel computers with significant latency, block sizes must not be too small, i.e., communication patterns should not be vector oriented but matrix oriented. This technique is called *message combining* (see Geist et al. [18], Gropp et al. [20]).

In general it can be said that in parallel computing large block sizes are more favorable than small ones, as long as the work load is reasonably balanced between processors.

## 11   Storage Efficiency

Symmetric and triangular matrices (as arising in Sections 5, 6, and 7) require special storage techniques to avoid significant waste of memory. A tradeoff between floating-point performance and storage efficiency influences the choice of the optimum blocking technique.

Triangular and symmetric matrices can be stored using only half the memory requirement of LAPACK routines using conventional storage and at the same time rely on Level 3 BLAS. Efficient storage schemes are discussed in Haunschmid, Kvasnicka [21], Kvasnicka [26] (see Fig. 6), and D'Azevedo, Dongarra [9]. The storage overhead of some of these schemes is given in Fig. 7.

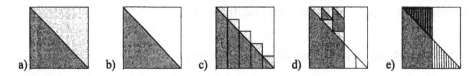

**Fig. 6.** Storage schemes for symmetric and triangular matrices: a) conventional (*unpacked*) storage, b) packed storage, c) block packed storage, d) and e) parallel packed storage scheme (d) $b = n/5$, e) $b = n/25$). The darker shaded areas indicate matrix elements whose values are stored and used; the lighter shaded parts indicate redundant storage.

Since the storage schemes c), d), and e) of Fig. 6 may lead to non-contiguous data layouts the respective algorithms become more complicated. In the algorithms, block sizes have to be used which are not larger than the block size of the underlying storage scheme.

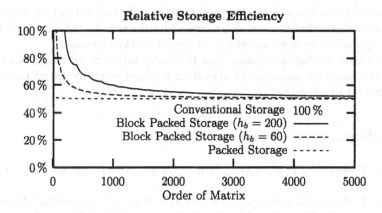

**Fig. 7.** Storage requirement of several storage formats. The ratio of required storage locations (words) to the total number of matrix elements (100%) is plotted.

## 12   Blocking Parameter Values

All blocking parameters have to be chosen in a problem and hardware dependent way. Typical values are

$$(32\ldots512) \geq b_{\text{parallel}} \geq b_{\text{storage}} \geq b_{\text{one-sweep}} \geq (32\ldots128) \geq b_{\text{two-sweep}} \geq (6\ldots16).$$

$b_{\text{parallel}}$ denotes the block size for data distribution, $b_{\text{storage}}$ the block size of the chosen storage scheme, $b_{\text{one-sweep}}$ the block size of the rank-$2k$ update of the one-sweep tridiagonalization (see Section 6), and $b_{\text{two-sweep}}$ the block size of the two-sweep tridiagonalization (see Section 7).

For local computations matrices only need to be blocked in one dimension. The other dimension can be as large as necessary, except for the complexity oriented blocking (see Section 9).

Level 3 BLAS routines are often called with rectangular matrices where one dimension is smaller by a factor of up to 100 than the other dimension. The smaller dimension should not be chosen below a certain hardware dependent level in order to avoid performance degradation.

Many block sizes will be chosen to be equal in practice. For instance, experience (made with SGI, DEC, HP, IBM, and SUN workstations) shows that values of

$$b_{\text{parallel}} = b_{\text{storage}} = b_{\text{lapack}} = 64 \quad \text{and} \quad b_{\text{two-sweep}} = 8$$

are a good choice for the tridiagonalization algorithms of Sections 6 and 7.

## 13   Conclusion

Choosing satisfactory blocking techniques and blocking parameters is not easy and straightforward. There are good reasons for choosing large blocks, for instance to better exploit the memory hierarchy. There are also good reasons for choosing small blocks, for instance, to reduce the number of floating-point operations which are not covered by Level 3 routines, or to reduce the total number of operations in some algorithm, or simply to save memory.

These conflicting reasons lead to totally different optimum block sizes, so that a portable *and* efficient algorithm requires several levels of blocking using different block sizes.

# References

[1] E. Anderson et al., LAPACK *Users' Guide*, 2nd ed., SIAM Press, Philadelphia, 1995.

[2] J. Bilmes, K. Asanovic, C.-W. Chin, J. Demmel, *Optimizing Matrix Multiply using* PHIPAC: *a Portable, High-Performance, ANSI C Coding Methodology*, Proceedings of the International Conference on Supercomputing, ACM, Vienna, Austria, 1997, pp. 340–347.

[3] J. Bilmes, K. Asanovic, J. Demmel, D. Lam, C.-W. Chin, *Optimizing Matrix Multiply using* PHIPAC: *a Portable, High-Performance, ANSI C Coding Methodology*, Technical report, LAPACK Working Note 111, 1996.

[4] C. H. Bischof, B. Lang, X. Sun, *Parallel Tridiagonalization through Two-Step Band Reduction*, Proceedings of the Scalable High-Performance Computing Conference, IEEE, Washington D. C., 1994, pp. 23–27.

[5] C. H. Bischof, B. Lang, X. Sun, *A Framework for Symmetric Band Reduction*, Technical report, Argonne Preprint ANL/MCS-P586-0496, 1996.

[6] C. H. Bischof, B. Lang, X. Sun, *The SBR Toolbox - Software for Successive Band Reduction*, Technical report, Argonne Preprint ANL/MCS-P587-0496, 1996.

[7] L. S. Blackford et al., SCALAPACK *Users' Guide*, SIAM Press, Philadelphia, 1997.

[8] S. Carr, R. B. Lehoucq, *Compiler Blockability of Dense Matrix Factorizations*, ACM Trans. Math. Software 23 (1997), pp. 336–361.

[9] E. F. D'Azevedo, J. J. Dongarra, *Packed Storage Extension for* SCALAPACK, Technical report, LAPACK Working Note 135, 1998.

[10] J. J. Dongarra, J. Du Croz, S. Hammarling, I. Duff, *A Set of Level 3* BLAS, ACM Trans. Math. Software 16 (1990), pp. 1–17.

[11] J. J. Dongarra, J. Du Croz, S. Hammarling, R. J. Hanson, *An Extended Set of* BLAS, ACM Trans. Math. Software 14 (1988), pp. 18–32.

[12] J. J. Dongarra, I. S. Duff, D. C. Sorensen, H. A. van der Vorst, *Linear Algebra and Matrix Theory*, SIAM Press, Philadelphia, 1998.

[13] J. J. Dongarra, S. J. Hammarling, D. C. Sorensen, *Block Reduction of Matrices to Condensed Forms for Eigenvalue Computations*, J. Comput. Appl. Math. 27 (1989), pp. 215–227.

[14] J. J. Dongarra, S. J. Hammarling, D. C. Sorensen, *Block Reduction of Matrices to Condensed Forms for Eigenvalue Computations*, Technical report, LAPACK Working Note 2, 1987.

[15] C. C. Douglas, M. Heroux, G. Slishman, R. M. Smith, GEMMW—*A Portable Level 3* BLAS *Winograd Variant of Strassen's Matrix-Matrix Multiply Algorithm*, J. Computational Physics 110 (1994), pp. 1–10.

[16] W. N. Gansterer, D. F. Kvasnicka, *High Performance Computing in Material Sciences. The Standard Eigenproblem - Concepts*, Technical Report AURORA TR1998-18, Vienna University of Technology, 1998.

[17] W. N. Gansterer, D. F. Kvasnicka, *High Performance Computing in Material Sciences. The Standard Eigenproblem – Experiments*, Technical Report AURORA TR1998-19, Vienna University of Technology, 1998.
[18] A. Geist, A. Beguelin, J. J. Dongarra, W. Jiang, R. Manchek, V. Sunderam, *PVM: Parallel Virtual Machine—A Users' Guide and Tutorial for Networked Parallel Computing*, MIT Press, Cambridge London, 1994.
[19] G. H. Golub, C. F. Van Loan, *Matrix Computations*, 3rd ed., Johns Hopkins University Press, Baltimore, 1996.
[20] W. Gropp, E. Lusk, A. Skjelum, *Using MPI*, MIT Press, Cambridge London, 1994.
[21] E. Haunschmid, D. F. Kvasnicka, *High Performance Computing in Material Sciences. Maximizing Cache Utilization without Increasing Memory Requirements*, Technical Report AURORA TR1998-17, Vienna University of Technology, 1998.
[22] High Performance Fortran Forum, *High Performance Fortran Language Specification, Version 2.0*, 1997.
[23] N. J. Higham, *Exploiting Fast Matrix Multiplication within the Level 3 BLAS*, ACM Trans. Math. Software 16 (1990), pp. 352–368.
[24] N. J. Higham, *Stability of a Method for Multiplying Complex Matrices with Three Real Matrix Multiplications*, SIAM J. Matrix Anal. Appl. 13-3 (1992), pp. 681–687.
[25] B. Kagstrom, P. Ling, C. Van Loan, *GEMM-Based Level 3 BLAS: High-Performance Model Implementations and Performance Evaluation Benchmark*, ACM Trans. Math. Software 24 (1998).
[26] D. F. Kvasnicka, *Parallel Packed Storage Scheme (P2S2) for Symmetric and Triangular Matrices*, Technical Report to appear, Vienna University of Technology, 1998.
[27] D. F. Kvasnicka, W. N. Gansterer, C. W. Ueberhuber, *A Level 3 Algorithm for the Symmetric Eigenproblem*, Proceedings of the Third International Meeting on Vector and Parallel Processing (VECPAR '98), Vol. 1, 1998, pp. 267–275.
[28] J. Laderman, V. Pan, X.-H. Sha, *On Practical Acceleration of Matrix Multiplication*, Linear Algebra Appl. 162-164 (1992), pp. 557–588.
[29] M. S. Lam, E. E. Rothberg, M. E. Wolf, *The Cache Performance and Optimizations of Blocked Algorithms*, Computer Architecture News 21 (1993), pp. 63–74.
[30] C. L. Lawson, R. J. Hanson, D. Kincaid, F. T. Krogh, BLAS for Fortran Usage, ACM Trans. Math. Software 5 (1979), pp. 63–74.
[31] V. Pan, *How Can We Speed Up Matrix Multiplication ?*, SIAM Rev. 26 (1984), pp. 393–415.
[32] R. Schreiber, J. J. Dongarra, *Automatic Blocking of Nested Loops*, Technical Report CS-90-108, University of Tennessee, 1990.
[33] V. Strassen, *Gaussian Elimination Is not Optimal*, Numer. Math. 13 (1969), pp. 354–356.
[34] C. W. Ueberhuber, *Numerical Computation*, Springer-Verlag, Heidelberg, 1997.
[35] R. van de Geijn, *Using PLAPACK: Parallel Linear Algebra Package*, MIT Press, 1997.
[36] R. C. Whaley, J. J. Dongarra, *Automatically Tuned Linear Algebra Software*, Technical report, LAPACK Working Note 131, 1997.
[37] S. Winograd, *On Multiplication of 2 × 2 Matrices*, Linear Algebra Appl. 4 (1971), pp. 381–388.

# HPF and Numerical Libraries*

Harald J. Ehold[1], Wilfried N. Gansterer[2], Dieter F. Kvasnicka[3], and
Christoph W. Ueberhuber[2]

[1] VCPC, European Centre for Parallel Computing at Vienna
ehold@vcpc.univie.ac.at
[2] Institute for Applied and Numerical Mathematics, Vienna University of Technology
ganst@aurora.tuwien.ac.at, christof@uranus.tuwien.ac.at
[3] Institute for Physical and Theoretical Chemistry, Vienna University of Technology
dieter@titania.tuwien.ac.at

**Abstract.** Portable and efficient ways for calling numerical high performance software libraries from HPF programs are investigated. The methods suggested utilize HPF's EXTRINSIC mechanism and are independent of implementation details of HPF compilers. Two prototypical examples are used to illustrate these techniques. Highly optimized BLAS routines are utilized for local computations: (i) in parallel multiplication of matrices, and (ii) in parallel Cholesky factorization. Both implementations turn out to be very efficient and show significant improvements over standard HPF implementations.

## 1 Introduction

High Performance Fortran (HPF [11]) is one of the most interesting approaches for high-level parallel programming. In particular, it provides very convenient ways for specifying data distributions and for expressing data parallelism. Development of parallel code using HPF is much easier and requires less effort than message passing programming, for example, using MPI.

However, in numerical applications the performance achieved with HPF programs is often disappointing compared to message passing code. This is partly due to the immaturity of HPF compilers, which can be explained by the difficulties to implement the advanced features of HPF efficiently. There is, however, another important aspect, which is often ignored. In order to achieve high performance it is crucial to integrate highly optimized library routines into HPF code.

Much effort has been spent on developing highly efficient implementations of the Basic Linear Algebra Subroutines (BLAS [12,7,8]) and on numerical libraries for dense linear algebra which use the BLAS as building blocks. Important examples are LAPACK [1], which is the standard sequential library for dense or banded linear algebra methods, and parallel libraries such as SCALAPACK [4] or PLAPACK [10]. Recently, code generation tools have been developed (see Bilmes

---

* This work was supported by the Special Research Program SFB F011 "AURORA" of the Austrian Science Fund FWF.

P. Zinterhof, M. Vajteršic, A. Uhl (Eds.): ACPC'99, LNCS 1557, pp. 140–152, 1999.

et al. [2], Whaley, Dongarra [15]) which automatically find the best choice of hardware dependent parameters for an efficient implementation of the BLAS.

The integration of such software into HPF is crucial for several reasons:

- In many scientific applications a lot of programming effort can be saved by using existing library routines as building blocks instead of (re-)coding them in HPF directly.
- A considerable amount of expertise has been incorporated into high quality software packages like the ones mentioned above. It is hardly possible to achieve comparable floating-point performance when coding in HPF directly. Even if a problem is well suited for an HPF implementation, the success of such an attempt heavily relies on the maturity of HPF compilers, which cannot be taken for granted at present (see Ehold et al. [9]).
- When standard numerical operations are coded in HPF the resulting code often suffers from poor local performance. Thus, one of the key issues is to optimize local performance in order to improve overall parallel performance.
- Highly optimized BLAS implementations are available for most target systems. Therefore, the use of BLAS routines for local computations ensures *performance portability*.
- Usually the main motivation for parallelization is performance improvement. It is essential to optimize the *sequential* code first in order to be able to measure the benefits of parallelization in a clean way. The sequential optimization typically involves restructuring the code, for example, by increasing the fraction of Level 3 (matrix-matrix) operations and by using appropriate BLAS 3 routines (or other high performance library routines) wherever possible. Thus, when the parallelization is done by using HPF, the necessity of combining the BLAS (and other numerical packages and libraries) with HPF arises quite naturally.

For all of these reasons the goal of this paper is to investigate ways to utilize high performance numerical libraries in an HPF context.

The basic facility provided by HPF for integrating procedures from other programming languages or models is the EXTRINSIC mechanism (HPF Forum [11]). This paper describes various methods, at different levels of abstraction, for calling existing library routines from HPF using this mechanism.

The techniques described are *portable* in the sense that they only rely on features from the HPF standard [11] and, additionally, on three of the HPF 2.0 Approved Extensions. In particular, the required HPF features are:

- an advanced form of the ALIGN directive, namely
  ALIGN A(j,*) WITH B(j,*) (replication of A along one dimension of B);
- the INHERIT directive;
- EXTRINSIC(HPF_LOCAL) subroutines;
- EXTRINSIC(F77_LOCAL) and EXTRINSIC(F77_SERIAL) subroutines.

Unfortunately, some HPF compilers do not yet support all of these features.

The main ideas and basic concepts presented in this paper are applicable to many numerical algorithms in dense linear algebra, but their implementation will differ in various technical details. For the purpose of illustration two operations, which arise very frequently at the core of numerous scientific applications, serve as prototypes: Matrix-matrix multiplication and, representing more complicated numerical algorithms, Cholesky factorization.

In Section 2 ways for integrating sequential library routines into HPF code are suggested for these two operations. Section 3 gives an overview of techniques that have been used to interface HPF to the SCALAPACK parallel library. In both cases experiments have been performed which demonstrate the considerable performance gains arising from the use of the techniques suggested.

# 2    Calling Sequential Routines from HPF

The topic of this section is the integration of extrinsic library routines into an HPF program for *local* computations only. All of the communication in the multiprocessor environment is organized by the HPF compiler.

A matrix-matrix multiplication routine (Section 2.1) and a Cholesky factorization routine (Section 2.2) were implemented. Both utilize the BLAS for local computations. Experiments were performed with the HPF compiler from PGI, *pghpf*, version 2.4, on a Meiko CS-2 and an IBM SP2. The CS-2 experiments used a standard Fortran implementation of the BLAS (LIBBLAS), compiled with the *pgf77* Fortran 77 compiler from PGI, and BLAS routines generated by the ATLAS package (Whaley, Dongarra [15]). On the SP2 the vendor optimized BLAS routines were utilized.

## 2.1    Multiplication of Matrices

As a first example, the computation of the product $C \in \mathbb{R}^{m \times n}$ of two matrices $A \in \mathbb{R}^{m \times l}$, $B \in \mathbb{R}^{l \times n}$ is considered where all matrices can have arbitrary distributions. An HPF routine called par_dgemm (*parallel* BLAS/dgemm) was developed, which performs the matrix-matrix multiplication. Internally, this operation is split up into local operations on subblocks, each of which is performed by calling the general matrix multiplication routine BLAS/dgemm.

In the general case, the multiplication of two distributed matrices involves *non-local* computations. Some of the data have to be replicated over several processors in order to localize all of the computations involved.

**Loop Orders.** The central operation in matrix-matrix multiplication is

$$C(i,j) = C(i,j) + A(i,k)B(k,j),$$

where $i = 1, 2, \ldots, m$, $j = 1, 2, \ldots, n$, and $k = 1, 2, \ldots, l$. Permuting the order of these three loops yields different algorithmic variants with different characteristics with respect to parallelization (and consequently, different requirements for expressing parallelism). The basic differences relevant to parallelization are as follows, where the variants are labeled by the index of the outermost loop.

**k∗ variant:** The two inner loops perform the outer product of a column of $A$ and a row of $B$. Since the entire matrix $C$ is updated in place at every step of the outer loop, this variant exhibits no communication requirements for elements of $C$. Computation of a local block of $C$ requires the corresponding parts of the columns of $A$ and of the rows of $B$. In order to localize all of the computations, the columns of $A$ have to be replicated in the direction of the rows of $C$, and the rows of $B$ have to be replicated in the direction of the columns of $C$.

**i∗ variant:** The two inner loops compute a row of $C$ by multiplying the corresponding row of $A$ with the columns of $B$. In this case it is usually most efficient to parallelize over both dimensions of the matrix $B$. This, however, requires distributed reduction operations for the elements of $C$.

**j∗ variant:** The two inner loops compute a column of $C$ by multiplying rows of $A$ with the corresponding column of $B$. In this case it is usually most efficient to parallelize over both dimensions of the matrix $A$. This again requires distributed reduction operations for the elements of $C$.

The $k*$ variant requires the smallest amount of local storage without extra communication for the elements of $C$. This variant is well suited to a two-dimensional distribution of array $C$, whereas for the other two variants either a one-dimensional data distribution is to be preferred (columnwise for the $i*$ variant, rowwise for the $j*$ variant) or extra communication needs to be performed for the elements of $C$.

For load balancing reasons two-dimensional data distributions are normally used in linear algebra algorithms. In such a setup, the $k*$ variant exhibits the best data locality, minimizes the amount of data to be replicated, and leads to the most efficient parallelization. Consequently, it was chosen for the implementation.

**Implementation.** For performance reasons, a *blocked* version of matrix-matrix multiplication (Ueberhuber [14]) was implemented. Two levels of wrapper routines are involved. The programmer calls the HPF routine par_dgemm, which takes the matrices $A$, $B$ as input and returns the product matrix $C$.

In this routine the HPF_LOCAL routine local_dgemm is called inside a loop. In addition to the blocks involved local_dgemm also takes their size (the block size of the algorithm) as an argument and performs the local outer products of a block column of $A$ and a block row of $B$ by calling the routine BLAS/dgemm.

In par_dgemm, work arrays for a block column of $A$ and for a block row of $B$ are aligned properly with $C$, and then the corresponding subarrays of $A$ and $B$ are copied there. This copying operation adjusts the distribution of the currently required parts of $A$ and $B$ to that of $C$. This copying is the only place where inter-processor communication occurs. It has the advantage of making the procedure fully independent of the prior distributions of $A$ and $B$. All of the necessary communication, which is entirely contained in the copying operation, is restricted to the outermost loop.

144     Harald J. Ehold et al.

**Experiments.** Fig. 1 illustrates the floating-point performance of `par_dgemm`
and the vendor supplied intrinsic (parallel) function MATMUL for multiplying
two $n \times n$ matrices on 16 processors of a Meiko CS-2. $A$ and $B$ were distributed
cyclically in both dimensions, and $C$ was distributed block-cyclically with block
size 20 in both dimensions.

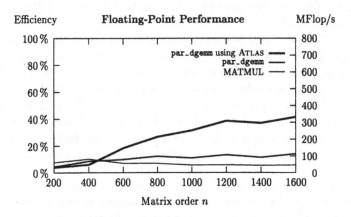

**Fig. 1.** Matrix multiplication on 16 processors of a Meiko CS-2. "Efficiency"
denotes the percentage of the peak-performance achieved.

A considerable performance improvement of `par_dgemm` over MATMUL is
evident for matrix orders $n > 500$ (roughly by a factor of 2 using the LIBBLAS,
respectively 5 using the ATLAS-BLAS). This holds for other distribution scenarios
as well.

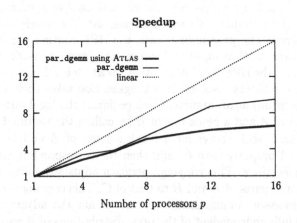

**Fig. 2.** Matrix multiplication on $p$ processors for $n = 1000$ on a Meiko CS-2.

Moreover, for large matrix orders $n$ the efficiency of par_dgemm improves, whereas the efficiency of the MATMUL routine decreases. The ATLAS generated BLAS/dgemm shows significant improvements in efficiency, especially for large matrices. It is important to keep in mind that 16 processors were used in the test run and that larger local blocks result in better utilization of the BLAS.

In order to evaluate parallel speedup (see Fig. 2), the runtime of the routine par_dgemm respectively par_dgemm using ATLAS on one processor was used as a reference value. The parallel speedup for up to 8 processors is quite the same for both par_dgemm routines. On more processors the ATLAS version does not scale so good anymore because the time for computation per processor is already so small that the communication becomes dominant.

Fig. 3 gives the efficiency on 1 to 16 processors of par_dgemm and MATMUL running on an IBM SP2. On the SP2 par_dgemm is based on the highly optimized ESSL/dgemm routine. The performance of par_dgemm scales reasonably well as long as the workload can be balanced well. MATMUL shows a very poor performance. MATMUL running on 16 processors of an IBM SP2 achieves a lower floating-point performance than ESSL/dgemm running on one processor. In this experiment it was not possible to test MATMUL running on one processor due to memory limitations.

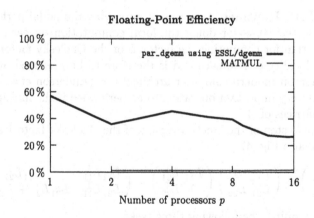

**Fig. 3.** Matrix multiplication on $p$ processors for $n = 2000$ on an IBM SP2.

## 2.2 Cholesky Factorization

Even if the HPF compiler does a good job with organizing data distribution and communication between processors, the performance achieved can be disappointingly low due to bad node performance. Experiments have been performed with several HPF implementations of the same algorithm for computing the Cholesky factorization $A = LL^\top$ of a real symmetric positive definite matrix.

As shown in Table 1, using the sequential BLAS-based Cholesky factorization LAPACK/dpotrf is up to 12 times faster than the best scaling HPF code (Version 1) on one processor and still 1.3 times faster than this code on 16 processors. The HPF code which yields best sequential performance (Version 2) does not show any parallel speedup.

**Table 1.** Cholesky factorization of a matrix of order 1000, using the vendor optimized routine LAPACK/dpotrf running on one processor only, and two HPF versions on a Meiko CS-2.

| $p$ | 1 | 2 | 4 | 8 | 16 |
|---|---|---|---|---|---|
| LAPACK/dpotrf | **9 s** | | | | |
| HPF Version 1 | 113 s | 57 s | 29 s | 18 s | 12 s |
| HPF Version 2 | 36 s | 43 s | 45 s | 45 s | 46 s |

**Utilizing BLAS Routines.** One way of improving the nodal performance in HPF is to use the BLAS for doing the local computation on each node. The blocked algorithm for computing the factor $L$ of the Cholesky factorization exhibits better locality of reference and is therefore to be preferred over the unblocked version on modern computer architectures (Anderson et al. [1], Ueberhuber [14]). A very important operation to be performed is the multiplication of certain submatrices of $A$.

The principle step of the blocked version of the Cholesky factorization is the following (see also Fig. 4):

$$\begin{pmatrix} A_{11} & A_{12} \\ A_{21} & A_{22} \end{pmatrix} = \begin{pmatrix} L_{11} & 0 \\ L_{21} & L_{22} \end{pmatrix} \begin{pmatrix} L_{11}^\mathsf{T} & L_{21}^\mathsf{T} \\ 0 & L_{22}^\mathsf{T} \end{pmatrix} = \begin{pmatrix} L_{11}L_{11}^\mathsf{T} & L_{11}L_{21}^\mathsf{T} \\ L_{21}L_{11}^\mathsf{T} & L_{21}L_{21}^\mathsf{T} + L_{22}L_{22}^\mathsf{T} \end{pmatrix}.$$

This step requires the following three tasks.

1. Factorize $A_{11} = L_{11}L_{11}^\mathsf{T}$ using LAPACK/dpotrf.
2. Solve the linear system $L_{21}L_{11}^\mathsf{T} = A_{21}$ for $L_{21}$ using BLAS/dtrsm.
3. Multiply $L_{21}L_{21}^\mathsf{T}$ using BLAS/dgemm and continue recursively with $A_{22} - L_{21}L_{21}^\mathsf{T} = L_{22}L_{22}^\mathsf{T}$.

In the HPF version the first task is done sequentially on one processor by using the HPF extrinsic kind F77_SERIAL, but the subsequent steps (BLAS/dtrsm and BLAS/dgemm) are parallelized. For simplicity it is assumed that the block size $b$ divides the dimension of the matrix $n$ evenly, where $n$ is the size of the leading dimension of matrix $A$. The distribution chosen for the symmetric matrix $A$ is (CYCLIC($b$),*). The one-dimensional distribution was chosen to avoid communication within rows.

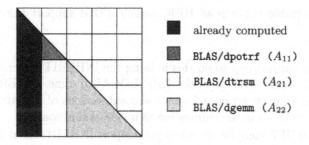

**Fig. 4.** Update areas of blocked Cholesky factorization.

**Using HPF_LOCAL.** The extrinsic kind HPF_LOCAL refers to procedures implemented in the HPF language, but in the *local* programming model. Its code is executed locally per processor and this gives the possibility to do local computations by calling the sequential BLAS routines.

In the parallel version, BLAS/dtrsm is called on each processor that owns local data of the block $A_{21}$ (see Fig. 5). Hence in the example shown in Fig. 5 (two processors $P1$ and $P2$) there are two calls to BLAS/dtrsm. In this example processor $P1$ owns blocks $B_1$ and $B_3$ and needs block $A_{11}$ and processor $P2$ owns block $B_2$ and also owns already block $A_{11}$. Since block $A_{11}$ is needed by all participating processors it is broadcast to all of them. The block size on distinct processors can differ. For that a mechanism for determining the correct size of the local array is needed. Inside the HPF_LOCAL routine the intrinsic function SIZE gives the desired information that can be used by the BLAS routine.

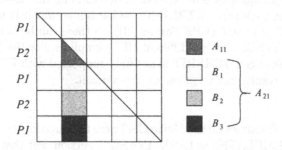

**Fig. 5.** Local blocks in parallel blocked Cholesky factorization.

The HPF_LOCAL version calls BLAS/dtrsm as an HPF_LOCAL routine by passing the appropriate array section to the subroutine. This section of the array needs to be passed to the HPF_LOCAL routine without changing the distribution of the array at the procedure boundary. In this paper two different methods were investigated to achieve this, using INHERIT or TEMPLATES, which are both portable between all HPF compilers that support the core HPF 2.0 language.

1. The natural way to do this is using the INHERIT directive in the interface and the declaration of the HPF_LOCAL routine. INHERIT tells the compiler that an array and hence also subsections of this array have the same distribution in the subroutine as in the calling routine.
2. If an HPF compiler does not yet support INHERIT then TEMPLATES can be used to ensure that the distribution of array sections is not changed at a procedure boundary.

The drawback of this second version is that two additional arguments occur which are only needed for specifying the proper size and distribution of the template in the interface. One parameter gives the size of the first dimension of the array in the calling routine. This value is used to declare a TEMPLATE in the interface of the callee that has the same size and distribution as the starting array. A second parameter gives the starting point of the array section passed to the subroutine. With this information the dummy argument can be aligned to the corresponding part of the starting array via the use of the TEMPLATE.

**Using F77_LOCAL.** The extrinsic kind F77_LOCAL refers to procedures that are implemented in Fortran 77 and in the local programming model. This suggests that F77_LOCAL is the natural choice for combining HPF with the BLAS, which use Fortran 77 calling conventions. However, since the Fortran 90 intrinsic function SIZE is typically not supported within a F77_LOCAL subroutine, different techniques have to be used for determining the size of local arrays. Another problem with this EXTRINSIC mechanism is that it is not clearly defined in the HPF standard (HPF Forum [11]), so there can be variations in the implementation details between different HPF compilers. For the HPF compiler used on the Meiko CS-2, INHERIT is not fully supported with F77_LOCAL, so no performance results can be given for this version.

**Experiments.** Runtimes on the Meiko CS-2 were the same for the HPF_LOCAL version using TEMPLATES as for the INHERIT version. For that reason Fig. 6 shows just one "HPF+BLAS" curve.

The sequential routine LAPACK/dpotrf was benchmarked on one processor. The efficiency values for $p \geq 2$ processors were determined by simply dividing the single-processor value by $p$.

On one processor the superiority of the HPF+BLAS versions over the pure HPF code can be seen. ATLAS-LAPACK/dpotrf yields very high efficiency on one processor. For an increasing number of processors a decreasing efficiency is to be observed. This phenomenon is due to the relatively small problem size ($n = 2000$) giving bad workload balancing for a one-dimensional processor distribution.

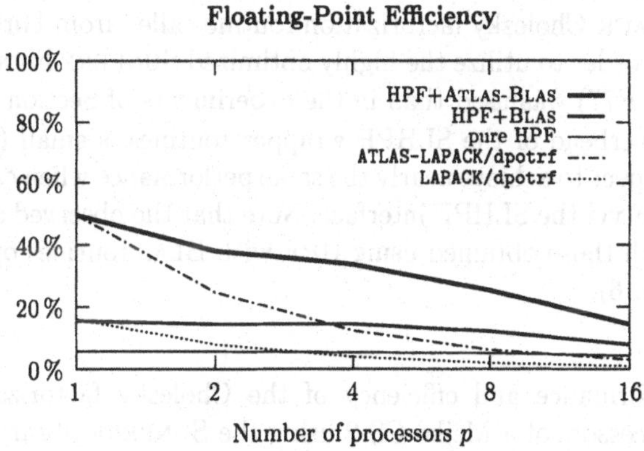

Floating-Point Efficiency

**Fig. 6.** Efficiency of the Cholesky factorization of a matrix of order $n = 2000$ on a Meiko CS-2.

# 3  Calling Parallel Libraries from HPF

As an alternative to the approach described in the previous section, extrinsic *parallel* library routines such as SCALAPACK [4] and PLAPACK [10] can be invoked from HPF for distributed computations. In this case, the extrinsic library procedure itself performs communication as well as computation, and HPF is used as a framework for conveniently distributing data and for organizing high-level parallelism. Calling distributed library routines from HPF mainly involves the issue of passing distribution information from HPF to the library routines.

This section gives a brief survey of techniques that have been used to interface HPF to the SCALAPACK parallel library.

## 3.1  Interfacing ScaLAPACK via EXTRINSIC Functions

Blackford et al. [5] have developed an interface called SLHPF from HPF to the SCALAPACK package [4]. This interface uses three layers of wrapper routines: (i) global HPF routines, which call (ii) HPF_LOCAL routines, which in turn call (iii) Fortran 77 routines, which take local assumed-size arrays as arguments. The first version of the SLHPF interface contains HPF wrapper routines for LU and Cholesky factorization, and for the level 3 BLAS routines BLAS/dgemm and BLAS/dtrsm, among others.

The wrapper routines do not have a significant influence on the execution time (see Blackford et al. [5]). Thus most of the total computation time is spent in SCALAPACK routines.

Unfortunately, at the time the present work was performed (August 1998), the SLHPF interface did not work in conjunction with the particular HPF compilers available on our Meiko CS-2 platform. Therefore, to give some indication of the expected HPF-SCALAPACK performance, Table 2 shows the performance of the SCALAPACK Cholesky factorization routine called from Fortran 77 on the Meiko CS-2. In order to utilize the highly optimized SUNPERF library, a different compiler (SUN F77) was used than in the experiments of Section 2.

Since the overhead of the SLHPF wrapper routines is small (see Blackford et al. [5]), we expect to achieve nearly the same performance when calling SCALA-PACK from HPF via the SLHPF interface. Note that the observed efficiencies are comparable with those obtained using HPF with BLAS routines optimized with ATLAS (see Fig. 6).

**Table 2.** Performance and efficiency of the Cholesky factorization on $p = 1, 2, \ldots, 16$ processors of a Meiko CS-2 using the SUNPERF library and SCALA-PACK. The matrix size was $n = 1000$.

| Processor | | | | Performance [Mflop/s] | | |
|---|---|---|---|---|---|---|
| $p$ | Grid | Time | Speedup | Peak | Observed | Efficiency |
| 1 | 1×1 | 11.9 s | — | 50 | 28 | 56 % |
| 2 | 2×1 | 6.3 s | 1.9 | 100 | 53 | 53 % |
| 4 | 2×2 | 3.8 s | 3.1 | 200 | 88 | 44 % |
| 8 | 4×2 | 1.9 s | 6.3 | 400 | 175 | 44 % |
| 16 | 8×2 | 1.3 s | 9.2 | 800 | 256 | 32 % |

Lorenzo et al. [13] developed another interface from HPF to SCALAPACK. In this interface, to call the PBLAS routine Pdgemm, for example, an HPF wrapper routine HPF_Pdgemm is called. This sets up the array descriptors of the BLACS (Basic Linear Algebra Communication Subprograms) using the HPF library function HPF_DISTRIBUTION, broadcasts this information to all processors, and then calls the PBLAS routine Pdgemm as an EXTRINSIC (F77_LOCAL) routine.

It should be noted that these interfaces from HPF to parallel libraries should also be useful for converting existing *sequential* numerical libraries to HPF. In many cases it should suffice to replace calls to sequential BLAS or LAPACK routines in the Fortran 77 code by calls to PBLAS or SCALAPACK routines in the HPF version.

There are, however, two possible sources of difficulty in interfacing HPF to parallel libraries:

- If the HPF compiler is not based on MPI, the initialization of the message passing library is difficult (Blackford et al. [4]). In this case, PVM may be used. PVM computation nodes may also be started and initialized during runtime, independent of the HPF runtime system.
- If the backend compiler is not the standard compiler of a given platform, incompatibilities with the optimized BLAS libraries may occur.

### 3.2 ScaLAPACK Implemented in an HPF Runtime System

The public domain HPF compilation system ADAPTOR (Brandes and Greco [6]) contains an interface to SCALAPACK. The interface is implemented directly in the HPF runtime system using the language C. This approach offers the following advantages.

- A more flexible redistribution strategy is possible.
- Subsections of arrays can be used without creating temporary arrays.
- Overhead resulting from the conversion of the array descriptors from HPF to PBLAS is reduced.

However, portability is reduced due to the proprietary implementation.

## 4 Conclusions

It has been shown that existing numerical high performance libraries can be integrated into HPF code in a portable and efficient way, using HPF language features only. In particular, it is possible to implement parallel BLAS on top of the sequential BLAS. This guarantees high local performance in HPF programs and yields significant performance improvements compared to pure Fortran implementations which do not take advantage of existing software packages and libraries.

Most high level linear algebra libraries are based on the BLAS. By invoking a parallelized BLAS version implemented along the lines suggested in this paper, these libraries can also be utilized within HPF programs.

**Acknowledgments.** We would like to thank John Merlin (VCPC, Vienna) for his helpful comments, his experienced interpretation of the HPF standard, and for proofreading the paper.

## References

1. E. Anderson, Z. Bai, C. Bischof, J. Demmel, J. Dongarra, J. Du Croz, A. Greenbaum, S. Hammarling, A. McKenney, S. Ostrouchov, and D. Sorensen: LAPACK *User's Guide*, 2nd ed. SIAM Press, Philadelphia 1995.
2. J. Bilmes, K. Asanovic, C.-W. Chin, J. Demmel, *Optimizing Matrix Multiply using* PHIPAC: *a Portable, High-Performance, ANSI C Coding Methodology*, Proceedings of the International Conference on Supercomputing, ACM, Vienna, Austria, 1997, pp. 340–347.
3. J. Bilmes, K. Asanovic, J. Demmel, D. Lam, C.-W. Chin, *Optimizing Matrix Multiply using* PHIPAC: *a Portable, High-Performance, ANSI C Coding Methodology*, Technical report, LAPACK Working Note 111, 1996.
4. L. S. Blackford, J. Choi, A. Cleary, E. D'Azevedo, J. Demmel, I. Dhillon, J. J. Dongarra, S. Hammarling, G. Henry, A. Petitet, K. Stanley, D. Walker, and R. C. Whaley: SCALAPACK *Users' Guide*, SIAM Press, Philadelphia 1997.

5. L. S. Blackford, J. J. Dongarra, C. A. Papadopoulos, and R. C. Whaley: *Installation Guide and Design of the HPF 1.1 Interface to* SCALAPACK, *SLHPF*, LAPACK Working Note 137, University of Tennessee 1998.
6. T. Brandes and D. Greco: *Realization of an HPF Interface to* SCALAPACK *with Redistributions*. High-Performance Computing and Networking. International Conference and Exhibition, Springer-Verlag, Berlin Heidelberg New York Tokyo 1996, pp. 834–839.
7. J. J. Dongarra, J. Du Croz, S. Hammarling, R. J. Hanson: *An Extended Set of* BLAS, ACM Trans. Math. Software 14 (1988), pp. 18–32.
8. J. J. Dongarra, J. Du Croz, S. Hammarling, I. Duff: *A Set of Level 3* BLAS, ACM Trans. Math. Software 16 (1990), pp. 1–17.
9. H. J. Ehold, W. N. Gansterer, and C. W. Ueberhuber: *HPF—State of the Art*, Technical Report AURORA TR1998-01, European Centre for Parallel Computing at Vienna, Vienna 1998.
10. R. van de Geijn: *Using* PLAPACK: *Parallel Linear Algebra Package*, MIT Press 1997.
11. High Performance Fortran Forum: *High Performance Fortran Language Specification Version 2.0*, 1997.
    URL: www.crpc.rice.edu/HPFF/hpf2/ or
    www.vcpc.unvie.ac.at/information/mirror/HPFF/hpf2/.
12. C. L. Lawson, R. J. Hanson, D. Kincaid, F. T. Krogh: BLAS *for Fortran Usage*, ACM Trans. Math. Software 5 (1979), pp. 63–74.
13. P. A. R. Lorenzo, A. Müller, Y. Murakami, and B. J. N. Wylie: *HPF Interfacing to* SCALAPACK, Third International Workshop PARA '96, Springer-Verlag, Berlin Heidelberg New York Tokyo 1996, pp. 457–466.
14. C. W. Ueberhuber: *Numerical Computation*, Springer-Verlag, Berlin Heidelberg New York Tokyo 1997.
15. R. C. Whaley, J. J. Dongarra, *Automatically Tuned Linear Algebra Software*, Technical Report, LAPACK Working Note 131, 1997.

# PARADEIS: An Object Library for Parallel Sparse Array Computation

Franck Delaplace and Didier Remy

Universit d'Evry, Laboratoire des Mthodes Informatiques,
91025 Evry Cedex, France
{delapla,remy}@lami.univ-evry.fr

**Abstract.** Parallel programming of sparse data structure has been recognized to be hard to debug and to require an excessive amount of development time. The PARADEIS library (PARAllel Data Environment for Irregular Structure) is an object library that aims at providing a data-parallel approach with object oriented languages to ease the treatment of the sparse data structure for SPMD-like programs on Massively Parallel Processors (MPPs) with distributed memory. This paper focuses on the core of communication layout: the conversion of global addresses to communications, and its relationship with partitioning strategies.

## 1   Introduction

In data-parallel languages, such as High Performance Fortran (HPF), the parallel data structures are arrays. Accesses to distributed arrays still refer to a unique global address space which provides an easy programming model. However, massively parallel distributed memory architectures do not always provide a mechanism to support a global address space. Each processor owns a local memory and data exchanges are performed by message passing. Thus, compiling data-parallel languages onto distributed memory architectures requires producing communication code which translates global addresses to communication instructions and local memory addresses. The core of the *compilation of communications* is to compute two kinds of information from a global address $M$: the address $p$ of the processor owning the corresponding data and its local memory address $m$ in the processor. The correspondence between a global address $M$ and the pair $(p,m)$ is performed by a closed form expression. From this formula, the generation phase of the communication code identifies on each processor the sent data, the received data and the computed data. This leads to the definition of three sets: *Send, Receive* and *Compute,*[1],[4]. These sets are computed according to rules which dictate the location of the computation. The *owner computes rule* establishes that the computation is performed by the processor which owns the stored values. Therefore, designing support for communication deduced from a global address requires the analysis of the intricate implications between distribution, communication support and data structure. We propose a run-time system support based on an inspector-executor scheme to manage these communications for sparse arrays. The proposed scheme emphasized the modularity to

P. Zinterhof, M. Vajteršic, A. Uhl (Eds.): ACPC'99, LNCS 1557, pp. 153–162, 1999.
© Springer-Verlag Berlin Heidelberg 1999

break the relations between the partitioning structure and the communication scheme. The paper is organized as follows. Section 2 introduces the main features in compilation of communication for irregular structures. Section 3 exploits this analysis and describes the choices made in PARADEIS. Section 4 details the inspector-executor scheme. Finally, section 5 reports experiments performed for the PARADEIS library.

## 2   Mapping, Communication, and Data Structures for Irregular Problems

Addressing irregular problem implies to deal with communication induced from irregular data structure partitioning and access. Irregular structure distribution can not be efficiently described with regular distribution scheme. Moreover, introducing a new distibution scheme implies a new array reference scheme. CHAOS [6] is an example of a run-time system that transforms global references into communications for indirect addressing and irregular partitioning. In order to understand the difficulties encountered when designing an inspector-executor scheme dealing with indirections [8], we survey the main features of this system:

*User-Defined Partitioning* The mapping that we consider is called a *user-defined partitioning* since it lets the users define their own mapping. It is defined by an array associated to a data structure. Given a one-dimensional array A, the mapping is stored in another array Map such that if $A(i)$ is mapped to a processor $p$ then Map$(i)=p$. The array Map having the same size as the array A is distributed according to a block distribution since it cannot in general be duplicated in every processor. The regular distribution is described by a function (e.g. $where_{Map}(i) = \lfloor i/b \rfloor$ for partitioning in b blocks). In addition to the mapping array Map, another array is required to describe the local memory location.

*Indirect array reference* A typical indirect array reference example is depicted in the following program. A run-time inspector-executor is needed to compute the communication induced from the array distribution.

```
forall(i=0:n)
      X(e2(i))=Y(e1(i))+1
```

*Inspector-Executor scheme* Every array in the previous programs is distributed. Y and X are distributed according to the indirect mapping respectively defined by the arrays MapY and MapX. The indexing arrays e1 and e2 are distributed according to a regular distribution (e.g. blocks for instance), so that the location of a specific indexing array can be computed from a formula. Consequently the other processors do not know which data are used for a particular computation in an given processor since they do not have these indexing arrays. The following five communications are due to the distribution of indexing arrays and mapping arrays. The processor computing the result must own all the references but the irregular partitioning prevents the use of the owner computes rule. The rule applied is almost the owner computes rule. The inspector-executor scheme proceeds as follows:

1. *Localize the information (processor and local memory) of the referenced data.* Given an iteration $i$, the processor owning $e1(i)$ must determine where $Y(e1(i))$ is mapped. This is done by a request to the processor owning $MapY(e1(i))$, $where_{MapY}(e1(i))$. The received message is composed of a pair $(p, l)$ which indicates where the value is located and its local address.
2. *Gather values.* Once the real location of a data item is found, a pair of communications is necessary to bring these values in the processor owning the index $e1(i)$.
3. *Perform computation and update values.* Since the computation follows almost the owner computes rule, the processor with the stored data is not always the processor where the computation is performed; therefore a communication is necessary to update the value.

PARADEIS addresses a subclass of this problem: the class of sparse linear algebra problems with user-defined partitioning, the most general description of a mapping. In sparse linear algebra problems, the indirect addressing is only due to the sparse storage format of the matrices. In other words, the program would have regular accesses if the matrices were not sparse.

## 3 PARADEIS

In PARADEIS, the choices of sparse data structure, accesses and mapping codification are guided by the possibility to compute the *Send*, *Receive* and *Compute* sets on each processor, which is the key to having only one communication for a global reference. The solution is based on two principles. First promotes the use of implicit sparsity for programming applications. Second defines a hierarchical data structure and associated algorithm for representing the sparsity pattern to various degrees of approximation. The size of the data may then be traded for the accuracy of the representation.

### 3.1  Implicit Sparse Data Structure

An implicit sparse data structure enables a programmer to write his/her program as if it operated on dense array formats. Implicit sparse storage eases the programmer's task since s/he does not have to deal with the sparsity. This approach has been introduced by Bik and Whishoft [3]. According to them, *sparse data structures (sparse arrays) are data structures that are logically contained in enveloping data structures (dense arrays)*. In PARADEIS, a parallel sparse data structure corresponds to an instance of a class which is accessed by several kinds of iterators.

CRS $M \times V$

```
for(i = 0; LIGNE[i] < nz; i++)
  for(j=RO[i]; j<RO[i+1]; j++)
    R[i] += M[j] * VA[CO[j]];
```

PARADEIS $M \times V$

```
Iterator i(M) ; Iterator j(i);
for(i.begin(); i.end(); i.next())
  for(j.begin(); j.end(); j.next())
    R[i] += M[j] * V[j]
```

**Fig. 1.** MV multiply

*Consequences for the inspector-executor scheme* The use of an implicit sparse data structure allows programs to be written as they are conceived for dense arrays. Therefore, the array references in the sparse linear algebra program are regular (affine). Regular accesses are described by functions which provide a global view of the communications. In PARADEIS, communications are specified by array sections of the form $[l : u : s]$, where $l$ stands for the lower bound of the section, $u$ stands for the upper bound of the section and $s$ stands for the stride. The local access is done by iterators. The implicit sparsity offers the ability to efficiently determine the *Compute* set on each processor. Figure 1 represents programs of a sparse matrix vector multiply. The program to the left is a sequential program with C.R.S format [5], the one to the right is a sequential program written in PARADEIS. The sparsity is hidden by the use of dedicated iterators (i,j) which give the taste of dense programming. These iterators scan the non-zero of the sparse matrices.

## 3.2  Mapping of Sparse Data Structure

In PARADEIS, the localization phase has been removed. In order to remove the communications combined with the localization phase (see Section 2), sufficient information must be kept on each processor to describe the mapping of all data. Since the mapping array Map (that define *only* the mapping of non zero values) cannot be duplicated because it is too large, we define an approximation of this mapping that requires less memory. A valid approximation is such that, given a non-zero value located at the coordinates $(x, y)$, if the value at $(x, y)$ is mapped onto $p$ then the approximation also gives this mapping. However, in the case of sparse arrays, the approximation can also define a mapping for coordinates which correspond to zeros in the sparse matrix. A *sparse tree descriptor* describes this approximation.

The mapping is described by a set of 3-tuples $([l_x : u_x], [l_y : u_y], p)$ signifying that the data items contained in this block are mapped to the processor $p$. Logically, the descriptor can be viewed as a tree.

– The root is the size of the matrix.
– The first level is the common global knowledge shared by every processor. It provides an approximation of the exact mapping

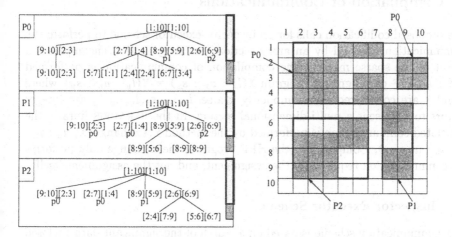

**Fig. 2.** Example of a descriptor codification

- The second level corresponds to local knowledge. It is the exact description of the values held by a processor. This information is only stored on the processor where the corresponding data are mapped.
- The leaves contain data if they are resident on the processor.

Figure 2 describes an example of this codification for a $10 \times 10$ matrix. Filled cells correspond to significant values whereas empty cells correspond to zeros.

### 3.3   Codification of the Partitioning by a Descriptor

Any partitioning can be applied providing it is described by a sparse array descriptor. For matrices, some partitioning algorithms such as BRD decomposition [2] provide a natural description of partition as a sparse tree descriptor. If the result cannot be directly codified by blocks, then it is decomposed into several independent blocks. These blocks are not necessarily contiguous but they still refer to a single partition. Once the mapping is defined the pieces of the sparse matrix are assigned to processors. The codification is performed in parallel. The part of the global view computed for one processor is broadcast to every processor.

The codification algorithm balances the accuracy of the description and the memory occupancy. The accuracy of the descriptor affects the amount of useless data sent in each messages. The number of blocks affects not only the memory occupancy but also the overhead induced by the inspector-executor phase, since the complexity of the inspector depends linearly on the number of blocks. The experiments (Section 5) have been performed with a "fine to coarse" algorithm which aggregates blocks to provide a larger block until a threshold is reached. The threshold represents a *density ratio* of the significant values in each block. Two different thresholds are defined: the first corresponds to the local level and the second to the global level.

# 4   Compilation of Communications

The proposed codification calls for an inspector-executor scheme to perform the communication involved by an array section. An array section encompasses a subset of the sparse matrix. The compilation of communication is performed as for an HPF paradigmatic program $X(l_x : u_x : s_x) = Y(l_y : u_y : s_y)$ where $X$ and $Y$ are considered to be implicitly sparse. $l_x, u_x, s_x, l_y, u_y, s_y$ are integer vectors to represent a multi-dimensional section. In the PARADEIS library, the associated communication is performed by the method $Y.\texttt{exchange}(X, l_x : u_x : s_x, l_y : u_y : s_y)$. Compared to the HPF program, the exchange only performs the communication required by the assignment, and not the assignment itself.

## 4.1   Inspector-Executor Scheme

The communication scheme is based on a search of the significant data enclosed in an array section by using a tree where levels are ordered by inclusion. The inspector-executor scheme consists of two phases.

**Approximation**   First, the search is performed according to the first level; it provides an approximation of the significant values owned by a processor.

**Refinement**   This search is performed at the second level by every processor, to send the non-zero values it owns.

We call an union of multi-dimensional array sections a *sparse block*. A *global sparse block* is a subset of blocks described at the global level whereas a *local sparse block* is a local subset of sparse blocks. $GS_p^X$ and $LS_p^X$ correspond respectively to the global and local description of $X$ located on the processor $p$. By extension, $GS_*^X$ represents the union of the global sparse blocks of all the processors. Let $X$ be a sparse array, and $me$ be the processor identifier, we define the following functions:

- $Own(X, s) = (P, B)$ gives the set $P$ of the processors owning a part of the sparse block $s$ according to $GS_*^X$. $B$ is a set of sparse blocks where each elements is indexed by a processor identifier such that for every $p \in P$, its corresponding sparse block is $B_p \in B$.
- $own(X, s) = P$ stands for the restriction of the previous function $Own$ to the set of processors.
- $\sigma(a, b, c)$ is a function which provides a subset of blocks $c' \subseteq a$ corresponding to $c \subseteq b$. This function computes the correspondence between the stored section and the read section.
  1. If $a, b, c$ are one-dimensional array sections such that:
     $a = [l_a : u_a : s_a]$, $b = [l_b : u_b : s_b]$, $c = [l_c : u_c : s_c]$, then the resulting section is:

$$c' = [l_a + s_a \frac{l_c - l_b}{s_b} : l_a + s_a \cdot \lfloor \frac{u_c - l_b}{s_b} \rfloor : \frac{s_c}{s_b} s_a]$$

2. If $a, b, c$ are multi-dimensional sections then $\sigma$ corresponds to the application of the previous formula to each of the dimensions.
3. If $a, b, c$ represent sparse blocks, then the result is the union of elementary array sections computed from the above formula.

The assigned part of an array may be determined for a given read sub-section using the $\sigma$ function. For the given following statement $A[0:21:3] = B[10:38:4]$, the corresponding stored part $B[18:34:8]$ is:

$$\sigma([0:21:3], [10:38:4], [18:34:8]) = [6:18:6]$$

Thus, $A[6:18:6]$ will be assigned by the values of $B[18:34:8]$.

**Inspector Phase** The first step is the computation of the expected *Send* and *Receive* sets. They are defined from the global knowledge.

**Send Inspector** The goal is to define an approximation of the blocks which are sent from the processor *me*. The inspector compares of the global knowledge and the array section to define the set of receiving processors and the data set to be sent.
   **Approximation**
   - $GR$ is a global set of blocks of X which correspond to the data of Y owned by the processor *me*:
     $GR = \sigma([l_x : u_x : s_x], [l_y : u_y : s_y], GS^Y_{me} \cap [l_y : u_y : s_y])$
   - $(PS, B)$ is the set of the processors which receive the expected value of $X$ owned by *me*. $B$ is the corresponding set of global blocks:
     $(PS, B) = Own(X, GR)$
   **Refinement**
   - $RB_p$ is the local set of blocks of $Y$ which will be sent to the processor identified by $p$:
     $\forall p \in P, RB_p = LS^Y_{me} \cap \sigma([l_y : u_y : s_y], [l_x : u_x : s_x], B)$

**Receive Inspector** The receive inspector computes the set of processors which send data to *me*. The receive inspector only performs an approximation.
   - $GW$ represents the set of addresses of X which need communications for their assignments. $GW = \sigma([l_y : u_y : s_y], [l_x : u_x : s_x], GS^X_{me} \cap [l_x : u_x : s_x])$
   - $PR$ is the set of the processors which are supposed to send values: $PR = own(Y, GW)$

**Executor Phase** Only significant values are transferred during a communication. Thus, the inspector phase transmits the local descriptor and the associated value. Values are stored in a dedicated temporary region of Y.

| *Send Executor* | *Receive Executor* |
|---|---|
| ```
for p ∈ PS − {me}
    send(RBp,value(RBp),tag,p)
end for
``` | ```
for p ∈ PR − {me}
    receive(B,D,tag,p)
    store (p,B,D) into LSY
endfor
``` |

## 4.2   Communication-Free Permutation

The descriptor represents the mapping function as well as the data structure. It enables some permutations to be performed on the data without communication. For example, the transposition of a matrix $A$ consists of exchanging values of $a_{ij}$ with $a_{ji}$. In our case, it suffices to exchange the array sections corresponding to rows and columns in the descriptor in every processor and to perform a local transposition of the values. This feature can be used for any permutation which does not induce an overlap of the resulting partition. For example, any permutation which preserves the Euclidean distance (such as a shift) can be treated this way. This property does not hold for a regular distribution because the mapping is determined by a function that cannot be changed.

## 5   Experiments

This section reports on experiments with PARADEIS. These experiments are performed on matrix vector multiply $R = M \times V$. where $R$ and $V$ are dense vectors whereas $M$ is a sparse matrix. $V$ and $M$ are distributed according to the BRD strategy. The experiments have been performed on an SP2 (8 nodes Power PC 66.6 Mhz, 256 MB/processor). PARADEIS has been developed in C++ with the PVM message passing library. The compiler is GCC 2.7.2 with the option -O. Tests have been performed on a large variety of sparse matrices coming from the Harwell Boeing collection. Only few results are reported here because of the lack of space.

Some $N \times N$ matrices are presented in Table 1. Their main characteristics are reported: number of non zero values, size $N$ and corresponding density.

|         | bcsstk13 | cavity09 | mcfe  | bcsstm13 | bcsstk19 |
|---------|----------|----------|-------|----------|----------|
| NZ      | 42943    | 32747    | 24382 | 11973    | 3835     |
| N       | 2003     | 1182     | 765   | 2003     | 817      |
| Density | 1.07%    | 2.34%    | 4.16% | 0.3%     | 0.57%    |

**Table 1.** Properties of the test matrices

**Scalability** We analyse the impact of PARADEIS on the scalability of the programs. The MV multiply is known to have a near linear speed-up. Figure 3 reports the results of the experiments. The efficiency is at least 75%, which shows that the management of the descriptor has little impact on the total execution time. The decrease of the speed-up is mainly due to communication.

**Object oriented language (C++) overhead** PARADEIS is implemented using C++ classes like SparseArray M in PARADEIS part of Figure 1. This leads to an overhead compared to a simple C array. Figure 1 describes two versions of

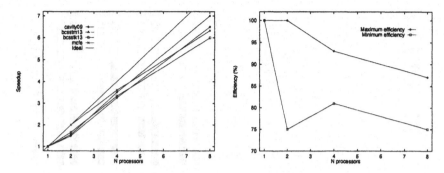

**Fig. 3.** Scalability and Efficiency versus Number of Processors

a sequential MV multiply. One can see that the PARADEIS version is very close to the dense version using C's array. To measure the impact of object layout (the use of a descriptor and iterators) we analyse the time spent to scan indexes. The criterion is the number of iterations per second. The performance ratio $r$ is always in favour of CRS code written in C. Yet the ratio is small and depends on the matrix: $1.2 \leq r \leq 1.8$ for the execution on a SUN SPARC 10, and $1.9 < r < 3$ for an RS6000. The overhead is mainly due to the descriptor part of the `SparseArray` structure, more precisely to an increase of the complexity of bound checking.

**Memory overhead** Two parameters influence the memory overhead: the density ratio and the structure of the matrix. Figure 4 reports two kinds of comparisons: The first one represents the evolution of the memory occupancy (in bytes) when the local density ratio increases. We also compare it with an estimation of the memory occupancy of the solution proposed in Vienna Fortran though BRD descriptors and local CRS arrays [7]. This first measure shows that the memory occupancy depends on the structure of the matrix. The second figure to the right are cumulative bars representing the ratio of memory occupancy for some fill values. From bottom to top of a bar, the first ratio (darkest grey) corresponds to nonzero elements, the next one represents the percentage of the zero element, the third one the size of the local descriptor and the last one (the top) the size of the global descriptor. An immediate conclusion is that the percentage of the global descriptor is negligible. The size of the local descriptor depends on the structure. The more the structure is scattered, the more the size of the local descriptor is important.

We compare the descriptor and non-zero value size of CRS and PARADEIS storage. CRS stores only non-zero values (NZ) in an array (as double-precision value for example). It also needs a descriptor of size equal to NZ plus the number of rows. In PARADEIS, the local descriptor is mainly between 10% and 50% of the used memory storage. It is the same amount as for CRS storage.

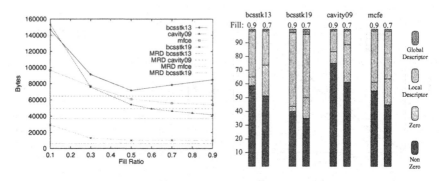

**Fig. 4.** Memory overhead

# 6  Conclusion

In this paper, we define the main features of an object library for data-parallel sparse matrices. The experiments show that PARADEIS and the object programming may be an efficient solution to ease programming of sparse problems. The PARADEIS library can be considered as a user library as well as a compiler backend library for high performance data-parallel languages. Two perspectives are considered: to extend the validation to more complex algorithm (like conjugate gradient); and to use another geometric basis as parallelograms instead of rectangular blocks in order to cover some particular structure like diagonal matrices.

# References

1. F. Andr, O. Chron, M. LeFur, Y. Maho, and J.L. Pazat. Programmation des machines mmoire distribue par distribution de donnes : langages et compilateurs. *T.S.I*, 12(5):563–597, 1993.
2. M.J. Berger and S.H. Bokhari. A partitioning strategy for nonuniform problems on multiprocessors. *IEEE Trans. Comput.*, 36(5):570–580, 1987.
3. A.J.C Bik and H.A.G Wijshoff. Automatic data structure selection and transformation for sparse matrix computation. *IEEE Transactions on Parallel and Distributed Systems*, 7(2):1–19, 1996.
4. F. Coelho, C.Germain, and J.L. Pazat. State of art in compiling HPF. In *The Data Parallel Programming Model*, volume LNCS 1132, pages 104–130. Springer, 1996.
5. I.S Duff, A.M Erisman, and J.K. Reid. *Direct Methods for Sparse Matrices*. Oxford Sciences Publications, 1986.
6. R. Ponnusamy, Y. Hwang, J.H. Saltz, A. Choudhary, and G. Fox. Supporting irregular distributions using data-parallel languages. *IEEE Parallel and Distributed Technology*, pages 12–24, Spring 1995.
7. M Ujaldon, E.L. Zappata, B.M. Chapman, and P.H. Zima. Vienna-Fortran/HPF extension for sparse and irregular problems and their compilation. *IEEE Trans. on Parallel & Distributed systems*, 8(10):1068–1082, October 1997.
8. R. von Hanxleden, K. Kennedy, C. Koelbel, R. Das, and J. Saltz. Compiler analysis for irregular problems in Fortran D. *Lecture Notes in Computer Science*, 757:97–111, 1993.

# Performance Analysis and Derived Parallelization Strategy for a SCF Program at the Hartree Fock Level

Siegfried Höfinger[1], Othmar Steinhauser[1], and Peter Zinterhof[2]

[1] Institute for Theoretical Chemistry, Molecular Dynamics Group,
University of Vienna, Währingerstr. 17, Ground Floor,
A-1090 Vienna, Austria
{sh, os }@mdy.univie.ac.at
http://www.mdy.univie.ac.at
[2] Institute for Mathematics, University of Salzburg,
Hellbrunnerstr. 34, A-5020 Salzburg, Austria
zinterhof@edvz.sbg.ac.at
http://www.mat.sbg.ac.at

**Abstract.** General performance measurement tools, as provided on a SGI-Power Challenge/R10000 (194 MHz) running IRIX 6.2, which are the shell-commands **perfex -a a.out** and **ssrun -[pcsampl, ideal, usertime] a.out**, have been used for profiling a new SCF-program based on the Hartree Fock theory. The main cpu-time-consuming subroutine was detected and a parallel version with PVM 3.3 at a host-node model has been deduced. Speed-up parameters are presented and discussed.

## 1 Introduction

Electronic structure determination has always been one of the key topics in computational chemistry and, due to the enormous progress in the performance of today's hardware architecture, this subject is becoming more and more important even in the fields of traditional bio-sciences, such as molecular biology, immunology, pharmaceutics, drug design and many more. As a result of this, a well designed algorithm for the calculation of molecular orbitals of large systems is still an interesting task one can pay attention to.

### 1.1 Time Independent Schrödinger Equation

The state of the art method for gaining some information about the electrons within a certain molecule is iteratively solving the *time independent Schrödinger equation* [1]

$$\left[ \sum_i^{electrons} -\frac{1}{2}\Delta_i + V(r_1, r_2...) \right] \psi_{el}(r_1, r_2...) = E \, \psi_{el}(r_1, r_2...) \qquad (1)$$

P. Zinterhof, M. Vajteršic, A. Uhl (Eds.): ACPC'99, LNCS 1557, pp. 163–172, 1999.

where the square bracket term on the left hand side of equation 1 is usually referred to as *the Hamiltonian Operator*, whose first part comprises *the Kinetic Energy* [1] of the electrons and whose second part describes *the Potential* for the electrons

$$V(r_1, r_2...) = - \sum_{i}^{electrons} \sum_{k}^{nuclei} \frac{Z_k}{|r_i - r_k|} + \sum_{i}^{electrons} \sum_{j}^{electrons} \frac{1}{|r_i - r_j|} \qquad (2)$$

consisting of *Nuclear Attraction* [2] and *Electron Repulsion*.

Furthermore, the $\psi_{el}(r_1, r_2...)$ of equation 1 is called *the Wavefunction of the Electrons* [3] and the *Eigenvalue E* on the right hand side of equation 1 stands for *the Total Energy* of the electrons and from all of these it may become clear that the whole Schrödinger equation represents an *Eigenvalue Problem*.

## 1.2    Decoupling of Electron Repulsion

The coupling of electrons as given from the last term of equation 2 leads to a correlated motion of electrons, that explicitly treated would result in tremendous computational efforts, which might be fairly released, if the following approximation is used,

$$\sum_{i}^{electrons} \sum_{j}^{electrons} \frac{1}{|r_i - r_j|} \approx \sum_{i}^{electrons} \left[ \int_{\acute{r}} \frac{\rho(\acute{r})d\acute{r}}{|r_i - \acute{r}|} + v_{eff}(r_i) \right] \qquad (3)$$

with $\rho(\acute{r})$ standing for the *Electron Density* around point $\acute{r}$ and $v_{eff}(r_i)$ being an *Effective Potential*.

With this approximation the general n-particle Schrödinger equation 1 reduces to its decoupled one-electron form of n [4] identical one-particle equations

$$\left[ -\frac{1}{2}\Delta - \sum_{k}^{nuclei} \frac{Z_k}{|r - r_k|} + \int_{\acute{r}} \frac{\rho(\acute{r})d\acute{r}}{|r - \acute{r}|} + v_{eff}(r) \right] \phi_k(r) = \varepsilon_k \, \phi_k(r) \qquad (4)$$

where now the $\phi_k(r)$ are one–electron wavefunctions or molecular orbitals (*MO* [5] ) and equation 4 once more reflects an eigenvalue problem.

The fundamental property characterizing a molecule, the so called one-electron density $\rho(r)$, is related to the n-particle wavefunction $\psi_{el}(r_1, r_2...r_n)$ by

$$\rho(r_1) = \int_{r_2} \int_{r_3} ... \int_{r_n} \psi_{el}^*(r_1, r_2...)\psi_{el}(r_1, r_2...)dr_2 dr_3...dr_n \qquad (5)$$

---

[1]   As usual, the $\Delta_i$ is used as an abbreviation for *the Laplacian Operator*.
[2]   the $Z_k$ stands for nuclear charge at center $r_k$.
[3]   Note, that for this very general formulation each electron has its own spatial coordinates $r_i$.
[4]   n stands for total number of electrons.
[5]   Molecular orbitals shall be designated $\phi_k$ and are said to form an orthonormal set.

and takes the special form

$$\rho(r) = \sum_i^{electrons} \phi_i^*(r)\phi_i(r) \tag{6}$$

for a *Slater Determinant.* [6]

## 1.3  Effective Potential

The practical representation of $v_{eff}(r)$ in equation 4 gives rise to two different main stream directions in quantum mechanics, where the one is called *Density Functional Theory* [5] and the other is referred to as *Hartree Fock Theory* [6] [7]. The latter has been used for the present program we are talking about.

In Hartree Fock theory ( HF ) the $3^{rd}$ and $4^{th}$ term of equation 4 are replaced by the following closed form

$$\left[\int_{\acute{r}} \frac{\rho(\acute{r})d\acute{r}}{|r - \acute{r}|} + v_{eff}(r)\right]_{HF} \triangleq \sum_j^{electrons} \left(J_j - \frac{1}{2}K_j\right) \tag{7}$$

with the J- and K-terms [7] given by [8]

$$J_j = J_j(r)\phi_k(r) = \left[\int_{\acute{r}} \phi_j^*(\acute{r})\frac{1}{|\acute{r} - r|}\phi_j(\acute{r})d\acute{r}\right]\phi_k(r) \tag{8}$$

$$K_j = K_j(r)\phi_k(r) = \left[\int_{\acute{r}} \phi_j^*(\acute{r})\frac{1}{|\acute{r} - r|}\phi_k(\acute{r})d\acute{r}\right]\phi_j(r) \ . \tag{9}$$

Thus the sum over J-terms together with equation 6 is precisely the *Coulomb Potential* ( $3^{rd}$ term of equation 4 ) and the sum over K-terms represents a *Nonlocal Effective Exchange Potential.*

So far we always talked about the electrons only and never spent a word on nuclear charge distribution, that theoretically should again be subject to another wavefunction and treated in the same quantum way the electrons had been. The neglect of an explicit wavefunction for the description of the nuclear charges is due to the fact, that the two wavefunctions, for electrons and nuclei either, may be separated into two independent ones, which is commonly known as the *Born-Oppenheimer Approximation* [9] [2].

---

[6] $\psi_{el}(r_1, r_2 ...) = \mathcal{P}[\phi_1(r_1)\phi_2(r_2)...\phi_n(r_n)]$ where $\mathcal{P}$ means Permutation and all is done in order to respect *the Pauli Principle* [3], that results from the *antisymmetric* character of the wavefunction.

[7] $J_j$ in equation 7 is called *Coulomb Operator* and $K_j$ in equation 7 is called *Exchange Operator.*

[8] Note, that the Index k refers to $\phi_k$ of equation 4 and thus shows up the exchange character of $K_j$.

[9] All nuclear centres are considered fixed in space with fixed partial charges $Z(r_k)$ located at those very points in space.

## 1.4  SCF-Procedure

As may be seen from equation 4 the decoupled one–electron Schrödinger equation depends on the electron density $\rho(r)$, which in its turn is a function of all MOs involved. Therefore all other electrons have a strong influence on the solution of the eigenvalue problem for a particular electron.

This leads to the paradox situation that one should already know all MOs in order to be able to determine one particular MO explicitly.

To overcome this principal problem, equation 4 has to be solved via a so-called *Self Consistent Field* procedure ( SCF ), which in a sketchy way may be described as

- Set up a first trial density $\rho_1(r)$, that naturally will be far away from the actual physical relevant one.
- Solve the eigenvalue problem 4 and thereby get access to a set of MOs, which in turn results in a better, more realistic new density $\rho_2(r)$ via equation 6.
- Resolve the next eigenvalue problem using the improved density $\rho_2(r)$ and again get another even more improved set of MOs and so on... until two subsequent sets of MOs will be almost identical to each other and differ only by a predefined small threshold value.

## 1.5  LCAO, Main Problem, and Method

According to *the LCAO – approach* [10], the MOs are again expanded into a linear combination of atomic orbitals( AO [11] ), where the latter might also simply be called *Basis Functions* [12]

$$\phi_k(r) = \sum_i c_{k,i} \, \varphi_i(r) \tag{10}$$

and the basis functions $\varphi_i$ are again expanded in a series over *Primitive Gaussians* $\chi_j$

$$\varphi_i(r) = \sum_j d_{i,j} \, \chi_j(r) \ , \tag{11}$$

which typically are *Cartesian Gaussian Functions* located at some place $(A_x, A_y, A_z)$ in space [13] [8] [9].

$$\chi_j(r) = N_j(x - A_x)^l(y - A_y)^m(z - A_z)^n \, e^{-\alpha_j(r-A)^2} \tag{12}$$

---

[10] Linear Combination of Atomic Orbitals.

[11] Atomic orbitals shall be designated $\varphi_i$.

[12] Basis functions are those widely known 1s, 2s, 2px, 2py, 2pz... orbitals for the description of one-electron atomic systems such as H, He$^+$, B$^{2+}$...

[13] An S-type basis function will consist of primitive gaussians with $l = m = n = 0$, a P-type however of primitives with $l + m + n = 1$, which may be solved at 3 different ways, either $l = 1$ and $m = n = 0$, or $m = 1$ and $l = n = 0$, or $n = 1$ and $l = m = 0$. D-type specification will likewise be $l + m + n = 2$ and similarly F-type $l + m + n = 3$.

The main problem for SCF-calculations is the evaluation of the ERIs, *the Electron Repulsion Integrals*, that are 6-dimensional, 4-center integrals over the basis functions $\varphi$.

$$ERI = \int\limits_{r_1} \int\limits_{r_2} \varphi_i(r_1)\varphi_j(r_1)\frac{1}{|r_2 - r_1|}\varphi_k(r_2)\varphi_l(r_2)dr_1 dr_2 \qquad (13)$$

There are various ways to compute ERIs [10] [11] [12], but the method used in the present program is the recursive method [14]described by *Obara and Saika* [12].

## 1.6 Shell Concept and Model-Molecule SO$_2$

Without any intention of going into further details, we just want to outline, at least basically, the principal scheme behind the recursive construction of ERIs due to *Obara and Saika* [12]. For all further report we present data for a simple molecule, that has been used as a certain kind of reference – SO$_2$ in particular. A schematic representation of the molecule is given in table 1. The applied basis function specification has been the standard 6-31g basis-set [13] [14]. The main advantage of this 6-31g basis-set is, that there exist cross-contracted shells, e.g. SP-contracted shells, which will utilize the same exponential factor ( $\alpha$ in equation 12 ) for S-type basis functions as well as P-type basis functions [15], and hence will offer a chance to calculate the more complicated ERIs, that contain P-type basis functions, in an recursive way from easily generated pure S-type ERIs.

As stated above, ERIs are 4-center integrals, and therefore we will have to combine 4 centres together with the according contracted shells to determine how many ERIs may be built in one subsequent, recursive subprocess. For example consider the center-quartett $\boxed{1}$ $\boxed{1}$ $\boxed{2}$ $\boxed{3}$, then one possible combination of contracted shells would be $(2)(3)(5)(9)$ [16], with the basic S-type ERI made up from basis functions 2 6 14 24, and after initial calculation of this basic S-type ERI, a total number of 63 related ERIs may be derived recursively, such as 2 6 14 25, 2 6 14 26, 2 6 14 27, 2 7 14 24, 2 7 14 25, 2 7 14 26 .... 5 9 14 27.

# 2  Performance Analysis

## 2.1  General Status

To shed some light onto critical regions of our program, we have used performance-measurement-tools, as provided on a SGI-Power Challenge/R10000 (194 MHz) running IRIX 6.2.

---

[14]  All complicated ERI-types ($l+m+n > 0$) may be deduced from the easier computed $(S_i, S_j|S_k, S_l)$ type.

[15]  For example basis functions 6, 7, 8 and 9 of contracted shell (3) at center $\boxed{1}$ will all have the same exponent $\alpha$.

[16]  There are 90 possible combinations of relevant contracted shells within this particular center-quartett and we have only picked one out of them.

**Table 1.** Shell concept explained at the model-molecule $SO_2$, that has been used for further analysis.

| Schematic Representation of $SO_2$ | | | | | | | | | | |
|---|---|---|---|---|---|---|---|---|---|---|
| $Center_{Element}$ | 1 $S$ | | | | 2 $O$ | | | 3 $O$ | | |
| Cntr. Shll.$_{Typ}$ | $(1)_S$ | $(2)_{SP}$ | $(3)_{SP}$ | $(4)_{SP}$ | $(5)_S$ | $(6)_{SP}$ | $(7)_{SP}$ | $(8)_S$ | $(9)_{SP}$ | $(10)_{SP}$ |
| Basisf.$_{Typ}$ | $1_S$ | $2_S$ | $6_S$ | $10_S$ | $14_S$ | $15_S$ | $19_S$ | $23_S$ | $24_S$ | $28_S$ |
|  |  | $3_{Px}$ | $7_{Px}$ | $11_{Px}$ |  | $16_{Px}$ | $20_{Px}$ |  | $25_{Px}$ | $29_{Px}$ |
|  |  | $4_{Py}$ | $8_{Py}$ | $12_{Py}$ |  | $17_{Py}$ | $21_{Py}$ |  | $26_{Py}$ | $30_{Py}$ |
|  |  | $5_{Pz}$ | $9_{Pz}$ | $13_{Pz}$ |  | $18_{Pz}$ | $22_{Pz}$ |  | $27_{Pz}$ | $31_{Pz}$ |

Some insight into the basic events of the r10000 processor for some specific program execution may be gained by using the shell-command **perfex -a a.out**. Then a table is given, which represents a listing of total counts of various events, such as $2^{nd}$ level data cache misses, or issued loads, or graduated loads, or ... that happened during program execution and thus [17] a relative weighting of critical events may be done to notice some basic bottlenecks. A listing of the top 6 events during our SCF-process together with some relative numbers of highly sensitive events is given in table 2 below.

**Table 2.** Relative timings of specific hardware counters on R10000, SGI for one SCF-process completion.

| Top 6 Events and Relative Timings | | | | |
|---|---|---|---|---|
| Event | Counts | Typical Event–Cost | Absolute Time [min] | Relative Time [%] |
| Cycles | $1.407 \ 10^{11}$ | 1.00 | 12.088 | 100.00 |
| Issued Loads | $7.402 \ 10^{10}$ | 1.00 | 6.359 | 52.61 |
| Graduated Loads | $6.125 \ 10^{10}$ | 1.00 | 5.262 | 43.53 |
| Issued Stores | $3.165 \ 10^{10}$ | 1.00 | 2.719 | 22.49 |
| Graduated Stores | $3.146 \ 10^{10}$ | 1.00 | 2.703 | 22.36 |
| Decoded Branches | $1.196 \ 10^{10}$ | 1.00 | 1.028 | 8.50 |
| Graduated FP Instruc. | $8.079 \ 10^{9}$ | 1.00 | 0.694 | 5.74 |
| Prim. Data Cache Miss | $3.280 \ 10^{8}$ | 9.03 | 0.254 | 2.10 |
| Mispredicted Branch | $1.317 \ 10^{9}$ | 1.50 | 0.170 | 1.40 |
| Sec. Data Cache Miss | $2.871 \ 10^{5}$ | 192.12 | 0.005 | 0.04 |

**Interpretation** At least no principal performance bottlenecks were encountered, especially when looking at highly sensitive events, such as data cache misses and mispredicted branches. So we don't think, that this application suffers from serious performance-problems, which could be solved at the programmer's level.

---

[17] After multiplying those absolute counts with specific event–cost–figures.

## 2.2   Main Time Consuming Modules

The other thing we were highly interested in, was to select those subroutines and functions, that play the main part during program execution as far as cpu-time is concerned. Therefore we did some performance-measurement on the SGI-Power Challenge/R10000 (194 MHz) running IRIX 6.2, namely **ssrun - [pcsampl, ideal, usertime] a.out**, which all produce ordered listings of the involved modules according to their fraction in the total cpu-time.

No matter which of the three optional parameters were selected in particular, it always turned out that 99.9 % of the total cpu-time was spent in *Subroutine FMAT*, which is our module for ERI–computation.


# 3   Parallelization

As has become clear from the previous section 2.2, our primary goal for parallelization purposes had to be subroutine *FMAT* — the ERI-calculation module.

For a first approach we used PVM 3.3 [18] for a host–node model due to the MPMD scheme [19], where the outermost loop [20] within subroutine *FMAT* had been split and partitioned over a certain range of nodes.

To begin with we present data resulting from a simple arithmetic, averaging partitioning scheme ( table 3 ), where each of the involved nodes got to work on a certain subset of center quartettes and the number of individual items within this subsets was derived by simply dividing the total number of all possible center quartettes through the number of nodes involved. ( partitioning scheme also shown on left side of table 4 )

**Table 3.** Wall clock times for parallel SCF on Dec 3000 cluster for different number of nodes and derived speed-up parameter. Arithmetic average partitioning scheme.

| Wall Clock Times and Speed–Up | | | |
|---|---|---|---|
| | Number of Nodes | Wall Clock Time [min] | Speed Up |
| Dec 3000 Cluster | 1 – sequential | 46:12 | |
| Dec 3000 Cluster | 2 | 32:03 | 1.44 |
| Dec 3000 Cluster | 4 | 21:00 | 2.20 |
| Dec 3000 Cluster | 8 | 13:34 | 3.46 |

---

[18]   Rel. 11 on the alpha-cluster made up of equal dec 3000 nodes, Rel. 10 on the SGI Power Challenge R10000.

[19]   Multiple ( different and executable ) Programs Multiple Data.

[20]   Loop over quartettes of center.

## 3.1 Load Balancing

After realizing the poor speed up factors obtained from the initial, simple arithmetic average partitioning scheme ( left side of table 4 ), we went one step further and introduced a pre-scanning subroutine, with which we could estimate the net work to be done and based on this information we built load-balanced pairlists representing the outermost loop over center quartettes. The latter scheme gave rise to the following pairlists of center [21] ( right side of table 4 ) and a summary of corresponding speed-up data for the likewise improved parallel version is given in table 5 for the Dec 3000 cluster as well as for the SGI-Power Challenge/R10000 (194 MHz).

**Table 4.** Arithmetic average and load balanced partitioning scheme of the outermost loop over center quartettes into node-specific pairlists of center for $SO_2$ — 4 nodes considered.

| Pairlist of Center for $SO_2$ Arithmetic Average | | | | Pairlist of Center for $SO_2$ Load Balanced | | |
|---|---|---|---|---|---|---|
| Node | Left Pair of Center | Right Pair of Center | | Node | Left Pair of Center | Right Pair of Center |
| I | 1 | 1 | | I | 1 | 1 |
| | 1 | 2 | | | 1 | 2 |
| | 1 | 3 | | | 1 | 3 |
| | 1 | 4 | | II | 1 | 4 |
| | 1 | 5 | | | 1 | 5 |
| II | 1 | 6 | | | 1 | 6 |
| | 2 | 2 | | | 2 | 2 |
| | 2 | 3 | $\Rightarrow$ | III | 2 | 3 |
| | 2 | 4 | | | 2 | 4 |
| | 2 | 5 | | | 2 | 5 |
| III | 2 | 6 | | | 2 | 6 |
| | 3 | 3 | | IV | 3 | 3 |
| | 3 | 4 | | | 3 | 4 |
| | 3 | 5 | | | 3 | 5 |
| | 3 | 6 | | | 3 | 6 |
| IV | 4 | 4 | | | 4 | 4 |
| | 4 | 5 | | | 4 | 5 |
| | 4 | 6 | | | 4 | 6 |
| | 5 | 5 | | | 5 | 5 |
| | 5 | 6 | | | 5 | 6 |
| | 6 | 6 | | | 6 | 6 |

---

[21] According to the specification done in table 1 pair 1 will be $\boxed{1}\,\boxed{1}$, 2 is $\boxed{1}\,\boxed{2}$, 3 is $\boxed{1}\,\boxed{3}$, 4 is $\boxed{2}\,\boxed{2}$, 5 is $\boxed{2}\,\boxed{3}$ and 6 is $\boxed{3}\,\boxed{3}$.

**Table 5.** Wall clock times for parallel SCF on either SGI-Power Challenge/R10000 (194 MHz) or Dec 3000 cluster for different number of nodes and derived speed-up parameter. Load balanced partitioning scheme.

| Wall Clock Times and Speed–Up | | | |
|---|---|---|---|
| | Number of Nodes | Wall Clock Time [min] | Speed Up |
| Dec 3000 Cluster | 1 – sequential | 46:01 | |
| Dec 3000 Cluster | 2 | 23:21 | 1.97 |
| Dec 3000 Cluster | 4 | 14:18 | 3.22 |
| Dec 3000 Cluster | 8 | 7:06 | 6.48 |
| SGI Pow.Chll. | 1 – sequential | 12:00 | |
| SGI Pow.Chll. | 2 | 6:03 | 1.98 |
| SGI Pow.Chll. | 4 | 3:52 | 3.10 |
| SGI Pow.Chll. | 8 | 1:57 | 6.15 |

### 3.2 Discussion

1. According to *Amdahl' s Law*

$$SpeedUp \leq \frac{1}{s + \frac{1-s}{N_{cpu}}} \tag{14}$$

with $s$ standing for the serial fraction and $N_{cpu}$ for the number of nodes, we should obtain speed up values almost equal to the number of nodes involved, if we assume that $s$ is sufficiently small and that communication may be neglected.

From section 2.2 we may conclude that $s = 0.001$, which leads to theoretical speed up values of 1.998 for 2 nodes, 3.988 for 4 nodes and 7.944 for 8 nodes. Nevertheless our measured speed up values as represented in table 5 differ quite a lot from those theoretical values, which is due to the fact, that despite serious efforts to enable a well balanced work distribution one cannot overcome the principal block structure for ERI-computation ( section 1.6 ). This means, that even after building load balanced pairlists all nodes will still have to work on slightly varying fractions of partial work, that are comparable to each other, but not exactly equal. [22]

So after 3 nodes for example have already finished their partial work, node IV may still be busy completing its last center quartette, which forces all other 3 nodes to stay idle in the meantime and thus induces an artificial enlargement of the serial fraction in program execution.

2. The small differences in speed up between the Dec 3000 Cluster and the SGI Pow.Chll. ( table 5 ) reflect the much faster processors of the SGI Pow.Chll., where for the latter the proportion computation : communication becomes

---

[22] Like the area of a puzzle can be divided into a number of almost equal partial areas, but not into exactly equal partial areas, without destroying some particular puzzle slice.

seriously smaller and hence system operations ( I/O, memory-management, communication ) gain importance.

3. Communication terms have been kept to an absolute mininum, so that only the varying density-matrices need have been sent and only the arising partial Fock-matrices need have been received within each of the iteration steps throughout the whole SCF-process ( section 1.4 ).

## 3.3   Conclusion

Recursive ERI-computation according to *Obara and Saika* [12] may not be done in 100 % parallel mode, but after introduction of some load balanced pairlists, a considerable amount of the global work may be parallel computed on distributed nodes.

## References

1. Schrödinger, E.: Quantisierung als Eigenwertproblem. Annalen der Physik. **79, 80, 81** (1926)
2. Born, M., Oppenheimer, R.: Zur Quantentheorie der Moleküle. Annalen der Physik. **84** (1927) 457
3. Pauli, W.: Exclusion Principle and Quantum Mechanics. Neuchatel, Griffon, 1st. Ed., Nobel Prize Lecture. (1947)
4. Slater, J.C.: The Self Consistent Field for Molecules and Solids: Quantum Theory of Molecules and Solids Mc Graw-Hill, New York, **4** (1974)
5. Parr, R.G., Yang, W.: Density Functional Theory of Atoms and Molecules. Oxford University Press, New York, (1989)
6. Hartree, D.R.: Proc. Camb. Phil. Soc., **24** (1928) 89
7. Fock, V.: Näherungsmethoden zur Lösung des Quantenmechanischen Mehrkörperproblems. Z. Physik, **61** (1930) 126 **62** (1930) 795
8. Davidson, E.R., Feller, D.: Basis Set Selection for Molecular Calculations. Chem. Rev., **86** (1986) 681–696
9. Shavitt, I.: The Gaussian Function in Calculations of Statistical Mechanics and Quantum Mechanics. Methods in Computational Physics, academic, New York, **2** (1963) 1–44
10. Saunders, V.R.: An Introduction to Molecular Integral Evaluation. Computational Techniques in Quantum Chemistry and Molecular Physics, Reidel – Dordrecht (1975) 347 – 424
11. McMurchie, L.E., Davidson, E.R.: One–and Two–Electron Integrals over Cartesian Gaussian Functions. J. Comp. Phys. **26** (1978) 218 – 231
12. Obara, S., Saika, A.: Efficient recursive computation of molecular integrals over Cartesian Gaussian functions. J. Chem. Phys. **84** (7) (1986) 3963 – 3974
13. Hehre, W.J., Ditchfield, R., Pople, J.A.: J. Chem. Phys. **56** (1972) 2257
14. Francl, M.M., Petro, W.J., Hehre, W.J., Binkley, J.S., Gordon, M.S., DeFrees, D.J., Pople, J.A.: J. Chem. Phys. **77** (1982) 3654
15. Geist, A., Beguelin, A., Dongarra, J., Jiang, W., Manchek, R., Sunderam, V.: PVM: Parallel Virtual Machine. A Users' Guide and Tutorial for Networked Parallel Computing MIT Press (1994)

# Computational Issues in Optimizing Ophthalmic Lens

Enric Fontdecaba *, José M. Cela, and Juan C. Dürsteler

[1] Universitat Politecnica de Catalunya enricf@ac.upc.es
[2] Universitat Politecnica de Catalunya cela@ac.upc.es
[3] Industrias de Óptica S.A. Dus@indo.es

**Abstract.** This paper reports the development of a non linear non constrained optimization code in a parallel platform. This code is used in a industrial design, different approaches are presented and the results obtained are shown.
*Keywords: Non linear non constrained optimization, Parallel Algorithms, Parallel Linear Solvers.*

## 1 Introduction

Nowadays, the competitive market demands a fast design cycle. The companies are forced to design products in short time. The new computational resources can help companies to gain competitive advantage. This paper explains the Progressive Addition Lens (hereinafter, PAL) design process, the key product for the ophthalmic lens manufacturers in Europe. We will show the complexity of the optimization algorithms involved and how a parallel approach can improve the design capability.

Industrias de Optica S.A. is the biggest Spanish lens manufacturer, its flagship product being the progressive lens. This kind of lens is used to compensate the presbiopya, resulting from the aging of the eye. This product is growing its market share and becoming the most important product for the ophthalmic lens manufacturers.

A progressive lens has three different vision zones. With the first zone the user can see distant objects. With the second zone the user can see at intermediate distances, due to a progressive change of optical power. The last zone is used for near vision. It is known that there is no analytical solution that gives the best possible progressive lens, so it is mandatory to use an optimization algorithm [2].

In addition to these three zones, used in phoveal vision, there is a fourth zone, the lateral zone. All the effort in the optimization process is devoted in reducing the astigmatism in this zone, improving the overall lens performance. In figure 1 the different zones can be observed.

---

* Also in Industrias de Óptica S.A.

P. Zinterhof, M. Vajteršic, A. Uhl (Eds.): ACPC'99, LNCS 1557, pp. 173–182, 1999.

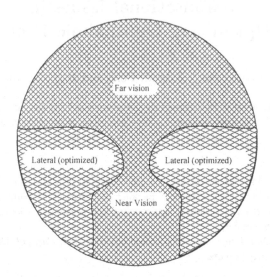

**Fig. 1.** Progressive Addition Lens vision zones

In the Progressive Addition Lens design process, it is necessary to optimize the lens surface for several sets of design parameters. This is an iterative process, so it is very important to use the fastest possible algorithm. This is the motivation that led us to a parallel approach.

## 2   Facing the Problem

In this section the different issues related to the model and the problem solving will be presented. First, the lens mathematical model will be explained. Second, the cost function and the optimization process will be detailed.

### 2.1   The Lens Surface

The issue in this optimization problem is to build an appropriate surface, thus a surface modeling tool is needed. We use B-Splines as the basis to model the optical surface. With this basis, two interesting properties are achieved:

**Local control.** When moving a control point of a B-Splines model, only the neighboring zones are modified. A control point of the surface is a degree of freedom in our optimization problem. The local control behavior leads to a quasi-regular sparse structure in the hessian matrix of the objective function.

$C^2$ **class surface.** The B-Splines are twice derivable. This property is needed in order to calculate the surface curvature. The optical power and the astigmatism, the lens properties measured, are related to the surface curvature. The computation of this curvature is the source of the non-linearity.

The B-Splines modeling allows enough flexibility. Also, as our algorithms improve, it is possible to obtain a more accurate model increasing the number of control points [3].

## 2.2   The Cost Function

It is necessary to develop a quality metric for the different lens surface designed. In our problem, we measure the ophthalmic power and the astigmatism, both of them computed with the surface curvature. We are interested in the whole surface behavior, so we integrate this properties in all the surface.

$$\sum_{i,j} Weight_{i,j} Property_{i,j}(x) \tag{1}$$

In equation 1 the vector $x$ represents the parameters of the optimization problems. Hereinafter, $n_p$ will denote the number of points involved in this sumatory and $n_c$ the vector dimension. In the problems tested, $n_p \approx 4n_c$.

The nonlinearity in this problem comes out when computing the $Property_{i,j}(x)$ function. This property, usually the ophthalmic power or astigmatism, is related to a non linear combination of the surface directional derivatives. And this surface is, as explained in the previous section, a sumatory of piecewise polynomial functions.

An important issue when choosing which optimization algorithm to use is the cost of the objective function. Let,

- $n_e$ , number of evaluation points needed in each partial derivative in order to construct the hessian matrix.
- $S$ , computing cost of the directional derivatives of the surface function in one evaluation point.
- $K$ , computing cost of the partial derivative of one of the terms of the sumatory of the cost function.

With these definitions, and taking the assumption that $n_p = 4n_c$, it can be shown that the cost of computing the objective function is:

$$C_{function} = 4Sn_c \tag{2}$$

The cost of computing the objective function gradient is:

$$C_{gradient} = 4Kn_c \tag{3}$$

The cost of computing the hessian with finite differences over the gradient:

$$C_{hessian} = 4K(n_c^2 + n_c) \tag{4}$$

And the cost of computing the hessian taking advantage of the special function structure is:

$$C_{Simpified-Hessian} = 8K(n_c + n_e) \tag{5}$$

Moreover, $n_e$ has a maximal value of 144 and a typical value of 90 while $n_c$ has a value of 1000 for the current designs. According to this result, computing a hessian with the simplified formula costs less than three times computing a gradient. This is the main reason that make the Newton approach so interesting to us.

It is also important to remark that, with the simplified hessian approach, the computational cost grows linearly with the number of control points, allowing us to face bigger problems. Nevertheless, the $LDL^t$ decomposition associated with the Newton algorithm remains with a cubic complexity.

## 2.3   Optimization Algorithms

After a complete review of the different non linear non constrained optimization algorithms available, the Newton algorithm was selected. The algorithms reviewed were, Polytope Method [7], Steepest Descent, Conjugate Gradient, Quasi-Newton and Newton [9].

In the numerical test it was clear the enormous possible gain we could achieve with an analytical derivative form of the cost function. We managed to obtain the derivative with exact formulas and had those analytical expressions programmed. With this modification the computing time for a Hessian was reduced enough to make the Newton algorithm preferable to a Quasi-Newton approach [6] [4].

In further developments we managed to compute the Hessian matrix with a simplified finite differences approach, taking advantage of the special cost function structure. This modifications reduced the Hessian computing cost to the equivalent of computing three to four gradients with the current problem size.

The last sequential algorithm implemented was a Newton algorithm with the modification proposed by Gill & Murray [4]. This modification improves the convergence rate, reducing the number of iterations. Gill & Murray use a $LDL^t$ decomposition to solve the Newton equation, but the diagonal term is modified in order to make the hessian positive definite. This technique is related with the trust region method. The reduction in the number of iterations performed ranges from 20 % for easy problems to 70 % for difficult problems.

We did not test Truncated-Newton or Limited Memory Quasi-Newton methods [8]. The computational costs of the hessian are low enough to expect no improving in using this methods.

# 3   Improving Performance

In an industrial product it is very important to reduce the design time. The company is very interested in improving the performance of its design process, so we decided to take advantage of the parallel computing platforms.

The targeted platform was a workstation cluster, so we chose PVM as the message passing environment for the new application [11].

We started the parallelization effort at the same time we were developing the sequential algorithms, the first algorithm ported to the parallel machine was the Pure Newton. As we conclude that the Gill & Murray modification was very important for the algorithm performance, this new algorithm was also ported.

## 3.1   Parallelization of Pure Newton Algorithm

In order to start the parallelization, a profile of the sequential version was performed. As a result of this profile it was clear that the biggest part of the CPU time was spent on building the Hessian. Those routines where the first ones to be parallelized. With the first parallel program, performance measurements were done to study its behavior. We used the analysis tools available on the CEPBA (European Center for Parallel Computing of Barcelona) [10], the Dimemas and Paraver tools, to perform those tests.

**Objective Function Parallelization** The numerical test showed that a very important part of the calculation time was spent in computing the objective function. Furthermore, the most important part is the Hessian computation. So, the first parallel approach faced the reduction of this time.

The Hessian is computed by finite differences of the gradient. In order to improve the performance, an analytical gradient routine was implemented. It is worth noting that symbolical mathematical packages like Mathematica or Maple failed to compute this analytic derivative.

In order to obtain a finite difference Hessian approach, it is necessary to calculate $n + 1$ ( $n$ is the problem dimension) function gradients. Those calculations are independent, so they are split among the different available processors. A master-slave approach is used. The other computations needed by the algorithm, the linear search and the linear equations system, are computed by the master. In table 1 the speed-up results of different problem sizes are shown. The tests were performed for 2,4,8,12 and 16 processors in order to study the algorithm scalability.

Studying the code and profiles, it was clear that as the function evaluation costs diminished due to its parallelization, the linear solver costs became more important in the total computing time. The traces obtained in our performance analysis tool corroborate this conclusion. In order to improve the scalability, the parallelization of the linear system solver was decided upon.

**Table 1.** Speed-Up with Function Parallelization

| Dimension | 2 Proc | 4 Proc | 8 Proc | 12 Proc | 16 Proc |
|-----------|--------|--------|--------|---------|---------|
| 70        | 1.69   | 2.63   | 3.75   | 3.38    | 3.95    |
| 140       | 1.85   | 3.16   | 5.00   | 6.13    | 6.96    |
| 390       | 1.92   | 3.56   | 6.12   | 8.10    | 9.18    |
| 1390      | 1.95   | 3.78   | 6.90   | 9.08    | 12.39   |

**Linear Solver Parallelization** In order to achieve a better scalability we parallelized the linear solver. We used preconditioned Krylov subspace iterative methods as linear solvers (GMRES(m)). The selected preconditioners are a set of different Incomplete Factorizations. The parallelization of the linear solvers is based on a Domain Decomposition data distribution. [1] The main bottleneck of the linear solver is the solution of the sparse triangular linear system arising from the preconditioner. The communication requirements of this operation depend on the block structure of the triangular factors. In order to minimize this bottleneck two strategies are used:

1. Control the fill-in at the block level with a different criteria than at the element level. Each new non-zero block implies a new communication between processors.
2. Perform a coloring of the domains which minimizes the fill-in at the block level and ensures the maximum parallelism.

Because the granularity of the Hessian assembly and the linear solver is quite different, we use a different number of processes in each phase. This means that additional communications are required to redistribute the data before and after the linear system solution phase. We must find for each problem size the optimum number of processes of each part in order to obtain the minimum execution time. In this way we can improve the scalability of the whole application.

The results are shown in table 2 and table 3. In the first one, only two of the available processors are used for the linear solver. In the other table, four processors are used. The results with the smaller data sets are not shown because due to their size they did not achieve any reasonable speed-up.

**Table 2.** Speed-Up with 2 processors for the Linear Solver

| Dimension | 2 Proc | 4 Proc | 8 Proc | 12 Proc | 16 Proc |
|-----------|--------|--------|--------|---------|---------|
| 390       | 1.84   | 2.91   | 4.08   | 4.63    | 4.20    |
| 1390      | 1.97   | 3.58   | 6.06   | 7.88    | 9.27    |

**Table 3.** Speed-Up with 4 processors for the Linear Solver

| Dimension | 2 Proc | 4 Proc | 8 Proc | 12 Proc | 16 Proc |
|-----------|--------|--------|--------|---------|---------|
| 390       | 1.73   | 2.58   | 3.45   | 4.02    | 3.34    |
| 1390      | *No convergence* | | | | |

In comparison with the results shown in table 1 this new parallelization introduces a slow-down in the total solving time. All the parallel performance is obtained by the first approach.

We found no way to avoid this behavior. In one hand, the condition number of the Hessian matrix is very high (even without any reordering) and this number rapidly increases with the problem size. In order to obtain convergence with any iterative method with this kind of matrix it is necessary to use a good preconditioner. A better preconditioner increases the filling on the Incomplete Factorization used. A more dense matrix implies more communication cost and longer execution time.

On the other hand, using several processors implies a data distribution, it is important to reorder the matrix in order to minimize the communication between processors but the reordering affects the numerical stability of the decomposition.

Summing up, without any reordering in the Hessian matrix the computing time is too big, due to the communication cost. With reordering in the matrix the iterative solver does not converge.

So, we achieve no increases in speed in parallelizing the linear solver. Analyzing the results and the code, we find two reasons for this behavior:

- As we use an iterative method, the number of iterations needed in order to solve the linear system is a key parameter. The parallelization, involving a matrix reordering increased the number of iterations. Furthermore, in the bigger case (when we expected some performance improvements), the reordering affected the algorithm convergence in such a way that made it diverge.
- With the solver parallelization, the number of communications is greatly increased. In the Hessian parallelization there are two communications, at the beginning and the end of the parallel phase. With the linear solver, there are communications in each linear solver iteration.

Summarizing, the linear system involved in the optimization algorithm is too small and too badly conditioned to be solved with a parallel iterative method.

### 3.2   Parallelization of the Gill & Murray Approach

When it became clear that the Gill & Murray approach leaded to better performance in the sequential algorithm we decided to parallelism this algorithm as well. Two improvements were included in this new parallel algorithm, the simplified hessian and the modified $LDL^t$ decomposition.

**Objective Function Parallelization** As we mentioned before, the Hessian is computed with a simplified finite difference scheme. In this scheme we take advantage of the special form of the objective function. This approach can be parallelized in a similar way to the parallelization performed for the Pure Newton Algorithm.

**Table 4.** Speed-Up of the G & M (problem dimension: 1390)

| Dimension | 2 Proc | 4 Proc | 6 Proc | 8 Proc |
|---|---|---|---|---|
| Speed-Up | 1.51 | 2.06 | 2.37 | 2.53 |
| Efficiency | 0.75 | 0.51 | 0.39 | 0.32 |

**Decomposition Parallelization** We are still working on this approach. But, taking into account the experience gathered with the Simple Newton parallelization, we expect no improvement in using several processors to perform the decomposition. Also the small size of the linear systems is a drawback for the parallelization of the linear solver.

# 4  Conclusion and Future Work

Comparing the three different approaches, it is clear that the best parallel behavior is obtained by the first one, the pure Newton with only the objective function performing in parallel. That's Amdahl's law. In the first approach we had about 95 % of the time spent in computing the hessian (Maximum ideal speed-up 20). This work could be performed in parallel with little overhead. In the last approach, the Gill & Murray modification with the simplified hessian computation, the hessian costs only 65 % of the total time. (Maximum ideal speed-up 2.8).

Nevertheless, as the sequential execution time is greatly reduced by the algorithmic improvements, the third approach is more performing in whatever computing platform you choose. The Gill & Murray, with the simplified hessian approach working in 4 processors is faster than the Pure Newton working with 16 processors.

The computing time obtained is satisfactory for the industrial process. It is not expected to use more than 12 machines at the same time. In fact INDO is installing a network of 6 DEC Alpha workstation with a Fast Ethernet switch. Taking the previous results into account, with the targeted platform, the third parallel approach is the most suitable for the company.

It is also interesting to remark the problems with the parallel linear solver. In our previous experience with linear systems from numerical simulations we have never found such a bad conditioned problem. In order to overcome this behavior we are thinking about new reordering methods.

The future work includes an upgrade of the basic sequential algorithm, and the changes needed by this improved approach. This new approach could consist in a non-linear least square formulation of the problem. We expect to follow this work with a constrained optimization formulation. Now, as the initial problem is successfully solved, we are trying to improve the model in order to obtain better designs. This initiative is partially supported by the EC under the OPTIPAR subproject inside the TTN project.

# 5  Acknockledgements

We would like to thanks all the R & D staff of INDO for their support in the development of this project. Specially to Roberto Villuela for their contribution to the tests and in developing key parts of the code.

# References

1. Cela, José M.; Alfonso, J. M.; Labarta, J.: *PLS: A Parallel Linear Solvers library for domain decomposition methods*: EUROPVM'96, Lecture Notes in Computer Science 1156, Springer-Verlag, 1996.
2. Dürsteler, Juan Carlos.: *Sistemas de Diseño de Lentes Progresivas Asistido por Ordenador.*: PhD Thesis. Universitat Politecnica de Catalunya, 1991.
3. Farin, Gerald.: *Curves and Surfaces for Computer Aided Geometric Design. A Practical Guide.*: Second Edition. Academic Press, 1990.
4. Gill, Philip E.; Murray, Walter & Wright, Margaret H.: *Practical Optimization.*: Academic Press, 1981.
5. Dennis & Schnabel.: *Numerical Methods for Unconstrained Optimization and Non Linear Equations.*: Prentice-Hall, 1983.
6. Nemhauser, G.L.; Rinnooy Kan, A.H.G. & Todd, M.J.: *Optimization.*: Elsevier Science Publishers B.V. 1989.
7. Nelder, J.A.; Mead, R. Computer Journal, vol 7 p308. 1965.
8. Liu, D.C.; Nocedal, J.: *On the Limited Memory BFGS Method for Large Scale Optimisation*: Mathematical Programming, 1989.
9. Nocedal, J.: *Theory of Algorithms for Unconstrained Minimization*: Acta Numerica, 1992.
10. Labarta, J., Girona, S., Pillet, V., Cortes, T., Cela, J.M.: *A Parallel Program Developement Environment*: CEPBA/UPC Report No. RR-95/02 (1995)
11. Gueist, A., Beguelin, A., Dongarra, J., Jiang, W., Manchel, R., Sunderam, V.: *PVM 3 User's Guide and Reference Manual*: Oak Ridge National Laboratory, TM-12187 (May 1994)

# Parallel Finite Element Modeling of Solidification Processes

Roman Wyrzykowski[1], Norbert Sczygiol[1], Tomasz Olas[1], and Juri Kanevski[2]

[1] Technical University of Czestochowa,
Dabrowskiego 73, 42-200 Czestochowa, Poland
roman@matinf.pcz.czest.pl, {sczygiol | olas}@imipkm.pcz.czest.pl
[2] Technical University of Koszalin,
Partyzantow 17, 75-411 Koszalin, Poland
kaniewsk@tu.koszalin.pl

**Abstract.** In the paper, parallelization of finite element modeling of solidification is considered. The core of this modeling is solving large sparse linear systems. The Aztec library is used for implementing the model problem on massively parallel computers. Now the complete parallel code is available. The performance results of numerical experiments carried out on the IBM SP2 parallel computer are presented.

## 1   Introduction

In the previous work [15], a basic formulation has been proposed for the mathematical description of solidification, based on the finite element method (FEM). This formulation gives the opportunity of creating effective computer software for the simulation of solidification. Taking into account the huge amount of computation needed by FEM in this application, such a software should be parallel in order to satisfy user demands with respect to the time of problem solving. This paper focuses on the parallel solution of large sparse linear systems occurred while using FEM for modeling solidification processes.

Direct methods for solving linear systems, like standard Gaussian elimination [3,5] lead to fill-in, and this makes these methods often expensive. For linear sparse systems, e.g. those arising in finite element approximations of partial differential equations, a promising alternative may be to use one of iterative methods [2], which work by repeatedly improving an approximate solution until it is accurate enough. These methods access the coefficient matrix $\mathbf{A}$ of the linear system only via a rather simple subroutine, based [3,19] on the matrix-vector product $\mathbf{y} = \mathbf{Ax}$. Hence the user needs only to supply this subroutine, which does not affect the matrix $\mathbf{A}$ in the course of computation. This advantage permits [3,20] exploitation of the sparsity or other special structure of $\mathbf{A}$, and simplifies parallel implementation of iterative solvers.

For sparse matrices, it is a common practice to store only the nonzero elements with information about their locations in a matrix. A variety of storage schemes are used [9] to store and process sparse matrices. These specialized

P. Zinterhof, M. Vajteršic, A. Uhl (Eds.): ACPC'99, LNCS 1557, pp. 183–195, 1999.
© Springer-Verlag Berlin Heidelberg 1999

schemes not only save storage, but also yield computational savings because unnecessary multiplication and additions with zero can be avoided.

In this paper, parallelization of finite element modeling of solidification is considered, for message passing paradigm of parallel programming. The core of this modeling is solving large sparse linear systems. One of possible approaches to this problem is to use dedicated parallel libraries. The P-SPARSLIB library [11] from the University of Minnesota, the Aztec package [8] from the Sandia National Laboratories, and the PETSc library [1] from the Aragonne National Laboratories are typical representatives of these parallel software packages built on the top of the MPI communication layer [6].

This paper is organized in the following way. Section 2 presents shortly the FEM-based approach used to numerical modeling of solidification, while in Section 3, some basic problems occurred in parallel implementation of the modeling are discussed, as well as possible alternatives to their solution. Section 4 describes our finite element parallel application based on the Aztec library. Section 5 gives conclusions.

## 2   Numerical Modeling of Solidification

Solidification is governed by a quasi-linear heat conduction equation containing the term of heat source $\dot{q}$, which describes the rate of latent heat evolution

$$\nabla \cdot (\lambda \nabla T) + \dot{q} = c\rho \frac{\partial T}{\partial t} , \tag{1}$$

where $T$ is the temperature, $t$ is time, $\lambda$ is the thermal conductivity coefficient, $\rho$ is the density, and $c$ is the specific heat. The heat source term is equal to

$$\dot{q} = \rho_s L \frac{\partial f_s}{\partial t} , $$

where $L$ is the latent heat, and $f_s$ is the solid phase fraction. The subscript $s$ refers to the solid phase. These two equations, with appropriate boundary and initial conditions, build together the temperature description of solidification. Taking into consideration the enthalpy, defined as

$$H(T) = \int_{T_{ref}}^{T} c\rho dT + \rho_s L(1 - f_s(T)) , \tag{2}$$

where $T_{ref}$ is the reference temperature, enables an enthalpy description of the solidification process. Differentiating the enthalpy given by Eq.(2) with respect to temperature, we obtain one of the enthalpy formulations [16], the so-called apparent heat capacity formulation

$$\nabla \cdot (\lambda \nabla T) = c^*(T) \frac{\partial T}{\partial t} , \tag{3}$$

in which $c^*(T) = dH/dT$ is called the "effective" heat capacity and accounts the latent-heat evolution.

Eq.(3) is transformed into FEM equations, for example, by the weight residue method. This method is based on finding an approximate solution of the differential equation, which describes the phenomenon of solidification in the region $\Omega$, by multiplying this equation by a certain weight function and next by integrating it over $\Omega$. The Galerkin method [10] is one of the most popular methods; it is used for further investigations. The application of the Green-Gauss theorem enables transformation of the above equation into an ordinary differential equation. Such a procedure is called semi-discretization because it is applied only to the discretization over the space.

The semi-discretization of Eq.(3) gives [15]:

$$\mathbf{M(T)\dot{T} + K(T)T = b(T)} , \tag{4}$$

where $\mathbf{M}$ is the mass matrix, $\mathbf{K}$ is the conductivity matrix, $\mathbf{T}$ is the temperature vector, and $\mathbf{b}$ is a vector whose values are calculated on the basis of boundary conditions. If $\mathbf{N}$ denotes the finite element interpolation function, the matrices $\mathbf{M}$ and $\mathbf{K}$ are expressed as

$$M(T) = \int_{\Omega^e} c^*(T)\mathbf{N}^T \cdot \mathbf{N} d\Omega , \tag{5}$$

$$K(T) = \int_{\Omega^e} \lambda(T)\nabla\mathbf{N}^T \cdot \nabla\mathbf{N} d\Omega . \tag{6}$$

There are some possibilities to approximate the "effective" heat capacity. The biggest problem here is to avoid any oscillations in the numerical solution. Del Guidice et al. [14] proposed to approximate $c^*$ as

$$c^* = \frac{\nabla H \cdot \nabla T}{\nabla T \cdot \nabla T} , \tag{7}$$

Many numerical simulations have shown the reasonable accuracy of this method.

The ordinary differential equation received as a result of semi-discretization is integrated over time using one of the linear k-step methods [18]:

$$\sum_{j=0}^{k} \alpha_j u_{n+j} = h \sum_{j=0}^{k} \beta_j f_{n+j}, \qquad n = 1, 2, \ldots ,$$

where $u$ is the analysed unknown, $f$ is the $u$- and $t$- dependent function, $h$ is the integration step, $\alpha$ and $\beta$ are certain parameters fulfilling the general stability conditions. For $k = 1$ we obtain a class of two-level integration schemes (e.g. Euler-backward), while for $k = 2$ we get a class of three-level integration schemes (e.g. Dupont II). The use of two-step schemes allows long time iterative computation in every time step to be avoided. The application of these schemes requires the use of an one-step scheme in the first calculation step for obtaining results at the time level $n + 1 = 2$. The initial conditions are applied at the time level $n = 1$.

The use of the Euler-backward and Dupont II schemes to Eq.(4) gives respectively

$$(\mathbf{M}^n + \Delta t \mathbf{K}^n) \, \mathbf{T}^{n+1} \; = \; \mathbf{M}^n \mathbf{T}^n + \Delta t \mathbf{b}^{n+1} \, , \tag{8}$$

$$\left( \mathbf{M}^0 + \frac{3}{4} \Delta t \mathbf{K}^0 \right) \mathbf{T}^{n+2} \; = \; \mathbf{M}^0 \mathbf{T}^{n+1} - \frac{1}{4} \Delta t \mathbf{K}^0 \mathbf{T}^n + \frac{1}{4} \Delta t \left( 3 \mathbf{b}^{n+2} + \mathbf{b}^n \right) \, . \tag{9}$$

The superscript $(^0)$ indicates that in calculation we use material properties evaluated for the temperature $T = (3/2) \, T^{n+1} - (1/2) \, T^n$.

## 3    Parallel Finite Element Modeling

According to the previous section, the core of the finite element modeling of solidification is solving repeatedly a system of linear equations. For example, in case of the Euler-backward scheme we have the following system:

$$\mathbf{A}^n \mathbf{T}^{n+1} \; = \; \mathbf{b}^n \, , \tag{10}$$

which corresponds to expression (8), for $n = 1, 2, \dots$. Before solving, the coefficient matrix $\mathbf{A}^n$ and the right-hand side vector $\mathbf{b}^n$ are calculated.

Parallel implementation of these computations is based [17] on the concept of dividing a large and computationally time consuming finite element problem into smaller and more manageable sub-problems which may be solved efficiently. Using this approach, a finite element mesh is divided into a finite number of sub-domains (or sub-meshes) such that the computational load per sub-mesh is approximately the same, these sub-domains are then solved concurrently over different processors.

The task of partitioning unstructured meshes for parallel finite element computations is not straightforward [17], since not only the load per subdomains has to be kept the same. Also, interprocessor communications have to be minimized. These communications are results of interaction (or coupling) between boundary nodes of sub-meshes.

Fig.1 shows an example of such an partitioning when a finite element mesh containing 13 nodes and 16 triangular elements is divided into four sub-domains each of them is mapped into a corresponding processor. Based on Fig.1, for each sub-domain, we distinguish between three types of nodes:

1. Internal nodes are those that are coupled only with nodes belonging to a given sub-domain.
2. Border nodes are those nodes in a given sub-domain which are coupled with nodes in other sub-domains. Internal and border nodes are called local ones.
3. External nodes are those nodes in other sub-domains which are coupled with local nodes.

In case shown in Fig.1, the sub-mesh corresponding to processor 2 consists of two internal nodes 8 and 10, as well as three border nodes 5, 6, 11. For this sub-mesh, its external nodes are 0, 1, 3, 4, 7, 9, 12.

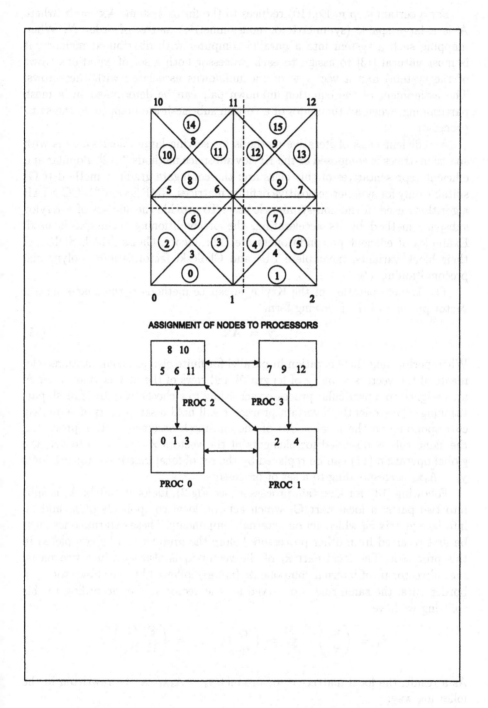

**Fig. 1.** Example of finite element mesh partitioning

For a certain step n, Eq.(10) reduces to the linear system $\mathbf{Ax} = \mathbf{b}$ where $\mathbf{A}$ is a large sparse (symmetric or nonsymmetric) matrix of order $N$. When mapping such a system into a parallel computer with distributed memory, it is most natural [13] to assign to each processor both a set of equations (rows of the system) and a vector $\mathbf{x}$ of the unknowns associated with these rows. The assignment of the equation-unknown pair can be determined by a mesh partitioning, when all the nodes of a certain sub-mesh are mapped to the same processor.

An efficient class of iterative methods for solving large linear systems with sparse matrices is composed of the Krylov subspace methods [2,12]. Popular and efficient representatives of this class are: the conjugate gradient method (CG) suitable only for symmetric positive definite matrices, GMRES and Bi-CG-STAB algorithms used in the non-symmetric case. To provide the success of a Krylov subspace method by its accelerating, the preconditioning technique is used. Examples of efficient preconditioners are [12]: Jacobi, Gauss-Seidel, SOR and their block variants, incomplete LU- and Choleski factorizations, polynomial preconditioning, etc.

The kernel operation of the Krylov subspace methods is the sparse matrix vector product of the following form:

$$\mathbf{y} = \tilde{\mathbf{A}}\mathbf{x}. \tag{11}$$

When performing this operation in parallel for finite element computations, elements of the vectors $\mathbf{x}$ and $\mathbf{y}$ of length $N$ and rows of the matrix $\tilde{\mathbf{A}}$ of order $N$ are assigned to a particular processor according to a chosen variant of mesh partitioning. Consequently, a certain processor will hold a set of rows of $\tilde{\mathbf{A}}$-matrix corresponding to the local nodes of the sub-mesh assigned to this processor; the same rule is observed for elements of the vectors $\mathbf{x}$ and $\mathbf{y}$. Moreover, the global operation (11) can be replaced by the set of local matrix-vector products $\mathbf{y}_i = \tilde{\mathbf{A}}_i \mathbf{x}_i$ corresponding to a set of processors.

Following [13], for a certain processor (see Fig.2), its local matrix $\tilde{\mathbf{A}}_i$ is split into two parts: a local part $\mathbf{C}_i$ which acts on local components of $\mathbf{x}$, and an interface matrix $\mathbf{E}_i$ which act on external components. These external ones must be first received from other processors before the product can be completed in this processor. The local part $\hat{\mathbf{x}}_i$ of the vector $\mathbf{x}_i$ is also split into two parts: the subvector $\mathbf{u}_i$ of internal components (nodes) followed by the subvector $\mathbf{v}_i$ of border ones; the same rule is observed for the vector $\mathbf{y}_i$. So, according to this splitting we have

$$\hat{\mathbf{x}}_i = \begin{pmatrix} \mathbf{u}_i \\ \mathbf{v}_i \end{pmatrix}, \quad \mathbf{y}_i = \begin{pmatrix} \mathbf{r}_i \\ \mathbf{q}_i \end{pmatrix}, \quad \mathbf{C}_i = \begin{pmatrix} \mathbf{F}_i\ \mathbf{G}_i \\ \mathbf{H}_i\ \mathbf{K}_i \end{pmatrix}.$$

As a result, the local matrix-vector product $\mathbf{y}_i = \tilde{\mathbf{A}}_i \mathbf{x}_i$ can be rewritten in the following way:

$$\begin{pmatrix} \mathbf{F}_i\ \mathbf{G}_i \\ \mathbf{H}_i\ \mathbf{K}_i \end{pmatrix} \begin{pmatrix} \mathbf{u}_i \\ \mathbf{v}_i \end{pmatrix} + \begin{pmatrix} 0 \\ \sum_{j \in NB_i} \mathbf{G}_{ij}\mathbf{v}_j \end{pmatrix} = \begin{pmatrix} \mathbf{r}_i \\ \mathbf{q}_i \end{pmatrix}. \tag{12}$$

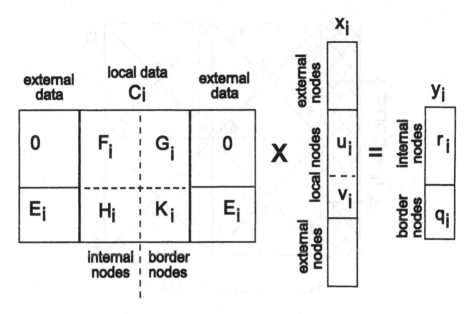

**Fig. 2.** Local matrix-vector multiplication

Here $\mathbf{G}_{ij}\mathbf{v}_j$ is a contribution to the local matrix-vector product (11) from the neighbouring sub-mesh number $j$, and $NB_i$ is the set of sub-domains that are neighbours to sub-mesh $i$. After grouping all the subvectors $\mathbf{v}_j$ of border components in the neighbouring sub-meshes into one vector $\mathbf{v}_{i,ext}$, expression (12) reduces to

$$\mathbf{C}_i\mathbf{x}_i + \mathbf{E}_i\mathbf{v}_{i,ext} = \begin{pmatrix} \mathbf{r}_i \\ \mathbf{q}_i \end{pmatrix}, \tag{13}$$

where

$$\mathbf{E}_i\mathbf{v}_{i,ext} = \sum_{j \in NB_i} \mathbf{G}_{ij}\mathbf{v}_j. \tag{14}$$

The key features of organizing data structures for implementing operation (13) efficiently are [13]: (i) to separate the border nodes from the internal ones; (ii) to order external nodes after local ones. This local ordering of nodes in a sub-mesh is completely different from the original global ordering (see Fig.1) and has important advantages, including more efficient interprocessor communication, reducing local indirect addressing during matrix-vector multiplication [11,8], and broad possibilities to overlap computations and communications. Hence, it is profitable to introduce a pre-processing stage into the parallel solution of a linear system, which is responsible for transforming the matrix $\tilde{\mathbf{A}}$ into its sparse distributed representation in accordance with the local ordering.

## 4   Parallel Application Based on the Aztec Library

The heat transfer problem corresponding to Eq.(1) without the heat source term $\dot{q}$ has been chosen as a model problem to verify and compare different approaches

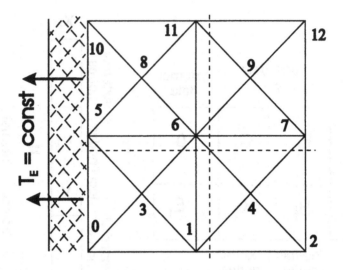

**Fig. 3.** Heat transfer problem

to parallelizing finite element computations. The numerical experiments have been carried out for a square plate illustrated in Fig.3. The left-hand side of this plate is subjected to a process of cooling which is described by the boundary condition of the third kind expressed as $-\lambda(\partial T/\partial n) = \alpha(T - T_E)$, where $T_E$ is the temperature of the environment.

To speedup as much as possible the way from formulation of the numerical model to its parallel implementation, it has been decided, for the starting period, to concentrate the efforts on developing the finite element code based on a parallel library dedicated for solving sparse linear systems occurred in problems of such a kind. The Aztec library [8] has been favoured before other possible candidates like P-SPARSLIB [11] and PETSC [1], for (i) possibility to use a wide spectrum of methods for solving sparse systems, (ii) moderate complexity, (iii) user-friendly documentation, (iv) relatively easy way of installing on both sequential and parallel machines, and (v) easy coupling with an advanced partitioning utility called Chaco [7]. The Aztec library and Chaco utility has been installed both on the INDY workstation in the Technical University of Czestochowa, and the IBM SP2 parallel MIMD computer. The main experiments have been carried out on the latter system located at the Wroclaw Supercomputing and Networking Center.

## 4.1    Aztec Library

Aztec is [8] an iterative library that greatly simplifies the parallelization process of solving the linear system of equations $\mathbf{Ax} = \mathbf{b}$ on distributed-memory message passing computers, where $\mathbf{A}$ is a user supplied $N \times N$ sparse matrix, $\mathbf{b}$ is a user supplied vector of length $N$, and $\mathbf{x}$ is a vector of length $N$ to be computed. The most complicated parallelization task for an Aztec user is the distributed matrix specification for a particular application. Aztec provides a collection of data transformation tools that allow creation of distributed sparse unstructured matrices for parallel solution. These tools accomplish also transition from the representation of data in the global distributed format, which is relatively easy to use, to the local distributed format, which is much more efficient when implementing iterative methods in parallel (see Section 3).

Aztec includes [7] a number of Krylov iterative methods such us CG, GMRES and Bi-CG-STAB to solve systems of equations. These methods are used in conjunction with various preconditioners such as polynomial ones or domain decomposition using LU or incomplete LU factorizations within subdomains. Aztec can use one of two different schemes for sparse matrix - either a point-entry modified sparse row format or block-entry variable block row format.

## 4.2 Parallel Finite Element Code

We started to develop the parallel finite element application on March 1998, beginning from the numerical model. Now a complete code written in the C language for the parallel solution of the model problem, as well as realization of some service functions, is available.

The kernel of this application are two pieces of code: the main program module and a set of routines, which are invoked from the main module to provide together the implementation of the following tasks:

1. describing the parallel machine (e.g. the number of processors),
2. initializing the matrix data structures, right hand side and initial guess,
3. choosing iterative methods, preconditioners and the convergence criteria,
4. performing in a loop whose index corresponds to time steps $n$ in Eq.(8) the following operations:
   - invoking the solver,
   - substituting the obtained solution $T^n$ into the right-hand side of Eq.(8) to update it.

Additionally, several software components have been developed to implement converters between different data formats used in the application. Indeed, the input information for this application is the description of a triangular mesh used to discretize the problem domain. This domain is meshed using the **Nuscas** package developed at the Technical University of Czestochowa. The Nuscas output is transformed into the data structures required by both the Chaco partitioning utility [7], and the main module of the application. Another converter is used to provide an interface between the Nuscas and the AVS Express visualization software. The AVS package is used here to visualize the results of domain partitioning in order to estimate their quality for eliminating a priori useless variants.

## 4.3 Performance Results

The parallel program was tested on the IBM SP2 machine. Tables 1-3 show the runtime, speedup and efficiency for different mesh sizes on SP2 with 1,4

| | N = 20201 | | | N = 100801 | | | N = 205441 | | |
|---|---|---|---|---|---|---|---|---|---|
| p | LSP | ILU | SymGS | LSP | ILU | SymGS | LSP | ILU | SymGS |
| 1 | 112.1 | 85.6 | 112.5 | 1136 | 877 | 1179 | 3506 | 2978 | 3450 |
| 4 | 31.9 | 45.7 | 52.7 | 228 | 329 | 369 | 699 | 878 | 979 |
| 8 | 29.7 | 38.3 | 47.0 | 163 | 217 | 263 | 439 | 541 | 641 |

**Table 1.** Runtime (in seconds) for the model problem on the SP2 machine as a function of number $p$ of processors and number $N$ of nodes in mesh, for different preconditioners

| | N = 20201 | | | N = 100801 | | | N = 205441 | | |
|---|---|---|---|---|---|---|---|---|---|
| p | LSP | ILU | SymGS | LSP | ILU | SymGS | LSP | ILU | SymGS |
| 1 | 1.00 | 1.00 | 1.00 | 1.00 | 1.00 | 1.00 | 1.00 | 1.00 | 1.00 |
| 4 | 3.51 | 1.87 | 2.13 | 4.98 | 2.67 | 3.20 | 5.02 | 3.39 | 3.52 |
| 8 | 3.77 | 2.23 | 2.39 | 6.97 | 4.04 | 4.48 | 7.99 | 5.50 | 5.38 |

**Table 2.** Speedup for the model problem on the SP2 machine

or 8 processors each with 64 MB RAM, for 10 time-steps of modeling. Since the efficiency of an iterative sparse linear solver depends significantly on a chosen variant of preconditioner, all experiments were carried out for different preconditioners, namely [8]: (i) least-squares polynomial one (or LSP shortly), (ii) non-overlapping domain decomposition preconditioner using the incomplete LU factorization (ILU), (iii) non-overlapping domain decomposition preconditioner using the symmetric Gauss-Seidel algorithm (SymGS). In all cases, the CG iterative method for solving linear systems was used. The numerical accuracy of modeling was verified by comparison of the computed temperatures with those obtained by the sequential **Nuscas** software.

The comparison of run times, speedups and efficiencies among different preconditioners (see Fig.4-5) shows that while in the sequential case the ILU preconditioner is the best, the LSP one is fastest for parallel modeling. The obtained performance results are promising and indicate that for sufficiently large meshes a significant reduction in runtime is achieved. For example, for our largest mesh with 205441 nodes, the use of the LSP preconditioner allows us to obtain the maximum speedup $S_p \approx p$ for $p = 8$ processors, and even a superlinear speedup

| | N = 20201 | | | N = 100801 | | | N = 205441 | | |
|---|---|---|---|---|---|---|---|---|---|
| p | LSP | ILU | SymGS | LSP | ILU | SymGS | LSP | ILU | SymGS |
| 1 | 100.0 | 100.0 | 100.0 | 100.0 | 100.0 | 100.0 | 100.0 | 100.0 | 100.0 |
| 4 | 87.8 | 46.8 | 53.3 | 124.6 | 66.6 | 79.9 | 125.5 | 84.8 | 88.1 |
| 8 | 47.2 | 27.9 | 29.9 | 87.1 | 50.5 | 56.0 | 99.8 | 68.8 | 67.3 |

**Table 3.** Efficiency (in %) for the model problem on the SP2 machine

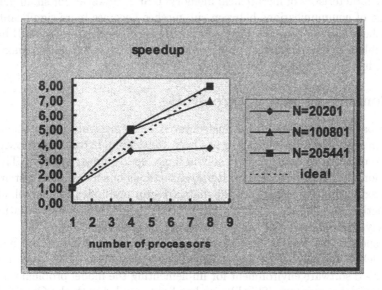

**Fig. 4.** Speedup for the polynomial preconditioner

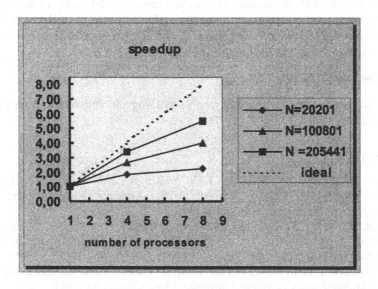

**Fig. 5.** Speedup for the domain decomposition ILU preconditioner

for $p = 4$. This anomalous phenomenon, when $S_p$ is greater than $p$, is probably due to characteristics of hierarchical memory in SP2. However, for small meshes, the time of communications between sub-meshes becomes more significant and makes the reduction decrease. For example, in case of the mesh with 20201 nodes, the doubling of the number of processors, from four to eight, allows maximum 11 % reduction in runtime (for the ILU preconditioner).

## 5  Conclusions and Future Works

In this paper, parallelization of unstructured finite element computations occurred in modeling of solidification has been considered. It has been shown that the core of these computations is solving large sparse linear systems. The use of the domain decomposition and iterative methods of solving such systems, as well as exploiting the features of the matrix-vector multiplication kernel, allows constructing parallel numerical algorithms feasible for their implementation on massively parallel processors as well as networks of workstations.

To speedup as fast as possible the way from formulation of the numerical model to its parallel implementation, the Aztec library has been used as a dedicated software environment for implementing the model problem on massively parallel computers. This library has been coupled with the Chaco utility responsible for static partitioning of finite element meshes. The performance results of numerical experiments carried out on the IBM SP2 parallel machine have been presented. The obtained results are promising and indicate that for sufficiently large meshes a significant reduction in runtime should be achieved.

Concurrently with developing the Aztec-based parallel code, an alternative approach using the from-scratch programming for networks of workstations running under the PVM [4] is being investigated. This approach is based on the object-oriented technique applied on the application level.

## Acknowledgements

This work was supported by the Woclaw Networking and Supercomputing Center under grant No. 05/98.

## References

1. Balay, S., Gropp, W., McInnes, L.C., Smith, B.: *PETSc 2.0 User Manual.* Tech. Report ANL-95/11 - Revision 2.0.22, Aragonne National Lab., 1998 (http://www.mcs.anl.gov/petsc/petsc.html)
2. Barret, R., Berry, M., Chan, T., Demmel, J., Donato, J., Dongarra, J., Eijkhout, V., Pozo, R., Romine, Ch., Van der Vorst, H.: *Templates for the Solution of Linear Systems: Building Blocks for Iterative Methods.* SIAM, Philadelphia (1994)
3. Demmel, J.W., Heath, M.T., Van der Vorst, H.: Parallel Numerical Linear Algebra. In: *Acta Numerica*, Cambridge University Press, Cambridge (1993)
4. Al Geist et al.: *PVM: Parallel Virtual Machine. A User's Guide and Tutorial for Networked Parallel Computing.* MIT Press, Cambridge MA (1994)
5. Golub, G., Ortega, J.M.: *Scientific Computing: An Introduction with Parallel Computing.* Academic Press, New York (1993)

6. Gropp, W., Lusk, E., Skjellum, A.: *Using MPI: Portable Parallel Programming with the Message-Passing Interface.* MIT Press, Cambridge MA (1995)
7. Hendrickson, B.: *The Chaco User's Guide.* Tech. Report SAND95-2344, Sandia National Laboratoires, 1995 (http://www.cs.sandia.gov/CRF/Aztec_pubs.html)
8. Hutchinson, S.A., Shadid, J.N., Tuminaro, R.S.: *Aztec User's Guide - Version 1.1.* Tech. Report SAND95-1559, Sandia National Lab., 1995 (http://www.cs.sandia.gov/CRF/Aztec_pubs.html)
9. Kumar, V., Grama, A., Gupta, A., Karypis, G.: *Introduction to Parallel Computing.* Benjamin/Cummings Publish. Comp., New York (1994)
10. Norrie D.H., de Vries, G.: *An Introduction to Finite Element Analysis.* Academic Press, New York (1978)
11. Saad, Y., Malevsky, A.V.: *P-SPARSLIB: A portable Library of Distributed Memory Sparse Iterative Solvers.* Tech. Report UMSI 95-180, Univ. Minnesota, 1995
12. Saad, Y.: *Iterative Methods for Sparse Linear Systems.* PWS Publishing, New York (1995)
13. Saad, Y., Lo, G.-L.: *Iterative Solution of General Sparse Linear Systems on Clusters of Workstations.* Tech. Report UMSI 96/117 & UM-IBM 96/24, Univ. Minnesota, 1996
14. Segal, A. Dalhuijsen, A.J.: Comparison of Finite Element Techniques for Solidification Problems. Int. J. Numer. Methods Eng. **23** (1986) 1807-1829
15. Sczygiol, N.: Numerical Modeling of Solidification of Castings. In: *Proc. Int. Workshop Parallel Numerics'97,* Zakopane, Poland, 1997, 255-262
16. Tamma, K.K., Namburu, R.R.: Recent Advances, Trends and New Perspectives via Enthalpy-Based Finite Element Formulations for Application to Solidification Problems. Int. J. Numer. Methods Eng. **30** (1990) 803-820
17. Topping, B.H., Khan, A.I.: *Parallel Finite Element Computations.* Saxe-Coburg Publications, Edinburgh (1996)
18. Wood, W.L.: *Practical Time-Stepping Schemes.* Clarendon Press, Oxford (1990)
19. Wyrzykowski, R.: On the Design of Systolic Array Architectures for the Iterative Solution of Linear Systems. In: *Proc. Int. Conf. Parallel Computing Technologies PACT'91,* Scientific World, 1991, 109-118
20. Wyrzykowski, R. Kanevski, J.: A Technique for Mapping Sparse Matrix Computations into Regular Processor Arrays. Lect. Notes in Comp. Sci. **1300** (1997) 310-317

# Architectural Approaches for Multimedia Processing

## (Invited Paper)

Sethuraman (Panch) Panchanathan, *Senior Member IEEE*

Visual Computing and Communications Laboratory
Department of Computer Science and Engineering
Arizona State University
P.O. Box 875406, Tempe, AZ 85287-5406
E-mail: panch@asu.edu

**Abstract.** The recent advances in VLSI technology, high-speed processor designs, Internet/Intranet implementations, broadband networks (ATM and ISDN) and compression standards (JPEG, MPEG, H.261, H.263 and G.273) are leading to the popularity of multimedia applications. Examples include, video over the internet, interactive TV, distance learning, telemedicine, and digital libraries. Multimedia refers to a combination of various media types including text, audio, 2D and 3D graphics, animation, images and video. Visual media (image, video and graphics) proliferation in multimedia applications demands high-powered compute engines, large storage devices, and high bandwidth networks for processing, storage, and transport of image/video data. Visual media processing poses challenges from several perspectives, specifically from the points of view of real-time implementation and scalability. There has been several approaches to obtain speedups to meet the computing demands in multimedia processing ranging from media processors to special purpose implementations. Note that a variety of parallel processing strategies are adopted in these implementations in order to achieve the required speedups. The objective of this paper to present a summary of the various architectural alternatives that exist for multimedia processing.

## 1. Introduction

Multimedia processing is becoming increasingly important because of the wide variety of applications. A variety of media processing techniques are typically used in multimedia processing environments to capture, store, manipulate and transmit multimedia objects such as text, handwritten data, audio objects, still images, 2D/3D graphics, animation and full-motion video. Example techniques include speech analysis and synthesis, character recognition, audio compression, graphics animation, 3D rendering, image enhancement and restoration, image/video analysis and editing, and video transmission. Multimedia computing presents challenges from the perspectives of both hardware and software. Each media in a multimedia environment

P. Zinterhof, M. Vajteršic, A. Uhl (Eds.): ACPC'99, LNCS 1557, pp. 196-210, 1999.
© Springer-Verlag Berlin Heidelberg 1999

requires different processes, techniques, algorithms and hardware. Hence, it is crucial to design processor architectures that meets the computing requirements of the various media types.

Visual media in a multimedia system contains a significant amount of information, and correspondingly involves a large volume of data in contrast to the other media types. Uncompressed digital video requires 250 Mb/s to support studio quality transmission of NTSC images (480 lines x 720 pixels/line x 24 bits/pixel x 30 frames/s). Even a simpler application such as video telephony (240 lines x 360 pixel/line x 16 bits/pixel x 10) requires 14 Mb/s to transmit the digital video signal in raw format. The bandwidth and storage requirements of visual information typically make it difficult to manage the data in its raw form. However, there is considerable redundancy in video data, both from an information theoretic view point as well as from the perspectives of structural content and human perception. A number of image and video compression standards, e.g., MPEG-1, MPEG-2 [1] and H.263[2] have been recently proposed to compress the visual data for a variety of transmission and/or storage applications. There is ongoing research and standardization efforts targeted towards future multimedia applications including MPEG-4 [3] and MPEG-7 [4] with the objective of integrating compression and content access functionality. These techniques and standards will involve execution of complex video processing tasks in real-time. The challenges can range from waveform coding implementations to scene modeling and understanding. For example, the principal objective of model-based image coding [5] or intelligent image coding is to understand the scene by modeling the objects to achieve a higher level representation. In addition, there is an increasing interest in 3-D image and stereoscopic video processing.

The complexity, variety of techniques and tools, and the high computation, storage and I/O bandwidths associated with visual processing pose several challenges, particularly from the point of view of real-time implementation. This is the principal reason for the slant of most media processor development towards visual processing. Several processing solutions ranging from multimedia extensions to general purpose processors such as the Intel MMX, programmable DSP architectures such as the TI-C6x series, Media processors like the Philips Trimedia processor, and special purpose architectures such as the C-Cube MPEG decoder chip-sets have been proposed to implement a variety of multimedia (particularly visual) processing operations.

The paper is organized as follows. The issues in multimedia hardware design is presented in section 2. The design trends in multimedia processor architectures are detailed in section 3. The conclusions are presented in section 4 followed by the references.

## 2. Issues in Multimedia Hardware Design

Multimedia applications require efficient VLSI implementations for various media processing algorithms. Emerging multimedia standards and algorithms will result in hardware systems of high complexity. In addition to recent advances in enabling VLSI technology for high density and fast circuit implementations, special investigation of architectural approaches is also required. The important issues in multimedia hardware design are listed below:

- Processor choices: DSP, RISC, etc.
- Parallelization and Granularity: MIMD, SIMD, coarse grain such as multiprocessor architectures and fine grain like superscalar and VLIW architectures, etc.
- Memory selection: SDRAM, VRAM, etc..

The issue of flexibility which corresponds to whether the architecture is dedicated or programmable is dealt with in section 3.

## 2.1   Processor Selection

Programmable architectures have several units in common. In general, every programmable architecture consists of data path, memory, input/output and control path. Data path is responsible for all the operations performed on data for the purpose of data input, manipulation, analysis, processing and output. Control path is generating all necessary signals to control the interaction between modules. There is always a contest between the complexity of these two parts in the design of a processor. The larger data path leaves less space for control path and vice versa.

In this section, we present the possible options for the design of the processor in a multimedia system as shown in    Figure 1. The categorization scheme is based on the format of instruction-set, available registers and the structure of data path.

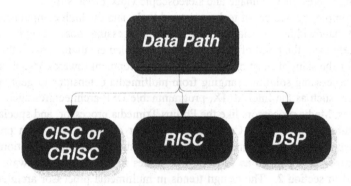

Figure 1- Data path selection

CISC (Complex Instruction Set Computing) microprocessors with more complex instruction sets provide better direct hardware support for the software developer than any other architecture. Instructions in a CISC processor are very powerful in their processing capability and support large numbers of registers and addressing modes. Control path in CISC processors is more complex in order to execute more powerful instructions. RISC (Reduced Instruction Set Computing) microprocessors offer faster execution of individual instructions by optimizing the processor for a smaller instruction set. DSP (Digital Signal Processing) microprocessors are optimized to perform digital processing operations such as filtering. Multiply and Accumulate (MAC) instruction is repeated in DSP algorithms and is performed in one cycle in DSP processors.

### 2.1.4  DSP

DSP processors are optimized for digital processing operations which include multiply and accumulate (MAC) operation. MAC operations $r = b + a \cdot x$, requires multiple clock cycles in CISC and RISC processors, while in a DSP data path it is executed in one clock cycle. Some characteristics of a DSP processor include:

- Multiple data and instruction buses.
- Parallel execution of MAC operation.
- Limited number of instructions.
- Efficient loop control.

## 2.2  Granularity

The granularity of a multimedia system defines the size of individual processing units by which tasks are executed. The granularity affects the number of processing units since, in any parallel architecture there is a tradeoff between the size and the number of processors.

Coarse grain systems are formed by a small number of large and complex processing units. In fine grain parallelism, there are large number of small processors. The intermediate possibilities between these two extremes can be referred to as medium-grain parallelism. Fine and medium grain parallelism have the potential of being faster, but they need more powerful compilers to divide small tasks between processing units efficiently. Most of multimedia processors are categorized as fine/medium grain processors. Each task is executed in parallel at the instruction level among several processing units. We now present the scheme of parallelism in these machines. Data level parallelism is presented in section 2.2.4.

### 2.2.1  Instruction Scheduling – Super Scalar

The objective of a super scalar system is to execute more than one instruction in each clock cycle. The basic idea is to build a processor whose data path includes multiple functional units and a modified control path to divide each task among the functional units and keep them busy as much as possible.

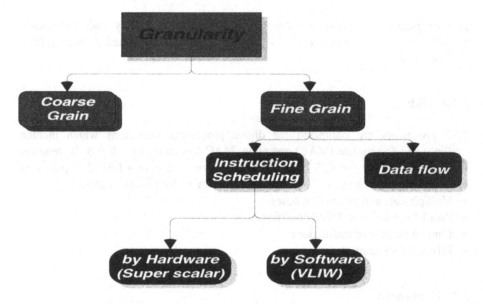

Figure 2 - Granularity issue in multimedia architectures

### 2.2.2  Instruction Scheduling – VLIW

VLIW (Very Long Instruction Word) processors achieve instruction level parallelism through software control in contrast to super scalar architectures. Parallel execution is arranged simply by setting the instruction bits that activate several functional units at the same time. VLIW architectures are data path intensive and require low control complexity.

### 2.2.3  Data Flow

Data flow architectures achieve parallelism based on the concept of executing program instructions as soon as their operands are ready, instead of following the sequence dictated by instruction code. The architecture hence can eliminate the need for a processor clock and hence the processor has extremely low power consumption in which the architecture itself has power management functions so that it operates only when data is present in the computational section.

### 2.2.4  Data Distribution

We recall from Section 2.2 that in fine/medium grain systems, parallelism can be achieved by either task distribution (instruction level parallelism) or data distribution. In data distribution parallelism, the data is distributed among several processing units which perform operations in parallel over different data segments. Processors are classified as SISD, SIMD, MISD and MIMID based on how they process the program instruction and data streams. It is clear that the last two classes employ data distribution for parallelism.

## 2.3   Memory Selection

Multimedia applications with large volumes of data require very large memory bandwidth and hence, high density, fast and low power storage is an essential part of each multimedia system. Also, the clock speed in processing units has been increased and fast memory is required to match the processing speed. In order to meet these requirements recently several approaches have emerged which increase the performance of DRAM memories. These techniques include extended data out (EDO) DRAM, synchronous DRAM (SDRAM), Rambus (RDRAM) DRAM and video (VRAM) DRAM.

### 2.3.1  EDO RAM

In EDO memories, output data can be maintained until the next CAS (column address) falling edge, resulting in fast continuos memory access. This DRAM has a two-stage pipeline, which lets the memory controller read data off the chip while it is being reset for the next operation.

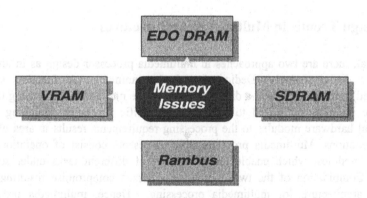

Figure 3 - Available DRAM options

### 2.3.2  SDRAM

SDRAM is another form of memory developed shortly after EDO. SDRAM performance is improved by introducing synchronous operation to DRAM. As a result of being in sync with the processor, it eliminates timing delays and makes the memory retrieval process much more efficient.

### 2.3.3  Rambus

RDRAM is an interface design to provide an optimized solution for data transfer between memory and processor. It adopts a 9-bit data bus, and there is no dedicated address bus. Instead packets including both command and address are sent to the chip first via the Rambus channel. Following the request packets, an acknowledge packet and a data packet are sent from the chip back to the controller. After initial latency, data access is achieved at high speed.

### 2.3.4  VRAM

Graphics memory must work very quickly to update, or refresh, the screen (60-70 times a second) in order to prevent screen flicker. At the same time, graphics memory must respond very quickly to the CPU or graphics controller in order to change the image on screen. With ordinary DRAM, the CRT and CPU must compete for a single data port, causing a bottleneck of data traffic.

VRAM is a dual-ported memory that solves this problem by using two separate data ports. One port is called the serial access memory (SAM) dedicated to the CRT, for refreshing and updating the image on the screen. The second port which is the random-access port is dedicated for use by the CPU or graphics controller, for changing the image data stored in memory.

## 3.  Design Trends in Multimedia Architectures

In general, there are two approaches in multimedia processor design as in any core processor design namely: Dedicated and Programmable. A function specific (dedicated) implementation is a direct mapping of the multimedia processing tasks to hardware modules optimized to execute the specific functions. Matching of the individual hardware modules to the processing requirements results in area efficient implementations. Multimedia programmable processors consist of operational and memory modules, which enable the processing of different tasks under software control. Combination of the two approaches offers a compromise resulting in an adapted architecture for multimedia processing.  Hence, multimedia processing architectures range from dedicated and adapted  modules to fully programmable media processors as shown in Figure 4. A brief description of each category is presented in the following sub-sections.

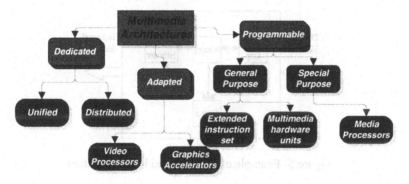

Figure 4- Multimedia Architecture Trends

## 3.1 Dedicated Architectures

Dedicated implementations could become the best choice based on available technologies, required computational bandwidth, production quantity and the target algorithm. For high volume consumer products, the optimization in silicon area and timing of the device are attractive features of dedicated architectures which result in a reduced production cost. In addition, designing a specific function architecture for a well defined and established (standard) algorithm may be the best alternative. Dedicated processors differ in terms of the computational potential - a compact module for a specific task such as a DCT chip to a complete MPEG-2 encoder as discussed below.

### Distributed (Chip-Set) Implementation

In a chip-set, each major processing module is configured as a separate chip such as a DCT chip, Huffman coder chip, motion estimator chip, etc. Each module is designed using a dedicated hardware architecture. In a distributed implementation, the designer is responsible for the interconnection of the chips. The advantage of this approach is the flexibility in selecting and connecting the different modules. The disadvantage is the increase in area and therefore the size of the system. A typical distributed implementation is shown in Figure 5. LSI Logic's L64735 DCT Processor Chip and L64765 Color and Raster/Block Converter Chip [6] are good examples of this approach.

### Unified Implementation

In this approach, the entire system is integrated on a single chip (or chip-set) which results in a low power dissipation and reduced silicon area. The main disadvantage of this approach is the lack of the flexibility. Figure 6 shows a typical unified implementation. An example of this approach is the C-Cube CL451 [7].

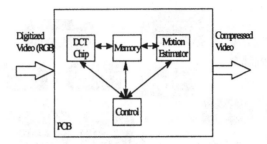

Figure 5- Example of a Distributed Implementation

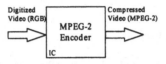

Figure 6- Example of Unified Implementation.

### 3.1.1 Adapted Architectures

We recall from section 1, that designing a more flexible architecture for multimedia applications is necessary because of the increasing number of multimedia applications. Dedicated architectures fail to respond to any change in the target algorithm. Most dedicated architectures for multimedia processing applications achieve an increase in flexibility by adapting a programmable architecture to the algorithmic requirements. Visual media being the most computationally intensive media in a multimedia environment, has been the main target of architecture adaptation. Examples include the graphics and video processor chips as discussed below.

### Graphics Accelerators

The graphics accelerator chip (or chip set) is designed to perform computationally intensive tasks by providing hardware acceleration forthe execution of low-level graphics operations. Hence, they often work as a coprocessor in workstations and personal computers. Recent chip set designs often include hardware assistance for displaying 3-D data and video streams. A typical system with a graphics accelerator is shown in Figure 7. An example of this category is the ViRGE/VX by S3 [8].

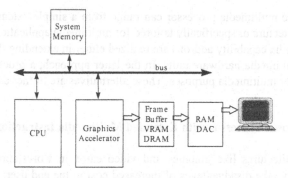

Figure 7- A typical graphics accelerator system

## Video Processors

Video processing tasks, such as DCT, motion estimation and variable length coding, demand a high profile processor. Most processing units accomplish higher speeds with the assistance of a video coprocessor which is capable of execution of these tasks. Several video processors have recently appeared in the literature. Figure 8 illustrates the utilization of a video processor in a system. An example of a video processor is the VCP by 8x8 [9].

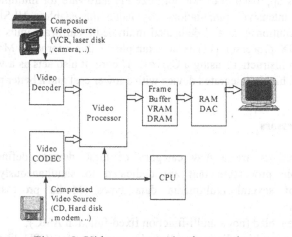

Figure 8- Video processor implementation

## 3.2 Programmable Architectures

In contrast to function specific approaches with limited flexibility, programmable architectures implement different tasks under software control. The main advantage of programmable architectures is the increased functionality. The design of a programmable multimedia processor can range from a simple extension to a general purpose architecture or specifically tailored for multimedia applications. In the former case, multimedia capability add-ons are realized either in extending the instruction set or adding multimedia hardware units. In the latter approach, a processor is designed specifically for multimedia purposes. These alternatives are further discussed below.

**General Purpose Processor with Extended Multimedia Instruction Set**

Adapted architectures like graphics and video cards in workstations and personal computers have the disadvantages of increased cost to the end user. General purpose processors including RISC and DSP have significant computing power but are not optimized for multimedia processing. Therefore, there is a strong desire among computer manufacturers to enhance existing architectures so that multimedia processing (video and graphics processing) is integrated into the next generation processors just as 2D graphics processing has been integrated into today's architectures. Extended multimedia instruction set which is introduced by Intel in MMX™ [10] processors is an example of this approach.

**General Purpose Processor with Multimedia Hardware Units**

The previous approach does not optimize the hardware for multimedia applications with highly intensive computations. By using the enabling VLSI technology the alternative solution is to add dedicated multimedia hardware units to the processor. The MediaGX processor [11] is an example of this approach. MediaGX not only executes x86 instructions using a Cyrix CPU core, it also acts as a virtual video card resulting in a highly integrated device with a lower cost and a better performance.

**Media Processors**

Media processors are a new category of logic devices defined as software-programmable processors that are dedicated to simultaneously accelerate the execution of several multimedia data types. Media processors meet three requirements:
1. software-enabled (not a multi-function fixed-function ASIC);
2. dedicated to multimedia (not multimedia extensions to a CPU, like MMX™;
3. capable of accelerating several multimedia functions simultaneously (not a DSP).

The increasing visual media based products and applications such as DVD (MPEG-2 video, Dolby Surround AC-3™ audio), 3D graphics, home movie editing (MPEG encoding), and video-phone make it expensive and difficult to design a dedicated chip or add-on boards for every new technology. Hence, media processors are the target of new multimedia hardware architects. A typical system implementation using a media processor is shown in Figure 9. There are several vendors now in the process of designing media processors. TriMedia [12] by Philips is an example of a media processor.

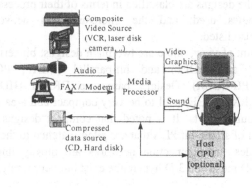

**Figure 9- Typical media processor system**

### 3.2.1  Embedded Memory-Based Processor Arrays

It has also been reported that the processor speed increases 60% while memory speed increases only 7% every year [13], and the memory capacity is quadrupled every three year [14]. The growing processor-memory bandwidth gap cannot be ignored as it results in a loss of performance. To narrow the processor-memory gap, the usual approach at the system level is to use large data/instruction caches. At the chip level, however, one approach is to embed memory modules to Application Specific Integrated Circuit (ASIC) processor arrays to remove the data transfer bottle neck at the I/O pins and the bus level. The other approach is to integrate modular logic circuits into the existing RAM's (Static RAM or Dynamic RAM) to take advantage of the inherently high memory bandwidth while exploiting parallelism across the memory columns. The former is sometimes referred to as *logic-with-memory*, while the later *logic-in-memory*.

### Logic-in-Memory MIMD Design

An example of the MIMD design is the EXECUBE [15]. It can be operated in either MIMD or SIMD mode. In MIMD mode, each of the 8 processing elements (PE's) fetches instructions and data from its own memory. The external control device can send instructions into an arbitrary subset of the 8 PE's through the broadcast logic and interface. When running at 25MHz, each chip delivers a performance of 50 MIPS. The design of this chip aims at the heavy digital signal processing (DSP) applications and not so much at the cost-performance issues. The 8 PE's are used to address the small number of taps/inputs in the Fast Fourier Transform (FFT) algorithms.

## Logic-in-Memory SIMD Design

In this section, the designs are classified in terms of their processing elements (PE), memory technologies used, and the interconnection network. The relative performances are also listed.

From the PE point of view, there are two main designs: bit-serial (C*RAM's, PC-RAM, SRC-PIM, and HDPP) and bit-parallel (IMAP, PC-RAM, CNAPS, EXECUBE, and PIP-RAM). Designs based on DRAM (HDPP, and the late C*RAM's, except EXECUBE) tend to be very compact and area efficient compared to their SRAM counter parts. It is noted that compact designs usually involve a number (usually 4) of single-bit PE's that are pitch-matched to the sense-amplifier of the DRAM modules. Interconnection networks are mostly linear with left-right communication, 2-D mesh and 3-D hypercube exist but very rarely.

### Table 3.1: Summary of logic-in-memory SIMD designs

| Name | N = No. of PE's | I = Bits/ALU | Memory per PE | R= Network | Perf. (Clock) |
|---|---|---|---|---|---|
| C*RAM92 [16] | 64 | 1 | 128b SRAM | 1-D or linear | --- |
| C*RAM93 [17] | 64 | 1 | 1Kb SRAM | linear | 3.2GIPS (25MHz) |
| IMAP [18] | 64 | 8 | 32Kb SRAM | linear | 3.84GIPS (40MHz) |
| PC-RAM [19] | 512 | 1 or 2n | 480b SRAM | linear | 25.6GIPS (25MHz) |
| SRC-PIM [20] | 64 | 1 | 2Kb SRAM | linear | 0.64GIPS (10MHz) |
| CNAPS [21] | 64(16), 16(4) | 16 | 4KB SRAM | linear | 3.2 GIPS (25MHz) |
| EXECUBE | 8 | 16 | 64KB DRAM | 3-D hypercube | 50MIPS (25MHz) |
| PIP-RAM [22] | 128(4) | 8 | 128Kb SRAM | linear | 7.68GIPS (30MHz) |
| HDPP [23] | 256 | 1 | 128b DRAM | 2-D mesh | 2.56GIPS (10MHz) |
| C*RAM96 [24] | 4096 | 1 | 4Kb DRAM | linear | --- |
| C*RAM97 [25] | 1024 | 1 | 16Kb DRAM | linear | --- |

A general trend towards large DRAM arrays is visible, together with a focus on image processing as key applications. The performance of the processor arrays very much depend on the clock rate and the number of available PE's, which, in turn, depends on the available silicon area and the ALU architecture.

## 4. Conclusions

The demands for processing multimedia data in real-time using unified and scalable architectures is ever increasing with the proliferation of multimedia applications. Multimedia hardware architectures have evolved from simple extensions of digital signal processors and small dedicated architectures to powerful parallel architectures. In this paper, we have presented a summary of the various architectural approaches for media processing. Different techniques and approaches ranging from dedicated modules to full programmable media processors have been presented. Based on available VLSI technologies, required computational achievement, production quantity and the target algorithm, it is the designer who will select the best VLSI implementation approach.

## Acknowledgements

The author acknowledges the work of his Ph.D. students Mr. Omid Fatemi and Mr. Thinh Le whose valuable contributions has made this paper possible.

## References

[1] D. L. Gall, "MPEG: A Video compression Standard for Multimedia Applications", Communications of the ACM, Vol. 34, No. 4, April 1991, pp. 59-63.

[2] ITU-T Recommendation H.263. "Video Coding for Low Bitrate Communication", October 1995.

[3] ISO/IEC JTC1/SC29/WG11, "Information Technology - Coding of Audio-Visual Objects. Visual ISO/IEC 14496 - 2 Committee Draft." Fribourg, October 1997.

[4] "MPEG-7 FAQ", http://drogo.cselt.stet.it/mpeg/faq/faq_mpeg-7.htm.

[5] K. Aizawa and T. S. Huang, "Model-Based Image Coding: Advanced Video Coding Techniques for Very Low Bit-Rate Applications", Proceedings of the IEEE, Vol. 83, No. 2, February 1995, pp. 259-271.

[6] "LSI Logic Consumer Products", http://www.lsilogic.com/products/unit5_6z.html.

[7] S. Bose, "A single chip multi-standard video codec", Proceedings IEEE Hot Chips V, Stanford CA, Aug. 1993.

[8] Phil Bernosky and Scott Tandy, "Bringing Workstation Graphics Performance to a Desktop Near You: ViRGE/VX", Proceedings IEEE Hot Chips 8, Stanford CA, Aug. 1996. http://infopad.eecs.berkeley.edu/HotChips8/9.2/.

[9] "8x8's Video Communication Processor", http://www.8x8.com/docs/chips/vcp.html.

[10] Michael Kagan, "P55C Micro-Architecture: The First Implementation of the MMX Technology", Proceedings IEEE Hot Chips 8, Stanford CA, Aug. 1996.

[11] "MediaGX Architectural System Overview", http://www.cyrix.com/process/prodinfo/mediagx/gxovervw.htm.

[12] Gerrit Slavenburg, Selliah Rathnam, and Henk Dijkstra, "The TriMedia TM-1 PCI VLIW Media Processor", Proceedings IEEE Hot Chips 8, Stanford CA, Aug. 1996.

[13] D. Patterson, T. Anderson, and K. Yelick, "A Case for Intelligent RAM: IRAM", Hot chip 8, 1996.

[14] J. Hennessy and D. Patterson, *Computer Organization and Design*, Morgan Kaufmann Publisher, 1994.

[15] T. Sunaga, H. Miyatake, K. Kitamura, P. Kogge, and Eric Retter, "A parallel processing chip with embedded DRAM macros", *IEEE Journal of Solid-State Circuits*, pp.1556-1559, vol.31, no.10, Oct.1996.

[16] D. G. Elliott, M. Snelgrove, and M. Stumm, "Computational-RAM: A Memory-SIMD Hybrid and Its Applications to DSPs", *IEEE Custom Integrated Circuits Conference*, pp.30.6.1-30.6.4, Boston, May 1992.

[17] C. Cojocaru, *Computational-RAM: Implementation and Bit-Parallel Architecture*, Master's dissertation, Department of Electronics, Carleton University, Dec. 1994.

[18] N. Yamashita, T. Kimura, Y. Fujita, Y. Aimoto, "A 3.84 GIPS Integrated Memory Array Processor LSI with 64 Processing Elements and 2Mb SRAM", *ISSCC'94*, pp.202-203, 1994.

[19] C. Cojocaru, *Computational-RAM: Implementation and Bit-Parallel Architecture*, Master's dissertation, Department of Electronics, Carleton University, Dec. 1994.

[20] M. Gokhale, B. Holmes, and K. Iobst, "Processing in Memory: The Terasys Massively Parallel PIM Array", *IEEE Computer*, 28(3), pp.23-31, Apr. 1995.

[21] D. Hammerstrom and D. Lulich, "Image Processing Using One-Dimensional Processor Arrays", *Proceedings of the IEEE*, vol.84, no.7, pp.1005-1018, Jul. 1996.

[22] Y. Aimoto, T. Kimura, Y. Yabe, "A 7.68GIPS 3.84GB/s 1W Parallel Image Processing RAM integrating a 16Mb DRAM and 128 Processors", *ISSCC'96*, pp.372-373, 1996.

[23] J. Gealow, F. Herrmann, L. Hsu, and C. Sodini, "System Design for Pixel-Parallel Image Processing", *IEEE Transactions on Very Large Scale Integration (VLSI) Systems*, vol.4, no.1, pp.32-41, Mar. 1996.

[24] R. Foss, "Implementing Application Specific Memory", *ISSCC'96*, Paper FP 16.1, 1996.

[25] R.N. McKenzie, W.M. Snelgrove, and D.G. Elliott, "A 1024 Processing-Element Computational RAM", TRIO, Kingston, May 1997.

# On Parallel Reconfigurable Architectures for Image Processing

Edwige Pissaloux

Laboratoire de Robotique de Paris (LRP), CNRS
10-14, avenue de l'Europe, 78 140 Vélizy, France
pissalou@robot.uvsq.fr
and
Université de Rouen, Faculté des Sciences
76 821 Mont Saint Aignan, France
Edwige.Pissaloux@univ-rouen.fr

**Abstract.** This paper proposes a short overview analysing the adequation between image processing requirements, parallel reconfigurable architectures dedicated to image processing and parallel computing. Through some examples of academic prototypes and commercialised computers the evolution of concepts useful for efficient and real time image processing is shown.

## 1. Introduction — Image Processing: A Challenge for a Universal Computing

Image Processing and Computer Vision were yesterday dedicated processings, today there are included in our PCs transformed in multimedia workstations. However, do the multimedia PCs support really and efficiently these new processings ? Besides technological progresses (very fast system clock, integration degree, superpipelinning,...), what else contributes to fast, error-free processing ? An answer to that question is : the convenient data structures for a given class of applications implemented at the hardware.

Indeed, from mathematical point of view, a computer is a hardware implementation of a set of algebras, algebras used in programming languages (such as Boolean ring, group of the residue classes modulo n in ⅃ and its image in Z, fields of reals and rationals). Some of these algebras are « naturally » implemented in hardware, others have to be obtained through convenient interpretation and composition of resources available in computers. Effectively, only the Boolean ring and group of the residue classes modulo n in ⅃ are naturally implemented at the computer hardware level.

Image processing has not its own algebra[1] implemented (not yet well defined ?) at the hardware level as well. The complexity of image processing data structures and operations is one reason for such situation. Traditionally, image processing is considered as a processing encompassing three levels :

---

[1] different from image algebra, [S82]

P. Zinterhof, M. Vajteršic, A. Uhl (Eds.): ACPC'99, LNCS 1557, pp. 211-225, 1999.

- image preprocessing ;
- image analysis ;
- image interpretation (scene analysis).

All these levels have their own complex data structures[2] and operations which can be applied to them. Data structures involved in operations can be regular and static (of the same size during the whole processing) (vectors, matrices, ...) and/or irregular and dynamic (i.e. data structures representations and sizes of which can vary during processing (trees, graphs, linked lists, ...)).

However, the spatial representation of data structures involved in operations require an efficient support for representation of relation concept inherent to complex data structures. Communication networks provide such support. Since the introduction of the reconfigurable communication network concept ([ML91]), the implementation in hardware of complex data structures dedicated to image/vision processings seems more efficient. Indeed, the reconfigurable networks, which are dynamic, suit naturally to dynamic data structures. Static data structure implementations on parallel reconfigurable computers can be most effective, and of lower cost.

Software implementations of image processing operations seem not be satisfactory any longer. This is especially true if the final applications require more or less hard real time constraints respect. Robot guidance, virtual reality system, videoconferencing, ... are a few examples of such applications ([MVA98]).

Consequently, in order to provide fast and efficient image processing/vision systems, it is necessary to design hardware systems (in fact, hybrid hardware/software systems) satisfying the final application requirements. Therefore, there is necessary to define and implement computer structures dedicated to image preprocessing, image analysis and image interpretation. Such implementations should consider both image data structure hardware representation and convenient operations ([Pis97]).

This paper tries to show the evolution of image processing/computer vision concepts implemented already at the hardware level through some architectural examples : parallel computers (laboratory prototypes, or commercial computers), and dedicated operators.

Section 2 addresses main data structures and different operations useful in image processing and image analysis. Section 3 gives some examples of dedicated architectures : systems and operators. Section 4 encompasses some concluding remarks.

## 2. Image Processing Data Structures

Three image processings levels mentioned above — image preprocessing, image analysis, image interpretation — can be defined through the complexity of data structures and operations applied to them. Image interpretation being strongly influenced by the final application and usually involving artificial intelligence

---

[2] By complex data structures we design data structures which elements are linked through a some relation between them such as a global order (sequences, for example) or a partial order (graphs, for example).

techniques ([NS93]), in the subsequent sections we will consider image preprocessing and image analysis only.

## 2.1. Mathematical Objects Useful for Image Processing

Sequences and graphs are two main objects supporting image processing data structures ([Pis87]).

Different image preprocessing operations imply the use of finite sequences s defined over a non empty set E such as :

$$s : [0, n-1] \to E$$

These sequences are usually referred as $s = s_0, ..., s_{n-1}$. Each index can be coded with m numbers referenced as a m dimensional index leading to m dimensional sequences. In computer science, m dimensional sequences are named (n-ary) hypercubes of dimension m (with n designing the number of elements per dimension). If m = 1 a n-ary hypercube (of dimension 1) is simply named a vector of n elements ; if m = 2 two dimensional arrays (meshes or matrices) are considered ; p dimensional arrays are used if m = p, ... .

Image analysis involves mainly finite graphs and weighted graphs.

A finite graph G defined over the set E is a function :

$$G : E \to P(E)$$

where $P(E)$ notes a set of partitions of E.

The G function is a forest if $\forall i \in E$ :

$$— G^{-1}(i) \text{ has one element at most}$$

$$— i \notin G^{p+1}(i), \quad \forall p \in \aleph.$$

A vertex $i \in E$ is :

— a root, if $G^{-1}(i) = 0$ ;

— a leaf, if $G(i) = 0$.

A forest having a unique root is a tree.

A weighted graph can be defined as follows :

Let : E be a non empty and finite set of states, and A a non empty finite set said alphabet. A partial function H :

$$H : E \times A \to P(A)$$

is a weighted graph.

A graph G defined for $\forall e \in E$ :

$$G : e \to U_{a \in A} \{H(e, a)\}$$

is said to be a frame of the weighted graph H.

## 2.2. Image Processing Computer Science Data Structures

Vectors, 2D arrays, lists (and self-organizing-lists — SOL— in particular); trees, graphs and pyramids are fundamental data structures for image processing.

In image preprocessings, the image is considered only as a set of ordered but not semantically related pixels. Consequently, vectors and arrays are the most suitable data structures for image representation at this level. Vectors encompass naturally lines/columns of images (acquired at video speed with some CCD camera, for example), while 2D arrays can represent whole images (total image ready for further real time, or not, processings, image analysis especially).

During the image analysis, the image can be considered as a set of its partitions, each partition being obtained through some segmentation algorithm (edge point segmentation, interest point segmentation, region segmentation, ...).

Lists are mainly used for image features extraction by segmentation algorithms (chain encoded representation of planar curves with Freeman's code, for example). In its queue or FIFO version, lists are also helpful for image displays.

Self organizing lists (SOLs) support efficiently acquired image lines/columns with additional information useful for image compression for example ([MW96]).

Trees and graphs support efficiently segmented image features (edge, region independent or co-operative detection), and allow for more effective and efficient, than the lists, analysis of information encompassed in images. However, graphs associated usually with images are represented with classical data structures such as 2D matrices, linked lists, region adjacency 2D matrix, ... .

There have been attempts to implement directly, at the hardware level, some new data structures useful for image analysis ([CLE93]), especially efficient for global communication in pyramids[3]. A C-graph (a concentration graph) is a sub graph of the graph of physical connections of the pyramidal network associated with a plan graph implemented in a pyramidal layer. A node p of a given C-graph is said to be a centralising node, if there is no node in a C-graph ancestor of p. The set of all centralising nodes pyramidally projected on a plan (a pyramidal layer for example) define the centralising graph. C-graphs are usually associated with image regions, while centralising graphs represent the connexity of connect components of a graph.

## 2.3. Image Processing Operations

### 2.3.1. Image Preprocessings

Image preprocessings are frequently assimilated with signal processings. Signal filtering is a basic operation, which usually precedes more sophisticated processings. Its role is to reduce the influence of errors related to real image quantification, i.e. a finite representation of an infinite (pixel) value. Indeed, the whole image acquisition procedure adds a noise related not only to image finite representation (sampling and quantification, truncation, rounding), but also due to the imperfection of used apparatus (camera lens, transmission links, ...).

From calculation point of view, the filtering is based upon convolution operation. Image convolution involves two matrices (cf. Fig. 1) : one representing the image (2D array of pixels), and another, a mask (frequently $3 \infty 3$), quantifying the relation which links variations of given pixel luminosity with the luminosity of its spatial

---

[3] A pyramid is a complex data structure encompassing matrices of different sizes linked usually by (binary or quaternary) tree.

neighbours (form mathematical point of view, this is an approximation of 2D luminosity gradient).

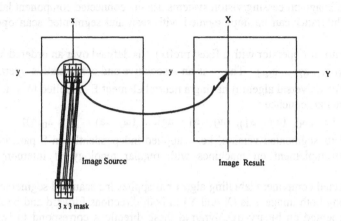

Figure 1. Image convolution calculation principle.

Although, the image convolution operation isn't a matrix multiplication, this latter is sometimes useful in image processing. The temporal complexity of $O(n^3)$ of the basic sequential algorithm for matrix multiplication can be reduced to $O(n)$ on a parallel reconfigurable 2D mesh - torus (cf. 3.1.).

Many other image preprocessings are performed iteratively, step-by-step, the global result being a sum of local partial results. The temporal complexity of these preprocessings is usually proportional to the image size ($O(n)$).

Some preprocessings involve data concentration (max, OR, ...), or distribution (thresholding, for example) operations, i.e. global operations, over the whole image. These operations calculate the global value through some reducing/broadcasting operators. The convergence of the algorithm depends on these global operation implementation, i.e. of the global communication performance (communication protocols, network topology, ...) ; it is suitable that the temporal complexity of global operation be at max of $O(\log n)$.

### 2.3.2. Image Analysis

Image analysis encompasses various algorithms, which are irregular and heavily depend upon data. Some of algorithms are : the instanciation of image/region global characteristics (medium or max value, moments, ...) resulting in a scalar data ; histogram, Fourier transform, pattern search, ... being a result of grey level analysis ; image segmentation corresponding to homogenous object extraction ; projections, Radon, Hough transforms, ... leading to the object localisation ; Freeman's code, edge polygonal coding, edge coding with a function, ... being different representations of edge coding ; quad-tree, multi-resolution representation, median axis transform, ... declining region coding operations ; region graph adjacency, Voronoi diagram, ... belonging to the class of neighbouring region processings.

These examples show that the image analysis involves more complex data structures, tree and graph, in general. For example, non-absorbing operators (binary addition, exclusive OR, ...) require operations on trees, tree embedded on graphs, on meshes or, in general, on hypercubes.

Almost all image processing/vision systems use the connected component labelling algorithm. This latter can be implemented with scan and segmented scan operators ([Ble89]).

Scan operator (or operator with a fixed prefix)   is defined over an ordered set S of n elements $a_0$, $a_1$, $a_2$, ... $a_{n-1}$ . The   is an associative and commutative operator (in the sense of the universal algebra) having a neutral element E . Applied to S it orders it in the following sequence :

$$( E , a_0, (a_0 \quad a_1), (a_0 \quad a_1 \quad a_2), ..., (a_0 \quad a_1 \quad ... \quad a_{n-2}))$$

Scan (and its segmented version, i.e.   applied independently on S partitions), is very easy to implement on machines with regular topology of interconnection network.

The connected component labelling algorithm applies the scan and segmented scan operators along both image axis (X and Y) in both directions forward and backward. When implemented on binary tree/pyramid these directions correspond to two data movements   up to the root, and down to the leaves (cf. below). With such implementation the temporal complexity of the algorithm drops to $O(log_2 N)$.

Image matching is another very useful operations for image analysis. Almost all vision guided systems (robotics, teleoperation, telesupervision, biological pattern recognition, ...) extensively use it. Image matching algorithms differ mainly by the primitives used for matching (it could be : raw pixels, interest points, edges, polygons, ...) and the way in which they find the best optimized (but very seldom optimal) match ; dynamic programming, geometric hashing, semantic networks, simulated annealing, ... are the most popular algorithms for matching.

For low level primitives (raw pixel, interest points, ...) image matching can be considered as (1-) graph matching problem. In this case, the search for images best matching can be obtained using orthogonal dynamic programming applied in X and Y directions independently (variables x and y separability) ([LPB97], [PLB98]).

## 3. Parallel Reconfigurable Computers

Because of the huge amount of data involved in processing, parallel computers seem a good support for real image based applications. Indeed, even on modern workstations typical image processing can take several tens of seconds what could be too slow for processing requiring the respect of real time hard constraints (such autonomous vehicle, some robotics processings, data compression/decompression, videoconferencing, ...). Therefore, dedicated reconfigurable parallel computers and/or dedicated reconfigurable operators bring a solution to image preprocessing and analysis.

A processing unit of a parallel computer (frequently referred as elementary processing unit or PE, in order to underlay the simplicity of effectively realised operations), and especially interconnection network are their key points.

The previous section has outlined some of the important operations in image processing. This section proposes an overview of existing PEs interconnection networks, thus the architectures.

Concept of reconfigurable architectures aims to bring a hardware support for image processing dynamic complex data structures. Indeed, the reconfigurability adapts easier to run time dynamic data structures. Moreover, it reduces the physical degree of networks, thus simplifies their design, realisation and cost. Furthermore, it allows the physical realisability of some theoretical topologies.

The interprocessor communications are of two types : local and global, performed usually on sets of connected pixels.

Local communications (with spatial neighbours, corresponding to a pixel most frequently defined neighbourhood in images, cf. Fig.1) are performed by local modifications of physical active network interconnections. For example, in the case of 2D PE arrays, only one between 4 topological neighbours (South, East, North, West) is accessed at the same time (the 3 other connections are physically disabled). Local reconfigurations take usually place during one PE instruction.

Global communications in reconfigurable architectures are mainly based on circuit switching. Circuit switching can be done in asynchronous or synchronous way. CLIP4 (Duff'76, United Kingdom), IBM polymorphic torus, Gated Connection Network, MAO (maille associative d'Orsay), $\mu$PD (circuit dedicated to dynamic programming), ... are few examples of reconfigurable architectures with asynchronous switches. PAPIA2 (Pavia University, Italy), Symphonie (linear processor, CEA, France), Sphinx (Système Pyramidal Hiérarchique pour le traitement d'Images Numériques), PICAP-3 (Linpöking University, Sweden), AIS-5000 (Applied Intelligent Systems), RVS (NEC), GAP(M.J. Irvin, Pennsylvania University, '95)... are examples of reconfigurable architectures with synchronous switches.

Global communications can be implemented with regular of irregular communication schemes. Regular communication schemes involve usually scans operations or use processor connections in power of 2 $(2^p)$. These latter, existing on some parallel processors, allow for global data exchanges with wormhole protocol. On a 2D mesh (MasPar MP-1, for example), these exchanges can be pipelined. Communications of every k-th processors require that intermediate nodes resend each bit of the message. Consequently, the global communication time drops from n $\infty$ k to n + k (in the case of the whole message memorisation).

A result of the mesh embedded on a hypercube is the « natural » existence of additional physical links between processors interconnected in power of 2 which do not exist on the mesh. These additional links can be used for implementation of global time effective communications on mesh.

On trees (such as CM—5, for example), connections in power of 2 exist by the definition of the topology. They lead to the global communication time of the order of $O(\log_2 N)$.

Concentration graphs — C-graphs — are important data structure for communication in pyramids. C-graph operations are : a parallel application of the same operation to all graph nodes (broadcasting), a parallel data exchange between two C-graphs (embedded in the same plan graph) through frontier pixels, mesh plan communications, tree communication (send-down (a message), send-up (a message)), with possible local reduction through an associative and commutative operator.

In the subsequent sections, few examples of existing computers (commercialised or prototypes) are given.

## 3.1. Reconfigurable Mesh Processors

2D arrays are naturally implemented on 2D mesh computers or hypercubes in general (where frequently, 2D mesh in embedded through virtual machine concept). First image processing computers have been linearly or mesh connected ; SYMPATI2 (linear), MPP, DAP, CLIP, ... are few example of them ([Uhr87], [PIS97]).

Matrix A and B multiplication can be very efficiently implemented on a 2D torus (Fig. 2). Starting with the « natural » implementation of data, i.e. where elements $a_{ij}$ and $b_{ij}$ are implemented on the processing element $PE_{ij}$, two shifts per iteration have to be executed in order to realize matrix multiplication on a torus :

- the k-th line of the matrix A to west by k steps,
- the k-th column of the matrix B to north by k steps.

After these shifts processors $PE_{ij}$ contain $a_{ik}$ and $b_{kj}$ ($k = (i + j)$ mod n), and both values contributing to the calculation of the element $c_{ij}$. Figure 3 gives the new data rearrangement after iterations on 4 ∞ 4 torus.

The implementation of the above algorithm (of the temporal complexity reduced to N(n), n number of elements per matrix dimension) is possible thank to mesh local communications, the network autonomy (reconfigurability), and torus links.

Almost all image processing mask-based operations (such as Sobel, Prewitt, Kirsch, Moravec, ...) take advantage from the additional torus links ; indeed, these links solve easily the image border problem (although, not in the perfect way).

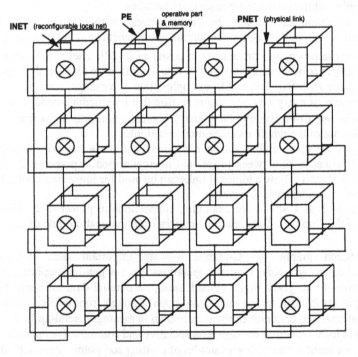

Figure 2. IBM polymorphic torus.

| a 00 b 00 | a 01 b 11 | a 02 b 22 | A 03 B 33 |
|---|---|---|---|
| a 11 b 10 | a 12 b 21 | a 13 b 32 | A 10 B 03 |
| a 22 b 20 | a 23 b 31 | a 20 b 02 | A 21 B 13 |
| a 33 b 30 | a 30 b 01 | a 31 b 12 | A 32 B 23 |

Figure 3. New matrix element rearrangement during matrix multiplication on torus.

### 3.2. Sphinx Pyramid

SPHINX (Système Pyramidal Hiérarchisé pour le traitement Numérique d'Images) (Fig. 4) is a pyramidal computer  developed by the Institute of Fundamental Electronics at Paris XI University, ETCA, a Defence Research Laboratory (France) and the Sodima Co. ([MER86], [PBME90]). The machine is organized as a set of stacked layers of PEs (Processing Element)  of decreasing size interconnected according to a dual network :
    • a mesh-based four-neighbour interconnection network within a layer;
    • a pure binary tree between layers.
    A PE is a bit serial processor. It has general purpose one-bit registers, and some functionally specialized registers, dedicated to pyramidal communications (Father, Left son and Right Son), and others, dedicated to in-layer communications (Neighbour register).
    On the i-th stage of the scan operation  implementation, k-th PE will read the partial result from the processor $2^k$ and combine it with its local value.
    Practical implementation of scan operation on binary pyramid requires two data movements over all pyramidal structure : UP (to the pyramid apex), then DOWN (to the pyramid base).
    During the Up data movement each processing element
    — executes the scan operation on its 2 sons, and send the result to the father ;
    — it memorises the left son's value.
    During the Down data movement each processing element  forwards :
    — to its left son the value received from its father ;

— to its right son the result of the scan operations applied to the data received from the father and the data memorised in Up movement.

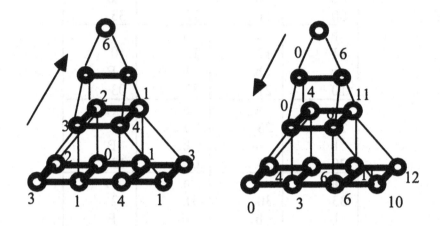

Figure 4. The max-scan operation implemented on the binary pyramid Sphinx.

Numerical values of the left pyramid (Fig. 4) represent, initial value distribution in the pyramid base, and data memorised in the UP movement for other layers ; numerical values of the right pyramid correspond, on edges, to values sent from fathers to their sons except to these attached to the pyramid base, which represent the final result, read from left to right, 4 by 4.

This algorithm shows the importance of inter-layer binary tree reconfigurable communications, and how easily the inter-PEs communications in power of 2 are realised on a binary pyramid.

### 3.3. MAO Mesh

The associative net architecture (or the MAO (Maille Associative d'Orsay), Fig. 5), [MER97]) implements a certain graph class at the hardware level. Several graph elementary operations are directly mapped onto a 2D mesh of P processors ($P_i$, $0 \cdot i \cdot$ (P-1)). Mao underlying concepts are : connect component, successor and ancestor.

An oriented graph $G = (P, E)$ of degree D is represented by P processors (vertices of G), and by its interconnection network (a sub-set of the physical interconnection network), which defines the set of graph's edges E.

G being oriented, there exist sets of ancestors and successors of $P_i$ in G. A set of processors of G is an equivalence class, if it is possible to define over G an equivalence relation $R_g$ such that $P_i$ and $P_j$ are equivalent if and only if $\exists$ k, $0 \cdot k \cdot$ (P-1), such that $P_i$ and $P_j$ belong to the k's ancestor set. Equivalence class define connected components of a given graph.

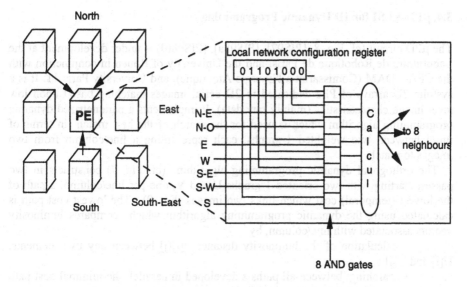

Figure 5. MAO reconfigurable mesh computer.

As a graph can be considered as a set of elements, it is possible to perform the usual set operations on elements of G (arithmetic and logic element-wise operations). The scan    (prefixed- or association-) operation involves a graph G considered as a parallel variable p of $p_i$ elements $(0 \cdot i \cdot (P-1))$, and c, one of its sub-graphs (or c-graph — the communication graph) which defines authorised data movements in G. The scan operation calculates the result $a_i$ (implemented on $P_i$) using the values $p_j$ of all successors of $P_i$.

Figure 6 gives examples of implementations of elementary operations on associative net.

The c-graph is an elementary parallel data structure. Its basic operations are the usual graph operations such as c-graph construction and deletion, c-graph topology redefinition, union /intersection/Cartesian product of two graphs, or insertion/deletion of one element. A associative net supports efficiently tree operations which has to be embedded on the graph on the MAO interconnection net before the execution of any operation.

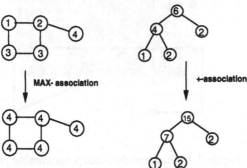

Figure 6. Implementation of associative operations on associative nets.

### 3.4. µPD VLSI for 1D Dynamic Programming

The µPD circuit (cf. Fig. 8, [PIS97], [PIS98a], [PIS98b]) is under development at the Laboratoire de Robotique de Paris, and the University of Rouen in conjunction with the CEA—DAM (Comissariat à l'Energie Atomique), and University Paris 11. It is a systolic 2D arrays of PEs dedicated to 2D aerial images matching based upon low level image characteristics (mainly raw data). It implements a parametrized dynamic programming algorithm which is used for the search of the best match (in terms of time) between two weighted 1-graphs, each representing a line/column from two images to match.

The orthogonal dynamic programming algorithm  (cf. Fig. 7) constructs in two passes, starting with two oriented 1-graphs U and V, one per line/column, a path of the lowest (temporal) cost which links two images (Fig. 9). The lowest cost path is calculated using the dynamic programming algorithm which  compares luminosity vectors associated with line/column, by

   • calculation of the luminosity distance d[i][j] between any two elements, U[i] and V[j] ;

   • searching, between all paths s developed in parallel, the minimal cost path $s_{min}$ associated with the cost *score* such as :

$$\text{score} = \min_{\text{paths } s} \left( \sum_{i, j} d[i][j] \times C(s) \right)$$

(the path s = • (i,j ) – where • is the concatenation operation, and i, j  { 1, ..., N}). The C(s) is a local cost of path s construction, i.e. the cost associated with the distance between two consecutive elements of the path. According to the fact that two vectors are equal if $\forall$ i ∈ {1,..., N}, U[i] = V[i], the cost of local diagonal path s is less than of an orthogonal one ; d[i][j] is said an elastic distance between compared elements.

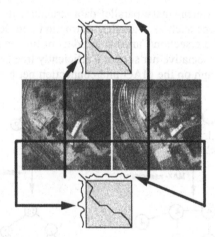

Figure 7. Orthogonal dynamic programming principle.

The μPD circuit mimes the parallel paths s development during two vectors comparison. It is a n-ary hypercube of dimension 2 (i.e. a 2D mesh with n processing elements per dimension). The whole μPD machine, a scalable massively parallel computer, is a MISD (systolic) data driven asynchronous computer. Each PE is 3 — connected to its three neighbours in the grid (east, south-east, south), and has its own local 2-bit memory (for PE direction activation code, important for the dynamic programming backtracking phase).

A parallel dynamic programming algorithm searches, in parallel, all possible paths s in $l_{opt}$ nibbles (asynchronously), where $l_{opt}$ is the length of the optimal path. On the k-th nibble, $0 • k • l_{opt}$, all active PEs have their spatial (machine) index of the form ( , n) or (n, ). Figure 9 gives an example of all paths s development in parallel ; the compared 1D signals are represented on X and Y axis. The shadowed area represents all parallely developed paths ; the bold path is the optimal path found during the backtracking stage. Results of real signals comparison are included inside the bold square (others are used for μPD simulator purposes).

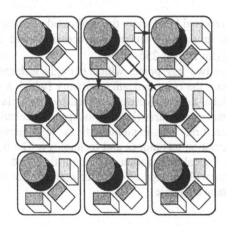

Figure 8. μPD circuit internal organization.

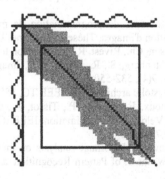

Figure 9. Path parallel development with the μPD circuit.

The μPD operator interconnection network is reconfigurable one : during the dynamic programming algorithm calculation, each PE of the machine reconfigurates locally network in all 3 directions (east, south, south-east) at convenient moment in systolic way. During backtracking phase the same physical links are re-oriented in opposite directions, in order to recover the minimal cost path for compared vectors. Consequently, the reconfigurability is a key for fast vector matching operation.

## 4. Conclusion

The design and implementation of most suitable architectures dedicated to image preprocessing and image analysis have been addressed. Starting from mathematical concepts and computer science data structures useful for vision, some hardware dedicated realisations of the convenient parallel reconfigurable computers as prototypes Sphinx, μPD and MAO or as commercialised computers (IBM polymorphic torus), have been briefly presented, and implementations of basic graph operations shown.

Experiments (not reported here) show that dedicated computers reduce by the order of 1000 the calculation time of image preprocessings and analysis compared to sequential computers (for images 256 ∞ 256).

However, there are two key points which strongly influences dedicated parallel computer performance : the communication network (which has to provide the most efficient local and global data exchanges required by given class of algorithms), and the convenient calculation structures (leading to a parallel calculation model).

The above architectural key implementations rely heavily on the computer execution mode, calculation techniques and technology. The biological approach for vision system : biological inspiration for algorithm and architecture design , biological components will perhaps offer better performance for the future systems. But when they'll become available ... .

## 5. Bibliography

[Ble89] Blelloch, G., Scans as parallel primitive operations, IEEE TC, 38 (11), 1989, 1526—1538.
[CLE93] Clérmond, Ph., Méthodes de programmation de machine cellulaire pyramidale : Applications en segmentation d'images, Thèse de Doctorat, Université PAris VII, 1993
[CLR90] Cormen, T., Leiserson, Ch., Rivest, R., Introduction to Algorithms, MIT Press, 1990
[HSZ87] Haralick, R., M., Stenberg, S., R., Zhuang, X., Image analysis using mathematical morphology, IEEE PAMI, 9(4), 532-550, 1987
[KU80] Hung, H., T., Why systolic architectures, IEEE TC, 15(1), 37-46, 1990
[LPB97] Le Coat, F., Pissaloux, E., Bonnin, P., Tissot, A., Garié, Th., Durbin, F., Parallel Algorithm for Very Fast Velocity Field Estimation, IEEE ICIP'97, Santa Barbara, CA, 26—29 Octobre, 1997, pp. II-179—182
[MZ92] Mérigot, A., Zavidovique, B., Image analysis on massively parallel computers : an architectural point of view, Int. J. of Pattern Recognition and Artificial Intelligence, vol. 6, N° 2 & 3 (1992), 387-393

[ML91] Maresca, M., Li, H., Polymorphic VLSI arrays with distributed control, *in* H. Li, Q. Stout, Reconfigurable Parallel Computers, Prentice Hall, 1991

[MER86] Mérigot, A., Bouaziz, S., Clérmont, Ph., Devos, F., Eccher, M., Méhat, J., Ni, Y., Sphinx, un processeur pyramidal massivement parallèle pour la vision artificielle, Proc. of the AFCET 7th RFIA Conf., Paris 1989, pp. 185-196

[MER97] Mérigot, A., Associative Nets : A Graph-Based Parallel Computing Model, IEEE TC, 46(5), May 1997

[MVA98] Proc. of the IAPR Machine Vision Applications (MVA)'98 Workshop, Makuhari, Chiba, Japan, Nov. 17—19, 1998.

[MW96] Myoupo, J., F., Wabbi, A., Improved systolic architecture for transpose heuristic, ICSA Conf. Parallel and Distributed Computer Systems, 64—68, 1996

[NS93] Nadler, M., Smith, E., Pattern Recognition Engineering, John Wiley, 1993

[PBMD90] Pissaloux, E., Bouaziz, S., Mérigot, A., Devos F., Co-programming : a tool for the development of software for massively parallel computers, The Euromicro Journal, Vol. 30, N° 1-5, 1990, pp. 569-676.

[PIS97] Pissaloux, E., Bonnin, P., On the Evolution of Parallel Computers Dedicated to Image Processing through Examples of Some French Computers, Int. J. on Digital Signal Processing, Academic Press, vol. 7, N° 1, January 1997, pp. 13—27

[PIS87] Pissaloux, E., Contribution a la conpcetion d'une machine orientee types de données, Thèse d'Etat, Université Paris 7, 1987

[PLB98a] Pissaloux, E., Le Coat, F., Bonnin, P., Tissot, A., Garié, Th., Durbin, F., Design and Realisation of a Parallel Systolic Architecture dedicated to Aerial Image Matching, Proc. of the IAPR MVA'98, November 17—19, 1998, Makuhari, Japon, pp. 402—405

[PLB98b] Pissaloux, E., Le Coat, F., Bonnin, P., Tissot, A., Garié, Th., Durbin, F., Design and Implementation of a Systolic Circuit dedicated to Dynamic Programming Algorithm Parallel Implementation, Proc. SPIE Int. Conf. on Configurable Computing : Technology and Applications, Boston, USA, 2—3 november 1998, vol. 3526, pp. 43 — 51

[S82] Serra, J., Image Analysis and Mathematical Morphology, Academic Press, London, 1982

[Uh87], Uhr, L., ed. Parallel Computer Vision, Academic Press, 1987

# Parallel Multiresolution Image Segmentation with Watershed Transformation*

Alina N. Moga

Albert-Ludwigs-Universität, Institut für Informatik
Chair for Pattern Recognition and Image Processing
Universitätsgelände Flugplatz, D-79085 Freiburg i.Br.
moga@informatik.uni-freiburg.de

**Abstract.** A parallel extension for the watershed segmentation is presented in this paper. By following regional minima, i.e. the seeds around which regions are grown, in images of lower resolution, a region merging criterion in the oversegmented watershed image is retrieved. Thus, regions which have a common ascendent at a certain resolution level unify, as well as neighboring regions which annihilate until that level. Fine until coarse segmentation results are obtained efficiently in parallel on a Cray T3E parallel computer. Additionally, the technique performs quasi-automatically and allows an easy incorporation of restrictive merging criteria within the pyramidal merging structure.

## 1 Introduction

Watershed transformation performs similar to region growing techniques, by labeling connected areas around regional minima in an image [2,7,8]. Its simple algorithmic design and efficient implementation, as the absence of user supplied parameters, make the watershed segmentation an attractive technique for many image processing applications. However, the method has a major shortcoming; for many natural images it produces oversegmentation. Solutions have been sought to reduce this effect, namely, to support the segmentation process by a priori information upon the meaningful objects in an image, and to further constrain the region growing only around these areas [6]. Also known as a marker based watershed transformation, the algorithm does not perform automatically. Human interaction and a parameter controlled procedure are required, as preprocessing stages, to provide the desired image cues [11].

In this paper, a quasi-automatic technique is proposed to improve the watershed segmentation result. The idea resides in the field of multiresolution image analysis, which was successfully employed in the segmentation process (see [1,3,4,12,13,14,15,16,17]). The multiresolution watershed segmentation algorithm that we propose is described in the next section. In Section 3, details about its parallel implementation follow. Experimental results, performance analysis, and conclusions finalize the paper.

---

* For our experimental results, we acknowledge the use of the Cray T3E parallel computer at the High-Performance Computing-Center Stuttgart, Germany.

P. Zinterhof, M. Vajteršic, A. Uhl (Eds.): ACPC'99, LNCS 1557, pp. 226–235, 1999.
© Springer-Verlag Berlin Heidelberg 1999

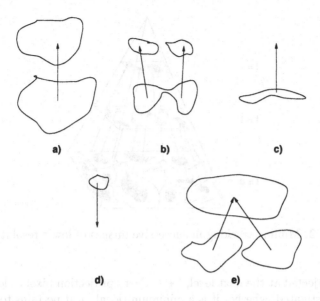

**Fig. 1.** Cases of matching minima a) single parent, b) splitting, c) annihilation, d) creation, and e) merging

## 2   Description of the Sequential Algorithm

A multiscale region merging technique is presented in this paper to improve the watershed segmentation. For this purpose, regional minima, i.e. the seeds around which the segmented objects are grown, are hierarchically organized across the scale space. Starting from the original input image, every image of lower resolution is the blurred version of the image at the previous level. The decimation ratio of the size of successive images is 4 : 1. The smoothing function is a Gaussian $5 \times 5$ kernel, $w$, which is chosen subject to four constraints [14]:

- separability: $w(m, n) = \hat{w}(m)\hat{w}(n)$,
- normalization: $\sum \hat{w}(m) = 1$,
- symmetry: $\hat{w}(m) = \hat{w}(-m)$,
- equal contribution: if $\hat{w} = \{c, b, a, b, c\}$, then $a + 2c = 2b$.

The solution of the combined constraints leads to $\hat{w} = \{\frac{1}{4} - \frac{a}{2}, \frac{1}{4}, a, \frac{1}{4}, \frac{1}{4} - \frac{a}{2}\}$, where $a$ is a free variable. In our implementation we used $a = 0.4$. If $I_0$ denotes the input image, the image $I_l$ at level $l > 0$ is obtained by the recurrence: $I_l(x, y) = \sum \sum_{m,n=-2}^{m,n=2} w(m, n)I_{l-1}(2x + m, 2y + n)$. The maximum level $l$ is given by the logarithm of the image diameter.

By progressively blurring and sampling an image, the image structure simplifies. Thus, regional minima move closer to other regional minima until they coalesce, or disappear [3,4,5]. The regions associated with these minima will consequently either merge, or annihilate. By mapping in a bottom-up manner regional minima from one level to the next, a hierarchy of child-parent relationships is established. More specifically, given a minimum $A$, at level $l$, all pixels

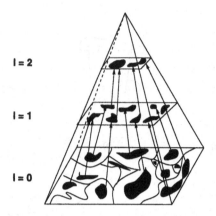

**Fig. 2.** Minima tracking in successive images of lower resolution

in $A$ are projected at the next level, $l + 1$. Every projection pixel at level $l + 1$ is further investigated, whether it is a minimum (local, or it pertains to a regional minimum), or a non-minimum. If all, or part, of the pixels in $A$, at level $l$, maps onto another minimum $B$, at level $l + 1$, then $B$ is the parent of $A$ (see Fig. 1.a). However, if $A$ maps onto more than one minimum, case encountered when $A$ splits, then $A$ has several parents (see Fig. 1.b). As a consequence, all parents of $A$ unify by a simple connected component operator, and, implicitly, all their children at level $l$ become brothers of $A$. Conversely, if the mapping of $A$ at level $l + 1$ does not intersect any regional minimum, $A$ does not have any parent and it annihilates (see Fig. 1.c). A reverse phenomenon is also possible, if a regional minimum at level $l + 1$ is created and it does not correspond to any regional minimum at level $l$ (see Fig. 1.d). In the algorithm presented in this paper, we do not however explore the newly created minima at intermediate levels. Moreover, at level $l + 1$, only regional minima which intersect the projections of the regional minima at level $l$ are investigated. The last case, minima merging, illustrated in Fig. 1.e, corresponds to the situation when two or more minima at level $l$ have the same parent at level $l + 1$.

Once the hierarchy of relationships between regional minima at consecutive scales has been built (see Fig. 2), one can retrieve a merging criterion between regions at level 0, i.e. in the oversegmented watershed image. Thus, considering any arbitrary node at a given level $l$ in the hierarchy, the underneath tree rooted at this node is parsed top-down until its leaves reached. Consequently, all regions, whose regional minima represent these leaves, coalesce. Additionally, neighboring regions which annihilate at a level smaller than $l$ are merged together. Different segmentation results are obtained, for various values of the level $l$. Thus, finer segmented objects are observed for low values of $l$, and the segmentation becomes coarser as $l$ increases.

Let us still underscore that, by progressively blurring the input image and linking the minima from level to level, we aim only to merge neighboring regions separated by "small" graylevel ridges. This automatically affects the watershed

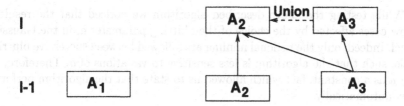

**Fig. 3.** Linking and unifying labels of regional minima at neighboring resolution levels

lines at the interface between the regions. We however do not intend to correct the remaining watershed lines, e.g. for edge dislocation, missing edge, false or spurious edges. Additionally, the final level $l$ in the pyramidal structure is subjectively chosen. For low values of $l$, fine structures are present, while oversegmentation still persists; on the other side, for high values of $l$, oversegmentation is reduced, but small structures are lost.

A compact description of the algorithm is next presented:

1. $l \leftarrow 0, I_0 \leftarrow I$;
2. $O_0 \leftarrow$ Watershed$(I_0)$;
3. Store $O_0$ as the segmentation result at level 0;
4. for every level $1 \leq l < l_{max}$ do
   (a) $I_l \leftarrow Gauss(I_{l-1}, w)$;
   (b) for each label $i$ in $O_{l-1}$, $lut_l(i) \leftarrow i$;
   (c) for each regional minimum $A$ in $I_{l-1}$ do:
      i. project each pixel $p = (y, x)$ in A onto $p'$ in $I_l$:
         A. $p' = (\lfloor y/2 \rfloor, \lfloor x/2 \rfloor)$ or $p' = (\lfloor y/2 \rfloor, \lfloor (x+1)/2 \rfloor)$ or $p' = (\lfloor (y + 1)/2 \rfloor, \lfloor x/2 \rfloor)$ or $p' = (\lfloor (y + 1)/2 \rfloor, \lfloor (x+1)/2 \rfloor)$;
         B. explore $p'$: if it is a minimum then
            – if the minimum to which $p'$ belong has not been labeled yet then label it with $O_{l-1}(p)$;
            – else UNION$(lut_l, O_l(p'), O_{l-1}(p))$;
      ii. if no pixel in $A$ projects onto a minimum, then store level $l$ as annihilation level for $A$ (i.e., $O_{l-1}(p)$);
   (d) for every regional minimum $A$ at level $l - 1$ which annihilates at level $l$ do:
      i. for every neighboring region $B$ of $A$ in $O_0$ which annihilates do: UNION$(lut_l, O_l(A), O_l(B))$;
   (e) Solve the connected components in $lut_l$;
   (f) for every pixel $p \in O_0$, $O_0(p) \leftarrow lut_l(O_0(p))$;
   (g) Store $O_0$ as the segmentation result at level $l$;

In Fig. 3, minimum $A_3$, at level $l - 1$, splits into two minima $A_3$ and $A_2$, at level $l$. Consequently, $A_3$ and $A_2$ unify, as belonging to the same class. However, $A_2$ and $A_3$ fuse not only because $A_3$ splits, but also because of $A_3$ and $A_2$, which, at level $l-1$, have the same parent at level $l$, namely $A_2$, and consequently merge.

While testing the above described algorithm we noticed that the result is in few cases affected by the choice of the blurring parameter $a$, in the Gaussian kernel. Indeed, only the regional minima are followed between successive blurring levels, such that the algorithm is less sensitive to variations of $a$. Therefore, we used $a$ as a constant, fact which allowed us to state that the algorithm performs quasi-automatically.

## 3   Parallel Implementation

Although the sequential implementation of the above described algorithm is not difficult, its integration into a multiprocessors environment is not straightforward. A Single Program Multiple Data (SPMD) paradigm with regular domain decomposition is next presented to parallelize the code for the watershed segmentation with multiresolution region merging. Thus, watershed segmentation of the original image is first executed in parallel as described in [9,11]. Next, region merging is performed by following regional minima across a multiscale Gaussian smoothing pyramid.

Since the first phase of the algorithm, namely parallelization of the watershed transformation has been extensively presented in [9,11], only the parallel implementation of the multiscale region merging is here explained. In order to obtain the subimage at the next lower resolution level, each input subimage is smoothed with the given Gaussian kernel, as explained in Section 2. The operation is local, but computation of the smoothed values for the pixels within the subimage edges requires the graylevels within a two pixels wide border around the subimage. These data are provided by each processor through a regular neighborhood communication pattern, imposed by the data distribution.

After the grayscale image at the next lower resolution level, $l+1$, has been distributed computed, i.e. each processor holds the smoothed version of its subimage from the previous level $l$, regional minima at level $l$ should be mapped at level $l + 1$. As in the sequential algorithm, only pixels within surviving minima are projected onto the subimage at level $l + 1$. Since the initial image distribution is maintained at every level of the multiscale pyramid (no data redistribution is performed), for every pixel, its projection pixel falls into the subimage owned by the same processor. Thus, no "vertical" communication between processors (to communicate data at consecutive smoothing levels) is needed. However, the analysis and diagnosis of a pixel as minimum or non-minimum is not entirely local. As already mentioned in [9,11], plateaus of connected pixels of constant altitude are not easily classified, as minima or non-minima, and hence appropriately labeled, when the image is distributed to several processors. Therefore, an a priori labeling of those plateaus shared by more than one processor as a regional minimum (RM) or not a regional minimum (NARM) is performed in parallel. The idea is that if a plateau has a lower brim, every pixel in the plateau should be labeled NARM; otherwise, RM. Local NARM/RM labeling is first performed in every processor, only in those parts of plateaus which extend to neighboring subdomains. In order to have a plateau uniformly classified, neighboring processors exchange the labels in the edges of their subimages. Thus, if a plateau has been labeled NARM in one processor and RM in the neighboring processor, the label in the latter part is corrected to NARM. Communication and label correction is iteratively executed until stabilization, i.e. no change has occurred in any processor due to communication.

After this step, pixels within each surviving regional minimum at level $l$ are projected onto the subimage at level $l + 1$. As in the sequential algorithm, for every surviving minimum pixel $p$, at level $l$, its projection pixel $p'$, at level $l + 1$, is investigated and diagnosed as a minimum or non-minimum. If after the local exploration of $p'$, and eventually of the plateau at which $p'$ pertains, $p'$ turns to be part of a regional minimum (all pixels on the plateau are labeled either RM, or INIT—the initialization label—, but there is no neighboring pixel of lower altitude than the plateau), the plateau is labeled with the label of $p$ at level $l$, if it has not been already labeled; otherwise, the two labels, of $p$ and $p'$, are united by a UNION operation.

A regional minimum survives if at least one of its pixels projects onto a minimum pixel at the next level, property which has a global nature. Therefore, due to the image distribution, it may be possible that parts of a plateau in some processors extinguish, while parts of the same plateau survive in other processors. Additionally, different parts of the same plateau at level $l$, but located into different processors, merge various other regional minima at level $l + 1$ (minimum splitting/merging); and conversely, onto a regional minimum at level $l+1$, shared by several processors, different regional minima from the level $l$ (minima merging) may project, leading to a non-uniform labeling of the regional minimum at level $l + 1$. Therefore, neighboring processors exchange the labels of pixels in the edges of the subimage, in order to unify those labels assigned to parts of the same plateau. Local label equivalences are further combined by a global connected operator (implementation details about the global connected component operator can be found in [9,11]) in order to provide for each label a global representative or annihilation level. Finally, neighboring regions of minima which annihilate are merged together by another global connected components operator. Alternatively, a special label can be assigned to every regional minimum which extinguish, in order to avoid the overhead of another global connected component operator. Annihilating regions which are smaller than 25 pixels merge any of the neighboring surviving regions.

Since the number of remaining regional minima at level $l+1$ is lower than at level $l$, a relabeling with consecutive labels, starting from 0 is performed in the output image of labels at level 0. Thus, the memory requirements for the lookup table in the connected components operator at level $l+2$ diminish. The resulted image of labels is stored as output of the multiresolution minima merging at level $l + 1$.

The afore described iteration is performed for every smoothing level. However, due to the constraint of keeping the initial image distribution constant across the whole processing, the number of levels for an image $N \times N$ when $P$ processors are used is bounded to $\log_4(N^2/P) - \log_4(25)$, i.e. the smallest

**Table 1.** Execution times of the multiscale watershed algorithm (in seconds)

| Image \P | 1 | 2 | 4 | 8 | 16 | 32 | 64 |
|---|---|---|---|---|---|---|---|
| Cermet(256 × 256) | 0.55844 | 0.29177 | 0.18419 | 0.12256 | 0.09649 | 0.08328 | 0.08011 |
| Peppers(512 × 512) | 1.93015 | 1.10929 | 0.61170 | 0.37135 | 0.25165 | 0.19355 | 0.18417 |
| Lenna (512 × 512) | 2.18953 | 1.36508 | 0.73339 | 0.43497 | 0.29626 | 0.23379 | 0.22226 |
| People (1024 × 1024) | 8.08802 | 4.88998 | 2.91150 | 1.68950 | 1.00565 | 0.66360 | 0.56577 |

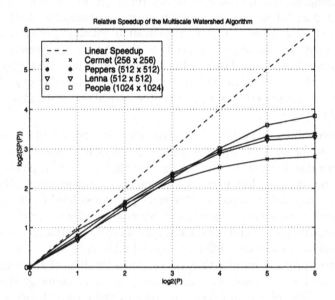

**Fig. 4.** Relative speedup

subimage is 5 × 5. Future work will focus on specializing the implementation to allow adaptive image redistribution at different resolution levels and incorporate load balancing reasonings.

## 4    Experimental Results and Conclusions

The algorithm has been implemented on top of *Message Passing Interface* (MPI) and tested on a Cray T3E parallel computer. Four images of different complexities and different sizes have been used to test the algorithm. The execution time was measured for each image in part and tabulated in Table. 1. Since the maximal height of the multiscale pyramid varies with the number of processors $P$ (see above), this value has been bounded to 5. Thus, the same pyramid structure, with 5 levels, is built for every number of processors used $1 \leq P \leq 64$. The running time drops significantly with increasing number of processors, but the gain stabilizes at 64 processors.

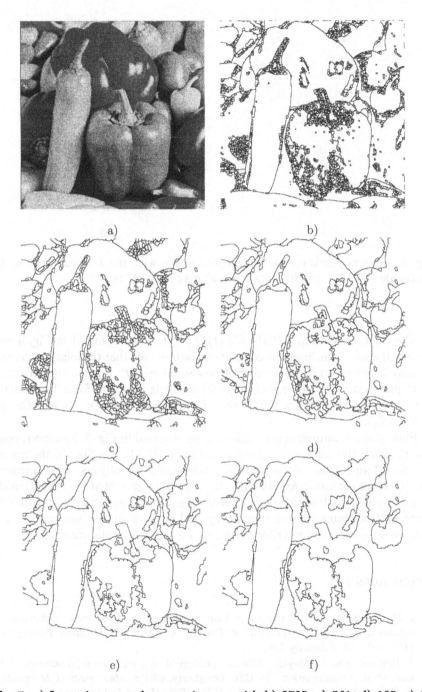

**Fig. 5.** a) Input image and output images with b) 2795, c) 561, d) 185, e) 97, and f) 46 regions

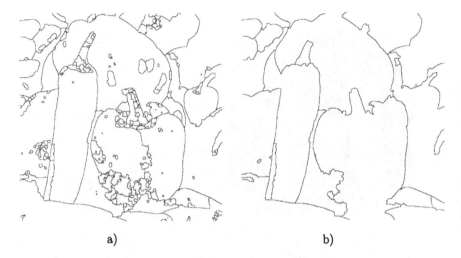

a)                                    b)

**Fig. 6.** *Peppers* Marker based segmented image with markers collected a) by multiscale morphology (264 regs), and b) manually (27 regs)

The relative speedup $SP(P) = T(1)/T(P)$ has been plotted in Fig. 4 and compared against the linear speedup. It can be observed that the relative speedup strongly increases with the number of processors used $P$, but the curve flattens at 64 processors. Moreover, better performance is obtained for a large image (see image *People* (1024 × 1024)) than for a small image (see image *Cermet* (256 × 256)).

Finally, good segmentation results can be observed in Fig. 5. Moreover, compared against the marker based segmentation in [10,11] (see Fig. 6), the results vary in a larger scale of granularity of the segmented objects, while preserving important contours. Additionally, the algorithm presented here is extremely simple, regular, and hence faster than the parallel marker based watershed algorithm, it performs quasi-automatically, and permits an easy incorporation of additional region merging criteria like size, average grayscale, variance.

# References

1. J.-M. Beaulieu: Hierarchy in picture segmentation: A stepwise optimization approach. *IEEE Transactions on Pattern Analysis and Machine Intelligence*, 11(2):150–163, February 1989.
2. S. Beucher and F. Meyer: The morphological approach to segmentation: The watershed transformation. In E.R. Dougherty, editor, *Mathematical Morphology in Image Processing*, pages 433–481, N.Y., 1993. Marcel Dekker Inc.
3. J. M. Gauch and S. M. Pizer: Multiresolution analysis of ridges and valleys in grey-scale images. *IEEE Transactions on Pattern Analysis and Machine Intelligence*, 15(6):635–646, June 1993.

4. L. M. Lifshitz and S .M. Pizer: A multiresolution hierarchical approach to image segmentation based on intensity extrema. *IEEE Transactions on Pattern Analysis and Machine Intelligence*, 12(6):529–540, June 1990.

5. T. Lindeberg: *Scale-Space Theory in Computer Vision.* Kluwer Academic Publishers, Dordrecht, The Netherlands, 1994.

6. F. Meyer: Minimum spanning forests for morphological segmentation. In J. Serra and P. Soille, editors, *Computational Imaging and Vision-Mathematical Morphology and Its Applications to Image Processing*, pages 77–84, Dordrecht, The Netherlands, September 1994. Kluwer Academic Publishers. Proceedings of the International Symposium on Mathematical Morphology, Fontainebleau, France.

7. F. Meyer: Topographic distance and watershed lines. *Signal Processing*, 38(1):113–125, July 1994.

8. F. Meyer and S. Beucher: Morphological segmentation. *Journal of Visual Communication and Image Representation*, 1(1):21–46, September 1990.

9. A. Moga and M. Gabbouj: Parallel image component labeling with watershed transformation. *IEEE Transactions on Pattern Analysis and Machine Intelligence*, 19(5):441–450, May 1997.

10. A. Moga and M. Gabbouj: Parallel marker-based image segmentation with watershed transformation. *Journal of Parallel and Distributed Computing*, 25(1):27–45, May 1998.

11. A. N. Moga: *Parallel Watershed Algorithms for Image Segmentation.* PhD thesis, Tampere University of Technology, Tampere, Finland, February 1997.

12. A. Montanvert, P. Meer, and A. Rosenfeld: Hierarchical image analysis using irregular tessellations. *IEEE Transactions on Pattern Analysis and Machine Intelligence*, 13(4):307–316, April 1991.

13. I. E. Pratikakis, H. Sahli, and J. Cornelis: Hierarchy determination of the gradient watershed adjacent groups. In *Proceedings of the 10th Scandinavian Conference on Image Analysis*, volume I, pages 685–692, Lappeenranta, Finland, June 1997.

14. A. Rosenfeld: *Multiresolution Image Processing and Analysis.* Springer-Verlag, 1984.

15. P. Schroeter and J. Bigün: Hierarchical image segmentation by multi-dimensional clustering and orientation-adaptive boundary refinement. *Pattern Recognition*, 28(5):695–709, 1995.

16. K. L. Vincken: Probabilistic multiscale image segmentation. *IEEE Transactions on Pattern Analysis and Machine Intelligence*, 19(2):109–120, February 1997.

17. S. G. Ziavras and P. Meer: Adaptive multiresolution structures for image processing on parallel computers. *Journal of Parallel and Distributed Computing*, 23:475–483, 1994.

# Solving Irregular Inter-processor Data Dependency in Image Understanding Tasks

Yongwha Chung and Jin-Won Park

Performance Evaluation Team, ETRI
{yongwha + jwpark}@computer.etri.re.kr

**Abstract.** It is challenging to parallelize the problems with irregular computation and communication. In this paper, we proposed a scalable and efficient algorithm for solving irregular inter-processor data dependency in image understanding tasks on distributed memory machines. Depending on the input data, in the scatter phase, each processor distributes search requests to collect the remote data satisfying certain geometric constraints. In the gather phase, the requested remote data are collected by using a DataZone data structure and are returned to each processor. For demonstrating the usefulness of our algorithm, we conducted experiments on an IBM SP2. The experimental results were consistent with the theoretical analyses, and showed the scalability and efficiency of the proposed algorithm. Our code using C and MPI is portable onto other High Performance Computing(HPC) platforms.

## 1 Introduction

Many problems with *regular* computation and communication have been parallelized on distributed memory machines by using explicit message passing paradigm or by using parallelizing compilers[5]. However, it is challenging to parallelize problems with *irregular* computation and communication. Such problems arise in intermediate and high-level image understanding tasks, for instance. In these problems, the nature of the irregularity depends on the input image and is unknown at the compile time[10]. To achieve high performance, the irregularity should be considered carefully and dynamic techniques may be required.

A key operation in intermediate and high-level image understanding tasks is to search feature data satisfying certain geometric constraints. For instance, in perceptual grouping, feature data satisfying certain geometric constraints are grouped to form structural hypotheses[7,8]. In another example, image matching, correspondences between feature data extracted from two different images are determined based on geometrical relationships[9].

Parallelizing such feature search is challenging since the operation is highly *data dependent*(the size and shape of the search is different for each data element). Thus, the inter-processor communication pattern is generally irregular, since the number of processors to communicate and the amount of data transferred depend on the distribution of input data. Our motivation for this research

P. Zinterhof, M. Vajteršic, A. Uhl (Eds.): ACPC'99, LNCS 1557, pp. 236–245, 1999.
© Springer-Verlag Berlin Heidelberg 1999

is to develop an efficient algorithm for solving irregular inter-processor data dependency on distributed-memory machines.

The irregular data dependency can be simply solved by using all-to-all broadcast operations to obtain all the feature data. By enabling each processor to receive all the feature data, the problem caused by the inter-processor data dependency is avoided. As the number of processors increases, however, the broadcast time grows linearly and becomes the dominant part of the total execution time. Furthermore, this simple solution requires a fixed amount of per-processor memory, and this makes it impossible to solve larger problems with a large number of processors.

In this paper, we proposed a scalable and efficient algorithm for solving irregular data dependency. A *DataZone* array was defined first such that only the remote data satisfying certain geometric constraints can be collected efficiently, instead of checking all the remote data. Then, depending on the input data, the search requests for remote data are distributed in the scatter phase. In the gather phase, the requested remote data are collected by using the DataZone array and are returned to each processor. Using a realistic abstract model of distributed memory machines, a tradeoff between computation and communication in solving irregular data dependency was also analyzed.

For demonstrating the usefulness of our algorithm, experiments were conducted on an IBM SP2. The experimental results were consistent with the theoretical analyses, and showed the scalability and efficiency of the proposed algorithm. Although we showed the experimental results for the feature search operation as an example, the idea can also be applied to many other irregular data dependency problems.

The organization of the paper is as follows. An overview of the irregular data dependency problem is given in Section 2. Section 3 discusses our approach for solving the problem. Experimental results are shown in Section 4. Concluding remarks are made in Section 5.

## 2 Irregular Data Dependency in Image Understanding

A key operation in intermediate and high-level image understanding tasks is to search feature data satisfying certain geometric constraints. Compared with parallelizing low-level image understanding operations[1,2,6], several issues must be considered in parallelizing such feature search operation.

- The computations are performed on *symbolic* data(point, line, and area).
- The search is *highly data dependent*(the size and shape of the search is different for each symbolic data).
- The search may need *non-neighbor* data.

For instance, the communication pattern, which is in terms of the processors to communicate and the amount of data to be transferred, depends on the distribution of input data and can be irregular. As shown in Fig. 1, the pattern

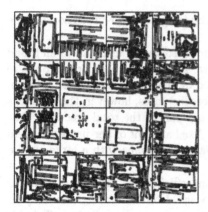
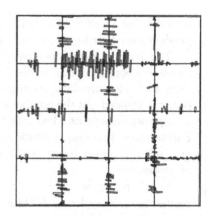

**Fig. 1.** Search windows generated by line segments have intra-processor data dependency(left) and inter-processor data dependency(right). The generated search windows are shown as small rectangles. The input line segments were extracted from an $1024 \times 1024$ Modelboard image on a 16-processor SP2.

of inter-processor data dependency is different for each processor. Such irregular communication patterns result in large parallelization overheads. Thus, the motivation for this research is to develop an efficient algorithm for solving irregular inter-processor data dependency. Another issue in parallelizing the feature search operation is related to unbalanced data distribution, and its solution was discussed in [3].

In the following sections, we explain how to solve the irregular data dependency in the feature search operation as an example. However, the idea can also be applied to solve irregular data dependency in other application domains.

## 3    Algorithm and Analysis

As we discussed in Section 1, the irregular data dependency in the feature search operation can be simply solved by using all-to-all broadcast operations to obtain all the feature data. However, this solution is not scalable and unsuitable for larger problems. An algorithm which does not employ the all-to-all broadcast operation is described first in this section. Then, the effect of using an efficient data structure is analyzed.

### 3.1    Notation

Let $P$ and $n \times n$ denote the number of processors and the image size, respectively. The input to each processor is an image block of size $\frac{n}{\sqrt{P}} \times \frac{n}{\sqrt{P}}$, along with the line segments extracted from the image block during the pre-processing phase(i.e., feature extraction). The output is a set of line segments collected from remote processors such that each processor can perform further operations

without any inter-processor data dependency. Feature extraction can be paral-
lelized by partitioning the $n \times n$ image array into $P$ blocks of size $\frac{n}{\sqrt{P}} \times \frac{n}{\sqrt{P}}$, and
then performing the contour detection and linear approximation operations for
a partitioned image block in each processor. A parallel implementation of the
feature extraction has been reported in [2,6].

Following the sequential algorithm for the feature search used in the perceptual grouping step[7], a search for each line segment is performed on both sides
of it(within a 4-pixel width region) to find other line segments satisfying certain
constraints(see Fig. 2). Following terminology is used in this paper:

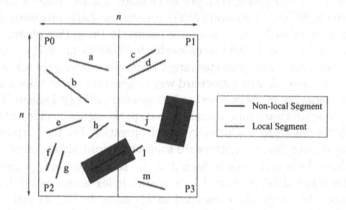

**Fig. 2.** An example illustrating the notations using 4 processors.

- *Token*: a data structure containing information about an input line segment,
  such as end-point coordinates, length, orientation, and average contrast of
  the line segment. In the following, line segments, segments, and tokens are
  used interchangeably. For the sake of explanation, a token performing a
  search is denoted as a *source* token, whereas a token to be examined in the
  search window is denoted as a *target* token. For a source token $s$, the search
  window generated by it is denoted as $W(s)$.
- *Owner(s)*: a processor which stores a source token $s$. In the example shown
  in Fig. 2, the owner of source tokens $e$, $f$, $g$, $h$, $i$ is processor $P_2$, and the
  owner of $j$, $k$, $l$, $m$ is $P_3$.
- *Domain($P_i$)*: an image block of size $\frac{n}{\sqrt{P}} \times \frac{n}{\sqrt{P}}$ assigned to a processor $P_i$ during the pre-processing phase. In the example shown in Fig. 2, Domain($P_2$) is
  the lower left block of size $\frac{n}{2} \times \frac{n}{2}$. We assume that each processor has information about the domain of the other processors as well as that for itself. For
  instance, each processor has information about Domain($P_0$), Domain($P_1$),
  Domain($P_2$), and Domain($P_3$).

- $Remote(W(s))$: the set of processors whose domain overlaps the search window $W(s)$, i.e., $Remote(W(s)) = \{P_i | Domain(P_i) \cap W(s) \neq \emptyset, P_i \neq Owner(s)\}$. In the example shown in Fig. 2, $Remote(W(i)) = \{P_3\}$.
- *Local, non-local* tokens: a token $s$ having empty (non-empty) $Remote(W(s))$ is denoted as a *local* (*non-local*) token.

## 3.2   Algorithm

The inter-processor data dependency can be managed in a *scatter-and-gather* fashion. In the scatter phase, each processor sends a source token $s$ having non-empty $Remote(W(s))$ to $Remote(W(s))$ processors. Each processor generates search requests for each non-local source tokens. In the gather phase, each processor $P_i$ sends back the target tokens located in $W(s) \cap Domain(P_i)$ to $Owner(s)$. Note that, each processor generates target tokens for each source token received in the scatter phase. A straightforward way to generate target tokens is to scan all the line segments to check whether it is located in $W(s) \cap Domain(P_i)$. However, this straightforward solution can degrade performance significantly because it scans all the line segments for every search operation. For the purpose of comparison, we denote this straightforward solution as *scan* algorithm(see Fig. 3).

To reduce the search time in Step 3 of the scan algorithm, pointers representing the token data are stored in a spatial index array, called *DataZone*. For instance, the image block assigned to $P_3$(shown in Fig. 2) and the corresponding DataZone are shown in Fig. 4. To find target tokens(i.e., $l$), only the neighboring area(i.e., the shaded locations on the DataZone array) needs to be searched; instead of scanning all the line segments(i.e., $j$, $k$, $l$, $m$) to check whether it is located within the requested area. The size of the DataZone array is $\frac{n}{\sqrt{P}} \frac{1}{SF} \times \frac{n}{\sqrt{P}} \frac{1}{SF}$, where $SF$ is a scale factor. By changing $SF$, we can control a tradeoff between the number of searched locations and the extra token data. For simplicity, $SF$ is chosen as 1(i.e., the size of the DataZone array is the same as that of an image block) in the following. The algorithm using the DataZone array is denoted as *spatial index* algorithm and is shown in Fig. 5.

**Procedure : Scan Algorithm**
  **Step 1:** Generate search requests for *non-local* line segments.
  **Step 2:** Scatter the search requests.
  **Step 3:** Scan all the line segments for each request.
  **Step 4:** Gather line segments located in the requested area.
**end**

**Fig. 3.** An outline of the scan algorithm for solving the irregular data dependency problem.

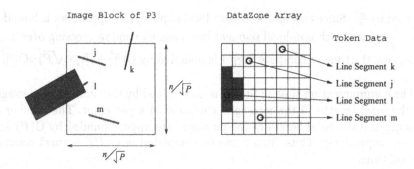

**Fig. 4.** An example illustrating the spatial indexing technique with the DataZone array.

**Procedure : Spatial Index Algorithm**
    **Step 1:** Create a *DataZone* array to store input line segments.
    **Step 2:** Generate search requests for *non-local* line segments.
    **Step 3:** Scatter the search requests.
    **Step 4:** Perform window search on the DataZone array for each request.
    **Step 5:** Gather line segments located in the requested area.
**end**

**Fig. 5.** An outline of the spatial index algorithm for solving the irregular data dependency problem.

## 3.3    Analysis

Let $P$ and $m$ denote the number of processors and the length of the message(bytes), respectively. The $n \times n$ image is assumed to be partitioned into $P$ blocks of size $\frac{n}{\sqrt{P}} \times \frac{n}{\sqrt{P}}$ during the pre-processing step. The communication time of each algorithm can be analyzed using the simple communication model described in [4]. According to the communication model, the cost of sending a message between two processors can be represented by two parameters: the *message startup time* $T$, which is the time required to initiate the communication, and the *transfer time per byte* $\tau$, which is determined by the physical bandwidth of the communication channel linking the source and destination processors. Then, the time required to send a message of size $M$ bytes is represented as $T + M\tau$. The details of the communication model can be found in [4].

If an all-to-all broadcast operation is used to solve the inter-processor data dependency, then the total sizes of the messages is bounded by $O(n^2 m)$. Thus, the communication time and per-processor memory requirement to solve the inter-processor data dependency are $O(PT + n^2 m\tau)$ time and $O(n^2 m)$ bytes, respectively.

In analyzing the scan algorithm shown in Fig. 3, the distribution of the input line segments over the partitioned image block of size $\frac{n}{\sqrt{P}} \times \frac{n}{\sqrt{P}}$ needs to be considered. The number of source tokens initially stored in a processor is upper bounded by $\frac{n^2}{P}$. Since the number of non-local segments in a processor is bounded by $4 \times \frac{n}{2\sqrt{P}}$ and each non-local segment has a search window crossing over $2\sqrt{P}$ processors, the time to compute Step 1 is bounded by $O(\frac{n^2}{P} + \frac{n}{\sqrt{P}} \times \sqrt{P})=O(\frac{n^2}{P})$, $P \leq n$.

The communication time in Step 2 is determined by the number of messages and the total lengths of the messages generated in a processor. The number of messages and the total length of the messages are upper bounded by $O(P)$ and $O(nm)$, respectively. Thus, Step 2 can be completed in $O(PT + nm\tau)$ communication time.

The time to perform Step 3 is determined by the number of source tokens received by a processor. Each processor can receive at most $4 \times \frac{n}{2\sqrt{P}}$ search requests, and each search request needs to check every segments initially stored in a processor. Since the number of segments initially stored in a processor is bounded by $O(\frac{n^2}{P})$, Step 3 can be completed in $O(\frac{n^3}{P\sqrt{P}})$ time.

Similar to Step 2, the communication time in Step 4 depends on the number of messages and the total length of the messages generated in a processor. The number of messages and the total length of the messages are upper bounded by $O(P)$ and $O(\frac{n^2m}{P})$, respectively. Thus, Step 4 can be completed in $O(PT+\frac{n^2m}{P}\tau)$ communication time. Finally, the inter-processor data dependency can be solved totally in $O(\frac{n^3}{P\sqrt{P}})$ computation time and $O(PT + \frac{n^2m}{P}\tau)$ communication time, $P \leq n$.

The per-processor memory requirement is determined by the number of target tokens received by a processor. Since the number of target tokens collected for each source token is upper bounded by $O(\frac{n}{\sqrt{P}})$, the additional memory requirement to solve the inter-processor data dependency is bounded by $O(\frac{n^2m}{\sqrt{P}})$ bytes.

The communication time of the spatial index algorithm shown in Fig. 5 is the same as that of the scan algorithm without it. However, the computation time can be reduced by using it. The time to create the DataZone in Step 1 is upper bounded by $\frac{n^2}{P}$, whereas the time to collect the target tokens in Step 4 is reduced to $O(\frac{n^2}{P})$ by using it. Thus, the total computation time is bounded by $O(\frac{n^2}{P})$. Note that the additional memory requirement which is proportional to $\frac{n^2}{P}$ does not increase the worst-case memory requirement.

In Table 1, the computation time, the communication time, and the per-processor memory requirement of each algorithm are compared.

## 4   Experimental Results

For the purpose of comparison, we implemented all three algorithms described in the previous section on an IBM SP2. The code was written using C and MPI message passing library. For each source token, a search was performed within the region on both sides of it within a 4-pixel width to find target tokens, using the algorithms shown in Fig. 3 and 5. The length of messages such as source

| | Computation Time | Communication Time | Memory Requirement |
|---|---|---|---|
| Broadcast | - | $O(PT + n^2 m \tau)$ | $O(n^2 m)$ |
| Scan | $O(\frac{n^3}{P\sqrt{P}})$ | $O(PT + \frac{n^2 m}{P} \tau)$ | $O(\frac{n^2 m}{\sqrt{P}})$ |
| Spatial Index | $O(\frac{n^2}{P})$ | $O(PT + \frac{n^2 m}{P} \tau)$ | $O(\frac{n^2 m}{\sqrt{P}})$ |

**Table 1.** Analysis of the computation time, the communication time, and the per-processor memory requirement of each algorithm to solve the inter-processor data dependency.

tokens and target tokens was 100 bytes. The input line segments were extracted from two $1024 \times 1024$ Modelboard images, and the number of input line segments was 3519 from Modelboard1 and 7825 from Modelboard2, respectively.

| | Broadcast | | Scan | | Spatial Index | |
|---|---|---|---|---|---|---|
| | Comm | Comp | Comm | Comp | Comm | Comp |
| 4 processors | 24 msec | - | 4 msec | 3783 msec | 4 msec | 211 msec |
| 16 processors | 35 msec | - | 5 msec | 2623 msec | 5 msec | 153 msec |
| 64 processors | 68 msec | - | 8 msec | 1083 msec | 8 msec | 76 msec |

**Table 2.** Execution time for Modelboard1 image on SP2.

| | Broadcast | | Scan | | Spatial Index | |
|---|---|---|---|---|---|---|
| | Comm | Comp | Comm | Comp | Comm | Comp |
| 4 processors | 50 msec | - | 7 msec | 32476 msec | 7 msec | 288 msec |
| 16 processors | 62 msec | - | 8 msec | 20819 msec | 8 msec | 194 msec |
| 64 processors | 103 msec | - | 16 msec | 9925 msec | 16 msec | 112 msec |

**Table 3.** Execution time for Modelboard2 image on SP2.

In Table 2 and 3, we show the execution times of each algorithm. The execution times of the broadcast algorithm increased with the number of processors used although it was very small for a 4-processor case. In contrast, the scan algorithm is scalable in that the execution times were reduced with the increased number of processors. This is because each processor is responsible for a smaller size of the image block with the increased number of processors. For the practical number of processors, however, the overhead to collect the target tokens can offset the gain from the scalability. Note that, the communication time observed in the scan algorithm was negligible. In the broadcast algorithm, there is no computation time because it simply performs the all-to-all broadcast operation regardless of the given pattern of the data dependency.

Compared with the scan algorithm, the effect of spatial indexing with the DataZone array is significant. The portion of computation times was reduced by an order of magnitude for Modelboard1 and two orders of magnitude for Modelboard2, respectively. The greater improvement was obtained from Modelboard2 image because the scan algorithm should scan the much longer list of the input segments for every search operation. Note that the amount of communication time of the spatial index algorithm was the same as that of the scan algorithm.

To evaluate the additional per-processor memory requirement of each algorithm, we show the average of the memory requirements over the processors in Table 4 and 5. The average of additional per-processor memory requirement of the broadcast algorithm was computed as *the total number of input line segments* $\times \frac{P-1}{P} \times 100 bytes$. For the scan algorithm, it was computed as $\sum_{i=1}^{P}$ *the total number of target tokens collected to processor* $i \times \frac{1}{P} \times 100 bytes$. In addition to this amount of per-processor memory requirement, the spatial index algorithm needs the memory space of size $\frac{n}{\sqrt{P}} \times \frac{n}{\sqrt{P}} \times 100 bytes$ for the DataZone array to reduce the search time. However, the additional requirement for the DataZone can be reduced by increasing the scale factor $SF$.

|  | Broadcast | Scan | Spatial Index |
|---|---|---|---|
| 4 processors | 263.9 KBytes | 0.2 KBytes | 1048.7 KBytes |
| 16 processors | 329.9 KBytes | 0.1 KBytes | 262.2 KBytes |
| 64 processors | 346.4 KBytes | 0.1 KBytes | 65.6 KBytes |

**Table 4.** Average of additional per-processor memory requirements for Modelboard1 image on SP2.

|  | Broadcast | Scan | Spatial Index |
|---|---|---|---|
| 4 processors | 586.8 KBytes | 0.3 KBytes | 1048.8 KBytes |
| 16 processors | 733.5 KBytes | 0.3 KBytes | 262.4 KBytes |
| 64 processors | 770.2 KBytes | 0.2 KBytes | 65.7 KBytes |

**Table 5.** Average of additional per-processor memory requirements for Modelboard2 image on SP2.

## 5   Concluding Remarks

We have presented a scalable and efficient algorithm for solving irregular data dependency in intermediate and high-level image understanding tasks, and shown the results of implementation on an IBM SP2.

A DataZone array was defined first such that only the remote data satisfying certain geometric constraints could be collected efficiently, instead of checking all the remote data. Then, depending on the input data, the search requests for remote data were distributed, and the requested remote data were collected by using the DataZone array.

The experimental results were very encouraging. Both the execution time and the additional per-processor memory requirement of the proposed algorithm were scalable with the number of processors used. The experimental results were also consistent with the theoretical analyses. The code written in C and MPI permits portability to various HPC platforms. We believe that our idea can also be applied to solve irregular data dependency in other application domains.

# References

1. D. Bader and J. JaJa, "Practical Parallel Algorithms for Dynamic Data Redistribution, Median Finding, and Selection", *Proc. of International Parallel Processing Symposium*, pp. 292-301, 1996.
2. Y. Chung, V. Prasanna, and C. Wang, "A Fast Asynchronous Algorithm for Linear Feature Extraction on IBM SP-2", *Proc. of Computer Architectures for Machine Perception*, pp. 294-301, 1995.
3. Y. Chung, J. Woo, R. Nevatia, and V. Prasanna, "Load Balancing Strategies for Symbolic Vision Computations", *Proc. of Conf. on High Performance Computing*, pp. 263-269, 1996.
4. I. Foster, *Designing and Building Parallel Programs: Concepts and Tools for Parallel Software Engineering*, Addison-Wesley, Co., 1995.
5. G. Fox, M. Johnson, G. Lyzenga, S. Otto, J. Salmon, D. Walker, *Solving Problems on Concurrent Processors: General Techniques and Regular Problems*, Prentice-Hall, Inc., 1988.
6. D. Gerogiannis and S. Orphanoudakis, "Load Balancing Requirements in Parallel Implementations of Image Feature Extraction Tasks," *IEEE Tr. on Parallel and Distributed Systems*, Vol. 4, No. 9, pp. 994-1013, 1993.
7. A. Huertas, C. Lin, and R. Nevatia, "Detection of Buildings from Monocular Views of Aerial Scenes using Perceptual Grouping and Shadows," *Proc. of Image Understanding Workshop*, pp. 253-260, 1993.
8. D. Lowe, *Perceptual Organization and Visual Recognition*, Kluwer Academic Press, 1985.
9. G. Medioni and R. Nevatia, "Matching Images using Linear Features", *IEEE Tr. Pattern Analysis and Machine Intelligence*, Vol. 6, No., 6, pp. 675-685, 1984.
10. C. Wang, P. Bhat, and V. Prasanna, "High Performance Computing for Vision", *Proceedings of IEEE*, Vol. 84, No. 7, pp. 931-946, 1996.

# A New Parallelism Management Scheme for Multiprocessor Systems

Xavier Verians[1], Jean-Didier Legat[1], Jean-Jacques Quisquater[1], and
Benoit Macq[2]

[1] Microelectronics Laboratory,Université Catholique de Louvain,
[2] Telecommunications Laboratory, Université Catholique de Louvain,
place du Levant, 3; 1348 Louvain-la-Neuve; Belgium

**Abstract.** Current multimedia and signal processing applications become more complex. This paper proposes a new parallelism management scheme that can explicitly deal with complex and general parallelism patterns. The parallelism description is based on a task flow graph
representation interlaced with control commands. A graph management
algorithm is proposed to extract eligible tasks and implement synchronization operations. We show that this management outperforms classical parallelism extraction in case of complex applications. Moreover, the
parallelism description can be directly inserted in sequential programs
without deep code modifications.
**keywords:** parallelism, task graph, task scheduling, multiprocessors

## 1  Introduction

Many applications have to be parallelized to be implemented on multiprocessor
systems. The parallelization task can be split into two steps: the first step consists
in the algorithm parallelization and the second step translates the extracted
parallelism into the system parallel syntax. This second part is not trivial and
often reduces the available parallelism to respect syntax constraints.

More particularly, new multimedia or image processing applications (MPEG-
4 codec, videophony, ...) are based on highly complex structures. Data are
divided into small blocks and taking into account some block statistics, a kernel
is selected and executed. Consequently, these applications are constituted by
kernels linked together by a complex control structure. Therefore, the available
parallelism moves from the kernel level where smaller data sets are used, to the
control level allowing to simultaneously execute several kernels and statistical
computations. Due to the close interaction between parallelism and program
control flow, it can only be exploited with a flexible description.

Finally, loops often are neither completely parallel nor fully sequential. Some
loop body parts need to be executed in the loop order while other parts are
completely independent. The corresponding task graph is close to an oriented
meshed structure (Fig. 1). Usually, this partial parallelism is ignored because
classical fork/join commands cannot easily describe such parallelism patterns. In
the best cases, loops are unrolled to partially exploit the available parallelism or

P. Zinterhof, M. Vajteršic, A. Uhl (Eds.): ACPC'99, LNCS 1557, pp. 246–256, 1999.

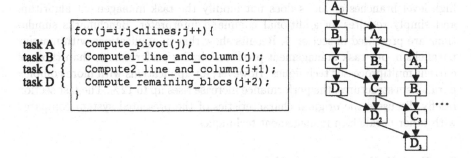

```
                for(j=i;j<nlines;j++){
    task A {       Compute_pivot(j);
    task B {       Compute1_line_and_column(j);
    task C {       Compute2_line_and_column(j+1);
    task D {       Compute_remaining_blocs(j+2);
                }
```

**Fig. 1.** Partially parallel loop (example taken from the LU algorithm). *left:* pseudo-C loop, *right:* corresponding task graph. $A_i$ is the $i^{th}$ instance of $A$.

an explicit pipelined implementation is programmed. Both solutions require that the programmer significantly modifies the program code to match the parallelism syntax.

These observations lead us to study a more general parallelism representation allowing to describe various parallelism patterns and simple enough to be handled at a fair cost by a multiprocessor system. The proposed parallelism management scheme is based on the following characteristics:

- A separation of the parallelism description and the main control structures from the computation code. It releases the processing units (PUs) of the parallelism management task.
- An original parallelism description with a complexity adapted to the extracted parallelism levels, able to describe general parallelism patterns. Complex parallelism patterns allow to schedule tasks sooner in the process. Moreover, it is possible to extract parallelism from structured sequential programs without major code modifications to compel with strict syntax.
- An original task management algorithm to quickly determine tasks ready to be executed and to perform efficient synchronization operations.

The parallelism management has been designed to be implemented on small-scale systems able to execute up to 64 simultaneous tasks. The system architecture is not imposed, provided that a centralized parallelism management and a shared memory are supported. The system can be a (simultaneous) multithreaded processor, where tasks are used to hide remote or memory access latency, or a multiprocessor system with simple RISC or complex superscalar processors.

This paper is structured as follows. The following section presents the multiple aspects of our parallelism description. It begins with some classical theoretical considerations before describing the parallelism representation. A study of the parallelism levels is presented in subsection 2.3. The proposed task management algorithm is developed in section 3. For the paper clarity, we suppose in section 2 and subsection 3.1 that the program does not contain branches or loops between tasks. We will show in subsection 3.2 that the generalization to high-level branches or loops does not modify the task management algorithm and simply requires an additional command interpreter. Benchmarks simulations are presented in section 4. Results show that in applications with massive parallelism, our task management produces similar performances than those of current multiprocessor techniques. However, for applications with more complex parallelism structures, the performance increase rises up to 12%. The last section briefly discusses the original characteristics of the presented system compared with other parallelism management techniques.

## 2   Parallelism Description

### 2.1   Theoretical Aspects

Program parallelism can be represented with a task graph. The program is divided into several tasks connected by some dependence links. A task is an instruction set which will be executed on a single processing unit. Tasks can communicate and interact together by using the shared memory. Their size and number of dependences is not limited. They can contain jumps, function calls and conditional branches, provided all jumps are referencing an address included in the same task.

A task graph [1] is a directed graph consisting of a set of vertices (tasks) $V$ and a set of arcs (dependences) on the vertices $A$. If the arc $(t_i, t_j) \in A$, then task $T_i$ must complete execution before task $T_j$ can start execution. We say that task $T_i$ is a *preceding task* of $T_j$. Arcs $(t_i, t_j)$ are also called *incoming arcs of $T_j$*. The notation used is the following: capitals to design tasks and lower-case letters for arcs. The *distance* between two tasks $d(T_i, T_j)$ is defined as the minimum number of arcs to follow to go from a task $T_k \in V$ to both $T_i$ and $T_j$.

The task graph defines a task ordering. This ordering is not strict as parallel tasks do not need to be executed one before another. A task $T_i$ is *eligible* for execution only if all tasks $T_j$ such that $(t_j, t_i) \in A$ have been completed.

At run-time, the task graph is unrolled to produce a directed acyclic graph (DAG), called *dynamic task graph*[1]. Due to its limited resources, the processor will only see a slice of this graph, called the *task window*. The task window holds all tasks already decoded by the system, but waiting for their execution to be launched or completed.

### 2.2   Task Program

Most current systems insert parallelism commands in the program code [2][3]. The processing units execute tasks and, when a parallel command is encountered, they execute it and update the eligible task queue. The time lost by the processing units to handle parallelism commands leads to an overhead becoming significant in case of complex parallelism structures. Moreover, the task creation order depends on the task execution, for example, when several connected tasks are created in different parallel threads. This can lead to complex protocols to identify synchronization edges between tasks[4].

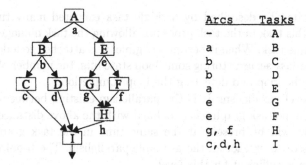

| Arcs | Tasks |
|------|-------|
| -    | A     |
| a    | B     |
| b    | C     |
| b    | D     |
| a    | E     |
| e    | G     |
| e    | F     |
| g, f | H     |
| c,d,h | I    |

**Fig. 2.** *left:* Task graph example, *right:* Corresponding task program

We propose another approach consisting in separating the parallelism information from the program code. This separate parallelism description, called the *task program*, is directly sent to the task manager, releasing the processing units from the parallelism decoding. The communications between the task manager and the processing units are reduced. Besides, the parallelism decoding can be performed in parallel with task execution. Finally, as the parallelism description syntax is independent of the task code, the syntax complexity can easily be adapted to the exploited parallelism level.

The multiple tasks defined in the program code create a static dependence graph containing loops and conditional parts. As the task program is also static, it needs to include some commands to implement loops and branches among tasks. At run-time, these commands are decoded to build the correct dynamic task graph. For the paper clarity, we will treat this point later in section 3.2. We will first suppose that the static task graph is already a directed acyclic graph. The static graph and the dynamic task graph are in this case identical.

The dynamic task graph decoding is performed by a sequential reading of the task program. A first constraint on the task program is that the task order has to be respected, i.e. for each task $T_i$, all preceding tasks $T_j$ are encoded before $T_i$ in the task program. Consequently, the task program is simply an ordered topological description of the task graph (Fig. 2). It allows to represent all the parallelism patterns which can be drawn as a directed acyclic graph. The syntax is simple: each line describes a task $T$ and its incoming arcs $(t_j, t)$. As $t$ is the same for all the incoming arcs, it is not noted.

## 2.3 Extended Parallelism Description

The limited size of task window can prevent the processor to extract all the available parallelism. The simple topological graph description has to be extended to allow an efficient parallelism extraction. An analysis of multimedia and signal processing applications shows three parallelism levels of increasing complexity:

1) Low-level parallelism: Signal processing applications are often based on small kernels easily parallelizable (DCT, quantification, filtering, ...). These ker-

nels can often be described by a single task executed many times. A tag identifies this task in the task program, allowing the task manager to handle it as a simple task. When it becomes eligible, it is attached to a dedicated resource, the *task server*, storing some loop state bits, the number of processors executing the loop and detecting the loop completion.

2) Medium-level parallelism: It is the parallelism existing between close tasks in the dynamic task graph, *close* defined with the above distance definition. Close tasks will be present at the same time in the task graph, allowing the processor to extract all the available parallelism. The topological graph description is sufficient for this level.

3) High-level parallelism: It often arises that multiple processing can be performed in parallel. If each processing is described by a large number of tasks, a sequential decoding cannot provide a comprehensive view on the global parallelism structure. A simultaneous decoding of different task graph parts is necessary. High-level commands, such as classical fork/join commands [5], are added to the task program. The key difference with other systems is that these commands are only used at a high level, where the command management overhead is negligible compared with the execution time of the multiple task sets.

# 3  Task Management

## 3.1  Parallelism Extraction Algorithm

The main task manager job is to quickly determine the eligible tasks. It is equivalent to identify the entry tasks of the task window graph. Moreover, when a task is completed, it has to remove it from the task window and has to compute the new eligible tasks.

We propose an algorithm based on a queue bank. The queue filling is performed in the task decoding order: if $T_j$ needs to be executed after $T_i$ completion, it will be added to a queue after $T_i$. The filling policy is simple: a task $T_i$ is added to the queue $q_j$ only if $T_k = tail(q_j)$ is such as $(t_k, t_i) \in A$. These two rules guarantee that the queues form a graph covering with each queue containing a chain of consecutive tasks. The task graph structure is explicitly conserved in the queue organization: different parallel branches are stored in different queues. It allows to implement intelligent task scheduling policies based on a dynamic analysis of the task graph. Moreover, thanks to the task ordering, eligible tasks will only be found in the queue heads.

The three basic operations on the task window graph are the following:

- add_task(T): Search a queue $q_k$ such as $tail(q_k) = T_j$, with $(t_j, t) \in A$. If no queue satisfies the condition, $q_k$ will point to an empty queue. Then all the incoming arcs $(t_j, t)$ are added to the queue, followed by the task identifier. If no empty queue is available, the decoding waits for the execution to proceed further until a queue is released.

- get_eligible_task(): If a task in a queue depends on other tasks of the task window graph, it is preceded by its incoming arcs. Only eligible tasks are not preceded by arcs nor tasks and are thus stored in the queue head. Finding an eligible task requires only to determine if the head of a queue contains an arc or a task identifier. When selected, the task is removed from the queue.
- remove_task(T): When a task T completes, all the arcs $(t, t_j)$ are removed from the queues. It is the most expensive operation. However, the critical operation is the removal of an arc preventing a new task to become eligible. These arcs are found in the queue heads. Thus the time between a task completion and a dependent task launching is determined by the delay needed to remove the arcs associated with T from the queue heads.

Tasks selected from the task graph are sent to an eligible task queue or to a task server in case of a low-level loop. Currently, no task scheduling policy is implemented. As the eligible queue is similar to the eligible queue of classical multiprocessor systems, except that the synchronization is already performed, classical task scheduling techniques can be implemented.

## 3.2 Generalization to High-Level Commands

Up to now, we have supposed that the task program was free of loops, branches and high-level parallel commands. We will now introduce these control instructions and show that they only require a small task interpreter to be handled by the task manager. The previous task management algorithm remains valid.

High-level commands can be considered as particular tasks executed by the task manager in place of the processing units. If these commands are executed when they are decoded, they are simply inserted in the task program. If they depend on some tasks, they are added to the task graph, modifying the task program consequently. The task graph management is not modified. The additional complexity is limited to the command interpreter which remains simple as the number of commands is limited. An example of a static task graph, and the derived task program is presented in figure 3.

High-level branches dynamically modify the task graph. They are considered as small tasks executed by the manager to modify the task decoding order. Branches are predicted to speculatively decode the most probable side. A branch tag is attached to the queues to allow a possible invalidation. When a branch becomes eligible, it is resolved. In case of wrong prediction, the tagged queues are invalidated and the decoding starts on the other branch side. The queue invalidation can be performed in parallel with the task decoding. The cost is thus hidden to the processing units. Several additional commands are defined to improve performance or to reduce some parallelism management operations.

The use of backward branches allows several tasks to be executed multiple times. In order to distinguish between the different task instances, tasks and arcs are renamed at the decoding stage to guarantee that each instance has a unique name in the task window. As the tasks in the task program are ordered, the arcs $(t_i, t_j)$ where $T_i$ has not been renamed are ignored. It guarantees the absence of dependence loops in the execution graph.

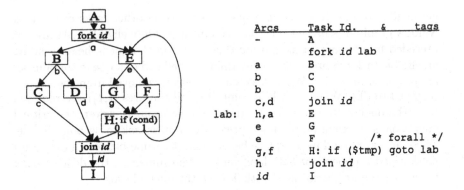

**Fig. 3.** *left:* Static task graph, *right:* Task program. The fork/join commands are not necessary as the two task graph parts are small. It is only set for example purpose.

## 4  Simulations

The task manager has been simulated in a simple multiprocessor architecture. The processing units are simple RISC processors. As the goal is to measure the efficiency of the parallelism extraction, the system has been supposed to have ideal buses and memories. The relative speedup is defined as the ratio between the measured execution time with the execution time on a single processor. The normalized speedup is defined as the ratio between the measured execution time with the execution time of an equivalent sequential program executed on a single processor. The compiler used is based on the SUIF compiler system [6], extended with the Harvard's MACHINE library [7]. Several passes have been added to handle the task and parallelism extraction.

Figure 4a displays the system performance with SPLASH and SPLASH-2 benchmarks. The overall performance is close to the ideal performance as given in [2][8], except for the LU benchmark where the speedup is increased. These SPLASH applications mainly have simple parallelism structures and the data sets are large enough to hide the parallelism management overheads.

The performance increase for the LU benchmark is achieved by exploiting the complex parallelism of the outermost loop. The gain is more visible when executing the program on smaller data sets (Fig. 4b). The parallelism extraction has been lead further than in the SPLASH benchmark, without modifying the underlying algorithm. Indeed, LU iterations are partially dependent. The SPLASH version does not extract this parallelism. The iterations have been sliced to extract the partial meshed-like parallelism (Fig. 1). The consequence is the "pipelining" of the multiple iterations across the system. Results show in all cases an improved performance. With a little parallelism available (>256 blocks), results are improved of about 12%. The instruction overhead is limited to less than 2%.

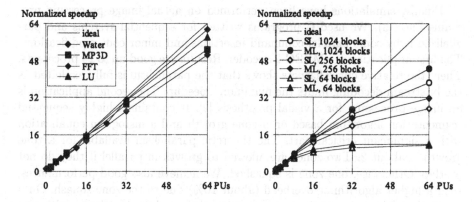

**Fig. 4.** *a)* SPLASH benchmark performances: WATER(64 mol.), LU(256x256 matrix), MP3D(3000 mol.), FFT($2^{14}$ points); *b)* LU performances with 64x64 matrix. SL is for classical single-level parallelism exploitation, ML is for the proposed multi-level parallelism exploitation.

The LU benchmark has also been simulated on a system with 2 cache levels (latency: 1/3/10 cycles) and an atomic bus (latency: 1 cycle) (Fig. 5a). Despite higher miss rates and no data locality optimizations, our parallelism management still produces better performance. The main overhead is due to the memory latency. The implementation of a classic data locality policy would sharply reduce this overhead. With more than 32 PUs, the locking overhead rises, as the critical parts become significant. However, the computation overhead remains negligible despite the deeper parallelism exploitation (idle time divided by 2.2).

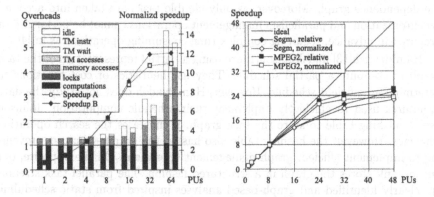

**Fig. 5.** *a)* left histogram and speedup A: classical parallelism management, right histogram and speedup B: with proposed parallelism management; *TM* is for Task Manager *b)* Image processing benchmarks

Finally, simulations have been performed on actual image processing programs(Fig. 5b). We used C programs written for sequential systems and parallelized it by some parallel command insertions and minor code modifications. The first application is a MPEG-2 decoder. Results are good up to 24 processors. Then, the relative curve flatness shows that the parallelism is fully exploited as the bitstream decoding limits the maximum speedup. The second application is an image processing for a visual prosthesis [9]. It contains a highly sequential segmentation algorithm based on a zone growth and a histogram equalization with a significant critical part. The theoretic parallelism available for a zone growth is about 2. Two zones are allowed to growth in parallel if they do not overlap. Otherwise, one zone is squashed. We achieve also good performances, with a higher algorithmic overhead (about 10%) due to the zone squash. On a classical multiprocessor system, these programs would need major rewriting to exploit some of the available parallelism.

## 5   Related Work

A large number of static and dynamic scheduling techniques have been proposed to exploit parallelism on multiprocessor systems. Static scheduling techniques[10] [11] are essentially based on a thorough analysis of the program dependence graph. The compiler uses its global view on the program structure to modify the graph and determine when tasks could be executed and to which processor they will be allocated. Their advantage is the ability to use complex algorithms to compile near-optimal solutions. However, they are not suited to data-dependent applications. Dynamic scheduling techniques [4][12] mainly use the eligible task queue to compute an efficient scheduling. They compute or read several statistics for each task and select a task matching some criteria. The drawback of these methods is that they do not use the information contained in the future tasks of the dependence graph. Moreover, as only eligible tasks are taken into account, synchronization and parallelism management operations cannot be performed or prepared in advance, requiring complex time-consuming operations as in [4].

Hamidzadeh[13] and Johnson[1] are using a partial representation of the task graph to compute a partial schedule. They use heuristics or cost functions to determine an efficient schedule. However, Hamidzadeh does not study the data structures used to efficiently implement their schedule computation. Johnson uses a hashing table to store the task graph. It increases the search operation efficiency. However, the hashing table also hashes the graph structure, preventing to implement efficient graph management operations. On the opposite, our queue bank stores the graph in a structured fashion. The parallel task streams are clearly identified and graph-based analyses inspired from static scheduling techniques can be adapted to extract a partial schedule. The eligible tasks can be sent to a pool implementing classical load balancing techniques[12].

# 6 Conclusion

We have presented a new parallelism management scheme to allow the exploitation of complex and general parallelism patterns in multiprocessor systems. The generality is achieved by representing the parallelism with a topological description. The separate parallelism description allows to implement a high pattern complexity with a limited dynamic overhead. Indeed, the parallelism is no more spread throughout the program code and is decoded in parallel to task execution. The advantages of the parallelism description are dynamically exploited by a co-designed task manager. It stores a structured representation of the task dependence graph allowing to implement efficiently graph management operations and task synchronization. Simulations validate the parallelism management on scientific and image processing applications. For highly parallel applications, we have similar performances than current multiprocessor parallelism exploitation. However, when the application presents more complex parallelism patterns, performances can be increased up to 12% for the LU SPLASH benchmark compared with classical parallelism management. Furthermore, the syntax generality does not require major program rewritings to be able to describe the parallelism.

Future work will be oriented to the study of the hardware/software implementation issue. Load balancing and data locality techniques will be added to enhance practical performances. A second research direction is the automation of the parallelization step.

# References

[1] Johnson, T., Davis,T., Hadfield, S.: A Concurrent Dynamic Task Graph. Parallel Computing, Vol 22, No 2, Ferbruary 1996, 327–333

[2] Singh, J.P., Weber W-D., Gupta, A.: SPLASH: Stanford Parallel Applications for Shared Memory. Computer Architecture News, 20(1), July 1994, 5–44

[3] Chandra, R., Gupta, A., Hennessy, L.: Integrating Concurency and Data Abstraction in the COOL Parallel Programming Language. IEEE Computer, February 1994

[4] Rost, J., Markus, F.-J., Li Yan-Hua: "Agency Scheduling" A Model for Dynamic Task Scheduling. Proc of the Int. Conf. on Parallel Processing, Euro-Par 95, Sweden

[5] Nikhil, R.S., Papadopoulos, G.M., Arvind: *T: A Multithreaded Massively Parallel Architecture. Int. Symp. on Computer Architecture, 1992, 156–167

[6] Stanford Compiler Group: The SUIF Library. Stanford University (1994)

[7] Smith, M.D.: The SUIF Machine Library. Harvard University (1997)

[8] Woo, S.C., Ohara, M., Torrie, E., Singh, J.P., Gupta, A.: The SPLASH-2 Programs: Characterization and Methodological Considerations. Proc. of the $22^{nd}$ Ann. Symp. on Computer Architecture, June 1995, 24–36

[9] Gilmont, T., Verians, X., Legat J-D., Veraart, Cl.: Resolution Reduction by Growth of Zones for Visual Prosthesis. Int. Conf. on Image Processing, 1996, 299–302

[10] Shirazi, B., Wang, M.: Analysis and Evaluation of Heuristic Methods for Static Task Scheduling. J. of Parallel and Distributed Computing, vol. 10, 1990, 222-232

[11] Park, G.-L., Shirazi, B., Marquis J., Choo, H.: Decisive Path Scheduling: A New List Scheduling Method. Proc. of the Int. Conf. on Parallel Processing, 1997, 472–480

[12] Dandamudi, S. P., Cheng, P. S.: A Hierarchical Task Queue Organization for Shared-Memory Multiprocessor Systems. IEEE Transaction on Parallel and Distributed Systems, vol. 6, no. 1, January 1995, 1–16
[13] Hamidzadeh, B., Lilja, D.J.: Dynamic Scheduling Strategies for Shared-Memory Multiprocessors. Proc. of the 16<sup>th</sup> Int. Conf. on Distributed Computing Systems, Hong Kong, 1996, 208–215

# A Flexible VLSI Parallel Processing System for Block-Matching Motion Estimation in Low Bit-Rate Video Coding Applications

Donglai Xu and Reza Sotudeh

Computer Architecture Research Unit, School of Science and Technology,
University of Teesside, Middlesbrough, TS1 3BA, UK
D.Xu@Tees.ac.uk

**Abstract.** In this paper, we design a flexible VLSI-based parallel processing system for an improved three-step search (ITSS) motion estimation algorithm that is superior to the existing three-step search (TSS) algorithm in all cases and also to the recently proposed new three-step search (NTSS) algorithm if used for low-rate video coding, as with the H.261 standard. Based on a VLSI tree processor and an FPGA addressing circuit, the proposed architecture can successfully implement the ITSS algorithm on silicon with the minimum number of gates. Because of the flexibility of the architecture, it can also be extended to implement other three-step search algorithms.

## 1 Introduction

It is well known that motion estimation algorithms play an important role in video sequence compression. Many fast block-matching algorithms for motion estimation have been proposed because of their lower computation overhead than that of full-search block-matching algorithm, such as the existing three-step search (TSS) algorithm [1] and the recently proposed new three-step search (NTSS) algorithm [2]. Recent studies show that the motion vector distribution of a real world image sequence within the search window is highly centre-biased. Based on this fact, we propose an improved version of the well-known TSS method, the ITSS algorithm [3], specifically aiming towards low bit-rate video coding applications. The ITSS has much better performance and faster speed than the original. Compared to the NTSS, its performance is better when applied to our target applications, such as videophone, and its speed is faster as well, without any direct or hidden costs.

Compared with TSS, the ITSS uses the same number of checking points in each step, but a different search pattern. This leads to better performance of ITSS while maintaining the same data flow of TSS. That means the architecture for TSS can also be used for ITSS. Many VLSI architectures [4], [5] have been proposed for TSS. However, these architectures either have low throughput, or high hardware cost, or too little flexibility. To avoid these drawbacks and to implement the proposed ITSS algorithm on silicon with the minimum number of gates, a low-latency and high-

P. Zinterhof, M. Vajteršic, A. Uhl (Eds.): ACPC'99, LNCS 1557, pp. 257-264, 1999.

throughput parallel pipeline computing architecture, based on a VLSI tree processor and an FPGA addressing circuit, is presented. Owing to its simple and modular properties, the tree processor is suitable for VLSI implementation, and because of the use of FPGA to implement addressing and control circuits, the architecture is flexible enough to implement different three-step search algorithms. Furthermore, the tree processor can be decomposed into sub-trees to reduce hardware cost and pin count. Memory interleaving and pipeline interleaving are also employed to enhance memory bandwidth and to raise pipeline utilisation to 100%.

In the next section, we will describe the ITSS algorithm. In section 3, we will present the computing architecture for ITSS. A sample design for videophone application will be discussed in section 4. Finally, the conclusion will be given in section 5.

## 2 ITSS Algorithm

The experimental results in [2] have shown that the block motion field of real world image sequences is usually gentle, smooth, and varies slowly. It results in a centre-biased global minimum motion vector distribution instead of a uniform distribution. This is particularly true in low bit-rate video applications, including videophone and video conferencing, where fast and complex movements are involved rarely. For most video sequences, there are nearly 80% blocks that can be regarded as stationary or quasi-stationary blocks and most of the motion vectors are enclosed in the central 5×5 area. For such a distribution, we have developed an improved three-step search algorithm (ITSS) which uses a centre-biased checking point search pattern adapted to the centre-biased motion vector distribution, hence its performance is expected to be far better than that of TSS which uses a uniformly distributed checking point search pattern. Additionally, the ITSS employs a smaller number of search points than TSS to speed up block matching. The details of the algorithm are given in the example described below.

Following earlier block-matching techniques, our example takes a block size of 16×16 pixels and a maximum search range of ±7 pixels in both horizontal and vertical directions. The mean absolute error (MAE) is used as an appropriate estimate of the block distortion measure (BDM). For a given $(x,y)$, the MAE between $block(m,n)$ of the current frame and $block(m+x,n+y)$ of the previous (reference) frame is defined as:

$$\text{MAE}_{(m,n)}(x,y) = \frac{1}{256} \sum_{i=0}^{15} \sum_{j=0}^{15} |f_k(m+i,n+j) - f_{k-1}(m+x+i,n+y+j)|, \qquad (1)$$

where $f_k(i,j)$ and $f_{k-1}(i,j)$ are the pixel intensities at position $(i,j)$ of the current frame $k$ and the previous frame $k-1$ respectively, and the $block(m,n)$ is the block with its upper left corner at position $(m,n)$ of a frame. The first step of the algorithm employs a centre-biased search pattern with nine checking points on a 5×5 window instead of the 9×9 window with TSS. The centre of the search window is then shifted to the point with minimum BDM. The search window size of the next two steps depends on the location of the minimum. If the minimum lies at the centre of the search window, the search will go to the final step (step 3) with a 3×3 search window. Otherwise, the search window size is maintained at 5×5 for step 2. In the final step, the search window is reduced to 3×3 and the search stops at this small search window. The three search steps of ITSS can be summarised as follows:

Step 1: The minimum BDM point is found from a nine-checking-points pattern on a 5×5 window located at the centre of the 15×15 searching area as shown in Figure 1(a). If the minimum BDM point is found to coincide with the centre of the search window, then go to step 3 else go to step 2.

Step 2: The search window size is maintained at 5×5 with a search pattern chosen after considering the two alternatives:

  a)  If the previous minimum BDM point is located at one of the corners of the previous search window, then five additional checking points are considered. An example is shown in Figure 1(b) where black circles and grey circles represent additional and previously evaluated pixels respectively.

  b)  If the previous minimum BDM point is located at the middle of any horizontal or vertical edge of the previous search window, then three additional checking points are considered. An example is shown in Figure 1(c) where black circles and grey circles represent additional and previously evaluated pixels respectively.

Step 3: The search window is reduced to 3×3 around the minimum BDM point found in step 2 as shown in Figure 1(d) and the direction of the overall motion vector is taken to be the minimum BDM point among these nine search points, of which eight are new.

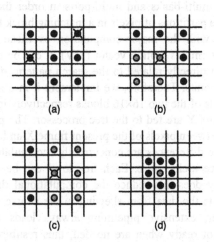

**Fig. 1.** Search pattern of the ITSS. (a) First step centred on centre pixel; (b) Second step centred on a corner pixel; (c) Second step centred on a middle pixel; (d) Third step

Because there are some overlapping check points on the 5×5 search window in the second step of ITSS, the total number of checking points will vary from a minimum of (9+8)=17, when step 2 can be skipped, to (9+5+8)=22 in the worst case which is three block matches less than TSS, eleven block matches less than NTSS.

## 3 Proposed Architecture

The system architecture of ITSS, which consists of memory subsystem, tree processor, and address generation circuit, is described in Figure 2.

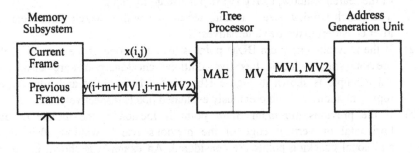

**Fig. 2.** Block diagram of system architecture for ITSS algorithm

The memory subsystem stores the current and previous video frames by which the tree processor is provided with input data. To enhance memory bandwidth, the memory system adopts multiple memory modules, such as an N×N module array, to apply memory interleaving for simultaneous accesses. In addition, the memory system is divided into multi-banks and multi-ports in order that the memory cycle time can afford the huge partitions of pixels in a matching block to be interleaved.

The tree processor, with the major computation overhead in this architecture, computes MAE and determines motion vector (MV). It can be a modular-processing engine because it is not directly related to the search pattern of ITSS algorithm and the position of search points. Therefore, we implement it in a single-chip VLSI. As shown in Figure 3, pixels of the two 16×16 blocks respectively in the present frame X and in the previous frame Y are fed to the tree processor. The processing element D computes differences between pixels of the present frame X and those of the previous frame Y in parallel. The differences are concurrently accumulated by the adders that comprise the binary tree architecture. Each tree level can be viewed as a parallel pipeline stage; not only does this reduce the computational data path length (tree height) but also prevents the data from skewing in the course of doing the parallel computations. However, extensive pipelining always leads to lost resources if computed results are not ready when are needed, thus resulting in hazards. In the pipelined computation of ITSS, during its three-step operations, the current step must complete its computation before the next step begins; therefore, it will lead to lost resources because the computed results of the current step are not ready when they are needed by the next step. To avoid performance degradation caused by these hazards, pipeline interleaving is employed. The main approach is to interleave the execution of adjacent block-matching tasks in a pipeline, that is the current task and the task next to it. They can be jointly processed without any interference because they are independent of each other. In other words, when a step of the current task awaits data, a step of the next task can be processed in advance to fill out the empty machine cycles caused by delays; thus, the computing machine can work at full speed.

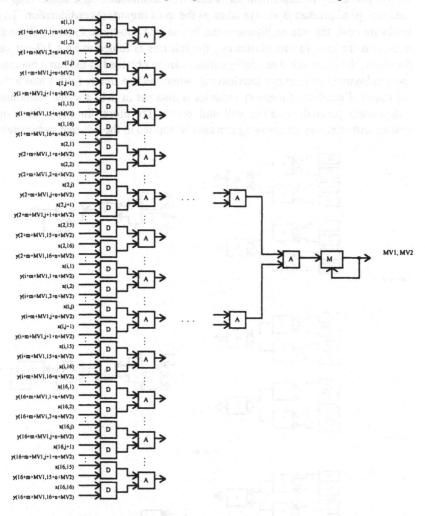

**Fig. 3.** Tree architecture processor for MAE and MV computing

The address generation unit will address memory for computing MAE and motion vector. It controls the search pattern of the ITSS algorithm and assigns search points by selecting the proper block address and pixel address within the block. By changing the design of this component, different hierarchical search algorithms could be implemented in this architecture. Therefore, FPGA implementation is used, to provide future flexibility.

## 4 Application Design

In VLSI implementation of the tree processor for specific applications, it is not always necessary to implement the whole tree architecture on a single chip because the area-speed product is always taken as the most important consideration. To reduce hardware cost, the tree architecture can be cut into sub-trees. Figure 4 illustrates the tree-cut technique. In this illustration, the full tree is folded to be a 1/2-cut sub-tree. Similarly, the other sub-tree configurations can be obtained in the same manner. With the employment of memory interleaving, whenever the tree is folded into a sub-tree, the required number of memory modules is also cut in half, and the phenomenon of independent pixel distribution will still occur. Therefore, the multi-bank memory system with memory interleaving can also be applied to the sub-tree architecture.

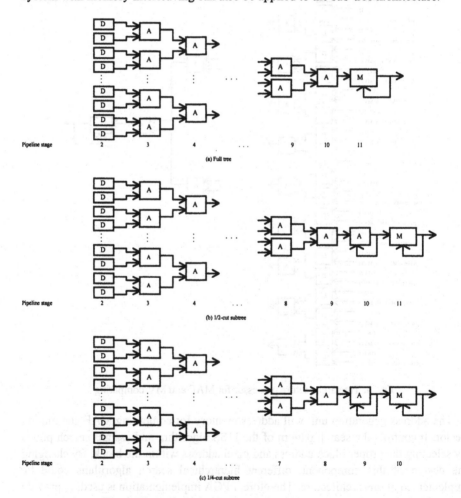

**Fig. 4.** Illustration of tree-cut technique for block size 16×16

In the following, we consider the hardware implementation of the tree architecture (N=16) for a videophone application. Suppose dynamic random access memories (DRAM's) with 200ns access time are used to implement the memory system and the time to latch a pixel from a memory module is simulated to be about 50ns. If a 1/C-cut sub-tree architecture and a B-bank memory system are adopted, then we have the following constraint:

$$\frac{200\text{ns} \times B}{(256 \text{ memory modules})/C} \geq 50\text{ns}; \text{ and thus } C \times B \geq 64. \tag{2}$$

Thus the 1/C-cut sub-tree architecture must connect with a memory system of at least 64/C memory banks to allow simultaneous accesses. On the other hand, the period time for each pipeline stage is also constrained by memory cycle time and is 200ns. For videophone application (288×352 frame size, 16×16 block size, and 10Hz frame rate), the throughput required by the processing system is

$$\frac{352 \times 288}{16 \times 16} \times 10 = 3.96\text{k blocks/second,} \tag{3}$$

so that 252.5us is allowed for each vector. With a 1/C-cut sub-tree, the number of time cycles required to compute a motion vector is 27×C. Thus, the sub-tree architecture for the application is constrained by:

$$200\text{ns} \times (27 \times C \text{ time cycles for 1/C - cut sub - tree}) \leq 252.5\text{us};$$
$$\text{and thus } C \leq 46.76. \tag{4}$$

Consequently, the 1/32-cut (C=32) sub-tree architecture with two memory bands (B=2) for this videophone application is optimal in speed or hardware cost for the realisation of ITSS algorithm.

The real throughput rate depends on the clock rate applied to it. For the above application, the clock cycles required to estimate a motion vector is 27×C=864. Suppose a clock rate of E MHz is applied to the sub-tree processor. Then, in order to estimate motion vector in real time for the application, there should be the following constraints:

$$\frac{E \times 10^6}{864} \geq 3.96 \times 10^3; \text{ then } E \geq 3.42 \text{ MHz.} \tag{5}$$

This means that the minimum clock rate to do real-time block matching is 3.42MHz. The single chip 1/32-cut sub-tree is currently synthesised and simulated in a 0.7um CMOS standard cell technology, requiring about 5000 equivalent gates.

## 5 Conclusions

In this paper, we describe a flexible and efficient architecture for implementation of the improved ITSS algorithm. In this architecture, simple and modular tree structures facilitate VLSI implementation, the FPGA designs for addressing and control circuits improve the flexibility of the system, and memory interleaving and pipeline interleaving enhance the overall system computing performance. Furthermore, the tree-cut technique is introduced to reduce the hardware cost of the tree processor; but still allowing real-time processing requirements to be met in the sub-tree architecture for specific applications. Finally, a videophone application is considered and its speed and hardware cost trade-offs are evaluated to find out an optimal VLSI implementation.

# References

1.  Koga T., Iinuma K., Hirano A., Iijima Y. and Ishiguro T.: Motion-compensated interframe coding for video conferencing, Proceedings of National Telecommunication Conference, pp. G5.3.1-G5.3.5, New Orleans, LA, USA (November 1981).
2.  Li R., Zeng B. and Liou M.L.: A new three-step search algorithm for fast block motion estimation, IEEE trans. on Circuits and System for Video Technology 4, 438-442 (1994).
3.  Booth W., Noras J.M.and Xu D.: A novel fast three-step search algorithm for block-matching motion estimation. In: Chin R., Pong T.C. (eds.): Computer Vision. Lecture Notes in Computer Science, Vol. 1352. Springer-Verlag, Hong Kong (1998) 623-630.
4.  He Z., Liou M.L., Chan P.C.H. and Li R.: An efficient VLSI architecture for new three-step search algorithm, Proceedings of the 1995 IEEE 38th Midwest Symposium on Circuits and Systems, pp. 1228-1231, USA (August 1995).
5.  Gupta G. and Chakrabarti C.: Architectures for hierarchical and other block matching algorithms, IEEE Transactions on Circuits and Systems for Video Technology 5, 477-489 (1995).

# Hierarchical Block Matching Motion Estimation on a Hypercube Multiprocessor*

Charalampos Konstantopoulos, Andreas Svolos, and Christos Kaklamanis

Computer Engineering and Informatics Department, University of Patras and
Computer Technology Institute, 70 G. Papandreou, GR 157 73, Athens, Greece
{konstant,svolos,kakl}@cti.gr

**Abstract.** Block matching motion estimation algorithms are widely
used in video coding schemes. In this paper, we design an efficient hierarchical block matching motion estimation (HBMME) algorithm on
a hypercube multiprocessor. Unlike systolic array designs, this solution
is not tied down to specific values of algorithm parameters and thus offers increased flexibility. Moreover, the hypercube network can efficiently
handle the non regular data flow of the HBMME algorithm. We also
assume that our multiprocessor is fine grained in contrast to most programmable architectures used in video coding where processors usually
have a large local memory. Apart from its practicality, the constraint of
limited local memory makes the algorithm design more challenging and
thus more theoretically interesting.

## 1   Introduction

Block matching motion estimation algorithms are widely used in video coding
schemes [3]. The basic idea is to divide the current frame into equally sized blocks,
and then to find for each block the best matching block in an available previous frame. This can be done by full (exhaustive) search within a search window
(optimal solution), or by using an intelligent non exhaustive search (suboptimal solution) in order to reduce the computation requirements. Additionally, a
multiresolution representation of video frames can be used for achieving higher
performance (hierarchical block matching algorithms). Next, the motion vector
of each block in the current frame is determined by the relative displacement of
the best matched block in the previous frame. As a measure of block similarity, the mean absolute difference between two blocks is typically used because
it requires no multiplication and has similar performance to the mean square
error.

Due to its high computational demands, video coding is usually implemented
in hardware. Hardware architectures can be split into application specific and
programmable [4]. In the first case, special purpose hardware is optimally designed. For example, a large number of architectures have been appeared for

---

* Work supported in part by the General Secretariat of Research and Technology of
  Greece under Project ΠΕΝΕΔ 95 ΕΔ 1623.

P. Zinterhof, M. Vajteršic, A. Uhl (Eds.): ACPC'99, LNCS 1557, pp. 265–275, 1999.

block matching motion estimation algorithms especially for the full search algorithm. Due to its highly regular data flow, most realizations of this algorithm are based on mesh-like systolic arrays. Despite their high efficiency, these application specific designs lack flexibility. A change in algorithm parameters or improvements in the coding scheme may lead to a costly hardware redesign. On the other hand, programmable architectures offer higher flexibility at a cost of reduced efficiency. The coding algorithms are developed in software and thus any change can be easily handled. In the literature, many designs have been reported that follow either of the two approaches (for a survey see [4, 5]).

In this paper, following the programmable architecture approach, we study the software based realization of the HBMME algorithm on a fine grained hypercube based multiprocessor. We also present an efficient full search block matching motion estimation (FSBMME) algorithm which is used as a subroutine in the HBMME algorithm. A basic assumption in our study is that the multiprocessor is fine grained. In contrast, most programmable architectures used in video coding are coarse grained multiprocessors where each processor has enough memory to keep all the data it will need throughout algorithm execution. Thus interprocessor communication is almost eliminated and processors can operate independently. In our fine-grained multiprocessor, the processors have limited memory and thus the interprocessor communication is inevitable. Clearly, in this kind of architecture the design of efficient algorithms is a more challenging task. In addition, the use of a powerful network such as the hypercube for the implementation of the HBMME algorithm is well justified because this algorithm has a non regular data flow and its inherent communication is not local. Thus it cannot be easily implemented on systolic arrays. Dedicated hardware designs for the HBMME algorithm[6, 7, 8], require high external memory bandwidth or relieve these requirements by using large on chip memory.

The rest of this paper is organized as follows. In Sect. 2, we briefly review the block matching motion estimation algorithms. In Sect. 3, we present the implementations of FSBMME and HBMME on the hypercube multiprocessor. Finally, in Sect. 4 we summarize our work in this paper.

## 2   Block Matching Algorithms

In the FSBMME algorithm, the current frame $N \times N$ is divided into blocks of size $M \times M$ and each block is compared with all the blocks of size $M \times M$ in the search window of size $(M + 2d) \times (M + 2d)$ in the previous frame (Fig. 1(a)). Here $d$ denotes the maximum displacement in each direction. We also refer to a block $M \times M$ with its top left corner at the picture element (pel) $(u, l)$ as block $(u, l)$. For all displacements $(x, y)$ $(x, y = -d, \cdots, d)$, the mean absolute difference (MAD) between the block $(u, l)$ of the current frame $X$ and the block $(u + x, l + y)$ in the search window of the previous frame $Y$ is given by:

$$MAD_{(u,l)}(x, y) = \frac{1}{M^2} \sum_{i=0}^{M-1} \sum_{j=0}^{M-1} |X(u+i, l+j) - Y(u+x+i, l+y+j)| \quad (1)$$

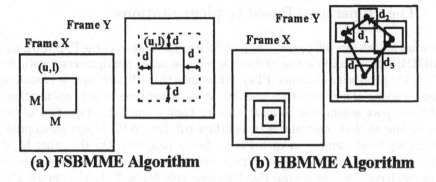

**(a) FSBMME Algorithm**     **(b) HBMME Algorithm**

Fig. 1. BMME algorithms

where $X(u+i, l+j)$, $Y(u+x+i, l+y+j)$ denote the intensity values of the corresponding pels. Now, the motion vector $v(u, l)$ of the block $(u, l)$ is given by

$$v(u, l) = arg \min_{(x,y)} MAD_{(u,l)}(x, y) \ . \qquad (2)$$

The choice of the right block size is important for the FSBMME algorithm. With either too large or too small block size, the algorithm might well yield false motion estimates [6]. The HBMME algorithm solves this kind of problems by using a multiresolution (hierarchical) representation of video frames in the form of a Laplacian pyramid [3]. The basic idea is to start the estimation of motion field from the lowest resolution level. At this level, the block size is relatively large in comparison with the frame size at that resolution level and the estimated motion vectors capture the large-scale movements existing in the scene. Then, these vectors are passed onto the next higher resolution level as an initial estimate. The higher resolution levels refine on the motion vector estimates and thus smaller block size should be used. The lower resolution frames in the pyramid are obtained by a series of low-pass filtering and subsampling operations.

There are a number of variations of the basic algorithm. The first variation is to skip the subsampling between successive levels of the pyramid. Alternatively, we can use only subsampling without low pass filtering. A third possibility is the use of overlapping blocks at each level. In this scheme, the motion vector of each block at one level is initialized as a linear interpolation of the motion vectors of its adjacent blocks at the previous lower resolution level. Finally, we can use either the FSBMME algorithm or a non exhaustive search BMME algorithm for motion estimation at each level.

An example of motion vector estimation by using a 3–level hierarchy is shown in Fig. 1(b). First the motion vector $d_3$ of the largest block is estimated (the lowest resolution level). Then at the next higher resolution level the vector $d_2$ is calculated around the point which $d_3$ points to. In the third level the vector $d_1$ is estimated using the smallest size block. The final motion vector is the vector sum $d$ of $d_1$, $d_2$, $d_3$.

## 3    The Hypercube Based Implementations

In what follows, we will present the parallel implementation of the FSBMME and HBMME algorithm in a fine grained hypercube based multiprocessor with $N^2$ ($N = 2^n$) processing elements (PEs). We assume that PEs are capable of sending and receiving along more than one link at the same time (n-port assumption). The one-port assumption is more easy to handle and the adaptation of our algorithms to that constraint is straightforward. Due to the n-port assumption, $\Omega (\log N)$ local memory in each PE is clearly necessary. On the other hand, the fine grained nature of the multiprocessor limits the amount of local memory PEs can have. Thus we assume that PEs have only $\Theta (\log N)$ local memory. This makes more difficult and hence interesting the design of efficient algorithms on the hypercube network. Finally, for convenience, we view the hypercube as a two dimensional $N \times N$ grid and thus when we use terms like row, column, block, we will actually mean the corresponding subhypercubes. Initially, the pixel values $X (i,j)$ and $Y (i,j)$ of the current and previous frame respectively are stored in PE $(i,j)$, that is, PE $j + iN$.

Before proceeding further, we must introduce the basic communication operations used by our algorithms. For more details see [1, 2].

- Shift($A$,$B$,$i$,$N$): $B (j) \longleftarrow A ((j + i) \, mod N)$ where $A (j)$, $B (j)$ are registers of PE $j$, $j = 0 \cdots N - 1$ and $N$ is the size of the hypercube. The complexity of the operation is $O (\log N)$ in general.
- Data_Sum($A$,$B$,$N$): the sum of the values of the register $A$ in all PEs is stored into the register $B$ of PE 0. The operation complexity is $O (\log N)$.
- Broadcast($A$,$B$,$N$): The PE 0 broadcasts the value of its register $A$ and this value is stored in the register $B$ of all PEs. The time complexity is $O (\log N)$.
- Random Access Read RAR($A$,$B$,$N$): Each PE reads the content of the register $A$ of some other PE and stores the data in its register $B$. The RAR operation involves a sorting step which determines the complexity of the operation. Considering only practical sorting algorithms, the complexity of the RAR is $O(\log^2 N)$ in the general case. Yet it is possible to lower the sorting step complexity by exploiting the special structure of the BMME algorithms.

A common alternative for handling data movements in an image grid is to use techniques of embedding mesh on hypecube [1, 2]. However, this solution is out of the scope of the paper as it turns the hypercube into a mesh network and thus techniques tailored to mesh networks should be used instead.

### 3.1    The FSBMME Algorithm

As has already been mentioned, in the FSBMME algorithm the current frame (frame $X$) is partitioned into $\frac{N^2}{M^2}$ non overlapping $M \times M$ blocks ($M = 2^m$). The basic operations in the algorithm are given by (1), (2). Clearly, each block corresponds to a subhypercube of size $M \times M = M^2$. As Fig. 1(a) shows, in order to find the motion vector of the block $(u, l)$, the MAD should be evaluated for all the $(2d+1)^2$ candidate vectors $(x, y)$. All pels of frame $Y$ required for these calculations are located inside the search window of the block $(u, l)$ and each PE in the block will only need $(2d+1)^2$ of these pels. These pels are transferred one at a time by a series of shift operations through the hypercube. After each shift, the MAD for the corresponding candidate vector is estimated by executing a Data_Sum operation inside the block $(u, l)$. A high level description of the parallel algorithm is given below:

```
Best_MAD[0,0]=infinity    /* In each block, initialize the best
  for r=-d to d do        /* so  far MAD value to infinity.
    for c=-d to d do
      shift(Y,temp,c+rN,N^2) /* temp is a temporary register
      diff[i,j]=|X[i,j]-temp[i,j]| for all i,j=0 ... N-1
      Data_sum(diff,MAD,M^2) within each block  /* calculate the MAD
      In each block if MAD[0,0] < BEST_MAD[0,0] then
                    BEST_MAD[0,0]= MAD[0,0]
                    best_vector[0,0]=(r,c)
                    endif
    endfor
  endfor
```

All operations in a single loop iteration are executed serially, one after another, while the iterations themselves are overlapped by initiating each iteration with one time step delay from the previous one. Rather than moving the frame $Y$ around the hypercube by a series of $(2d+1)^2$ horizontal shifts of $\pm 1$ and $(2d+1)$ vertical shifts of the same distance, we maintain the initial placement of $Y$ by transferring copies of this frame. If we used shifts of $\pm 1$, then a shift could not start before all its previous shifts had finished. Instead, in our algorithm each PE transfers the required pel values from their original positions by a series of shifts of $c + rN$ (where $r, c = -d, \cdots, d$) through a hypercube of size $N^2$. This dramatically reduces the interdependence between loop iterations. Now, an iteration needs the result of the previous ones only when it compares its MAD value with the best MAD estimated so far. But the scheduling of the operations is such that this information is always available on time.

When overlapping operations in a network, collisions may arise among operations trying to send or receive over the same link. In our algorithm, most operations are performed without collisions. We prove the following lemma:

**Lemma 1.** *Collision arises only between Data_Sum and shift operations. At most 2 packets try to pass through the same link at the same time.*

*Proof.* We first prove that collisions between shift operations cannot arise. In a single invocation of the shift operation, hypercube dimensions are visited one at a time in descending order, starting from the most significant one. Since in our algorithm each shift operation is initiated with one step delay from its previous one, all the concurrent shifts use different hypercube dimensions at any time and so no collisions occur.

Following the same reasoning, we can prove that Data sum operations are also performed without collisions among themselves. However it can be easily seen that there are collisions between Data Sum and shift operations. Yet, at most two packets may try to pass through the same link at the same time and thus the time complexity order of the algorithm does not increase.          □

Using the previous lemma, we can easily prove that the FSBMME algorithm can be executed on a hypercube network with $O\left(d^2 + \log N + \log M\right)$ delay. It should be noted that this complexity shows only the delay of communication operations which are the most costly part of each parallel algorithm. At each pipeline stage of the above scheme $O\left(\log M\right)$ partial sums are calculated and thus the total arithmetic complexity remains $O\left(d^2 \log M\right)$ the same as without pipeline. Yet, these operations are integer additions so their overhead is expected to be much smaller than that of communication operations.

The maximum number of partial sums calculated by processors at each pipeline stage can be greatly reduced by slightly modifying the above scheme. The Data sum operation in its simplest form uses a fixed spanning tree of the hypercube network and the final sum is obtained at node 0, the root of this spanning tree. It is almost straightforward to alter the Data sum algorithm so as to collect the final sum at an arbitrary hypercube node. We now make use of this modified Data sum operation. In the first invocation of this operation, the final sum is obtained at node 0. In the second invocation, the sum is collected at node 1. Generally, in the $i^{st}$ invocation the sum arises at node $i - 1$ where $i = 1 \cdots (2d + 1)^2$. Using this scheme, the maximum number of partial sums calculated by processors at each pipeline stage is reduced to $(\log \log \log M)$ a nearly constant number in practice. Due to space limitation, the proof of this fact is omitted. After this series of Data sum operations, there are $(2d + 1)^2$ sums distributed to the first $(2d + 1)^2$ processors of each $M \times M$ block. The minimum of this sums can be found in $O\left(\log M\right)$ by using a data reduction operation, an operation almost identical to Data Sum except that at each time step the operation executed at each processor is the minimum in place of sum.

## 3.2  The HBMME Algorithm

In this section we will study the realization of the HBMME algorithm on the hypercube network. First we describe the implementation of a simplified hierarchical scheme (basic scheme) where there is neither subsampling nor low-pass filtering between successive layers. Then we will show how to adapt the basic scheme in order to handle the other variations of the HBMME algorithm. Due to space limitation, we will not consider overlapping blocks and non exhaustive search algorithms. However, the techniques which we develop in the following subsections can be easily adapted to these cases. An increase in time complexity should be expected in the case of overlapping blocks as PEs in overlapping regions must execute all the operations relevant to the overlapping blocks. The slowdown factor is proportional to the degree of block overlapping.

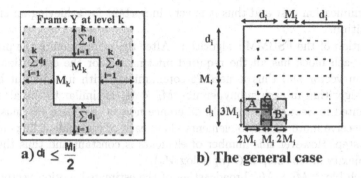

**Fig. 2.** Efficient techniques for the basic scheme of the HBMME algorithm

**The Basic Scheme.** We assume that there are $k$ levels in the hierarchy, $k$ being the lowest and 1 the highest (initial) resolution level. Due to the absence of subsampling, the dimensions of the video frame at each level remain the same and equal to $N \times N$. We introduce some useful notation. $M_i \times M_i$ will denote the dimensions of the blocks at the level $i$ where $i = 1, \cdots, k$ and $M_i$ is a power of 2. $d_i$ will denote the maximum vertical and horizontal displacement that a block can have at level $i$. We also make the realistic assumption that $M_i > M_{i-1}$.

The values of $d_i$ greatly affect the methods we adopt in implementing the HBMME algorithm. First, we will study the case $d_i \leq \frac{M_i}{2}$ $(i = 1, \cdots, k)$ which is often met in practice and then we will deal with the general case. In the first case, the design of an efficient algorithm requires much less effort than in the general case as will become clear in the next paragraphs.

*The case $d_i \leq \frac{M_i}{2}$.* The execution of the algorithm starts from the $k^{th}$ level and ends up at the first level. The calculations at a particular level cannot start before the motion estimates from the lower resolution levels are available. So, there is no parallelism across the levels. On the other hand, the computations inside each level can be easily parallelized by following a data parallel approach.

Now, we will describe the execution of the algorithm at the lowest resolution level $k$. Exactly the same techniques are used at the other levels as well. Figure 2(a) shows a block at level $k$. The area enclosed by the dashed line contains all the pels of frame $Y$ that could possibly be demanded by the PEs of this block at all levels of the hierarchy. The basic steps at level $k$ are:

- transfer of pels of the shaded region of frame $Y$ inside the block $M_k \times M_k$. Two horizontal shifts by $\sum_{i=1}^{k} d_i$ and two vertical shifts by the same displacement can perform this transfer in $O(\log N)$ time. Obviously, these operations are performed in parallel for all the blocks of the frame. After these movements each PE will hold 9 pels at most. This is due to the values of parameters $d_i$ $(d_i \leq \frac{M_i}{2}, i = 1, \cdots, k)$. Later, when we examine the general case, it will become clear that this initial concentration greatly reduces the communication cost and thus it is very important for achieving an efficient algorithm.

- execution of the FSBMME algorithm. After the completion of the previous step, each block has all the required information for the estimation of its motion vector and thus it needn't communicate with its adjacent blocks. The algorithm in each subhypercube $M_k \times M_k$ is similar to the algorithm presented in Sect. 3.1. The only difference now is that each PE must send and receive more than two elements along each of its incident links at each time step. However this number of elements is constant and thus the total complexity of this step is $O\left(d_k^2 + \log M_k\right)$.
- In each block $M_k \times M_k$, broadcasting of the estimated motion vector to all the PEs of the block ($O(logM_k)$ delay).

After the estimation of the motion field at level $k$, the algorithm visits the other levels of the hierarchy and it repeatedly executes the three steps above. For example, at level $k - 1$ the first step is the collection in each block $M_{k-1} \times M_{k-1}$ of all the pel values of frame $Y$ which are within $\sum_{i=1}^{k-1} d_i$ pels around the new position of the block. Clearly, the new position of the block at level $k - 1$ is determined by the motion vector of its "parent" block $M_k \times M_k$ at level $k$. The complexity of the first step is $O\left(\log M_k\right)$ ($O\left(\log M_{i+1}\right)$ at level $i$) and not $O\left(\log N\right)$ as at the level $k$. This is because the step can now be performed by using only subhypercubes $M_k \times M_k$ in contrast to the level $k$ where the whole hypercube $N \times N$ is used. The second and third step at level $i$ have complexity $O\left(\log M_i + d_i^2\right)$ and $O\left(\log M_i\right)$, respectively. Thus, for the case $d_i \leq \frac{M_i}{2}$ ($i = 1 \cdots k$) the total complexity of the HBMME algorithm is $O\left(\log N + \sum_{i=1}^{k}(\log M_i + d_i^2)\right)$.

*The general case.* In the general case the parameters $d_i$ have arbitrary values, thereby complicating the design of an efficient algorithm. If we try to apply the methods of the previous subsection, we will soon find out that after the first step of the data collection in each block, all the PEs have neither constant nor bounded load. This violates our assumption that the memory of each PE is at most $O\left(\log N\right)$. Thus, we have to devise different techniques to handle the general case.

Let us see in more detail how difficulties arise due to the absence of the initial data collection step. The execution at level $k$ is not affected, and the motion field can efficiently be estimated by simply using the FSBMME of the Sect. 3.1. The situation becomes complex after leaving level $k$; the vectors of the motion field that have been estimated at level $k$ might well have different values and directions. As a result, the search windows at level $k - 1$ are not at the same distance and in the same direction from their corresponding $M_{k-1} \times M_{k-1}$ blocks. This lack of regularity makes the use of shift operations problematic as a large number of operations are required to transfer all the pel values asked by PEs at that level. Even worse, this variation in the values of the motion vectors becomes larger and larger as the execution moves to higher resolution levels.

Thus we have to resort to more general communication operations such as RAR operations. As has been mentioned, the time complexity of a RAR operation on a hypercube of size $N$ is $O\left(\log^2 N\right)$ when using practical sorting algorithms. As each PE needs the value of $(2d_i + 1)^2$ pels at level $i$, the same number of RAR operations is required and thus the total delay is $O\left(d_i^2 \log^2 N\right)$. Unfortunately, this complexity cannot be lowered by overlapping RAR operations. The practical sorting algorithms usually employed in RAR operations, when pipelined, cause a large and non constant number of packets to contend for each link, and thus greatly increase the local memory requirements.

We will now present a technique for increasing the number of concurrent communication operations as well as decreasing the complexity of the employed sorting step in the RAR operations. The basic idea is to take advantage of the special structure of the HBMME algorithm. Although it is not possible to initially transfer all the pel values needed by the PEs of a block inside the block, these values can be transferred in batches. Figure 2(b) shows the search window of a block $M_i \times M_i$ at the level $i$. With at most 9 RAR operations, the pels of the region $A$ are transferred inside the block $M_i \times M_i$. After this transfer each PE will have 9 pel values in its local memory. Clearly, the block $M_i \times M_i$ has now all the necessary information for the estimation of the MAD at $(2M_i + 1)^2$ possible displacements. As has been shown in Lemma 1, the communication operations required by these estimations (shift and data_sum operations) can be easily overlapped and thus these calculations can be completed in time $O\left(M_i^2 + \log M_i\right)$.

After this set of computations, the next batch of pel values should be transferred inside each block $M_i \times M_i$. In Fig. 2(b), these pels are located inside the region $B$. In fact, since the regions $A$ and $B$ intersect at region $C$, only the pels outside this region need to be transferred. In general, following the route of Fig. 2(b), we can calculate the MAD for all the candidate vectors inside the search window.

Besides the above improvement, the time complexity can be further reduced by performing the sorting step of each RAR operation with delay lower than $O\left(\log^2 N\right)$. We prove the following lemma:

**Lemma 2.** *The sorting step of a RAR operation at level $i$ can be performed with $O\left(\log^2 N - \log^2 M_i\right)$ delay.*

*Proof.* In each invocation of RAR operation, each PE $(u + i, l + j)$ $(i, j = 0 \cdots M_i - 1)$ of a block $(u, l)$ must read the value of the pel $(u + i + x, l + j + y)$ where $(x, y)$ is a displacement vector. Due to this special reading pattern, the input of the sorting step of each RAR operation consists of $\frac{N^2}{M_i^2}$ sorted lists of at most $M_i^2$ elements each. These lists correspond to the blocks of the frame and the elements of each list are sorted by the coordinates of the pels to be read. This ordering is revealed, if we consider a different labeling of the hypercube nodes: the node $(i, j)$ $(i_{n-1}i_{n-2}\cdots i_{m_i}i_{m_i-1}\cdots i_0 j_{n-1}\cdots j_{m_i}j_{m_i-1}\cdots j_0, M_i = 2^{m_i})$ is labeled as $i_{n-1}i_{n-2}\cdots i_{m_i}j_{n-1}\cdots j_{m_i}i_{m_i-1}\cdots i_0 j_{m_i-1}\cdots j_0$. The new graph is an isomorphism of the original network and thus is also a hypercube network. In this network the sorting step can be executed with delay $O\left(\log^2 N - \log^2 M_i\right)$ by recursively merging larger and larger sorted lists [2]. □

Now it is clear that the estimation of the motion field at the level $i$ takes $O\left(\max\left(1, \frac{d_i^2}{M_i^2}\right) \cdot \left(\log^2 N - \log^2 M_i\right) + d_i^2\right)$ time and thus the total delay of the HBMME algorithm is $O\left(\sum_{i=1}^k \left(\max\left(1, \frac{d_i^2}{M_i^2}\right) \cdot \left(\log^2 N - \log^2 M_i\right) + d_i^2\right)\right)$. This complexity can be further reduced by using an already familiar technique: partition the $N \times N$ hypercube into $\frac{N^2}{D^2}$ $D \times D$ blocks where $D = 2^{\lceil \log\left(\sum_{i=1}^k d_i\right)\rceil}$ and then transfer into each $D \times D$ block all the values of the pels located inside its adjacent $D \times D$ blocks ($O\left(\log N\right)$ delay). After this transfer, the PEs of each $D \times D$ block will have inside their block all the information they could possibly need at all the levels of the hierarchy. Now, all the communication required by the algorithm is confined inside these $D \times D$ blocks and thus the total complexity is reduced to $O\left(\log N + \sum_{i=1}^k \left(\max\left(1, \frac{d_i^2}{M_i^2}\right) \cdot \max\left(1, \log^2 D - \log^2 M_i\right) + d_i^2\right)\right)$.

**Subsampling and Low-Pass Filtering** We now consider the inclusion of subsampling and low-pass filtering in the basic scheme of the HBMME algorithm. We first deal with the operation of subsampling. In the most practical subsampling scheme, the frame at level $i$ results from the frame at level $i-1$ by keeping every second pixel along both vertical and horizontal direction, namely the pels $\left(k2^{i-1}, l2^{i-1}\right)$ where $k, l = 0 \cdots \frac{N}{2^{i-1}} - 1$. The PEs that store these pels clearly form a sybhypercube of size $2^{2(\log N - i + 1)}$. Thus at level $i$ the algorithm operates entirely within this subhypercube and all the techniques presented in previous subsections can also be applied with minor modifications.

Even if, in addition to subsampling, low-pass filtering is employed in the multiresolution scheme, the techniques developed for the basic scheme will still apply. The most commonly used low pass filter takes the form of a small template and the value of each pel is given by a linear function of the values of the pels in its neighborhood. This kind of operation can be implemented on the hypercube by using the Template matching algorithm in [1]. As a result of low pass filtering, the values of some pels change across the levels of the hierarchy. Thus the corresponding PEs must keep more than one values in their local memory (at most $k$, the number of levels). On the other hand, due to subsampling the number of "active" PEs is reduced by a factor of 4 as we are moving from one level to the next lower resolution level. Hence it is possible to balance the load of PEs by distributing the extra load to "inactive" PEs. Specifically, the values of the pel $(i, j)$ at various levels are stored in those PEs which are neighbors in the hypercube with the PE $(i, j)$. At the beginning of the execution at one level, this PE can take the value of its pel for that level from one of its neighbors in the hypercube with only $O(1)$ delay. Apart from this initial step at each level, the rest of the algorithm uses the techniques developed for the basic scheme. The reader should be able to fill the technical details. It is also obvious that in the case of subsampling and low-pass filtering the total complexity of the HBMME algorithm is greatly lowered because the subhypercubes involved in the algorithm execution are getting smaller and smaller at lower resolution levels.

# 4 Conclusions

We have presented efficient hypercube-based algorithms for full search and hierarchical block matching motion estimation. A basic feature of our design is that it remains valid for the whole range of the values of block matching algorithm parameters. This is very important in such a rapidly evolving research field as video coding where the optimal parameters of coding algorithms have not been fixed yet. Our algorithms present theoretical interest too. The non regular data flow of HBMME in combination with the fine grained nature of the employed multiprocessor make the design of an efficient algorithm a difficult task. Under these limitations, we devised a number of techniques for achieving high performance by exploiting the rich interconnection structure of the hypercube network. Although the constraint of $O(\log N)$ local memory at each processor may seem unrealistic in terms of practicality, the techniques in this paper can be easily adapted to the case where each processor has some but limited memory, namely $O(k)$ local memory where $k$ is a free parameter. In a future work, we shall give the details of how these techniques are modified in this case.

# References

[1] Ranka, S., Sahni, S.: Hypercube Algorithms with Applications to Image Processing and Pattern Recognition. Springer-Verlag, 1990
[2] Leighton, T.: Introduction to Parallel Algorithms and Architectures: Arrays-Trees-Hypercubes. Morgan Kauffman Publishers, San Mateo, California, 1992
[3] Tekalp, M.: Digital Video Processing. Prentice Hall Signal Processing Series, 1995
[4] Pirch, P., Demassieux N., Gehrke, W.: VLSI Architectures for Video Compression-A Survey. Proceedings of the IEEE Vol. 83 No.2 February 1995, 220-246
[5] Cheng, S., Hang, H.: A Comparison of Block-Matching Algorithms Mapped to Systolic-Array Implementation. IEEE Transactions on Circuits and Systems for Video Technology, Vol.7, No. 5, October 1997, 741-757
[6] Gupta, G., Chakrabarti, C.: Architectures for Hierarchical and Other Block Matching Algorithms. IEEE Transactions on Circuits and Systems for Video Technology, Vol.5, No. 6, December 1995, 477-489
[7] Komarek, T., Pirsch, P.: VLSI Architectures for Hierarchical Block Matching Algorithm. IFIP Workshop, Grenoble, France, Dec. 1989, 168-181
[8] De Vos, L.: VLSI Architectures for the Hierarchical Block-Matching Algorithms for HDTV Applications. Proc. SPIE Visual Commun. Image Processing '90, vol. 1360, 398-409

# Classification Based Speed-Up Methods for Fractal Image Compression on Multicomputers

Jutta Hämmerle and Andreas Uhl

RIST++ & Department of Computer Science and System Analysis
University of Salzburg, AUSTRIA
{jhaemm,uhl}@cosy.sbg.ac.at

**Abstract.** Since fractal image compression is computationally very expensive, speedup techniques are required in addition to parallel processing in order to compress large images in reasonable time. In this paper we introduce a new parallelization approach for fractal image compression algorithms which employ block classification as speedup method suited for multicomputers.

## 1 Introduction

Fractal image compression [3, 11, 15] has generated much interest in the image compression community as possible competitor [4] to well established compression techniques (e.g. DCT-JPEG) and newer technologies (e.g. wavelets) and as extension [2, 19] to such methods. One of the main drawbacks of conventional fractal image coding is the high encoding complexity (whereas decoding complexity is much lower) as compared to e.g. transform coding. On the other hand fractal image compression offers interesting features like resolution-independent and fast decoding, and good image quality at low bit-rates which makes it an interesting candidate for off-line applications (e.g. video-on-demand (VoD), photo or video CD-ROMs, etc.).

However, speedup techniques are necessary in order to accelerate the encoding phase of fractal compression. Two different approaches can be distinguished:

- Sequential techniques [18],
- High Performance Computing.

In this paper we introduce a new parallelization approach for fractal image compression on multicomputers which employs the sequential speedup method "block classification" and which is especially designed for the use in memory critical applications (i.e. if large images are processed). In contrary to to existing parallel algorithms for this type of task we achieve reasonable performance and may even use dynamic load balancing. The concept discussed in this work is not restricted to fractal compression but may be applied to the general block matching algorithm (which is the fundamental basis of vector quantization or block-based motion estimation techniques).

P. Zinterhof, M. Vajteršic, A. Uhl (Eds.): ACPC'99, LNCS 1557, pp. 276–285, 1999.
© Springer-Verlag Berlin Heidelberg 1999

## 2    Fractal Image Compression

Fractal image compression exploits similarities within images. These similarities are described by a contractive transformation of the image whose fixed point is close to the image itself. In our implementation, the image transformation consists of block transformations which approximate smaller parts of the image by larger ones using contractive affine transforms. The smaller parts are called ranges and the larger ones domains. All ranges together (range pool) form a partition of the image. The domains (having twice the height and length of the ranges) can be selected freely within the image and may overlap. For each range an appropriate domain must be found. The domain blocks are transformed to match a given range block as closely as possible.

To compare image blocks (i.e. range and domain) usually rms (root mean square error) is used. The transformation applied to the domain classically consists of the following parts:

- Geometrical contraction (usually implemented by averaging the domain).
- Affine motion (modeled by using the 8 isometries of the square block).
- Gray value adaptation (a least square optimization is performed in order to determine the best values for the parameters describing contrast and brightness modification, respectively).

In a non-adaptive algorithm for each range the block transformation with the smallest rms-error becomes part of the image transformation - no matter how good the range can be covered. In an adaptive algorithm the error of a single block transformation is compared to a predefined maximum error. If the calculated error is larger the specific range (which is located at the "first quadtree level") is split into smaller blocks (located at the "second quadtree level") which are then covered independently or coded directly in the case of the lowest quadtree level. So - in contrast to the non-adaptive algorithm - a specific image quality can be guaranteed. We denote this algorithm (which may allow several quadtree levels, in our concrete examples three) "adaptive quadtree algorithm" [3, 15].

The search for appropriate transformations is computationally enormous expensive and is the reason for the need for accelerating the encoding speed of fractal coding.

In literature several sequential complexity reduction schemes are presented (for a summary of such techniques see [18]). In this paper we focus on discrete feature extraction methods. The domain pool is divided into several classes according to specific discrete features within the domain blocks. The features of the range determine the class in which optimal matches should be located (i.e. only within this single class the search for the optimal match is performed).

For our experiments we use a grey-value classification scheme which is introduced by Fisher [3]. This scheme uses the ordering of mean and variance values in the four block-quadrants as classification feature and results in 72 classes. This is only one possibility for a discrete feature extraction method. Many other techniques exist which may result in an either higher or lower number of classes (see e.g. [12, 13]).

## 3  Fractal Image Compression on Multicomputers

A significant amount of work has been already devoted to fractal compression algorithms for MIMD architectures preserving sequential coding quality – we identify two major groups of algorithms suited for MIMD computations which trade off memory versus communication demand and are therefore applied according to the relation between memory capacity of one PE and image size [20]:

- **Algorithms Class A** (parallelization via ranges): Within this class it is necessary that at least the entire image can be stored in the memory of each PE. Out of the image-data the complete domain pool can be produced. To each PE a subset of the range pool is assigned – either statically or dynamically – and the PE calculates the best transformations for its range subset. Algorithms of this class may therefore include a dynamic load balancing. According to memory capacities and communication costs further distinctions can be made. For more details see e.g. [5, 10, 16, 21, 22].
- **Algorithms Class B** (parallelization via domains): Within this class of algorithms the domain pool cannot be stored in the memory of one PE - therefore the domain pool is distributed evenly among the PEs. The most efficient algorithms transfer the ranges among the PEs in a pipelined manner in order to carry out the range-domain comparisons of all sub-domain pools. According to synchronization levels and available architectures further distinctions can be made. For more details see e.g. [1, 8, 9, 14].

We have also investigated algorithms employing both parallel processing and sequential speedup techniques (e.g., using block based classification for class A algorithms [6] and geometric searching [7] on parallel architectures).

## 4  Discrete Feature Extraction Speed-Up on Multicomputers

Parallel algorithms of class A are very efficient and show nearly linear speedup in the order of the number of processors [20]. They may include dynamic load balancing, and the number of PEs can be larger than with algorithms of class B before a bottleneck in the host-process occurs. Therefore, this class is a good basis for combining parallel algorithms and sequential speedup techniques in an efficient way [6].

When processing large images on distributed memory architectures unfortunately these algorithms cannot be applied due to memory restrictions. Consequently, the domain pool has to be distributed among the PEs (algorithms of class B). Algorithms of this type can be distinguished whether the assignment of parts of the domain pool to each PE is done prior to or after any precalculations. In the first case the domain pool is distributed evenly among the PEs (fixed distribution), in the second case the distribution is performed according to the results of the precalculations (adaptive distribution).

Parallel precalculations for discrete feature extraction methods with predefined classes include the following steps:

1. Distribution of the domain pool to the nodes in a way that each node receives an equal sized part.
2. On each node: Precalculations for specific sub-domain pool.
3. Redistribution of the domain pool including results of precalculations which can either be organized by a host process or by the nodes themselves.

Obviously, these procedures require a significant amount of communication and synchronization. Therefore, it is often more efficient to perform the precalculations in sequential since the small computational effort does not justify the parallelization effort (e.g., this is true for the classification applied in our experiments).

## 4.1 Fixed Domain Pool Distribution

The domain pool is distributed evenly among the PEs. Each PE performs necessary precalculations (i.e. classification, downsampling) for its private domain pool. If a range is processed by a PE it is first classified and the corresponding class of the private domain pool is being searched. Subsequently, the range is passed on to the next PE in the pipeline.

For a specific range a specific class provides domains for an optimal match and only this class is being searched. The sizes of the classes in the private domain pools on the PEs will generally be very different. Therefore, the processing time for a given range might be very different on different PEs which leads to a severe load balancing problem.

Consider the following bad case: On each PE exists one class which contains far more domains of the private pool as compared to the other classes. The PE which holds the large class where the range belongs to has a significantly higher processing time than all other PEs. This effect may occur for almost each range at different time instances. Since the "slowest" range determines the speed of the pipelined execution the speedup is limited and - in a worst case - equal to or even smaller than the speedup of the parallel algorithm without additional sequential speedup techniques.

The requirement for optimal speedup concerning discrete feature extraction methods is therefore a uniform size distribution of the domain classes on each PE. The size distribution of the domain classes can be controlled to some extent if discrete feature extraction methods with image adapted classes are being used – if fixed or heuristic classes are used a different strategy is needed for avoiding the load balance problem. However, if fixed distribution is used, it is important not to execute the pipeline in synchronized manner – the unsynchronized pipelined processing mode is capable of compensating the load balancing problem to some extent.

## 4.2   Adaptive Domain Pool Distribution

For this type of algorithms it is necessary to have knowledge about the class structure and size of the classes of both range and domain pool, respectively. These precalculations are done prior to any adaptive data distribution. In contrary to fixed distribution we do not perform a distribution of the domain pool itself but we distribute the classes among the PEs – both domains **and** ranges of a given class are assigned to a given PE which calculates the optimal match. This means that we neither apply data parallelism via ranges (class A) nor via domains (class B) but via classes.

An optimal distribution of the classes to the PEs therefore depends on:

- – Size distribution of range pool,
- – Size distribution of domain pool,
- – relation $\frac{c}{p}$ (c=number of classes, p=number PEs).

Distributing the classes evenly among the PEs would again lead to a load balancing problem since

a) the size of the classes (i.e. the number of domains in the classes) is generally very different and
b) the amount of computation inside one class is again very different, since a different number of ranges belongs to each class.

According to the precalculations we know the number of ranges and domains belonging to a given class (for example see Figure 1).

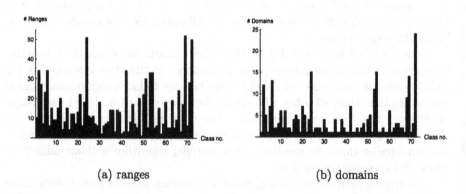

(a) ranges      (b) domains

**Fig. 1.** Number of ranges and domains for all classes of the first quadtree level (Lena image).

The use image-adaptive classes showing uniform class sizes would save the computation of the class complexities at the cost of the costly adaptive class generation. However, the number of range-domain comparisons within a class can be determined in any case – we denote this number as "class complexity" (see Figure 2). The total of all class complexities gives the overall complexity of the algorithm. Obviously, we have to distribute the class complexities evenly among the PEs.

These observations lead to the following optimization problem: Given the number of PEs and the precalculated class complexities. Distribute the class complexities as uniformly as possible among the PEs. This type of optimization can be interpreted as simple knapsack problem with several knapsacks of equal size. An additional constraint of this optimization may be the memory capacity of the PEs – in the case of very small capacity it may not be possible to distribute the class complexities at once but a step-by-step assignment (possibly dynamically) might become necessary.

An additional issue to be considered is the fact that the calculated distribution of class complexities is valid only for the first quadtree level of the adaptive quadtree decomposition. The distribution of classes of the domain pool of the subsequent quadtree levels among the PEs is different from the distribution of the first level as well (for a comparison of class complexity distribution of three quadtree levels see Figure 2). Therefore, a class complexity calculation and distribution is required at each quadtree level. These procedures involve a synchronization and centralization step at each quadtree level which limits parallelization efficiency.

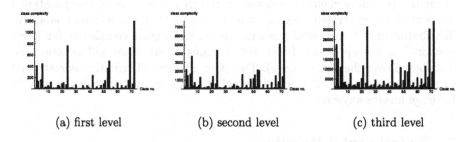

(a) first level            (b) second level            (c) third level

**Fig. 2.** Class complexities of different quadtree levels (Lena image).

The subsequent procedure has to be performed at each quadtree level after the class complexities have been calculated. Preferably, entire classes are distributed among the PEs according to their calculated complexities. The decision whether a class complexity is distributed or not is controlled by a tolerance parameter which specifies the allowed deviation from the uniformity of the final complexity distribution. Theoretically a completely uniform distribution is achievable by setting this parameter to zero, but in this case precalculation takes too long and communication and memory demand rises since many small complexities (the minimal size in our case is one range and the domain pool of one class) have to be distributed.

Although sophisticated techniques may be used for distributing the class complexities among the PEs the following simple technique has proven to be sufficiently precise. The class complexities are sorted and subsequently assigned to the PEs starting with the largest one. If a class complexity exceeds one PE's computed quota it is assigned to the PE with the next lower complexity level, and so on. If a class complexity exceeds all PE's quotas it is assigned to the PE with the current lowest complexity level. The amount of complexity exceeding the PE's quota is assigned again to the set of not yet distributed class complexities. This is done by assigning all domains of this class to the PE, whereas only the corresponding share of ranges is assigned to it. Unfortunately, this technique has to be paid with a slight increase of memory demand on the PEs since the domains of a single class may be assigned to different PEs for several times.

This technique is applied until all class complexities have been distributed. Depending upon the choice of the tolerance parameter we result in a more or less uniform load distribution.

Note that if for any reason dynamic load balancing is required (e.g. due to changing load conditions in multi-user environments or memory constraints) it is straightforward to create a task pool of any desired structure using the technique described above. Load balancing is not possible in a sensible way in the case of a fixed distribution.

A final remark on memory requirement: in memory critical systems the fixed domain pool distribution is done by distributing the image itself evenly. On each PE the necessary domains are calculated from this image part, which may dramatically reduce memory demand in return for increase of computational demand if an overlapping domain partition is used. In such a case adaptive distribution is not successful since the technique "higher complexity for lower memory" is not applicable (since the domains of one class will generally be distributed over the entire image). Therefore adaptive distribution is restricted to non-overlapping or "little-overlapping" domain partitions (which is a must for large images anyway).

## 5    Experimental Results

For our experiments we use a FDDI interconnected NOW consisting of 8 DEC AXP 3000/400 workstations, a Parsytec CC-48, and a SGI POWERChallenge GR. PVM is employed for implementing a message passing based host/node configuration.

We apply the adaptive quadtree algorithm (starting with range-size $16 \times 16$ pixels we partition until size $4 \times 4$ is reached, subsequently the range is coded directly) to the following images with 256 grayvalues: the well known Lena image ($512 \times 512$ pixels for the Parsytec, $1024 \times 1024$ pixels on the NOW) and a large satellite image ($2048 \times 2048$ pixels on the SGI). The domains in the domain pool are chosen to be non-overlapping (as it is done in most fractal coders in order to keep the complexity reasonably low). The necessary precalculations (i.e. classification) are not performed in parallel due to reasons explained earlier.

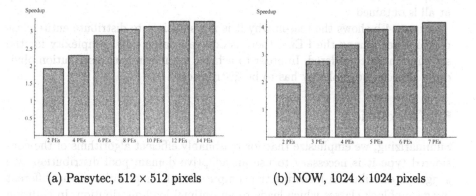

(a) Parsytec, 512 × 512 pixels          (b) NOW, 1024 × 1024 pixels

**Fig. 3.** Performance of adaptive distribution applied to the image Lena.

For the smallest image considered we notice a speedup saturation at 3.3 employing 12 PEs on the Parsytec (see Figure 3.a, the overall processing time with 12 PEs is 7 seconds), whereas we already achieve speedup 4.17 with 7 PEs on the NOW for the 1024 × 1024 pixels image (see Figure 3.b).

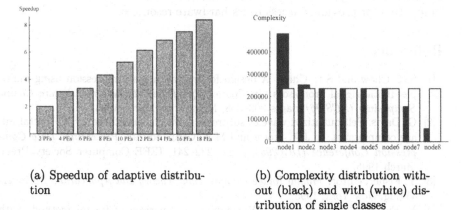

(a) Speedup of adaptive distribution          (b) Complexity distribution without (black) and with (white) distribution of single classes

**Fig. 4.** Performance of adaptive distribution on the SGI (satellite image, 2048 × 2048 pixels).

For the 2048 × 2048 pixels satellite image we observe increasing speedup up to 18 PEs (where speedup 8.4 is achieved - Figure 4.a). This clearly shows that we achieve reasonable speedup and even moderate scalability with the adaptive distribution technique provided the datasize is not too small. This stands in complete contrast to results using fixed distribution where almost no speedup at all is obtained.

Figure 4.b shows the reason why it is not sufficient to distribute entire class complexities among the PEs – there is one very large class complexity for the satellite image considered. In order to achieve a uniform load distribution obviously this class complexity has to be distributed.

# 6    Conclusion

Summarizing, we emphasize that for reasonably efficient algorithms of the considered type it is necessary to use an adaptive domain pool distribution. We suggest a distribution following the computational complexities of the different image-subblock classes which leads to an optimal load-distribution. In addition to pre-computed load-distribution the proposed technique may be used for dynamic load balancing in a multi-user environment with varying load-conditions. Future research will be conducted on how to employ multidimensional nearest neighbour search techniques [17] (which is currently the most efficient sequential speedup technique) within parallel algorithms.

## Acknowledgements

The first author was partially supported by Österreichische Nationalbank, Jubiläumsfonds project no. 6900. We thank the Paderborn Center for Parallel Computing for providing access to its hardware resources.

## References

[1] S.K. Chow and S.L. Chan. A design for fractal image compression using multiple digital signal processors. In *Proceedings of the International Picture Coding Symposium (PCS'96)*, pages 303–308, Melbourne, March 1996.

[2] G. Davis. Self-quantized wavelet subtrees: A wavelet-based theory for fractal image compression. In J.A. Storer and M.A. Cohn, editors, *Proceedings Data Compression Conference (DCC'95)*, pages 232–241. IEEE Computer Society Press, March 1995.

[3] Y. Fisher, editor. *Fractal Image Compression: Theory and Application.* Springer-Verlag, New York, 1995.

[4] Y. Fisher, T.P. Shen, and D. Rogovin. A comparison of fractal methods with DCT (JPEG) and wavelets (EPIC). In *Neural and Stochastic Methods in Image and Signal Processing III*, volume 2304-16 of *SPIE Proceedings*, San Diego, CA, USA, July 1994.

[5] M. Guggisberg, I. Pontiggia, and U. Meyer. Parallel fractal image compression using iterated function systems. Technical Report Technical Report CSCS-TR-95-07, Swiss Scientific Computing Center, May 1995.

[6] J. Hämmerle. Combining sequential speed-up techniques and parallel computing for fractal image compression. In R. Trobec, M. Vajtersic, P. Zinterhof, J. Slic, and B. Robic, editors, *Proceedings of the International Workshop on Parallel Numerics (Parnum'96)*, pages 220–233, 1996.

[7] J. Hämmerle and A. Uhl. Fractal compression of satellite images: Combining parallel processing and geometric searching. In E.H. D'Hollander, G.R. Joubert, F.J. Peters, U. Trottenberg, and R. Völpel, editors, *Parallel Computing: Fundamentals, Applications and New Directions*, number 12 in Advances in Parallel Computing, pages 121–128. North Holland, 1998.

[8] D.J. Jackson and T. Blom. A parallel fractal image compression algorithm for hypercube multiprocessors. In *Proceedings of the 27th Southeastern Symposium on Sytem Theory*, pages 274–278, March 1995.

[9] D.J. Jackson and W. Mahmoud. Parallel pipelined fractal image compression using quadtree recomposition. *The Computer Journal*, 39(1):1–13, 1996.

[10] D.J. Jackson and G.S. Tinney. Performance analysis of distributed implementations of a fractal image compression algorithm. *Concurrency: Practice and Experience*, 8(5):357–380, June 1996.

[11] A.E. Jacquin. Fractal image coding: A review. *Proceedings of the IEEE*, 81(10):1451–1465, October 1993.

[12] C.K. Lee and W.K. Lee. Fast fractal image block coding based on local variances. *IEEE Transactions on Image Processing*, 7(6):888–891, 1998.

[13] S. Lepsøy. *Attractor Image Compression - Fast Algorithms and Comparisons to Related Techniques*. PhD thesis, The Norwegian Institute of Technology, Trondheim, June 1993.

[14] H. Lin and A.N. Venetsanopoulos. Parallel implementation of fractal image compression. In *Proceedings of Canadian Conference on Electrical and Computer Engineering*, pages 1042–1045, Montreal, September 1995.

[15] N. Lu, editor. *Fractal Imaging*. Academic Press, San Diego, CA, 1997.

[16] A. Oswald and J. Ball. A parallel quadtree approach to fractal image compression. In *Proceedings of the International Conference on Parallel and Distributed Processing Techniques and Applications (PDPTA'96)*, pages II/914–917, 1996.

[17] D. Saupe. Accelerating fractal image compression by multi-dimensional nearest neighbor search. In J.A. Storer and M.A. Cohn, editors, *Proceedings Data Compression Conference (DCC'95)*, pages 222–231. IEEE Computer Society Press, March 1995.

[18] D. Saupe and R. Hamzaoui. Complexity reduction methods for fractal image compression. In J.M. Blackledge, editor, *Proc. IMA Conf. on Image Processing; Mathematical Methods and Applications (1994)*, pages 211–229. Oxford University Press, September 1995.

[19] N.T. Thao. A hybrid fractal-dct coding scheme for image compression. In *Proceedings of the IEEE International Conference on Image Processing (ICIP'96)*, volume I, pages 169–172, Lausanne, September 1996. IEEE Signal Processing Society.

[20] A. Uhl and J. Hämmerle. Fractal image compression on MIMD architectures I: Basic algorithms. *Parallel Algorithms and Applications*, 11(3–4):187–204, 1997.

[21] G.D. Veccia, R. Distasi, M. Nappi, and M. Pepe. Fractal image compresson on a MIMD architecture. In H. Liddel, A. Colbrook, B. Hertzberger, and P. Sloot, editors, *High Performance Computing and Networking. Proceedings of HPCN Europe 1996*, volume 1067 of *Lecture Notes on Computer Science*, pages 961–963. Springer, 1996.

[22] P. Zinterhof and P. Zinterhof jun. A parallel version of an algorithm for fractal image compression. In *Workshop Paragraph 1994*, number 94-17 in RISC - Linz Report Series, 1994.

# Accurate Motion Estimation in Image Sequences: Massive vs. Distributed Parallelism

Laurent Gatineau and François Meunier

Dept INF - Institut National des Télécommunications
9 rue Charles Fourier - 91011 Evry Cedex - France
{gatineau, meunier}@etna.int-evry.fr

**Abstract.** Motion estimation for image sequences has several important applications (vision, tracking, 3D reconstruction, indexing...). When high quality motion vector fields are required for large images, the computation load is huge: parallelism gives results in acceptable time. Which architecture to use depends on time constraints, architecture suitability to the algorithm, etc. This paper reports current work on parallelizing a multi-resolution, multi-attribute method, on two different architectures, one a massively parallel computer, the other a network of workstations. Although of older a technology, the parallel computer has various communication possibilities and a large bandwidth. On the other hand, networks of workstations are a possibly emerging standard for low-cost super-computing. Comparison of both shows that, for this kind of algorithm, even with some load balancing they provide less gain if communicating only via a classical network. The next stage of this work is to address high bandwidth networks.

## 1 Introduction

Many applications: robotics, guidance, video sequence compression, video indexing, 3D reconstruction... require to determine the motion of visible points within a sequence of images $\mathcal{I}_1, \mathcal{I}_2 \ldots$. In this paper, two parallelizations are presented, on two different architectures of a multi-resolution, multi-attribute method yielding dense and high quality motion fields for large images.

To determine motion, many methods exist, according to various hypothesis (homogeneous motion, points' intensity constancy...) and requirements (accuracy, real-time...). One interesting qualitative comparison is in [2]. For low accuracy requirements, motion estimation may be done by *block-matching* methods (eg. in video compression MPEG2, 4). *Differential methods* aim at more accuracy. They use the *optical flow* equation, and overcome under-determination by various ways (hypothesis of locally smooth displacement field, robust estimators [1], additional weighted equations [9], multi-grid relaxation algorithm [6], etc). Many further enhancements and other sophisticated methods attempt to deal with occlusions and discontinuities (Markov fields [7], etc). *Multi-resolution methods* cope with large displacements and obtain good quality results. To each image is associated a pyramid of images reduced at smaller scales. At the coarser

P. Zinterhof, M. Vajteršic, A. Uhl (Eds.): ACPC'99, LNCS 1557, pp. 286–295, 1999.

level, correct matches at large distances are found; from them finer matches at lower levels can be derived.

Up to now the methods which provide dense, high-quality displacements fields, although very computation demanding (in the order of 40 Giga floating point operations per pair of successive images), are mostly sequential. Parallel computing is needed if wanting results in less than hours, but it is not that obvious to achieve. For example a parallelization of a Markov field, multi-resolution method [8], using a workstation network under PVM [4], shows dramatic gains in computation-time but important losses in communication-time. The method we chose would allow to process each pair of successive images $\mathcal{I}_t, \mathcal{I}_{t+1}$ separately on a dedicated processor. However, this trivial parallelization is excluded because preventing future improvements of the method; moreover our aim in the middle term is to deal with a sequence as much as possible on the fly.

For all that, some super-computing capability seems in order. The first architecture we use is a massively parallel computer, a MasPar MP1. Although this SIMD computer was issued in 1990, it has interesting features. First it has a large power (1.2 G-Flops) in full configuration; its many elementary processors are weak (4 bits, 12.5 MHz), but they may amount to many workstations. Second, this power benefits of three different communication facilities. For regular communication schemes, the bandwidth is 23 G-bytes/s. In addition, irregular schemes are possible, as well as some emulation of a CREW-PRAM capability. Finally, the machine is well suited to image processing, and to test parallel algorithm prior to hardware integration; it may be equipped with parallel I/O.

On the other hand, the ubiquitous PC/workstation networks are potential challengers to supercomputers, if high bandwidth networks indeed provide communication capabilities in accordance: low latency, broadcast facility, etc. In order to acquire an experience on that, this work is a first stage, comparing a parallelization on some massive parallel computer, with a parallelization on a network of stations communicating through a classic Ethernet network. The next ongoing stage is to compare with a high bandwidth, low latency network under installation.

The rest of the paper is organized as follows. The motion computation algorithm is summarized in title 2. The parallelization on the massive parallel computer, and on the workstation network are presented in title 3 and 4. Title 5 reports the experimental results. Title 6 concludes by a discussion and the perspective.

## 2   The Differential Multi-resolution Method

The method of Weng and al. [11] for dense motion estimation is a global minimization of differences of attributes between matched pixels in images $\mathcal{I}_t$ and $\mathcal{I}_{t+1}$. The attributes considered for each point $\mathbf{u}$ in image $\mathcal{I}_t$ are: $i(\mathbf{u})$ grey level intensity, $e(\mathbf{u})$ edgeness, $p(\mathbf{u})$, $n(\mathbf{u})$ positive and negative cornerness, see [11] for details. Two points $\mathbf{u}$ and $\mathbf{u} + d(\mathbf{u})$ in $\mathcal{I}_t, \mathcal{I}_{t+1}$ respectively may be matched in inverse proportion of their attribute differences. The differences are iteratively minimized by progressively varying the vector field $d(\mathbf{u})$. To cope with large displacements, this minimization is applied on a multi-resolution basis. Images and pixel attributes are successively averaged in increasing levels $l = 0, 1\ldots$ of coarsening: if $\mathcal{I}_t^0$ is a $W \times H$ image, $\mathcal{I}_t^1$ is $W/2 \times H/2, \ldots \mathcal{I}_t^l$ is $W/2^l \times H/2^l$. The set of all levels for an image is termed a "pyramid".

## 2.1  Multi-resolution Motion Estimation

Computing displacement vectors from $\mathcal{I} = \mathcal{I}_t$ to $\mathcal{I}' = \mathcal{I}_{t+1}$ involves working on both their pyramids, level by level. When turning to the next image $\mathcal{I}$, first a new pyramid is built for it: a *bottom-up stage* computes attributes for level 0 and derives upper levels by averaging. Then, using both the new and the previous pyramids, displacements are computed in a *top-down stage*. This computation starts at the highest level, then projects the displacement vectors of level $l$ on level $l - 1$, where they provide initial, approximate values for an iterative minimization[1].

## 2.2  Computation of Displacement Vectors at Level $l$

At each level $l$, the displacement estimate $d(\mathbf{u})$ for a point $\mathbf{u}$ in $\mathcal{I}_t^l$ (level $l$ of image $t$) determines its match, i.e. the point $\mathbf{u} + d(\mathbf{u})$ in the next image $\mathcal{I}'^l = \mathcal{I}_{t+1}^l$. The goal is a global minimization, for all current match-pairs, of the differences of all their attributes. It is achieved by iterative improvements, which minimizes residual differences:

$r_i(\mathbf{u}, d(\mathbf{u})) = i'(\mathbf{u} + d(\mathbf{u})) - i(\mathbf{u}), \quad r_e(\mathbf{u}, d(\mathbf{u})) = e'(\mathbf{u} + d(\mathbf{u})) - e(\mathbf{u}),$
$r_p(\mathbf{u}, d(\mathbf{u})) = p'(\mathbf{u} + d(\mathbf{u})) - p(\mathbf{u}), \quad r_n(\mathbf{u}, d(\mathbf{u})) = n'(\mathbf{u} + d(\mathbf{u})) - n(\mathbf{u})$

Besides, two other residual differences are used, in order to smooth the displacement vector variations, in magnitude and in orientation: $r_d(\mathbf{u}, d(\mathbf{u})) = d(\mathbf{u}) - \bar{d}(\mathbf{u})$, $r_o(\mathbf{u}, d(\mathbf{u})) = cross\{d(\mathbf{u}), \bar{d}(\mathbf{u})\}/\|\bar{d}(\mathbf{u})\|$, with $cross\{(a, b), (c, d)\} = ad - bc$, and $\bar{d}(\mathbf{u})$ is the average displacement vector at $\mathbf{u}$ over a radius $r$.

To a field $(d(\mathbf{u}))$ of displacement vectors corresponds a field of residual difference vectors $r = (r_i, r_e, r_p, r_n, r_o, r_d^t)^t$, one for each point $\mathbf{u}$. The goal is to find a displacement field globally minimizing $\|\Lambda r\|^2$, $\Lambda = diag\{1, \lambda_e, \lambda_p, \lambda_n, \lambda_o, \lambda_d, \lambda_d\}$, for all $\mathbf{u}$, with $\lambda_e, \lambda_p, \lambda_n, \lambda_o, \lambda_d$ weighting parameters to be set dynamically. Starting with null initial displacement vectors, successive variations $\delta_d(\mathbf{u}) = d^{it+1}(\mathbf{u}) - d^{it}(\mathbf{u})$ are computed at iteration $it + 1$, according to:

$$r(\mathbf{u}, d(\mathbf{u}) + \delta_d(\mathbf{u})) \triangleq r(\mathbf{u}, d(\mathbf{u})) + J(\mathbf{u})\delta_d(\mathbf{u}) + o(\|\delta_d(\mathbf{u})\|)$$

with $J(\mathbf{u})$ the partial derivative of $r(\mathbf{u}, d(\mathbf{u}))$ with respect to $d(\mathbf{u})$. Minimizing the quadratic form $\|\Lambda(r(\mathbf{u}, d(\mathbf{u})) + J(\mathbf{u})\delta_d(\mathbf{u}))\|^2$ by zeroing its derivative with respect to $\delta_d$ amounts to take:

$$\delta_d(\mathbf{u}) = - \left(J(\mathbf{u})^t \Lambda^2 J(\mathbf{u})\right)^{-1} J(\mathbf{u})^t \Lambda^2 r(\mathbf{u}, d(\mathbf{u})) \tag{1}$$

Several iterations are achieved, each in two steps: a) estimate $\delta_d(\mathbf{u})$ for each $\mathbf{u}$ using (1), b) update the displacement for each point: $d(\mathbf{u}) = d(\mathbf{u}) + \delta_d(\mathbf{u})$.

---

[1] Actually, to further improve the quality, several up and down stages are applied.

**Fig. 1.** Estimated displacements from two $512 \times 512$ size images from a sequence with large displacements (1 vector among 32 displayed).

# 3   Parallelization on a Massively Parallel Computer

SIMD massively parallel computing is suitable here: the massive amount of computation is of the data-parallel type (same operations for each pixel), with neat temporal dependencies between the different stages/iterations. This section describes the global scheme, then an enhancement to best cope with the irregularities of the displacements while keeping the communication load moderate.

## 3.1   Overall Parallelization Scheme

The SIMD machine we use is a MasPar MP1 with 2k=2048 Processor Elements (PEs) with 64 Kb memory each. The PEs are arranged in a torus grid, the *xnet*, but have several ways to communicate. Communication between neighbours (regular scheme) via the *xnet* allows about 3 G-Bytes/s bandwidth. Irregular communication schemes between non neighbour PEs are possible through a switching device, delivering about 170 M-bytes/s[2]. These figures derive from the documentation and measurements. They are well scalable in proportion of the number of PEs[3]. Because SIMD architectures in general provide equivalent types of communication facilities, our parallelization is not specific to the MasPar.

**Data Mapping.** Since the same processing applies to each pixel, each data level is partitioned among the PEs, on a geometric basis because PEs periodically need some pieces of data from neighbour domains [4]. For $512 \times 512$ image and 2k PEs, the domain devoted to a PE is $8 \times 16$ at level $l = 0$, and shrinks progressively at upper levels. In the highest levels, although the domain size becomes $1 \times 1$, not all PEs may be kept active. We use a sub-grid of neighbour PEs: at the expense of some auxiliary moves, the many latter communication steps are more efficient.

---

[2] for a random communication pattern. Besides there is some synchronous CREW-PRAM facility, which allows to emulate broadcasts at about 20 G-bytes/s.

[3] a configuration may include 1k, 2k, ..., 16K PEs. MP1 series has 4 bits PEs, realizing 32/64 bits operations by sequencing. The MP2 series has 32 bits PEs.

[4] this has no drawback, since the load is well distributed.

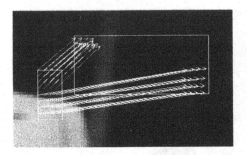

**Fig. 2.** Dynamic domain (*large rectangle*) corresponding to the proper domain of a PE (*small rectangle*) at level 0, the figure shows the upper right corner of the computer in fig. 1, with 5 times zooming. *Arrows* figure the displacements.

**Communications.** We use two types of communication. Regular communications (via the grid) occur after each iteration within a level's processing, for exchanging new values of displacements among neighbours. Also they occur after a level $l$ is processed, to project values on an upper/lower level $l'$. If $l' = l + 1$ and at level $l$ the $n_l^2$ active PEs had domain size $1 \times 1$, the data are gathered in a $(n_l/2)^2$ sub-grid at level $l'$. Conversely, when $l' = l - 1$ there may be some spreading of data. These moves through the grid are possible in $O(W_l)$ steps, with $W_l$ the dimension of the level $l$ image, by neighbour communications [5].

Irregular communications, via the switch, occur when a PE needs data far out of its domain. Estimating displacement $d(\mathbf{u})$ for a pixel $\mathbf{u}$ in image $\mathcal{I}$ requires information in image $\mathcal{I}'$ at pixel $\mathbf{u} + d(\mathbf{u})$. Since $\mathcal{I}'$ is mapped like $\mathcal{I}$, if the displacements are large, all PEs don't dispose of all data. The adequate way to get these foreign data according to an irregular scheme is to use the switch. Because on most of SIMD architectures these communications are much costly, we introduce a notion of *dynamic domain*, in order to reduce their frequency.

### 3.2   Dynamic Domains

The *dynamic domain* of a processor $P_k$ at level $l$ is this part of image $\mathcal{I}'$ and corresponding attributes it should need for computing the displacements $d^l(\mathbf{u})$, for pixels of its own domain. Displacements previously computed at level $l+1$ are initial estimates of the $d^l(\mathbf{u})$. From the $\mathcal{I}'^l(\mathbf{u} + d(\mathbf{u}))$ values each PE determines the rectangular part of $\mathcal{I}'$ it surely will need, see fig. 2. This estimate may prove too small during the following iterative improvements of the $d^l(\mathbf{u})$. So we make each PE in order to overestimate the size of its dynamic domain, and request the largest surrounding area fitting its available memory. To provide each PE with its dynamic domain, the level $l$ image and all attributes are shifted on the whole

---

[5] communications at distance greater than 1 could provide $O(\lceil \log_2 W_l \rceil)$ steps, a minor gain because gathering is necessary only when $W_l$ is smaller than the grid's dimension.

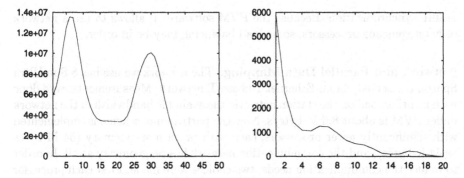

**Fig. 3.** Number of accesses out of the PEs' domains with respect to the distances (in pixels), for example 512 × 512 images and large displacements. *Left/right:* without/with dynamic domains.

processor grid: each PE takes the data it needs when passing. This procedure is achieved in $O(N)$ steps[6] by communications within the grid.

**Benefits Provided by Dynamic Domains.** When provided with its dynamic domain, the probability a PE could nevertheless need some extra data is dramatically reduced, see example figure 3. The number of accesses out of the memories of the PEs decreases from 248 millions, without dynamic domains, to 18 thousands, with dynamic domains. These extra accesses out of local memory obey an irregular pattern: they are satisfied by communication via the switch. Less numerous, they concern smaller pieces of data. If displacement vectors are medium-sized, using dynamic domains allows to avoid no such extra access out of local memory.

Because regular communications via the grid offer a very large bandwidth (3 G-bytes/s with 2K PEs) the procedure for acquiring dynamic domains is fast. For 512 × 512 images (with four 1 byte attributes) and large displacements, it is about 5 s at level 0. For a pair of 512 × 512 images, the time for the whole process (6 levels, with repetitions) is about 65 s, about 10 % of the total processing time (7 min). Without dynamic domains, every time a PE needs foreign data a communication takes place and the total time is more than 10 hours. In other words, dynamic domains amount to anticipate the foreign communication requirements and arrange them by larger chunks, to make the irregular communications as efficient as possible. All details about this massively parallelization are in [3].

## 4    Parallelization on a Workstation Network with PVM

The principle of the parallelization on a workstation network is again data partitioning. The large amounts of available memory enables the use of fewer communications. The architecture is now an asynchronous one: communication will

---

[6] $N$ is the number of processors used.

ensure synchronization. Because the PVM software [4] allows to use a network of heterogeneous processors, some load balancing may be in order.

**Network and Parallel Data Mapping.** The network we use has 8 Sun Ultra Sparcs, connected via an Ethernet 10 base T network. Measurements were done with no other load on the station network: the available bandwidth of the network under PVM is about 8.6 M-byte/s. Now the partitioning scheme is implemented with significantly fewer processors, each with much more memory (64 M-bytes vs 64 K-bytes), and the communication network has no geometry at all. In order to reduce the communication needs, two choices were made. First each processor has the knowledge of the whole images. Second each processor works on a band instead of a rectangle. Therefore, getting data within a neighbour border now requires two messages instead of four.

**Communications.** Iterative improvements of the displacements minimize differences of attributes in corresponding pixels in $\mathcal{I}^l(\mathbf{u})$ and $\mathcal{I}'^l(\mathbf{u} + d(\mathbf{u}))$. Meanwhile, attribute values in $\mathcal{I}^l$ are constant. Now, a processor $P_k$ which needs *some* parts of these data outside its domain, has enough memory to get them all *once*[7]. Thus plenty of memory renders the caching principle of previous dynamic domains pointless. The only need for communication is, after each iteration improving the displacement vectors $d^l(\mathbf{u})$ for its points, each $P_k$ must send some new values to neighbour processors. These neighbours have to compute averaged displacement vectors, on a width-one neighbourhood: $P_k$ sends new values of its top/bottom lines. While waiting for these values a processor cannot compute the newly averaged displacement vectors for its border. But meanwhile, it can compute them for the interior of its domain: computation and communication may largely overlap.

**Possible Load Balancing.** The load balancing strategy used is a GDGM one (Global Decision Global Migration), as opposed to Local Decision Local Migration (LDLM), see [5]. This strategy works well if the processors are of different types, or if they have different loads during long periods. The load balancing process can be divided into four phases [10]:

1. **Load Evaluation:** Because Global Decision is used this phase is executed every $l + 2$ iterations ($l$ the current level). Each processor broadcasts to all others the computation time it has spent since the previous load balancing stage.
2. **Profitability Determination:** If the difference of elapsed times for the lowest and fastest processors is greater than a threshold, then a load balancing stage is initiated.
3. **Work Transfer Vector Calculation:** Each processor derives the size and position of its band from the elapsed times measured.

---

[7] with a small bandwidth, it is even better *each* $P_k$ computes the whole attribute part of the new pyramid, rather than each computing a part and then multi-casting.

**Table 1.** Computation times for 1 image pair (including I / O ).

| Type of sequence | MasPar MP1 1k PEs | Sun Ultra Sparc 1 | Gain |
|---|---|---|---|
| $512 \times 512$ large displacements | 7 min 14 s | 58 min | 8.0 |
| $512 \times 512$ medium displacements | 5 min 51 s | 58 min | 9.9 |
| $512 \times 512$ small displacements | 5 min 8 s | 58 min | 11.3 |
| $256 \times 512$ small displacements | 1 min 31 s | 14 min 20 s | 9.5 |
| $128 \times 512$ small displacements | 31 s | 3 min 28 s | 6.7 |

4. **Data Migration:** Each processor gets the complete field of displacement vectors of the current level.

This method is simple to implement and works very well on a heterogeneous network of workstations: with 6 Ultra Sparc and 2 Sparc 5 stations, elapsed time is 39 min without load balancing and 19 min with one load balancing stage. On a homogeneous network of workstations (8 Ultra Sparc) and different loads, results are medium. Time is 19 min without load balancing and 17 min with 2 load balancing stages (including 16 s for phase one and 1 min for phase four). Those results are medium for two reasons. First, the load balancing strategy can't deal with the load of the network. Second a LDLM strategy could be better for sporadic extra loads, but needs some iterations to achieve the balance [5,10]. An improvement could be to refine the load balancing of a global method, by combination with a local one.

# 5  Experimental Results

The results for the two parallelizations will be discussed further in the conclusion. The measures on the network of workstations were done during an idle period, so no load balancing was needed.

## 5.1  Results for the Massively Parallel Computer

Table 1 compares the times obtained with the massively parallel machine (Mas-Par MP1 with 2K PEs), to those with a sequential one (a Sun Ultra Sparc1). Those times include the whole process including computation of attributes at level 0, which takes 1 s on the MasPar and 7 s on the Sun. The gain ratio of 11.3, for small displacements and large images, decreases to 8.0 with large displacements[8]. A gain ratio of about 10 means the parallel machine amounts to as many Sun Ultra Sparcs. As the performance of the machine scales well, with a full configuration (16384 PEs) this figure would be about 90, and 400 with the MP2 type (32 bits processors). This confirms that the many, weak processors communicate through efficient devices, mostly the grid, then the switch.

---

[8] and so it does for smaller images: on the whole less processors are active, see [3].

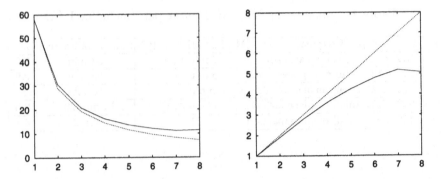

**Fig. 4.** *Left:* Time (min) with respect to $N$, the number of stations. *Right:* Speed-up with respect to $N$. *Solid lines:* Measurements, for one $512 \times 512$ image pair. *Dashed lines:* Best theoretical bound (assuming null cost communication)

These results also show that dynamic domains provide a good solution to the restricted amount of memory, by anticipated caching (without dynamic domains, the computation time is 10 hours). This also summarizes the challenge: to have communication devices providing a functional equivalent to a PRAM, without paying too much for delays and bottlenecks.

### 5.2  Results on the Workstation Network

Figure 4 shows that the time, obtained with the workstation network described in § 4, decreases with respect to the number of processors, up to a point. The cost of communication explains the difference between theoretical and experimental curves. As opposed to the massively parallel machine, the bandwidth does not scales with the number of stations. With more stations the communications times increase faster than the computation time decreases. This occurs quite soon for this algorithm, which requires a linearly increasing number of communications, of relatively small amounts, at each iteration; this kind of classical behaviour appears also in [8].

## 6  Conclusion: Discussion and Future Work

The configuration of the massively parallel computer we use delivers an effective power that should be available with about ten Sun Ultra Sparc1 stations. Although benefiting from simplifications, thanks to much more memory, the network parallelization cannot efficiently use this number of stations (it provides a best speed-up of only 4.5). This comes from the insufficient performance of the network. Moreover, because the performance of the MasPar scales well, a full configuration (16k PEs) of the MasPar MP1 should provide a time-gain of about 90, and four time more this figure for the MP2 series. This means that the time to deal with a pair of $512 \times 512$ images could be of about 9 s. Although still far from real-time, such a performance is indeed interesting in several situations.

Current high bandwidth networks such as ATM, should allow better performance. With the various requirements of different applications, the question is, however, to have not only high bandwidth, but also low latency, and possibilities such as broadcast and, if possible, multi-cast. Such networks clearly could allow to efficiently use more powerful stations all at the same time. In daily practice of course these are used by many other users; so load balancing, as reported above, would be mandatory.

Another (similar) architectural trend consists in connecting powerful processors (PCs) by point-to-point high speed links and dedicated I/O interfaces. We are currently considering the logical continuation of the present work, eg. porting the same parallelization on a set of 450 MHz PCs, fully connected by links enabling an effective bandwidth of 80 M-Byte/s. A configuration with four PCs should enable the evaluation of scalability, thus to determine at which rate sequences could be computed with larger configurations. Another direction of future work is to parallelize other methods, including other computations related to sequences, for example segmentation of the motion field in view of object recognition or object reconstruction, or indexing.

# References

1. A. Bab-Hadiashar, D. Suter. Robust Optic Flow Estimation Using Least Median of Squares. In Int. Conf. on Image Processing, volume 1, pages 513-516, Lausanne, Switzerland, September 1996.
2. J.L. Barron, D.J. Fleet, SS. Beauchemin. Performance of Optical Flow Techniques. International Journal of Computer Vision, 12(1):43-77, 1994.
3. L.Gatineau, F.Meunier. A Massive Parallelization of a Multi-resolution Algorithm for Motion Estimation. Technical report INF/98/01, Institut National des Télécommunications, May 1998.
4. A.Geist, A.Beguelin, J.Dongarra, W.Jiang, R.Mandchek, V.Sunderam. PVM: Parallel Virtual Machine. A Users' Guide and Tutorial for Networked Parallel Computing. The MIT Press, 1994.
5. R. Lüling, B. Monien, F. Ramme. A Study on Dynamic Load Balancing Algorithms. Technical report $PC^2$/TR-001-92, Paderborn Center for Parallel Computing, Germany, June 1992.
6. E. Mémin, P. Pérez. Robust Discontinuity-preserving Model for Estimating Optical Flow. In Int. Conf. on Pattern Recognition, pages 920-924, 1996.
7. T.D. Nguyen, K. Fazekas. Mean Field Approximation to Multimodal Motion Estimation Problem. In $8^{th}$ European Signal Processing Conference, Trieste, September 1996.
8. P.Piscaglia, B.Macq, E.Mémin, P.Pérez, C.Labit. Parallelized Robust Multiresolution Motion Estimation. In Int. Conf. on Image Processing, volume I, pages 945-948, Lausanne, Switzerland, September 1996.
9. M. Tistarelli. Multiple Constraints to Compute Optical Flow. IEEE Transactions on Pattern Analysis and Machine Intelligence, 18(12):1243-1250, December 1996.
10. J. Watts, S. Taylor. A Practical Approach to Dynamic Load Balancing. IEEE Transactions on Parallel and Distributed Systems, 9(3):235-248, March 1998.
11. J. Weng, S.T. Huang, N. Ahuja. Motion and Structure from Image Sequences. Springer-Verlag, 1993.

# A Real-Time Distributed Video Image Processing System on PC-Cluster

Daisaku Arita, Yoshio Hamada, and Rin-ichiro Taniguchi

Department of Intelligent Systems, Kyushu University,
6-1 Kasuga-koen, Kasuga, Fukuoka, Japan
{arita,yhamada,rin}@limu.is.kyushu-u.ac.jp

**Abstract.** This paper describes a PC-cluster system for real-time distributed video image processing. The PC-cluster consists of eight PCs connected by a very high speed network. The key issue of this system is synchronization of distributed video data. Frame synchronization module is introduced to realize three kinds of synchronization: forward synchronization, barrier synchronization and backward synchronization. Forward synchronization is to notify of timing to start processing. Barrier synchronization is to wait for all data that are processed at the same time. Backward synchronization is to cancel processing and transferring useless data. Experimental results are also shown to confirm the performance of the PC-cluster.

## 1 Introduction

Computer vision researches have been expanded from small-scaled, or toy world oriented, experimental studies to large-scaled, real world oriented, practical studies. CDV (Cooperative Distributed Vision) project[2, 3] in Japan aims to establish scientific and technological foundations to realize real world oriented practical computer vision systems, and to develop practical computer vision applications. One of the research issues of the CDV project is observation of objects/environments by distributed multiple sensors. To observe objects or environments with multiple sensors, it is almost impossible that only one computer handles all of stream data from multiple cameras. According to this consideration, we are developing a distributed real-time image processing system as a base architecture of the CDV project.

The problems of such distributed systems for computer vision is the performance of the network and the cost of the system construction. Fortunately, the cost of commercially available workstations and PCs is getting cheaper and their performance is getting much higher. In addition, recently several very high speed networks, whose throughput is more than 1 Gbps and is high enough to transfer video images without compression in real-time, are commercially and economically available. Based on this background, we are researching for real-time distributed video image processing on a PC cluster, which consists of multiple off-the shelf PCs connected via Myrinet, whose throughput is about 1.28 Gbps/full duplex.

P. Zinterhof, M. Vajteršic, A. Uhl (Eds.): ACPC'99, LNCS 1557, pp. 296–305, 1999.

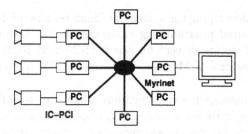

**Fig. 1.** PC-cluster

The advantages of this system are as follows.

- low cost — no need to develop any special hardwares.
- flexibility — components can be easily replaced by new ones.
- scalability — additional PCs can be easily installed.
- availability of peripherals — variety of peripherals are available.
- choice of OS — many types of operating system, Windows, UNIX, or real-time OS, are available.

In this research, we are focusing on software issues rather than hardware issues. When the system only deals with static data, or static image, the problem is not so difficult, i.e., we can use several standard software environments for distributed/parallel computers such as PVM[1] and MPI[4]. However, in real-time image processing, the synchronization within the system is quite important. This is because, we aim at parallelizing not only low level, or very primitive, image processing algorithms, but also higher level, or rather complicated and non-homogeneous, image processing algorithms, which may cause imbalance of computation loads of PCs in the system. Because such environments as PVM and MPI do not provide synchronization mechanisms suitable for real-time image processing, we cannot use them. In this paper we will outline our system, and propose frame synchronize module to realize three kinds of synchronization. And we will also show preliminary experimental results.

## 2   System Overview

The PC-cluster consists of eight PCs(Fig. 1), whose CPUs are Pentium IIs or Pentium Pros[1]. All PCs are connected by Myrinet, a crossbar-switch based switching network, whose features are full-duplex 1.28+1.28Gbps throughput, less than 500ns latency and cut-through switching. Three of PCs have a video capture board, ICPCI, which can capture uncompressed video images from a CCD camera in real-time. Since three CCD cameras are synchronized by a sync-generator, video images captured by those cameras are completely synchronized.

---

[1] To achieve optimal performance, the choice of PC mother board, particularly, the choice of chip set, is quite important.

We have been developing the system on Linux because of device driver availability and distributed programming environment. We have adapted a freely-distributed ICPCI device driver[2] for our NTSC-based RGB camera systems. For the low level control of Myrinet, we have used PM library[6] developed by RWCP, Japan.

PM provides message passing functions which can restart from a blocked state, i.e., a waiting state for a message, in response to a message arrival interrupt. As a result, no busy waits are necessary to receive data and signals from other PCs.

## 3   Representation of Time

In distributed real-time processing systems, it is important to handle the time of the system. In our system, a time quantum signal of 1/30 second cycle is generated from an input video signal. We call it Frame Synchronize Signal (FSS). FSS has a sequence number represented in integer which indicates the order of image frame capturing. FSS is transfered via the network, originating in an image capturing component, from upper components to lower components of the data flow, and timing control of each component is achieved referring to the FSS.

When such mechanism is employed, synchronization of sequence numbers generated from multiple cameras becomes a problem. That is, although all of the cameras are synchronized by the common external sync signal, it is not guaranteed that the start time of image capture is the same on all of the PCs. To solve the problem, the internal clocks of all of the PCs are synchronized by NTP[5] and all of the PCs start capturing at the same time based on the internal clocks. We have examined that no problem occurs by taking an image sequence of an LED counter driven by vertical sync signal. In this examination, when the counter is taken by multiple cameras, image frames in which the counter indicates the same value have the same time stamp, i.e., the same sequence number.

## 4   Modules

We consider that the following parallel processing schemes and their combinations are executed on the PC cluster. From the viewpoint of program structure, each PC corresponds to a component of a structured program of image processing. We call it distributed component (DC). In order to distinguish hardware and software, we use the word DC hereafter[3].

**Data gathering** Video images captured by multi cameras are processed by DCs and integrated on the third processing stage.

---

[2] URL=http://www.gom-online.de/devicedriver/icpci-english.html

[3] Practically, there is no difference between PC and DC in the current implementation. However, in principle, it is possible to map multiple DCs on a PC.

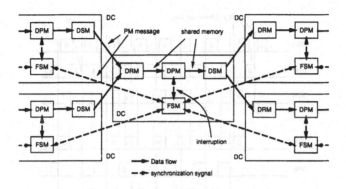

**Fig. 2.** Modules and data flow

**Pipeline parallel processing** The whole function is divided into sub-functions. Each sub-function corresponds to a DC.

**Data parallel processing** Video image is divided into sub-images. Each sub-image is processed in a DC.

**Function parallel processing** From the first DC video images are multicast to two DCs. Different Functions are executed on two paths.

To realize the above schemes efficiently, in each DC, the following modules, each of which is implemented as a UNIX process, are activated(Fig. 2).

**Data Processing Module(DPM)** This module is the main part of the image processing algorithms, and is to process its input data. It receives data from a DRM and sends data to a DSM via UNIX shared memory.

**Data Receiving Module(DRM)** This module is to receive data from other DCs via PM messages, and has buffers for queuing data. When a data request demand arrives from its succeeding DPM, it returns pointers to data.

**Data Sending Module(DSM)** This module is to send data to other PCs via PM messages, and has buffers for queuing data. When processed data arrives from its preceding DPM, it sends the data to the succeeding DCs.

**Frame Synchronization Module(FSM)** This module is introduced to make executions of different DPM synchronize with each other. FSM sends FSSs to the succeeding FSM, and/or receives FSSs from the preceding FSM. An FSS originates in a V-SYNC signal from a video capture board. FSM also sends start signals to activate the DPM in the PC and cancelation signals to neighbor FSMs. Details about FSM are described in the next section.

## 5   Synchronization Mechanism

Distributed video image processing requires to receive, process and send image data in real-time continuously. To realize this, synchronization among DCs is necessary. In our PC-cluster system three kinds of synchronizations are provided.

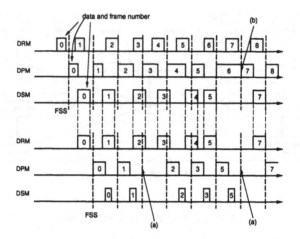

**Fig. 3.** Forward synchronization: This is a time table of two connected PCs. Data are transfered from upper to lower. Usually one data from the preceding PC comes between two FSSs, and DPM can process data between two FSSs. At point (a) data receiving is late for FSS. At point (b) data processing is late for FSS.

|                          | data processing is in time | data processing is late |
|--------------------------|----------------------------|-------------------------|
| data receiving is in time | no error                   | error type 2            |
| data receiving is late   | error type 1               | error type 3            |

**Table 1.** Three error types of forward synchronization

## 5.1 Forward Synchronization

Forward synchronization is to notify a DPM of timing to start processing. To realize this synchronization, FSS is transfered from the FSM in the preceding DC. When FSM receives FSS, it sends FSS to the next FSM(s) and sends a signal to a DPM to start its computation. When a DPM receives the start signal, it requests the next data to its DRM and starts computation. If a DPM has not finished computation of the previous data, or the next data is not received by its DRM when FSS arrives, an error recovery function is invoked(Fig. 3). Table 1 shows error types of forward synchronization and Fig. 4 shows typical examples of error recoveries and their typical execution time tables.

## 5.2 Barrier Synchronization

When one DC gathers data from two or more DCs, it should wait for all data that are processed at the same time. Barrier synchronization realizes this function. In Fig. 5, the DC receives data and FSS from three preceding DCs, (a), (b) and (c). The timings of FSSs' arrival are different from one another because the

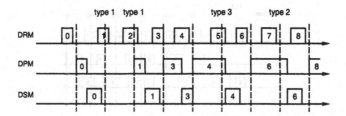

(a) **Data missing**: Recovery for error type 1 is that DPM doesn't work till the next FSS. Recovery for error type 2 and 3 is that DPM continues working and at the next FSS it processes the newest data. In this case only complete data are sent to the following DC but some data are lost.

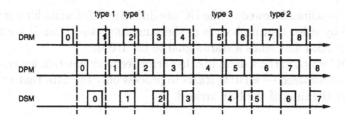

(b) **Incomplete data transfer**: Recovery for error type 1 is that DPM waits for data. Recovery for error type 2 and 3 is that DPM stops processing and DSM sends incomplete data. In this case no data are lost but incomplete data are sometimes sent.

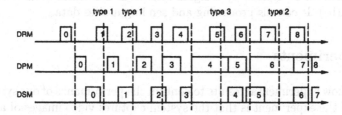

(c) **Complete queuing**: Recovery for error type 1 is that DPM waits for data. Recovery for error type 2 and 3 is that DPM continues working and as soon as it finishes processing it starts processing next data. In this case no data are lost and all data are complete. But it is possible that latency of data increases and also a certain size of queue is required.

**Fig. 4.** Examples of error recovery by forward synchronization: Type 1, type 2 and type 3 in figures mean that the FSS can detect the type of synchronization error.

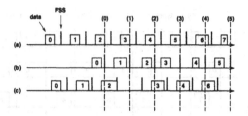

**Fig. 5.** Barrier synchronization: This figure shows a time table of a DRM in a DC which receives data and FSS from three DCs, (a),(b) and (c). When FSS from (b) is received((0),···,(5)), FSM sends signals to the succeeding DCs and the DPM.

lengths of pipelines followed by the DC are different. FSM waits for a set of FSSs generated by video capture boards at the same time, and, then, it sends FSS to succeeding FSM and sends a start signal to the DPM.

In a DC which gathers data, another error recovery function for error type 1 should be considered — a DPM starts to process only the data that are already received at the time of the FSS arrived.

## 5.3 Backward Synchronization

In case that overtime processing makes the next data being lost such as Fig. 4(a), a DPM on a DC, which gathers data may process few data or, in the worst case, none(Fig 6(a)). To avoid it we introduce backward synchronization. Backward synchronization is realized by a signal sent from DC which loses data to others(Fig. 4(b)). It cancels processing and sending useless data.

## 6  Experiments

We will show experimental results to confirm the performance of our system. The outline of the experiment is that the system captures video images of an human arm with three colored markers by two cameras, calculates the 3D positions of markers, and displays wire-frame images of the arm seen from any view points in real-time.

In this experiment each PC works as follows(Fig. 7).

1. PC1 and PC3 capture color video images and shrink them into half-sized (320 × 240) images.
2. PC2 and PC4 extract each marker by thresholding[4], find 2D coordinates of each marker center on the image plane, and calculate the 3D line equation through each marker center and the view point of camera.

---

[4] To make extraction easy, the markers are colored red, green and blue respectively.

3. PC5 gathers two 3D line equations and finds 3D coordinates of the point of their intersection for each marker.
4. PC6 generates and displays the sequence of wire-frame images of the arm.

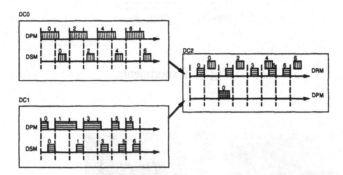

(a) **No backward synchronization**: In this case (maybe a very unfortunate case), DC2 can process only data 0 and 6.

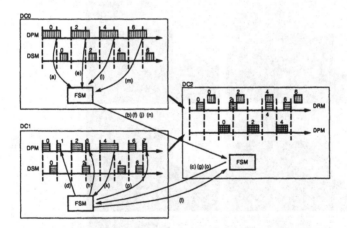

(b) **Backward synchronization**: Backward synchronization informs lost data to other DCs to avoid useless processing. (a) informs overtime processing, (b) informs that data 1 is lost, (c) informs that it is useless to process data 1 and (d) cancels processing data 1. (e)(f)(g)(h) are similar to (a)(b)(c)(d). (i)(j) are similar to (a)(b), but (k)(l) inform that data 5 is lost on DC1, then no cancelation signal is sent. (m)(n)(o)(p) are similar to (a)(b)(c)(d). In this case DC2 can process data 0, 2, 4 and 6.

**Fig. 6.** Backward synchronization: DC2 gathers data from DC0 and DC1.

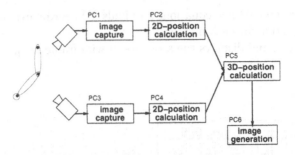

**Fig. 7.** Experiments

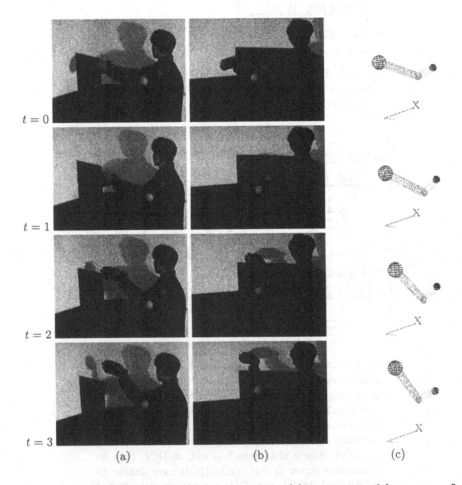

**Fig. 8.** Experimental results: Images in column(a) are captured by camera 0. Images in column(b) are captured by camera 1. Images in column(c) are wireframe images from the view point whose 3D coordinates are the same as those of the camera 0.

Fig. 8 shows experimental results. The frame rate in this experiment is 30 fps, NTSC frame rate. And In this experiment data-missing and complete-queuing are used for error recovery function(Incomplete-data-transfer is not suitable for this experiment). We have proved that synchronization mechanism works well.

However, in the current implementation, it is not possible that the system processes full-size(640 × 480) video image at 30 fps. To solve this problem, we are now implementing an improved system, which uses DMA to transfer video data between main memory and Myrinet buffer.

# 7 Conclusions

This paper describes a PC-cluster system for real-time video image processing. To process video images in real-time, we have employed a very high speed network connecting PCs and introduce three kinds of synchronization mechanisms. From our experiments, the synchronization mechanisms work effectively.

Our future works are more detailed analysis of the synchronization mechanism, development of programming library for this system, and performance evaluation with various applications.

# Acknowledgments

This work has been partly supported by "Cooperative Distributed Vision for Dynamic Three Dimensional Scene Understanding (CDV)" project (JSPS-RFTF96P00501, Research for the Future Program, the Japan Society for the Promotion of Science).

# References

[1] Al Geist, Adam Beguelin, Jack Dongarra, Weicheng Jiang, Robert Manchek, and Vaidy Sunderam. *PVM: Parallel Virtual Machine – A Users' Guide and Tutorial for Networked Parallel Computing.* The MIT Press, Cambridge, Massatusetts, 1994.

[2] Takashi Matsuyama. Cooperative distributed vision. In *Proc. of 1st International Workshop on Cooperative Distributed Vision*, pages 1–28, 1997.

[3] Takashi Matsuyama. Cooperative distributed vision – integration of visual perception, action, and communication –. In *Proc. of Image Understanding Workshop*, 1998.

[4] Message Passing Interface Forum. MPI: A message-passing interface standard. *International Journal of Supercomputer Applications*, 8(3/4), 1994.

[5] D. L. Mills. Improved algorithms for synchronizing computer network clocks. *IEEE/ACM Trans. Networks*, 3(3):245–254, 1995.

[6] Hiroshi Tezuka, Atsushi Hori, Yutaka Ishikawa, and Mitsuhisa Sato. Pm: An operating system coordinated high performance communication library. In Peter Sloot Bob Hertzberger, editor, *High-Performance Computing and Networking*, volume 1225 of Lecture Notes in Computer Science, pages 708–717. Springer-Verlag, 1997.

# Modeling and Scheduling for MPEG-4 Based Video Encoder Using a Cluster of Workstations

Yong He[1], Ishfaq Ahmad[2], and Ming L. Liou[1]

[1] Department of Electrical and Electronic Engineering
{eehey, eeliou}@ee.ust.hk
[2] Department of Computer Science
iahmad@cs.ust.hk
Hong Kong University of Science and Technology
Clear Water Bay, Kowloon, Hong Kong

**Abstract.** In this paper, we first present an Object Composition Petri Nets (OCPN) based model methodology for describing the dynamic behaviour of the multiple video objects and user interactions during the entire MPEG-4 video session; then, a Group of Video Object Plane (GOV) based periodical scheduling algorithm is proposed to assign the encoder tasks to a cluster of workstations with load balancing guarantee. The scheduling scheme can allocate the tasks efficiently according to the timing constraints as well as user interactions. The performance of the encoder can scale according to the number of workstations used. The experiment results indicate that a real-time encoding rate can be achieved for the sequences with multiple video objects.

## 1 Introduction

Most current multimedia applications can be viewed as the merging of traditional computer, communications, and broadcasting industries. In order to support the functionalities, such as content-based interactivity, high compression and random access, a new international standard, MPEG-4, is currently being developed by MPEG (Moving Picture Experts Group) and expected to be finalized towards the end of 1998 [1]. With a flexible toolbox approach, MPEG-4 is capable of supporting diverse new functionalities and satisfy various application requirements and hence will cover a broad range of multimedia applications [5].

MPEG-4, due to its content-based representation nature and flexible configuration structure, is considerably more complex than previous standards such as MPEG-1, MPEG-2 and H.263. In addition, MPEG-4 enables user to manipulate and process the objects with various dedicated tools. Any MPEG-4 hardware implementation is likely to be very much application specific. Therefore, software-based implementation is a natural and viable option. The main problem with such an approach is the requirement of a huge amount of computing power to support real-time encoding. With the developments in parallel and distributed systems, a higher degree of performance at an affordable cost (such as a network of workstations or PCs) can be achieved, provided the parallelism

P. Zinterhof, M. Vajteršic, A. Uhl (Eds.): ACPC'99, LNCS 1557, pp. 306–316, 1999.

from the application at hand is effectively extracted. While although MPEG-4 encoding is highly suitable for implementation using parallel and distributed systems, it is nevertheless a non-trivial task because of the unpredictable nature of MPEG-4 workload.

In this paper, we use a Petri net based modeling technology to describe the temporal relationships between various video objects and user interactions. An efficient scheduling algorithm is proposed to improve the performance of the video encoder with load balancing guarantee. With the proposed approaches, our encoder can allocate the tasks efficiently according to the timing constraints as well as user interactions. The performance of the encoder can scale according to the number of workstations used, With 20 workstations, the encoder can achieve a real-time encoding rate on some multiple objects sequences.

The rest of this paper is arranged in the following manner: Section 2 gives a brief overview of MPEG-4 video verification model. Section 3 describes the proposed implementation approach in detail, including the Object Composition Petri Net (OCPN) based modeling and a dynamic scheduling algorithm. Section 4 provides the experimental results and the last section presents the conclusion.

## 2   Overview of MPEG-4 Video Verification Model

MPEG-4 aims at providing the standard technological elements enabling the integration of the production, distribution and content access paradigms of the multimedia environment [4]. MPEG-4 video is an object-based hybrid natural and synthetic coding standard which specifies the technologies enabling the functionalities such as content-based interactivity, efficient compression and error resilience [8]. Fig.1 illustrates MPEG-4 video coding and composition procedure with user interactions. Both encoder and decoder are based on the concept of video object planes (VOPs).

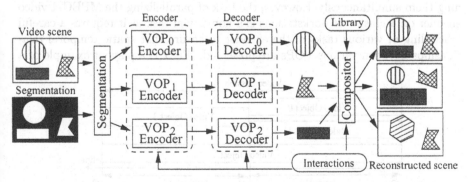

**Fig. 1.** MPEG-4 video scene coding and composition.

The video encoder is composed of a number of VOP encoders. Each object is segmented from the input video signal and goes through the same encoding scheme separately. The bitstreams of different VOPs are then multiplexed and transmitted. At the decoder, the received bitstream are demultiplexed and decoded by each VOP decoder. The reconstructed video objects are then composited by the composition information (which is sent along with the bitstream) and presented to the user. The user interaction with the objects such as scaling, dragging and linking can be handled either in the encoder or in the decoder.

Each VOP encoder consists of three main parts: shape coder, motion estimation/compensation, and texture coder. Shape coder is used to compress the alpha plane information which indicates the object region and contour within the scene. Motion estimation and compensation (ME/MC) are used to reduce temporal redundancies, and the techniques such as unrestricted ME/MC, advanced prediction mode and bidirectional ME/MC are supported to obtain a significant quality improvement. The texture coder which deals with the intra and residual data after motion compensation of VOPs includes algorithms that are similar or identical to the ones used in H.263.

MPEG-4 also supports scalable coding of video objects in both spatial and temporal domains, and provides error resilience across various media. In addition to the above basic technologies used in the encoder structure, the toolbox approach of MPEG-4 video makes it possible to achieve more improvement for some special cases by dedicated tools. Further details on the coding and syntax of MPEG-4 video can be found in [6].

## 3   MPEG-4 Video Encoder Parallelism

In our MPEG-4 based multimedia project, we have implemented an video encoder on a cluster of dedicated workstations that collectively work as a virtual parallel machine. The architecture of MPEG-4 video encoder as shown Fig.1 also happens to be very suitable for distributed computing. Each input VOP is encoded separately and efficient performance can be achieved by decomposing the whole encoder into separate tasks with individual VOP encoders and running them simultaneously. However, the task of parallelizing the MPEG-4 video encoder on a cluster of workstations is a non-trivial task as it requires a careful scheduling of various tasks of the encoder to ensure that spatio-temporal relationships between various VOPs are preserved.  Fig.2 is a playout example of a

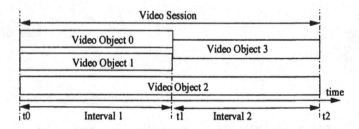

**Fig. 2.** Playout scenario of a MPEG-4 video session in time chart.

MPEG-4 video session in time chart. The presentation scenario begins with concurrent playout of $VO_0$ and $VO_1$, then followed by a new video object $VO_3$ at the time $t_1$. While $VO_2$ can be treated as the background video object throughout the whole session. The distinct characteristic of each video object, such as the shape size, playout duration and spatial allocation, can be different from each other, and the temporal relationship among the objects may also be designed synthetically. For example, we can specify that $VO_0$ and $VO_1$ are synchronized during the presentation from $t_0$ to $t_1$, and $VO_2$ and $VO_3$ are synchronized from $t_1$ to $t_2$; the playout frame rate of $VO_2$ and $VO_3$ is 5 frames/sec. and the frame rate of $VO_0$ and $VO_1$ is 10 frames/sec. According to the states of the video objects, the entire session can be partitioned into a number of presentation intervals as Fig.2 shows.

For MPEG-4 based video presentation, the system must observe and obey the temporal relations among various video objects at the time of playout. The coordination of real-time presentation of video sequences and maintenance of the timing-order among video objects are known as temporal synchronization. There are two styles of synchronization, which can be identified as intra-object synchronization and inter-object synchronization. Intra-object synchronization refers to the maintenance of real-time constraints across a continuous video object sequence and is required to meet the respective playout deadlines within the sequence. Inter-object synchronization refers to the maintenance of real-time constraints across more than one video object sequences and has to be satisfied since the playout of video objects may be dependent to each other in the scene.

The maintenance of temporal synchronization is the most crucial requirement in a distributed multimedia system. To enable a perfect playout manner at the client site, the entire system must be able to operate in a synchronized fashion. Since the encoder is the most computational intensive component of the entire system, we address the problem of designing efficient scheduling algorithms for encoding multiple video objects on the fly that guarantee real-time continuous presentation at the client site. Here, we assume that the network can provide a certain level of Quality of Service (QoS) throughput for the issues of networking communication is beyond this paper.

In real-time applications, the video encoder should be fast and efficient enough to meet the explicit playout timing requirement. It is impossible to achieve such real-time performance with a single PC or workstation, and parallel processing seems to be a viable solution for such problem as mentioned above. Due to the limited number of available processors, we have to determine the processing schedule of the video object tasks and manage the system resources carefully according to the playout requirements of the tasks, so as to guarantee a smooth playout at the presentation clients.

## 3.1 Modeling MPEG-4 Video with Object Composition Petri Nets

To elaborate a feasible scheduling scheme, a model specifying the timing constraints among different video objects is necessary. Various modeling techniques have been proposed and the Petri net based model has been shown to be a useful technique for specifying object level synchronization requirements [10]. Here, we employ such a model known as the object composition Petri net (OCPN) to represent the occurrence of multiple video objects due to its ability to explicitly capture all necessary temporal relations. In general, an OCPN refers to a Petri net graph, each place represents the playout of a video object while each transition represents a synchronization point. However, in order to model the flow of MPEG-4 based video objects, more definitions are essential to allow the dynamic behaviours such as user interactions and object attributes variation. Therefore, We introduce an augmented OCPN model which is capable of supporting interactive operations while satisfying the complex synchronization requirement. With this model, one can precisely describe temporal information of the video object, user interaction, as well as scheduling procedure involved in the entire real-time video session.

The OCPN is 9-tuple $OCPN = \{P, T, A, M, D, TS, PR, PC, PS\}$ where:
$P = \{p_0, p_1, \ldots, p_m\}$ is a finite set of places with $m \geq 0$;
$T = \{t_0, t_1, \ldots, t_n\}$ is a finite set of transitions with $n \geq 0$ and $P \cap T = \Phi$;
$A = \{P \times T\} \cup \{T \times P\}$ is a mapping arcs between places and transitions;
$M : P \to I$ represents the tokens distributed in the place where $I$ is the integer;
$D : P \to R^+$, represents the augmented duration $D$ of the place where $R^+$ is a non-negative real number;
$TS : P \to \{IntraS, InterS\}$ defines different transition types. $IntraS$ is the intra-object synchronization and $InterS$ is the inter-object synchronization;
$PR : P \to R$ represents the request of the clients;
$PC : P \to C$ defines the model construction operation;
$PS : P \to S$ defines the scheduling operation in the encoder.

The above $P$, $T$ and $A$ definitions are the same as that of the Petri net. Number of tokens to be deposited at a given place $p$ is indicated by $M$. The firing condition of both $IntraS$ and $InterS$ are determined by the token states at the input place.

Generally, a MPEG-4 based video session is much complicated and enforces different characteristic at various syntax levels, such as video session (VS), video object (VO) and video object plane (VOP) as specified by the standard. It is important that such hierarchical levels of synchronization can be depicted by the model to enable the handling of large and complex MPEG-4 scenarios. For OCPN, such hierarchical modeling capabilities can be achieved through subnet replacement: at the highest level abstraction, OCPN can be represented by a set of abstract place, and each abstract place can be decomposed into an OCPN subnet, the abstract places in the sub-OCPN indicate finer-grained data units and timing constraints; similarly, the sub-OCPN places can in turn be an abstraction of the lower OCPN subnet, and such replacement process can be recursively applied until to the lowest level where each place can not be decomposed any more. In our approach, we define three levels OCPN, namely VS-OCPN, VO-OCPN and VOP-OCPN. The abstract place of OCPN at different level is defined as:
$PVS : P \to VS$ represents the entire video session;
$PVO : P \to \{VO_0, VO_1, \ldots, VO_{N-1}\}$ represents a set of video objects;
$PVOP : P \to \{VOP_0, VOP_1, \ldots, VOP_{M-1}\}$ represents video object planes;

VS-OCPN is the highest level with an abstract place $PVS$ which indicates the entire video session. VO-OCPN is the middle level with a set of places $PVO$ representing the video objects, and the transition known as $InterS$ describes the temporal precedence and synchronization between the video objects. The lowest level VOP-OCPN is composed of places $PVOP$ which represent frames of the video object. Both intra-object synchronization $IntraS$ and inter-object synchronization $InterS$ can be depicted by the VOP-OCPN. Fig.3 shows a decomposition process of an OCPN, and Fig.4 is the global VOP-OCPN representation equivalent to the example of Fig.2. The time instance $t_1$, $t_2$ and $t_3$ are set to 0 second, 3 second and 6 second respectively.

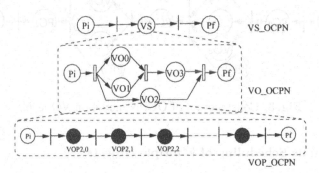

Fig. 3. Decomposition of an OCPN.

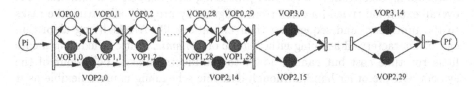

Fig. 4. Global VOP-OCPN example.

For most pre-orchestrated video sequences, as the timing constraints are known a prior, it is possible to determine the structure of a OCPN beforehand. For live video data, however, since related knowledge for constructing the OCPN cannot be verified a prior, a model has to be generated along with the video session. In addition, either pre-orchestrated or live video data can be ceased and new video data can be introduced at any time by the user interactions, and the temporal relationships between the video objects may also be manipulated by the user. Due to such unpredictable behaviours of the user interaction, the OCPN model should be able to support both deterministic as well as imprecise events on the fly.

Fig.5 illustrates the generation strategy of the OCPN (using Fig.3 as an example) which allows user participate actively. Corresponding to the user request as we assumed, the OCPN model can be generated at run-time. At the beginning of time $t_0$, we can obtain the attributes such as playout deadlines and data dependency of $VO_0$, $VO_1$ and $VO_2$ initially, and construct the OCPN as place $PC_0$ indicated. Such model may remains the same if all VOs status are stable. Being caused by user's interaction $PR_0$ at time $t_1$, $VO_0$ and $VO_1$ are halted and a new $VO_3$ is succeeded and synchronized with $VO_2$. Thus, the model should be changed by $PC_1$. The same approach $PC_2$ is performed at time $t_2$ and the video session is stopped by the request $PR_1$. With such dynamic construction process, external interrupts can be quickly responded and scheduling decision can be carried out consequently.

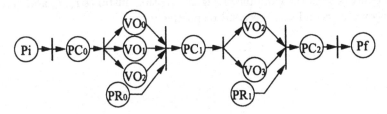

**Fig. 5.** Construction of OCPN model on VO level.

## 3.2 Dynamic Scheduling of Multiple Objects

Using the information generated by the model, the video objects need to be scheduled to multiple workstations for concurrent encoding. The objective of a scheduling algorithm in a parallel processing environment is to minimize the overall execution time of a concurrent program by properly allocating the tasks to the processors and sequencing their executions [3]. A scheduling algorithm can be characterized as being either *static* or *dynamic*. Static scheduling incurs little run-time cost but cannot adapt to the indeterministic behaviour of the system. On the other hand, although dynamic scheduling is more flexible as it can be adjusted to system changes, it incurs a high run-time cost. In a MPEG-4 video session, any static scheduling method may cause load imbalance due to the unpredictable behaviour of the video objects with varying computation requirement.

In order to solve such problem, we have to observe object variation and adjust the processor distribution of each object dynamically, while this may also introduce high inter-processor communication cost which may outweigh the benefit achieved. To compromise such trade-off in a natural way, we propose a dynamic scheduling algorithm, called *GOV-Adjusting Scheduling* (GAS) algorithm where GOV stands for Group of VOP, which periodically detects the workload information of the processors and performs the scheduling. In order to minimize the inter-processor communication cost, the period is based on the Group of VOP (GOV). GOV is an optional syntax level specified by the standard for random access and error recovery purpose. The GOV header usually is followed by the I-VOP performing Intra-coding which is independent to the previous VOPs and no data exchange needed. Therefore, the re-assignment of the processors will not introduce additional inter-processor communication.

Some of the additional notations used in the algorithms are presented below:

$V$: is scheduling sequence of the video session

$P_k$: is the $k$-th processor of the system, and total number of processors is $K$

$R_k$: is the scheduler on $P_k$

$L_x$: is the $x$-th GOV intervals

$G_i : M \rightarrow g_i$ where $g_i \in I$, is a mapping from the set of tokens to a set of groups $G_i$ containing $g_i$ processors each. Basically, such mapping describes the processors distribution with the scheduling scheme

$v_k$: is the partitioned data area of the VOP to $P_k$

$\tau_i$: is the synchronization interval of the video object $VO_i$.

## GAS Algorithm:

1. Initialize the GOV interval $L_x$
2. Sort VOPs of each existing video object $VO_i$ in $V$ using EDF rule
3. Measure the shape size of the first I-VOP along the $L_x$ as $S_i$
4. Initialize $R_k = \{\}$ and calculate the $g_i$ for processors distribution

$$
g_i = \begin{cases} \left\lfloor K \times \frac{S_i/\tau_i}{\sum_{i=0}^{N-1} S_i/\tau_i} \right\rfloor & \text{if } \left\lfloor K \times \frac{S_i/\tau_i}{\sum_{i=0}^{N-1} S_i/\tau_i} \right\rfloor > 1 \\ 1 & \text{otherwise} \end{cases}
$$

where $i = 0, 1, \ldots, N-2$, and

$$
g_{N-1} = K - \sum_{i=1}^{N-2} g_i
$$

5. Decide whether task merging should be performed or not
6. Point to the first $VOP_{i,j}$ of $V$
7. Schedule $R_k$ to the pointing $VOP_{i,j}$ with $P_k \in G_i$ and $\{R_k = R_k \cup VOP_{i,j}\}$
8. Partition the VOP and map the data area $v_k$ to $P_k$ where $v_k \approx \frac{S_i}{g_i}$
9. Advance the $VOP_{i,j}$ to the next $VOP_{i,j+1}$ along the $V$
10. Repeat last three steps until the end of the GOV
11. Goto step 2 and start the next GOV scheduling until the end of the video session

GAS algorithm divides the existing processors into a number of groups and each group handles single video object concurrently. Such allocation is performed periodically on the basis of the GOV. Within each group, a balanced data partitioning method [2] is employed for further encoding speedup.

An earliest-deadline-first (EDF) rule is applied in the second step of GAS algorithm, it specified that the tasks with earlier deadlines are assigned higher priorities and are executed before tasks with lower priorities. In our implementation, VOPs with the earlier playout deadlines or synchronization constraints are encoded and delivered first.

The criterion of processor allocation is tied to the shape size and playout duration as Step 4 shows, namely, the larger object size, the more processors assigned; the shorter the playout duration, the more processors assigned. While this may cause load imbalance between the groups since the distribution of the processors may not be proportional to the size of the video objects due to the excessive difference between the object size. One feasible solution is to merge the smallest object tasks together recursively as Step 5 indicates.

Such task merging is usually performed whenever the number of tasks is greater than the number of physical processors by merging a pair of tasks into a single co-task. There are various merging approaches for different purpose. While In our approach, such task merging step is to guarantee the load balancing among the processors. We define the GOV of each video object as a task, and find a pair of tasks which have minimum workload among all the clusters; then the pair of tasks are merged as one task and such operation is recursively executed until the load balancing can be met, namely, the distribution ratio of the processors are nearly equal to the size ratio of the video objects.

## 4    Experimental Results

We have tested the GAS algorithm on several composed sequences to demonstrate the performance achieved. All the video objects, such as *Akiyo*, *News1* and *Weather* as labeled in Fig.6, are obtained from the MPEG-4 standard test library with QCIF format and represent various characteristics in terms of spatial detail and movement.

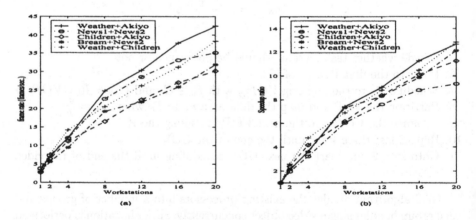

**Fig. 6.** Encoder performance

Our software-based implementation is applied on the MPEG-4 video verification model (VM8.0) encoder. The parallel platform is a cluster of 20 Sparc Ultra 1 workstations connected by a ForeSystems ATM switch (ASX-1000) which provides fast communication among the workstations. For inter-processor communication and synchronization, we use *Message Passing Interface* (MPI) [9], which ensures the portability of our MPEG-4 video encoder across various platforms. Furthermore, various additional software optimization, such as fast motion estimation algorithm, *Visual Instruction Set* (VIS) and Solaris C compiler optimizations, have been made to further speedup the encoder computation.

Fig.6(a) shows the encoding frame rate achieved by the GAS algorithms for the sequences with two video objects using various number of workstations. We can achieve frame rate higher than the real-time performance (30 frames/second) on most standard test sequences. Fig.6(b) is the overall speedup, a near-linear speedup relationship demonstrates that the performance of the encoder can scale according to the number of workstations used.

## 5  Conclusions

In this paper a software-based parallel implementation of MPEG-4 video encoder using a cluster of workstations has been proposed. The contribution of our work includes the use of an OCPN model to capture the spatio-temporal relations between multiple objects of MPEG-4 video, and a dynamic scheduling algorithm to implement MPEG-4 video encoder for multiple video objects. In our future work, we will explore MPEG-4 decoders and interactive methodology for supporting multimedia applications.

## Acknowledgments

This work was supported by the Hong kong Telecom Institute of Information Technology. The authors would also like to thank Dr. Ya-Qin Zhang of Sarnoff Corporation for technical support.

## References

1. L. Chiariglione, "MPEG and Multimedia Communications," *IEEE Transactions on CSVT*, vol. 7, no. 1, pp. 5-18, Feb. 1997.
2. Y. He, I. Ahmad and M. L. Liou, "Real-Time Distributed and Parallel Processing for MPEG-4," *Proceedings of the 1998 International Symposium on Circuits and Systems*, vol. 3, pp. 603-606, 1998.
3. Y. K. Kwok and I. Ahmad, "Dynamic Critical-Path Scheduling: An Effective Technique for Allocating Task Graphs to Multiprocessors," *IEEE Trans. on Parallel and Distributed Systems*, vol. 7, no. 5, pp.506-521, May 1996.
4. ISO/IEC, "MPEG-4 Version 1 Overview," JTC1/SC29/WG11 N2323, July 1998.
5. ISO/IEC, "MPEG-4 Applications Document," JTC1/SC29/WG11 N2322, July 1998.
6. ISO/IEC, "MPEG-4 Video Verification Model 8.0," JTC1/SC29/WG11 N1796, July 1997.
7. T.D.C. Little and A. Ghafoor, "Synchronization and Storage Models for Multimedia Objects," *IEEE Journal on Selected Areas in Communication*, vol. 8, pp. 413-427, Apr. 1990.
8. T. Sikora, "The MPEG-4 Video Standard Verification Model," *IEEE Transactions on CSVT*, vol. 7, no. 1, pp. 19-31, Feb. 1997.

9. D. W. Walker, and J. J. Dongarra, "MPI: a Standard Message Passing Interface," *Supercomputer*, vol. 12, no. 1, pp. 56-68, Jan. 1996.
10. M. Woo, N. U. Qazi, and A. Ghafoor, "A Synchronization Framework for Communication of Pre-orchestrated Multimedia Information," *IEEE Network*, vol. 8, pp. 52-61, Jan. 1994.

# Fractal Video Compression on Shared Memory Systems

Andreas Pommer *

Research Institute for Software Technology RIST++
University of Salzburg, Austria
apommer@cosy.sbg.ac.at

**Abstract.** Fractal image compression is known to be very demanding with respect to CPU-power, and demands are higher again for fractal video compression. Methods to speed up the encoding process can be divided into two groups: one contains sequential speedup-methods like classification, transformation into other search spaces, ... The other possibility is the use of parallel systems. Here we like to show the possibilities and experiences with shared memory MIMD machines.

## 1 Introduction

With the ever increasing demands of internet users regarding images and video in combination with the limited bandwidths the need for good compression methods seems to be clear. Various methods have been proposed, among them DCT-based methods like JPEG[27], MPEG[7] and H.261[13] and derived standards. Another approach is compression by wavelets[28]. And yet another method is fractal compression[15, 5, 19].

Fractal Compression is the attempt the reduce storage space by detecting self-similarities within the image and by storing just the references. However the task of detecting these similarities is computationally very intensive. Now there two ways for speedup (without loosing quality) are possible, the first is the group of sequential speedup methods (like classification, transformation into better search spaces, ... ). The other possibility is the move towards high performance computing (like dedicated hardware, clusters of general-purpose processors, ... ). Depending on the algorithms used combinations of both are possible, too. Here we will focus on the parallelization without sequential speedup-methods.

The field of fractal coding and parallel computation is very broad and previous work has been done for example for still images using SIMD (a MasPar) in [12]. In [10] Hämmerle writes about speedup methods for fractal coding of still images using MIMD architectures (a cluster of workstations) and in [14] Jackson shows another MIMD approach (on a nCube system), Andonova [1] describes fractal coding of video using a MIMD architecture (a transputer network).

* This work is supported in partial by the Austrian Science Foundation FWF, project number P11045-ÖMA, and by RIST++

P. Zinterhof, M. Vajteršic, A. Uhl (Eds.): ACPC'99, LNCS 1557, pp. 317–326, 1999.
© Springer-Verlag Berlin Heidelberg 1999

In the next section we will describe fractal compression of still images and video in more detail. Afterwards there is a section about different approaches of parallelization for a shared memory MIMD system, especially a SGI Power Challenge with 20 R10000 processors. Then a short discussion of results follows with relative speedup results, absolute execution times as well as different scheduling strategies. Performance results concerning the quality of the compression lie outside of the focus of this paper.

## 2    Image and Video Coding by Fractal Methods

### 2.1    The Basics of Fractal Image Coding

Fractal compression tries to exploit intra-image redundancies, especially visual similarities by using transformations between parts of the image. The transformations used are usually affine, and have to be contractive or, at least, eventually contractive [5]. The aim of fractal compression is the find a transformation which has its fixed point very close to the original image. If you want to go into detail read [15, 24].

The transformation $T$ consists of a geometric part (translation, rotation, reflection and size reduction) and of an optical part (contrast and brightness change). The image $x$ is usually partitioned into non-overlapping square parts which are called *"range blocks"* $r_i$. Another set of squares (which are allowed to overlap and which have usually double side length of range blocks) forms the *"domain blocks"* $d_j$. For each of the previous mentioned transformations a domain block is chosen, together with the appropriate geometric and visual parameters. The result should match a range block as close as possible. Usually the rms-metric is used to calculate the distance between the original and the coded block, this metric is a method to measure the quality.

$$T : \{d_j\} \rightarrow \{r_i\} \text{ with} \tag{1}$$

$$T = \bigcup T_k \text{ and } T_k : d_j \rightarrow r_i \tag{2}$$

The goal is to find good $T_i$ so that $\text{rms}(x, Tx) < \Delta$ where $\Delta$ is some predefined collage error. Usually for every $T_i$ an error is calculated, if it exceeds the threshold the range is divided up into 4 squares in a quadtree fashion, as well as the domains in the pool, and the process repeats until the error is below the threshold or the maximum depth of quadtree-subsectioning is reached.

### 2.2    Fractal Video Coding

To be able to estimate the complexity, or the amount of calculations involved in fractal coding, consider the following example (a worst case calculation): To find the best transformation for each range block the set of all domain blocks has to be searched and the transformation parameters have to be calculated using least-squares optimization. Consider a still image of size $256 \times 256$ with range

blocks of size 8 × 8 and a full domain pool. Then a total of 475 799 552 least
squares computations has to be performed — quite a big number for a small
image, and now imagine that there is no still image, but a video stream with 25
frames per second ...

Video compression using fractals works very similar to still image coding.
There exist two different approaches: the first one (called *"2D approach"*) is to
code each frame of a video sequence on its own or in combination with neighboring frames (fractal inter-/intra-frame coding). It is common to all methods that
the first frame has to be intra-coded like a still image. Afterwards the methods
differ, another aspect many methods have in common is that they use motion
estimation, a well-known technique in video coding to reduce the bit-rate. For
more information about 2D fractal video coding consider [6, 21, 16, 8].

The other approach ( *"3D approach"*) is to use three-dimensional range and
domain blocks and perform the transformations on these cubes. The first two
dimensions are the same is in the previously explained 2D case, the third dimension is the time: parts of some consecutive frames are grouped together to one
cube – a 3D block [18].

In the case where the optimal transformation for a cube is not below a quality
threshold the cube is partitioned into 8 sub-cubes — this is also called oct-tree
subsectioning. Other approaches, like Lazar [17], are more adaptive and partition
the block either in the time-direction into 2 parts or in the spatial dimensions
into 4 parts and may be implemented in the future.

A concept similar to motion estimation in the 2D-case has been proposed
for the 3D-case where the restriction of domain *cubes* has been relaxed, each
slice in the temporal direction is allowed to be shifted by a small amount[3]. A
short overview on various approaches to fractal video compression is given in
[22]. Mainly the "Claire"-sequence has been used for out experiments.

## 3 Different Methods of Parallelization

Besides attempts to accelerate fractal coding by parallelization using SIMD[9, 12]
systems there has been much research on the field of MIMD systems[29, 26, 10,
25].

The research for different methods to parallelize and to use different parameters for parallelization were performed on a SGI Power Challenge System
equipped with 20 Mips R10000 processors, the software was written in Power
C [4, 20]. This is a C derivative which is able to control parallelization at the
compile/link-phase via #pragma-statements (together with parameters).

The algorithm which was used for parallelization is a traditional sequential
fractal video coder which was coded following the ideas in [6] in the 2D case
and [17] in the 3D case. The only difference (or enhancement) is that a loop
which calculates the least squares approximation for a large number of block
is executed in parallel by a number of PEs. The parallel loop is either to loop
covering all ranges or the loop covering all domains for a certain range, this will
be detailed in the next subsections.

Some of the more commonly used statements are listed in the following:

- #pragma parallel: a parallel region starts
- #pragma pfor: the following for-loop can be performed in parallel
- #pragma critical: protects the access to critical statements, to avoid race conditions

### 3.1   Parallelization via Ranges

As every block is independent from all other it is possible to parallelize the encoding of the ranges. Each PE receives a number of blocks which it has to encode. The amount and order of blocks for each PE has to be determined and is usually architecture-dependent.

**Pseudo-code:**

```
do initialization stuff
do in parallel for each range r:
  find best match for r searching through all possible domains
  if the best match is not sufficient then
    perform quadtree-subsectioning recursively within this thread
  else
    write parameters to a structure associated with r
parallel section ends
collect results from all ranges and write to file
```

### 3.2   Parallelization via Domains

Another approach to parallelize is the search for a suitable domain block. For every given range all PEs receive a subset of the pool of domain blocks, each PE determines the best transformation for its subset, then the global optimum has be calculated.

**Pseudo-code:**

```
do initialization stuff
do in sequence for each range r:
  find best match for r searching in \
      parallel within all possible domains, that is:
    for each isometry calculate the optimal parameters and the PSNR
    store parameters which result in the best PSNR up to now
  if the best match is not sufficient then
    perform quadtree-subsectioning recursively \
            (and search again in parallel all domains)
  else
    write parameters to a structure associated with r
collect results from all ranges and write to file
```

## 3.3  Parallelization in the 2D Case

Again, several different approaches are possible. One is to assign every frame
to a new PE, however, latency will be high and it is possible that PEs are idle
for a large amount of time. The next approach is the parallelization by ranges
as explained earlier: Each PE calculates the optimal parameters for its ranges
(motion estimation, fractal transform, ...), afterwards all results are collected.

The third approach is an extension of the domain-parallelization method:
For a given range some PEs can perform the search for the optimal parameters
of motion estimation while some other PEs search the domain pool for their
optimal parameters. Afterwards it will be decided which gives the best overall
result using some rate-distortion criterion because usually less bits are needed
to encode the parameters of motion estimation (typically in the range of 10-15
bits) than to encode the fractal transformation (about 30 bits).

Another possibility is to perform these two searches in sequence because
motion estimation (ME) is far less computationally expensive than the fractal
search, if the result of ME lies below a threshold it can be considered good enough
so that no (time consuming) fractal search is necessary. This approach was taken
for the time measurements in the 2D case: motion estimation is performed by
one PE, if it does not succeed, a parallel section for fractal coding follows.

## 3.4  Parallelization in the 3D Case

In the 3D case the concept of motion estimation does not exist. Therefore interframe redundancy has to be exploited by fractal encoding, by the range-to-
domain transformations. The parallelization can be done either via ranges or via
domains. Another approach which is considered less favorable could be to assign
a group of frames to every single PE. However latency is even higher than in the
2D-one-frame-per-PE approach.

The algorithm under investigation uses a very simple approach, it uses cubes,
and partitions them in an oct-tree manner if the error threshold is not reached.
The domain pool is restricted to blocks which lie prior to the range block in the
time-dimension or at least occur at the same time.

# 4  Results

## 4.1  Speedup Observations

The numbers are obtained using the gss-scheduling strategy, for an explanation
of various scheduling strategies please read the next sections. As can be seen
if figure 1 parallel computation for the range blocks performs better than for
the domain blocks. It can also be observed that the 2D versions do not scale
as good as the 3D versions. This can be explained easily: in the 2D case there
is a significant amount of computation which lies outside the parallel section
and which is therefore not affected by the number of PEs. This is the part
of the motion estimation which uses a minor amount of computational time

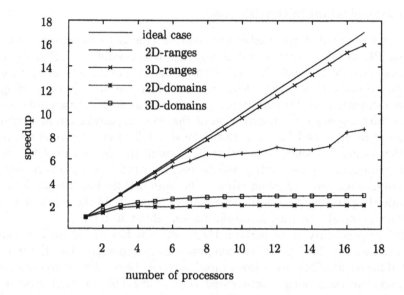

**Fig. 1.** Speedup results for coding in 2D/3D and parallelization via ranges/domains

as compared to parallel fractal coding, and if the motion estimation already achieves a result below a certain quality threshold then the routine for parallel fractal coding is not called at all.

The parallelization of the domains is worse because the critical section to store the optimal transformation parameters is still within a fast loop, although it is moved outwards one loop already.

| 2D/3D | parallel | #PE | seconds |
|-------|----------|-----|---------|
| 2D | ranges | 2 | 861 |
| 2D | ranges | 16 | 205 |
| 2D | domains | 2 | 30392 |
| 2D | domains | 16 | 19946 |
| 3D | ranges | 2 | 39931 |
| 3D | ranges | 16 | 5177 |
| 3D | domains | 2 | 53639 |
| 3D | domains | 16 | 28462 |

**Table 1.** Absolute times for compression

Figure 1 in combination with Table 1 shows that the 2D versions perform faster than the 3D versions but they do not scale as good as the 3D versions.

## 4.2 Experiments with Different Scheduling Strategies

Documentation [4, 20] for the examined system, a SGI Power Challenge, lists several scheduling strategies for loops and other parallelizable code built into the C compiler/runtime system:

**simple:** tells the run time scheduler to partition the iterations evenly among all the available threads/PEs

**dynamic:** tells the run time scheduler to give each thread *chunksize* iterations of the loop. *chunksize* should be smaller than $\frac{\text{number of total iterations}}{\text{number of threads}}$.

**interleave:** tells the run time scheduler to give each thread *chunksize* iterations of the loop, which are then assign to the threads in an interleaved way.

**gss:** (guided self scheduling) tells the run time scheduler to give each PE a varied number of iterations of the loop. If $I$ iterations remain and $P$ threads are working on them, the piece size is roughly $\frac{I}{2 \cdot P} + 1$.

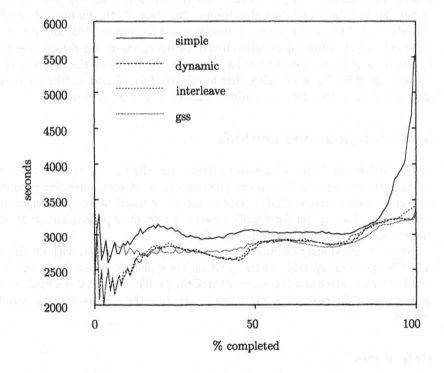

**Fig. 2.** Expected total time for different scheduling strategies

Figure 2 shows a down-sampled plot of the expected time to complete while processing the 3D range blocks in parallel. 14 PEs were used for all four types of scheduling strategies. The expected time is a linear extrapolation based on the previously needed time for that part of the work. To the left some initial oscillations can be seen, and the value at the right edge represents the actual time. The *simple* scheduling scheme fails miserably, the increase in the expected time results from the wrong assumption that all threads which are distributed evenly across the available PEs have an equal amount of work to do, they all got the same amount of cycles, but in the center of the frames the quadtrees had to be subsectioned more often since there was more activity.

### 4.3  Different Block Sizes

For the "dynamic"- and the "interleave"-scheduling type it is possible to provide the size of the block which each thread receives for computation. The calculations for this block have to be done completely by the assigned PE, after finishing it a new block is assigned to it. Therefore the size of the block is a parameter for load balancing. A block is a certain amount of iterations in the parallel loop. Various tests for different scenarios have been examined, in the example in Fig. 3 it is the case where 3D video blocks are coded, the loop of the range blocks runs in parallel and the scheduling algorithm is "interleave", the number of parallel threads is 14. The total amount of iterations in this case in 1188, and it can be observed on Fig. 3 that block with sizes of 24 iterations or less achieve the best results. It is obvious that the results become worse for block sizes greater than the average of 84, but it surprises that for values of 64, 48 and 32 (the latter two are not shown in the plot) the optimal balance still is not achieved.

## 5  Conclusions and Outlook

In this article the basic principles of fractal encoding of video have been explained, 2D coding and 3D coding. The approaches to parallelize this encoding step on shared memory MIMD systems include parallelization via ranges and via domains. Various vendor-supplied methods to control parallelization are explained and plots of related results are shown.

Since the previously mentioned calculations were performed without the so-called "sequential" speedup methods which are explained in [23, 2, 11] and others, like classification and transformation of the problem into better suited search spaces, one of the next steps will be to incorporate these ideas into the parallel version.

## References

[1] S. Andonova. *Parallel Fractal-Based Predictive Coding of Moving Images.* PhD thesis, Institute of Automation Technology, University of Bremen, Düsseldorf, Germany, 1995.

[2] B. Bani-Eqbal. Speeding up fractal image compression. In *Proceedings from IS&T/SPIE 1995 Symposium on Electronic Imaging: Science & Technology*, volume 2418: Still-Image Compression, pages 67–74, 1995.

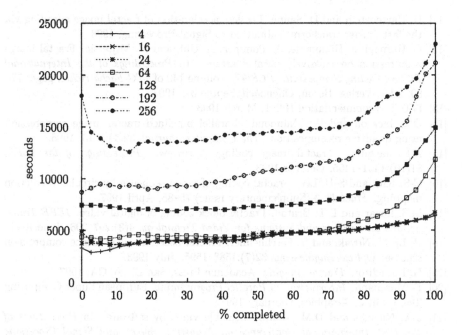

**Fig. 3.** Expected total time with different block sizes in the 3D/range/interleave case

[3] M. Barakat and J.L. Dugelay. Image sequence coding using 3-D IFS. In *Proceedings of the IEEE International Conference on Image Processing (ICIP'96)*, Lausanne, September 1996. IEEE Signal Processing Society.

[4] Barr E. Bauer. *Practical Parallel Programming*. Academic Press, New York, NY, USA, 1992.

[5] Y. Fisher, editor. *Fractal Image Compression: Theory and Application*. Springer-Verlag, New York, 1995.

[6] Y. Fisher, T.P. Shen, and D. Rogovin. Fractal (self-VQ) encoding of video sequences. In A.K. Katsaggelos, editor, *Visual Communications and Image Processing '94*, volume 2308 of *SPIE Proceedings*, Chicago, IL, USA, September 1994.

[7] D.J. Le Gall. The MPEG video compression algorithm. *Signal Processing: Image Communications*, 4(4):129–140, 1992.

[8] M. Gharavi-Alkhansari. *Fractal-Based Image and Video Coding Using Matching Pursuit*. PhD thesis, University of Illinois, 1997.

[9] S. Giordano, M. Pagano, F. Russo, and D. Sparano. A novel multiscale fractal image coding algorithm based on SIMD parallel hardware. In *Proceedings of the International Picture Coding Symposium (PCS'96)*, pages 525–530, Melbourne, March 1996.

[10] J. Hämmerle and A. Uhl. Approaching real-time processing for fractal compression. In J. Biemond and E.J. Delp, editors, *Visual Communications and Image Processing '97*, volume 3024 of *SPIE Proceedings*, pages 514–525, San Jose, February 1997.

[11] H. Hartenstein and D. Saupe. Lossless acceleration of fractal image encoding via the fast fourier transform. submitted to Signal Processing, 1998.

[12] C. Hufnagl, J. Hämmerle, A. Pommer, A. Uhl, and M. Vajtersic. Fractal image compression on massively parallel arrays. In *Proceedings of the International Picture Coding Symposium (PCS'97)*, volume 143 of *ITG-Fachberichte*, pages 77–80. VDE-Verlag, Berlin, Offenbach, September 1997.

[13] ITU-T recommendation H.261, March 1993.

[14] D.J. Jackson and W. Mahmoud. Parallel pipelined fractal image compression using quadtree recomposition. *The Computer Journal*, 39(1):1–13, 1996.

[15] A.E. Jacquin. Fractal image coding: A review. *Proceedings of the IEEE*, 81(10):1451–1465, October 1993.

[16] C-S. Kim and S-U. Lee. Fractal coding of video sequence by circular prediction mapping. *Fractals*, 5 (Supplementary Issue):75–88, April 1997.

[17] M.S. Lazar and L.T. Bruton. Fractal block coding of digital video. *IEEE Transactions on Circuits and Systems for Video Technology*, 4(3):297–308, June 1994.

[18] H. Li, M. Novak, and R. Forchheimer. Fractal-based image sequence compression scheme. *Optical Engineering*, 32(7):1588–1595, July 1993.

[19] N. Lu, editor. *Fractal Imaging*. Academic Press, San Diego, CA, 1997.

[20] B. Marchand. *Introduction to Parallel Programming*. European Super Computing Team, Silicon Graphics, February 1997.

[21] J.A. Nicholls and D.M. Monro. Scalable video by software. In *Proceedings of the 1996 International Conference on Acoustics, Speech and Signal Processing (ICASSP'96)*, Atlanta, May 1996.

[22] A. Pommer. A survey of fractal video coding. In B. Zovko-Cihlar, S. Grgić, and M. Grgić, editors, *International Workshop on Systems, Signals and Image Processing*, pages 55–58, Zagreb, Croatia, June 1998. online version follows.

[23] D. Saupe and R. Hamzaoui. Complexity reduction methods for fractal image compression. In J.M. Blackledge, editor, *Proc. IMA Conf. on Image Processing; Mathematical Methods and Applications (1994)*, pages 211–229. Oxford University Press, September 1995.

[24] D. Saupe, R. Hamzaoui, and H. Hartenstein. Fractal image compression – an introductory overview. In D. Saupe and J. Hart, editors, *Fractal Models for Image Synthesis, Compression and Analysis ACM SIGGRAPH'96 Course Notes 27*, New Orleans, Louisiana, August 1996.

[25] A. Uhl and J. Hämmerle. Fractal image compression on MIMD architectures I: Basic algorithms. *Parallel Algorithms and Applications*, 11(3–4):187–204, 1997.

[26] G.D. Veccia, R. Distasi, M. Nappi, and M. Pepe. Fractal image compresson on a MIMD architecture. In H. Liddel, A. Colbrook, B. Hertzberger, and P. Sloot, editors, *High Performance Computing and Networking. Proceedings of HPCN Europe 1996*, volume 1067 of *Lecture Notes on Computer Science*, pages 961–963. Springer, 1996.

[27] G.K. Wallace. The JPEG still picture compression standard. *Communications of the ACM*, 34(4):30–44, 1991.

[28] M.V. Wickerhauser. *Adapted wavelet analysis from theory to software*. A.K. Peters, Wellesley, Mass., 1994.

[29] P. Zinterhof and P. Zinterhof jun. A parallel version of an algorithm for fractal image compression. In *Workshop Paragraph 1994*, number 94-17 in RISC - Linz Report Series, 1994.

# The Split-Proxy Approach:
# A New Architecture for Parallel Video Servers§

Giuseppe De Pietro[1], Marco Lerro[1]

[1] IRSIP – CNR, Via P. Castellino 111 - 80131 Naples (Italy)
depietro.g@irsip.na.cnr.it

**Abstract.** Parallel architectures are often used for implementing Video Server in Video on Demand systems. In this paper, we propose an architectural model well suited for SMP parallel computers able to reduce the communication overhead related to video-stream data transfer from storage servers to the video clients. Experimental results obtained on a Quadrics CS-3 parallel computers are presented and discussed

## 1    Introduction

The possible diffusion of an interactive and multimedia television systems opens new interesting economic prospective for producers and end-users. For this reason, many telecommunication industries and research groups are employing a lot of their resources to study the feasibility of this new video service, nowadays known as *Video on Demand* (VoD). Generally speaking, the Video on Demand has been defined as a collection of services like

- **Movies on Demand**, where customers can select, play, and control movies with full and more VCR capabilities.
- **Interactive News Television**, where customers navigate inside a virtual newspaper selecting and deepening only the desired news.
- **Distance learning**, where customers subscribe to courses being taught at remote sites. Students tailor courses to individual preferences and time constraints.
- **Video conferencing**, where customers can negotiate with each other, by using several type of integrated data like audio, video, text, and graphics.

One of these services on which more research efforts are lavished is the *Movie on Demand* (MoD). It can be defined as the service of *interactive* delivery of a video-stream chosen from a *large* set of available programs. This brief definition sins of various ambiguous aspects that need to be clarified. For interactive delivery is intended the ability to control the stream flow through VCR-type controls (i.e. stop,

§ This research has been supported by EC FESR Project "Sviluppo di tecnologie Digitali di Grafica Avanzata per Applicazioni Industriali e Commericiali".

P. Zinterhof, M. Vajteršic, A. Uhl (Eds.): ACPC'99, LNCS 1557, pp. 327-336, 1999.

start, rewind, fast forward, and pause commands); besides, the waiting time between the control request and the reaction must be limited to a prefixed range of time. When this range is not longer than few seconds are, the MoD is said *Pure Movie on Demand* (P-MoD), otherwise *Near Movie on Demand* (N-MoD). The waiting time will largely determine the take-up of the service. About the available set of programs, to be successful a MoD services must give the appearance of an infinite availability of programs (the ideal case). However, the MoD system can exploit statistical information to offer at last the most popular videos of the moment, while alternatively storing techniques may be adopted for less required videos. Finally, a MoD service should support different audio and video resolutions in according with the user-bandwidth requirements. The fundamental innovation of this new service is that it will allow users to select the programs they want to view at the time they want to view them. Therefore, MoD is fundamentally different from broadcast video, cable television, or even *Pay Per View* (PPV) television. In fact, in these television systems, the only user's choice consists of selecting from a number of pre-programmed channels, which are oriented towards the *average user* within an *average time* of day for watching TV.

In the following sections, we explain some VoD architectures present in literature and after show our solution: the split-proxy approach. We discuss about the hardware environment adopted, the software architecture behind this approach and some results about a global performance test. Finally, we expose the conclusions and our future work.

## 2    Parallel Video Server Architecture

Parallel architectures have been introduced for VoD applications in order to overcome problems related to the lack of performance, fault tolerance, and efficient mass storage systems. The main work of a VoD server is to *pump* data streams from a mass storage system and to deliver these data to the final clients through a communication network. In a parallel video server, the data streams are striped across disks that usually belong to the same file system. The data striping can be provided as facility of the operating system running on the machine, can be provided by special devices (e.g. disk arrays) or can be managed at higher abstraction level by programmers. A combination of these three approaches is also possible.

Clearly, in order that the video client displays the originally video stream, the data split across the disks has to be re-ordered and merged for the final destination. This work is performed by software or hardware module called *proxy*, which knows the system's configuration (number and address of data servers, data locations, and striping policy).

There are three main solutions to implement the proxy [1]:

1. At the server computer - proxy at server
2. At an independent computer - independent proxy
3. At the client computer - proxy at client

Note that in this context the term computer refers to the hardware performing the proxy function that may not be a computer in the general sense. In the following, we describe these three approaches.

## Proxy-at-Server

In this architecture, one proxy is placed on each storage server. All servers are locally connected through an interconnection network (see Fig. 1). Each proxy manages the data retrieving from its storage system and requests to the other proxies the data placed elsewhere. Finally, the proxy combines the gathered data, and delivers them to the connected client. Clearly, in this approach, clients know nothing about system configuration and do not play any role in the data-retrieving phase: therefore, a client transparency is obtained.

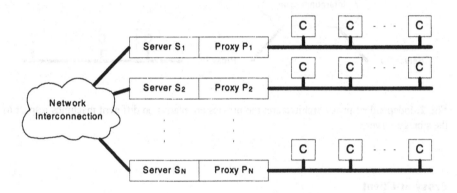

**Fig. 1.** Proxy-at-server architecture: the proxies are placed on each storage server.

Two main drawbacks are related to proxy-at-server solution. First, because the same processor is shared by the proxy and retrieve processes, there's an overhead due the competition for the same resource. Second, there is a communication overhead whenever data do not reside on the local storage; in this case, the required data have to be read from other server's local storage, processed and then transmitted to the client. Therefore, the data transmission requires in general a double step to reach the destination, and then a double bandwidth is needed.

## Independent-Proxy

In this case, proxies run on different processors respect to the storage server processes. The back-end storage servers and proxies processors are connected through an internal interconnection network (see Fig. 2), while the clients are connected to the proxies by another external network, usually with different characteristics from the internal one. The main advantage of this approach with respect to proxy-at-server is the reduction of the processing overhead. Clearly, this approach requires additional hardware (processors and communication links); moreover, the concept of data locality is lost, because storage server are *anonymous* for a proxy (as matter of fact, the topology of the internal communication network, whenever is not bus oriented, can strongly influence the time spent from retrieve data from a given server). Like the proxy-at-server solution, the independent-proxy architecture requires a double bandwidth.

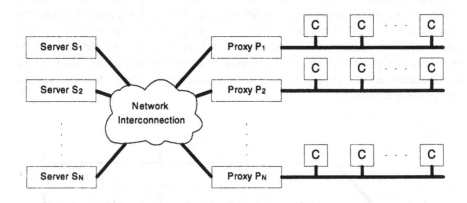

**Fig. 2.** Independent-proxy architecture: the proxies are placed on different machines respect to the storage servers.

**Proxy-at-Client**

In this third architecture, all the proxy functionalities are moved to the client side (see Fig. 3). Clients need to be equipped with additional software modules and/or hardware in order to combine data coming from storage servers into a single video stream suitable for the video-client application. In this case, there is a reduction of communication overhead with respects the proxy-at-server and independent-proxy architectures, because data are directly transmitted to the client. Another advantage is that a proxy failure involves only one client, while in the others approaches it involves all the clients connected to that proxy.

However, this approach presents some disadvantages related to the absence of client transparency. First, there is a processing overhead for the client: this implies that the client hardware must provide suitable hardware and software to satisfy additional requirements due to the implementation of proxy functionalities. Second, while the proxy-at-server and independent-proxy architectures essentially require the client to be able to decode a video stream compressed in a standard format (e.g. MPEG 2) and do not care about what is the client hardware (a PC or a Set Top Box), for the client-at-proxy architecture the proxy implementation is strongly related to the particular client hardware and, in general, there is no portability among different clients.

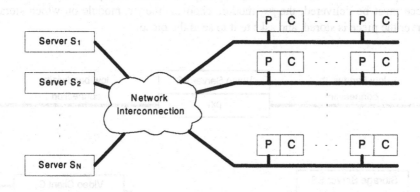

**Fig. 3.** Proxy-at-client architecture: the proxies are placed directly on the video clients.

## 3  The Split-Proxy Approach

As described in the previous section, each of three main approaches known in literature and used to develop most of video-server prototypes presents some vantages and disadvantages. In an efficient video-server system, the use of the proxy should hide the video-server implementation and configuration details to the client; at the same time, the video stream delivery should not require additional communications that can reduce the system efficiency. To reach these two goals, we propose a new approach named *split-proxy* architecture. It is based on the independent-proxy architecture where the storage server processes run on a different processors respect to the proxy processes. In our explanation we'll refer to the machines where the physical video streams are stored as *storage servers*, while the machines that interacts among clients and storage servers will be named *video servers*. In the independent-proxy architecture, the video stream flow from the storage servers to the video server, on which the proxy first merges the information and after sends them to the client. The basic idea behind our proposal consists of skipping the first data transfer, by directly sending the stream information from the storage servers to the client[1].

To reach this goal, we split the proxy into two modules (we'll call them $px_1$ and $px_2$): the first ($px_1$) running on a video server processor, the latter ($px_2$) running on the storage server processor. Clearly, the two modules have different functionalities. The $px_1$ module acts as the entry point for the client; that is, it shows to the client the available services, interacts for choosing the desired video stream and accepts the video-transport requests issued by the client. When a video stream request is accepted, the $px_1$ module activates the $px_2$ modules on all the storage servers that can

---

[1] We assume that each video stream is striped on more storage servers and that a data redundancy is provided in the manner that a single stream's piece is replicated on several storage servers depending on the fault-tolerance degree needed.

332    Giuseppe De Pietro and Marco Lerro

participate to the stream delivering. During the stream transport, when a stream's piece must be delivered, the $px_1$ module chooses[2] the $px_2$ module on which storage server the piece is stored, and asks to it to send the piece.

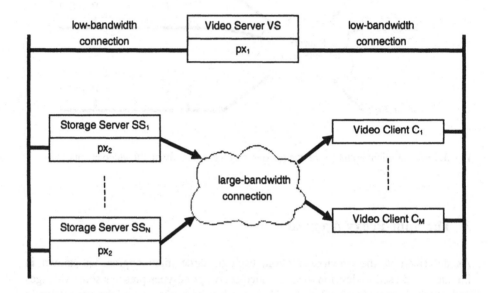

**Fig. 4.** Split-proxy architecture: the proxies are split in two modules. One ($px_1$) to interact to the client and another one ($px_2$) to send stream data from the storage server to the client.

Each $px_2$ module listens for transmission requests coming from the $px_1$ module and works to send the stream's pieces required directly to the client. When the piece has been sent, the $px_2$ module communicates this event to the $px_1$ module and waits for a new request. Therefore, the $px_2$ module can be viewed as a *bridge* that directly connects the storage server to the corresponding client. This implies that a client must have multiple network connections: one to the $px_1$ module to send requests and one to each $px_2$ module involved to receive the video stream. This doesn't necessary mean that several distinct physical links must connect the client to the proxy modules, but in many circumstances a low-bandwidth connection[3] can be used to communicate to the $px_1$ module (like a telephone line), while a large-bandwidth connection can be adopted to receive the stream from the $px_2$ modules.

---

[2] The storage server choice aims to balance the load across the storage servers.
[3] Consider that communications among the client and the video server (where the $px_1$ module is placed) consist of messages exchanges and requests deliveries coded by less bytes than a video stream flow. Moreover, these communications are occasional compared to the continuos flow for the video transmission.

## 4  Implementation Details

For the hardware implementation of both video and storage server we have used a preliminary version of the CS-3 machine produced by Quadrics Supercomputers World (QSW). This parallel computer can be viewed as a cluster of SMP nodes connected through a Fast Ethernet network[4]. Each SMP node is essentially provided of four Ultra2 at 250 MHz SPARC processors equipped with 512 Mbytes of central memory, and a storage system formed by twenty disks with a capacity of 4 Gbytes for each one, stripped by RAID level 5. About the client hardware, we have used Intel Pentium 2 processors equipped with 64 Mbytes and a board for hardware decoding of MPEG-2 video streams. Moreover, in order to increase the client number for the simulations, a client has also been implemented for Unix platform: it doesn't decode the received stream but simply throw away the data consuming the same rate required by the real clients.

About the software development, we have adopted the distributed object approach for both video server and client. In particular, a CORBA[5] compliant environment has been used as middle-ware layer, and the modules have been coded using C++. Moreover, the overall software architecture for the VoD service and the functionalities of each module has been developed according to the DSM-CC[6] standard. This standard essentially defines the core interfaces for the user-to-user communications at the ISO-OSI layer 7 (application specific layer). The DSM-CC primitives are based on the MPEG delivery system, and works are in progress to make the DSM-CC as an integral part of the MPEG-2 standard. The interfaces are described by the OMG IDL[7] [3] in the perspective of using the CORBA standard as development environment to produce interoperable applications in a client/server scenario. The most important interfaces defined in the DSM-CC standard are:

- **Service Gateway interface.** This interface provides the methods for browsing and discovering other services, authentication of end-user, registration of new service and new users, resolution of connections between clients and servers, and managing, opening and resuming of sessions.
- **Directory interface.** This interface provides the methods for the management of a name space where objects and data are registered and subsequently retrieved.
- **Stream interface.** This interface defines the essential methods for controlling continuous media stream playing. The methods extend the typical command offered by a VCR.

---

[4] This solution is being substituted by a very fast internal switching network.

[5] CORBA stands for Common Object Request Broker Adapter.

[6] DSM-CC stands for Digital Store Media-Command & Control

[7] IDL stands for Interface Definition Language. It is a language very similar to the C++ about the statement for declaring the structure of classes (here named interface). The interfaces defined by the IDL represent the common start point for both client and server development. Since the IDL is independent from implementation details, the client and server applications can be developed on different platforms and languages preserving the compatibility for the interfaces.

For the storage server, we used available software modules developed by other people within the activities related to the same VoD project [2]. Simply speaking, the streams are placed into a stripped file system, and retrieved by a database that exploits a whole time-based index of each stream in order to retrieve the needed frames.

About the network, our VoD system is for the moment based[8] on multiple Fast Ethernet networks that assure a physical bit rate of 100 Mbits per second. In particular, we have grouped several clients and one storage server on more isolated networks. In this way, we have obtained more clusters of VoD subsystem all controlled by the same video server, dividing the data stream traffic to each cluster.

## 5  Preliminary Results

Some tests have been performed in order to evaluate the effectiveness of the split-proxy approach. We used three *Symmetric Multi Processor* (SMP) modules: one for clients, one for the video server on which run the $px_1$ module and one for the storage server on which run the $px_2$ module, all connected through a Fast Ethernet network (see figure 4). Different operating conditions have been tested, by changing the number of active clients and the bit-rate of the video-streams. Several measurements have been taken by some software monitors inserted into all critical points of the system. As we expected, the main bottleneck is represented by the network. Due to the characteristics of the Fast Ethernet networks[9], in absence of overhead related to proxy server communications, we have estimated the number of clients that can be simultaneously served equal to

$$N_C = E_I E_N N_S / S_{BR} \qquad (1)$$

where

- $E_I$ is the network interface efficiency,
- $E_N$ is an efficiency factor taking into account the protocol overhead,
- $N_S$ in the bit rate provided by the network,
- $S_{BR}$ is the stream bit rate.

By means of speed test, we have estimated that the overhead factor $E_I E_N$ is equal to 0.8. As we used for the performance measurement a Fast Ethernet network, Ns is 100 Mbits/s. If we suppose to deliver all the streams at same bit rate, by applying (1), we have that the maximum number of clients that can be simoultaneously served is

$$Nc \cong 27 \qquad \text{for } S_{BR} = 3\text{Mbit/s}$$
$$Nc = 20 \qquad \text{for } S_{BR} = 4\text{Mbit/s}$$

---

[8] In the next future, the passage to ATM networks is programmed.
[9] where all the connected SMP modules share the same physical medium.

In figures 5, 6 is shown the maximum number of clients for typical video transmission rates. Results show that the limit of the number of clients that can be simultaneously served with a costant bit rate (25 for 3Mbit/s streams and 17 for 4 Mbit/s streams) are very close to the estimated numbers by applying the (1).

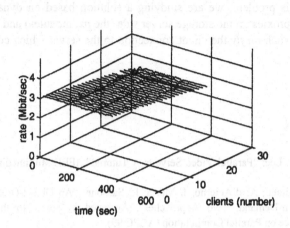

**Fig. 5.** Bit rate sustained by the network for each client in a split-proxy architecture. The bit rate requested is of 3 Mbits/sec.

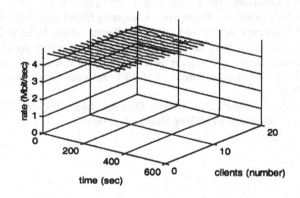

**Fig. 6.** Bit rate sustained by the network for each client in a split-proxy architecture. The bit rate requested is of 4 Mbits/sec.

## 6 Conclusions

In this paper, we have proposed the split-proxy architecture for implementing parallel video servers. Experimental results show the effectiveness of the approach for SMP machines. As explained in the previous sections, both the proxy-at-server and independent-proxy solutions require a double bandwidth due to a double step for the video-data transmission: the first from the storage servers to the video server, the second from this to the video client. Our solution aims to avoid this last passage using a direct connection from the storage server to the client. However, when the video data are stripped across more storage servers in order to achieve a load balancing across all the storage servers, there is the problem for connection switching. To overcome this problem, we are studying a solution based on dynamic connections between the proxies on the storage server side (the $px_2$ modules) and the video clients, capable to switch *on fly* the client connection to the server which contain the needed data.

# References

1.  Jack Y.B. Lee: "Parallel Video Servers: A Tutorial". IEEE Multimedia, April-June 1998, pp. 20-28
2.  F. Amalfitano, A. d'Acierno, I. Marra, L. Sansone: "An Object-Oriented DataBase for Movies-on-Demand: Two Approaches". Accepted as poster to the 4th International Conference on Parallel Computation (ACPC 99).
3.  IDL ISO/IEC 14750 Specification.
4.  Baldonado, M., Chang, C.-C.K., Gravano, L., Paepcke, A.: "The Stanford Digital Library Metadata Architecture". Int. J. Digit. Libr. 1 (1997), pp.108–121
5.  Bruce, K.B., Cardelli, L., Pierce, B.C.: Comparing Object Encodings. In: Abadi, M., Ito, T. (eds.): Theoretical Aspects of Computer Software. Lecture Notes in Computer Science, Vol. 1281. Springer-Verlag, Berlin Heidelberg New York (1997), pp. 415–438
6.  Van Leeuwen, J. (ed.): Computer Science Today. Recent Trends and Developments. Lecture Notes in Computer Science, Vol. 1000. Springer-Verlag, Berlin Heidelberg New York (1995)
7.  Michalewicz, Z.: Genetic Algorithms + Data Structures = Evolution Programs. 3rd edn. Springer-Verlag, Berlin Heidelberg New York (1996)

# A Wavelet Toolbox for Large Scale Image Processing

Geert Uytterhoeven, Dirk Roose, and Adhemar Bultheel

Department of Computer Science, Katholieke Universiteit Leuven,
Celestijnenlaan 200A, B-3001 Heverlee, Belgium,
Geert.Uytterhoeven@cs.kuleuven.ac.be
http://www.cs.kuleuven.ac.be/~geert/

**Abstract.** The wavelet transform has proven to be a valuable tool for image processing applications, like image compression and noise reduction. In this paper we present a scheme to process very large images that do not fit in the memory of a single computer, based on the software library WAILI (Wavelets with Integer Lifting). Such images are divided into blocks that are processed quasi independently, allowing efficient parallel programming. The blocking is almost completely transparent to the user.

## 1 Introduction

Wavelet transforms have good decorrelating properties. Wavelets are localized in both the spatial domain and the frequency domain, and are based on a multi-resolution analysis. They can have vanishing moments, which means that a smoothly varying signal can be represented with a small set of basis functions. The combination of this properties makes wavelets successful for image compression.

In application areas like Geographical Information Systems (GIS), image sizes are measured in gigabytes (GB) and even terabytes (TB). For example, an aerial color image of Belgium, where each pixel corresponds to one square meter, merely consumes 90 GB of storage. It is obvious that such images need to be stored and handled in a compressed form. Wavelets can provide for this since wavelet-based techniques offer better compression rates than other techniques. Moreover, many image processing operations can be done in the wavelet domain. Some of them can even be done better in the wavelet domain than in the spatial domain.

We have developed the software library WAILI, which allows the combination of image compression and processing in one package. In this paper we describe the extension of WAILI to block-based processing, which allows for a parallel implementation.

P. Zinterhof, M. Vajteršic, A. Uhl (Eds.): ACPC'99, LNCS 1557, pp. 337–346, 1999.

## 2    Wavelets and the Lifting Scheme

### 2.1    Predict and Update

The wavelet transform of a 1D signal is a multi-resolution representation of that signal where the wavelets are the basis functions which at each resolution level give a highly decorrelated representation. At each resolution level, the signal is split into a high pass and a low pass part and the low pass part is split again etc. These high pass and low pass parts are obtained by applying certain wavelet filters.

The lifting scheme is an efficient implementation of these filtering operations. Several introductions to the lifting scheme are available [7, 6]. Suppose that the low resolution part of a signal at level $j + 1$ consists of a data set $\lambda_{j+1}$. This set is transformed into two other sets at level $j$: the low resolution part $\lambda_j$ and the high resolution part $\gamma_j$. This is obtained first by just splitting the data set $\lambda_{j+1}$ into two separate data subsets $\lambda_j$ and $\gamma_j$ (e.g. the even samples and the odd samples). Such a splitting is sometimes referred to as the *lazy wavelet transform*. Then these two sets are recombined in several subsequent pairs of lifting steps which decorrelate the two signals.

A *dual lifting* step can be seen as a prediction: the data $\gamma_j$ are 'predicted' from the data in the subset $\lambda_j$. When the signals are still highly correlated, then such a prediction will usually be very good, and thus we do not have to keep this information in both signals. That is why we can keep $\lambda_j$ and store only the part of $\gamma_j$ that is not predictable (the prediction error). Thus $\gamma_j$ is replaced by $\gamma_j - \mathcal{P}(\lambda_j)$ where $\mathcal{P}$ represents the prediction operator. For smooth signals, the prediction error is small. This is the real decorrelating step.

However, the new representation has lost certain basic properties, which one usually wants to keep, like for example the mean value of the signal. To restore this property, one needs a *primal lifting* step, whereby the set $\lambda_j$ is updated with data computed from the (new) subset $\gamma_j$. Thus $\lambda_j$ is replaced by $\lambda_j + \mathcal{U}(\gamma_j)$ with $\mathcal{U}$ some updating operator.

In general, several such lifting steps can be applied in sequence to go from level $j + 1$ to level $j$. To recapitulate, let us consider a simple lifting scheme with only one pair of lifting steps.

**Splitting** (*lazy wavelet transform*) Partition the data set $\lambda_{j+1}$ into two distinct data sets $\lambda_j$ and $\gamma_j$.

**Prediction** (*dual lifting*) Predict the data in the set $\gamma_j$ by the data set $\lambda_j$.

$$\gamma_j \leftarrow \gamma_j - \mathcal{P}(\lambda_j).$$

**Update** (*primal lifting*) Update the data in the set $\lambda_j$ by the data in set $\gamma_j$.

$$\lambda_j \leftarrow \lambda_j + \mathcal{U}(\gamma_j).$$

These steps can be repeated by iteration on the $\lambda_j$, creating a multi-level transform or multi-resolution decomposition.

The inversion rules are obvious: revert the order of the operations, invert the signs in the lifting steps, and replace the splitting step by a merging step. Thus, inverting the three step procedure above results in:

**Inverse update**      $\lambda_j \leftarrow \lambda_j - \mathcal{U}(\gamma_j),$

**Inverse prediction**   $\gamma_j \leftarrow \gamma_j + \mathcal{P}(\lambda_j),$

**Merge**                $\lambda_{j+1} \leftarrow \lambda_j \cup \gamma_j.$

## 2.2  Integer Transforms

In practice, discrete signals are represented by integers. Doing the filtering operations on these numbers however will transform them in rational or real numbers because the filter coefficients need not be integers. To obtain an efficient implementation of the discrete wavelet transform, it is of great practical importance that the wavelet transform is represented by a set of integers as well, while the transform should still be invertible. This is easily achieved within the lifting framework.

We round the intermediates of each lifting step to integers (for example the nearest integer) and indicate this operation by square braces. Thus, we actually compute rounded values:

$$\gamma_j \leftarrow \gamma_j - [\mathcal{P}(\lambda_j)], \qquad \text{and} \qquad \lambda_j \leftarrow \lambda_j + [\mathcal{U}(\gamma_j)].$$

It is not difficult to verify that each step of the lifting scheme with rounding is perfectly invertible and thus the whole signal is perfectly reconstructible, whatever the rounding rule we use, on condition of course that the rounding is deterministic [1, 8].

## 2.3  Example: Cohen-Daubechies-Feauveau

The popular family of classical biorthogonal wavelets constructed by Cohen, Daubechies and Feauveau [2] fits in the above scheme. Especially its member with two vanishing moments for both the primal and dual wavelet (hence named CDF $(2, 2)$ wavelet) is widely used.

Thanks to the lifting scheme, the accompanying wavelet transform can be implemented in an efficient way [8]). From the second generation viewpoint, one transform step of a discrete signal $x = \{x_k\}$ looks like:

**Splitting** Split the signal $x$ (i.e. $\lambda_{j+1}$) into even samples (i.e. $\lambda_j$) and odd samples (i.e. $\gamma_j$):

$$s_i \leftarrow x_{2i}, \qquad \text{and} \qquad d_i \leftarrow x_{2i+1}.$$

**Prediction** Predict the odd samples using linear interpolation:

$$d_i \leftarrow d_i - \frac{1}{2}(s_i + s_{i+1}).$$

**Update** Update the even samples to preserve the mean value of the samples:

$$s_i \leftarrow s_i + \frac{1}{4}(d_{i-1} + d_i).$$

As a result, the signal $s = \{s_k\}$ is a coarse representation of the original signal $x$, while the signal $d = \{d_k\}$ contains the high frequency information that is lost when going from resolution level $j + 1$ to resolution level $j$.

A whole family of lifting schemes can be constructed in this way, of which the above example is just the simplest possible case. Since lifting steps use coefficients from regions near the coefficients that are updated, one needs special care at the boundaries. This is also important for a parallel implementation.

Note that this transform works on one-dimensional data. For two-dimensional data, like images, it can be applied row and columnwise, resulting in a tensor product transform at each step. Instead of with 2 subbands (low pass and high pass), one will end up with 4 subbands: LL — low pass in both the horizontal and vertical directions, LH — low pass in the vertical, high pass in the horizontal direction, HL and HH. When iterated on the LL subband, the result is a multi-resolution decomposition as shown in Figure 1. This ordering of the subbands at the different resolution levels is called the "Mallat" ordering [3].

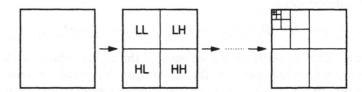

**Fig. 1.** The two-dimensional wavelet transform: iteration on the LL subband, showing the wavelet coefficients in "Mallat" ordering

## 3 WAILI: Wavelets with Integer Lifting

WAILI is a software library — written in C++ — providing wavelet transforms and wavelet-based operations on two-dimensional images [8]. WAILI is available in source form[1] for research purposes.

WAILI implements various integer wavelet transforms, using lifting technology. It includes some image processing operations, such as:

- Crop and merge on wavelet transformed images. This allows to cut a rectangular subimage out of a large wavelet transformed image, or to replace a rectangular area in a large wavelet transformed image.
- Noise reduction, using thresholding based on generalized cross validation.
- Simple compression, using thresholding of subbands.

---

[1] WWW: http://www.cs.kuleuven.ac.be/~wavelets/

# 4   Very Large Images and Tiling

We define an image to be *very large* if it does not fit in the memory of the computer system that is processing it. Note that by this definition, whether an image is very large or not depends not only on the image but also on the computer system.

## 4.1   Tiling

In order to process very large images, they have to be divided in blocks (tiling).

Fortunately wavelets are localized in both the spatial and frequency domains. For the division in blocks only the spatial domain is important, and this means that one needs only a limited subset of the input image to calculate a subset of the wavelet coefficients. The same is true for the inverse transform. From Figure 2, it is clear that only a fraction of the blocks in the right part of the figure are needed to reconstruct the rectangle in the left part.

Since not all blocks have to be present in main memory at the same time, this allows for the successful processing of very large images on small uni-processor machines with limited memory. Moreover, the processing of the blocks can be distributed among multiple machines and/or CPUs.

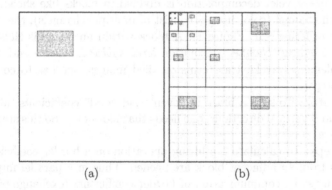

(a)                                  (b)

**Fig. 2.** Locality in the (a) *space* and (b) *wavelet domain*. The dotted lines in (b) indicate block boundaries

Within WAILI, each subband is divided separately in blocks. Blocks are square and block sizes must be powers of two (except near the borders of the image). Because small blocks cause more administrative overhead, it is recommended to use the same block size at different resolution levels.

Since we want to process both complete images and subimages that are part of a large image, the upper left corner of a (sub)image is not always located at the origin $(0, 0)$, but may have arbitrary coordinates (see Figure 3). To facilitate the wavelet transforms, the internal boundaries of the blocks must always be aligned to the block size. Hence, not only the blocks at the right and bottom borders of the image may be rectangular and smaller than the block size, but also at the left and upper borders.

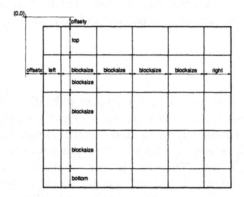

**Fig. 3.** Division of a subband in blocks. The margins (top, bottom, left and right) are chosen such that the coordinates of the block boundaries are always multiples of the block size

The whole wavelet decomposition is divided in blocks like shown in fig. 4. Within each channel (color images consist of multiple channels), the tree forms a "Mallat" ordering (cfr. Figure 1). A block within an image is thus uniquely defined by 5 indices: *channel* , resolution *level, subband, column* and *row*.

To implement wavelet transforms on tiled images, one has to consider the following steps:

**Splitting step:** To split a block in "even" and "odd" coefficients, in both the horizontal and vertical directions, one needs that block only. So this step is trivial to parallelize.

**Lifting steps:** To calculate the lifting operation on a border coefficient, some coefficients from an adjacent block are needed. Thus in a parallel implementation there must be communication of border coefficients (exchange of "overlap regions", with a width depending on the wavelet transform). Since in a primal lifting step the "even" blocks (i.e. the blocks with the even coefficients) are updated by the "odd" blocks, and vice versa in a dual lifting step, these updates can be done independently in each block, except for the communication just mentioned. Hence, this step can be parallelized easily. The parallel efficiency grows with increasing block size due to the "perimeter effect".

## 4.2   Block Management

We extended WAILI to support very large images that are divided in blocks. This is nearly completely transparent to the user: all traditional operations can be performed, just like on small non-blocked images.

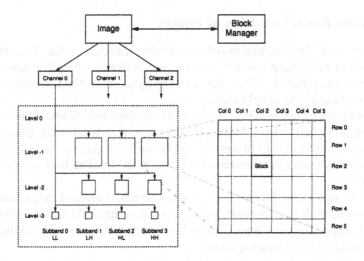

**Fig. 4.** Subdivision of an *image* into *channels*, *levels*, *subbands* and *blocks*

The "block manager" is responsible for managing the blocks of which a very large image is composed. To access a part of the (transformed) image, the block manager is asked for blocks and retrieves them, e.g. from disk or via a network. If a block is modified, it has to be updated on the storage device. Blocks are passed to the compression module to remove redundancy among the wavelet coefficients within each block. The block manager for sequential processing also keeps a cache of recently used blocks to speed up processing. An overview of the complete system is given in Figure 5.

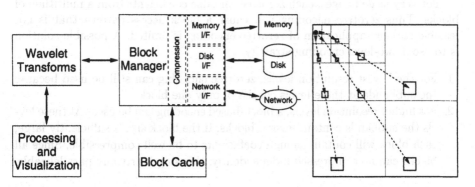

**Fig. 5.** Design of the large scale image processing system     **Fig. 6.** Zerotree encoding

To obtain a parallel version, we only need to adapt the block manager and introduce communication of the border coefficients during the lifting steps.

## 4.3   Block-Based Denoising of Images

WAILI uses soft-thresholding to reduce additive, stationary noise. The threshold is selected at each transform level and for each subband separately, using generalized cross validation [8]. Since this is an asymptotical method, it only performs well for sufficiently large subbands.

For wavelet transformed images that are divided into blocks, the denoising technique is applied independently on each block of each subband. Because only the highpass subbands are processed and blocks belonging to lowpass subbands are left alone, it does not become visible where the borders between the blocks are located.

Experiments showed that one gets better results, compared to the case where complete subbands are denoised, because now the threshold is more adapted to the local properties of the image. Thus denoising is not only easily parallelizable, it also yields a better image quality.

# 5   Image Compression

Due to their decorrelating properties, wavelets are well suited for the compression of images and signals. If a sufficiently smooth image is transformed, most high pass coefficients will be zero or very small. Compression algorithms based on wavelets often use a variant of "zerotree encoding" [5, 4]. Most compression methods are targeted at "small" images. In that case the individual subbands are too small to get a sufficient compression gain from compressing them separately, hence zerotree encoding combines the subbands to exploit the correlation that is still present among spatially related coefficients at different resolution levels. (see Figure 6).

For very large images, each zerotree contains coefficients from a multitude of blocks. Thus zerotree encoding will cause a block access pattern that is not feasible and the application of zerotree encoding is limited. A possible solution is to use a two-level encoding strategy:

1. For the lowest resolution levels, zerotree encoding can still be used because the subbands at these levels consist of only one block.
2. For higher resolution levels, a block-based encoding can be used. At these levels the subbands contain several blocks. If the block size is sufficiently large, each block will contain enough coefficients to be well compressible. Since all blocks can be compressed independently, this step is trivially parallelizable.

# 6   Results

We have no parallel implementation of WAILI yet. To get an impression of the performance of our wavelet transform code, we transformed an image of $3584 \times 3584$ pixels, using a block size of $128 \times 128$ pixels, for a total of $28 \times 28$ blocks. We used the Cohen-Daubechies-Feauveau (2, 2) biorthogonal wavelets.

All calculations were done on 16 bit integer data on a single processor machine (Intel Pentium II, 350 MHz) with sufficient memory to avoid paging. Note that after 2 transform levels, the subband size is no longer divisible by the blocksize and one has to take care of rectangular blocks.

Each level of the CDF $(2, 2)$ wavelet transform of $n$ blocks consists of 5 steps:

- Lazy wavelet transform in both the horizontal and vertical directions. In order to maintain the blocksize, 4 blocks are jointly transformed into 4 new blocks. (in total $n$ blocks are transformed)
- Lifting steps on the columns of the image: a dual lifting step ($n/2$ blocks) and a primal lifting step ($n/2$ blocks).
- Lifting steps on the rows of the image: a dual lifting step ($n/2$ blocks) and a primal lifting step ($n/2$ blocks).

Timings are shown in Table 1. An additional overhead of about 0.6 seconds was caused by the supply of blocks by the block manager. However, in a parallel implementation we can assume that the blocks are distributed over all processors, so this overhead can be spread over the processors too.

In a parallel implementation, the lazy wavelet transform requires no communication if the 4 neighboring blocks are transformed by the same processor. If we would use $14 \times 14 = 196$ CPUs, the lazy wavelet transform for the first level could be calculated in 1.33 ms and the calculation time for the 4 lifting steps would be 28.5 ms. The lifting steps require communication of the border coefficients. For the CDF $(2, 2)$ transform, in each lifting step 2 strips of 128 coefficients must be exchanged between processors. On current parallel systems (e.g. IBM SP2), the communication cost for such short messages is dominated by the startup time ($\mathcal{O}(50\ \mu s)$), which is negligible compared to the calculation cost.

However, the amount of data to be transformed is divided by 4 at each level. Hence, at the second level, we can keep the blocksize equal to 128, but then only 1/4 of the CPUs can be kept busy. Thus the execution time for the second level will also be approx. 30 ms. A similar reasoning holds for the other levels, but in our example rectangular blocks will appear at level 3, causing an additional overhead.

We can conclude that a parallel implementation can be efficient if the number of transform steps is small, and the number of processors is not too large, relative to the image size.

# 7 Conclusion

In this paper, we have described a strategy to perform wavelet-based image compression and processing on very large images. The images are split in blocks, which can be handled nearly independently.

The (integer) wavelet transform is implemented using the lifting scheme, which can be parallelized easily and efficiently, since only communication of pixels at the borders of the blocks is required.

| Transform Level | Lazy Transform # Calls | Time | Lifting Steps # Calls | Time | Total Time | Cumulative Total Time |
|---|---|---|---|---|---|---|
| 1 | 784 | 0.26 | 1568 | 5.59 | 5.85 | 5.85 |
| 2 | 196 | 0.02 | 392 | 1.34 | 1.22 | 7.07 |
| 3 | 49 | 0.01 | 128 | 0.33 | 0.34 | 7.41 |

**Table 1.** Timings for the wavelet transform of an image of size 3584 × 3584. Each call corresponds to the processing of one elementary block. All timings are in seconds

The method is implemented as an extension to the software library WAILI. Blocks are accessed and handled under supervision of a block manager. Although the software has not been parallelized yet, the limited communication and the existing block manager allow an easy and efficient parallelization.

The division in blocks also has a positive influence on the performance of our denoising algorithm. Because denoising works independently on the blocks, it adapts better to the local properties of the image.

# 8    Acknowledgements

This paper presents research results of the Flemish Information Technology Action Program ("Vlaams Actieprogramma Informatietechnologie"), project number ITA/950244. The scientific responsibility rests with its authors.

# References

[1] R. Calderbank, I. Daubechies, W. Sweldens, and B.-L. Yeo. Wavelet transforms that map integers to integers. *Appl. Comput. Harmon. Anal.*, 5(3):332–369, 1998.

[2] A. Cohen, I. Daubechies, and J. Feauveau. Bi-orthogonal bases of compactly supported wavelets. *Comm. Pure Appl. Math.*, 45:485–560, 1992.

[3] S. G. Mallat. Multifrequency channel decompositions of images and wavelet models. *IEEE Trans. Acoust. Speech Signal Process.*, 37(12):2091–2110, 1989.

[4] A. Said and W. A. Pearlman. Image compression using the spatial-orientation tree. In *Proc. IEEE Int. Symp. Circuits and Syst., Chicago, IL*, pages 279–282, May 1993.

[5] J. M. Shapiro. Embedded image coding using zerotrees of wavelet coefficients. 41(12):3445–3462, 1993. IEEE Trans. Signal Process.

[6] W. Sweldens. The lifting scheme: A new philosophy in biorthogonal wavelet constructions. In A. F. Laine and M. Unser, editors, *Wavelet Applications in Signal and Image Processing III*, pages 68–79. Proc. SPIE 2569, 1995.

[7] W. Sweldens. The lifting scheme: A construction of second generation wavelets. *SIAM J. Math. Anal.*, 29(2), 1997.

[8] G. Uytterhoeven, F. Van Wulpen, M. Jansen, D. Roose, and A. Bultheel. WAILI: A software library for image processing using integer wavelet transforms. In K.M. Hanson, editor, *Medical Imaging 1998: Image Processing*, volume 3338 of *SPIE Proceedings*, pages 1490–1501. The International Society for Optical Engineering, February 1998.

# Hardware and Software Aspects for 3-D Wavelet Decomposition on Shared Memory MIMD Computers

Rade Kutil[1] and Andreas Uhl[2]

[1] Department of Computer Science and System Analysis
University of Salzburg, AUSTRIA
[2] RIST++ & Department of Computer Science and System Analysis
University of Salzburg, AUSTRIA
{rkutil,uhl}@cosy.sbg.ac.at

**Abstract.** In this work we discuss hardware and software aspects of parallel 3-D wavelet/subband decomposition on shared memory MIMD computers. Experimental results are conducted on a SGI POWERChallenge GR.

## 1 Introduction

In recent years there has been a tremendous increase in the demand for digital imagery. Applications include consumer electronics (Kodak's Photo-CD, HDTV, SHDTV, and Sega's CD-ROM video game), medical imaging (digital radiography), video-conferencing and scientific visualization. The problem inherent to any digital image or digital video system is the large amount of bandwidth required for transmission or storage.

Unfortunately many compression techniques demand execution times that are not possible using a single serial microprocessor [13], which leads to the use of general purpose high performance computers for such tasks (beside the use of DSP chips or application specific VLSI designs). In the context of MPEG-1,2 and H.261 several papers have been published describing real-time video coding on such architectures [1, 3].

Image and video coding methods that use wavelet transforms have been successful in providing high rates of compression while maintaining good image quality and have generated much interest in the scientific community as competitors to DCT based compression schemes in the context of the MPEG-4 and JPEG2000 standardization process.

Most video compression algorithms rely on 2-D based schemes employing motion compensation techniques. On the other hand, rate-distortion efficient 3-D algorithms exist which are able to capture temporal redundancies in a more natural way (see e.g. [10, 6, 5, 15] for 3-D wavelet/subband coding). Unfortunately these 3-D algorithms often show prohibitve computational and memory demands (especially for real-time applications). At least, prohibitive for a common microprocessor. A shared memory MIMD architecture seems to be an interesting choice for such an algorithm.

P. Zinterhof, M. Vajteršic, A. Uhl (Eds.): ACPC'99, LNCS 1557, pp. 347–356, 1999.
© Springer-Verlag Berlin Heidelberg 1999

348    Rade Kutil and Andreas Uhl

As a first step for an efficient parallel 3-D wavelet video coding algorithm the 3-D wavelet decomposition has to be carried out (followed by subsequent quantization and coding of the transform coefficients). In this work we concentrate ourselves on the decomposition stage.

A significant amount of work has been already done on parallel wavelet transform algorithms for all sorts of high performance computers. We find various kinds of suggestions for 1-D and 2-D algorithms on MIMD computers (see e.g. [16, 14, 12, 7, 4, 8] for decomposition only and [9, 2] for algorithms in connection with image compression schemes). On the other hand, the authors are not aware of any work (except [11]) focusing especially on 3-D wavelet decomposition and corresponding 3-D wavelet based video compression schemes.

In this work we discuss hardware and software aspects of parallel 3-D pyramidal wavelet decomposition on shared memory MIMD computers.

## 2   3-D Wavelet Decomposition

The fast wavelet transform can be efficiently implemented by a pair of appropriately designed Quadrature Mirror Filters (QMF). A 1-D wavelet transform of a signal $S$ is performed by convolving $S$ with both QMF's and downsampling by 2; since $S$ is finite, one must make some choice about what values to pad the extensions with. This operation decomposes the original signal into two frequency-bands (called subbands), which are often denoted coarse scale approximation and detail signal. Then the same procedure is applied recursively to the coarse scale approximations several times (see Figure 1.a).

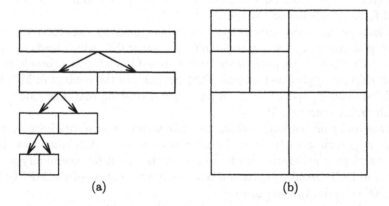

(a)                              (b)

**Fig. 1.** 1-D (a) and 2-D (b) pyramidal wavelet decomposition.

The classical 2-D transform is performed by two separate 1-D transforms along the rows and the columns of the image data $S$, resulting at each decomposition step in a low pass image (the coarse scale approximation) and three detail images (see Figure 1.b). To be more concise, this is achieved by first convolving the rows of the low pass image $S_{j+1}$ (or the original image in the first decomposition level) with the QMF filterpair G and H (which are a high pass and a low pass filter, respectively), retaining every other row, then convolving the columns of the resulting images with the same filterpair and retaining every other column. The same procedure is applied again to the coarse scale approximation $S_j$ and to all subsequent approximations.

By analogy to the 2-D case the 3-D wavelet decomposition is computed by applying three separate 1-D transforms along the coordinate axes of the video data. The 3-D data is usually organized frame by frame. The single frames have again rows and columns as in the 2-D case ($x$ and $y$ direction in Figure 2, often denoted as "spatial coordinates"), whereas for video data a third dimension ($t$ for "time" in Figure 2) is added. As it is the case for 2-D decompositions, it does not matter in which order the filtering is performed (e.g. a 2-D filtering frame by frame with subsequent temporal filtering, three 1-D filterings along $y$, $t$, and $x$ axes, e.t.c.). After one decomposition step we result in 8 frequency subbands out of which only the approximation data (the gray cube in Figure 2) is processed further in the next decomposition step. This means that the data on which computations are performed are reduced to $\frac{1}{8}$ in each decomposition step.

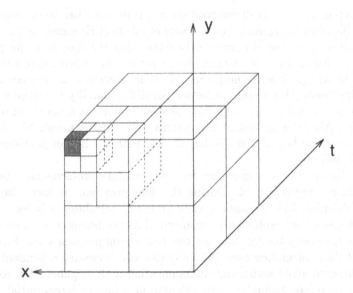

**Fig. 2.** Classical 3-D wavelet decomposition.

In our implementation we have chosen to apply the very natural frame by frame approach. A pseudo code (using in-place transforms) of such a 3-D wavelet transform with max_level decomposition steps applied to a video S[x,y,t] looks as follows:

**Sequential 3-D Wavelet Decomposition**

```
for level=1...max_level {
    for t=1..t_max {
        for x=1...x_max {
            for y=1...y_max {
            convolve S[x,y,t] with G and H } }
        for y=1...y_max {
            for x=1...x_max {
            convolve S[x,y,t] with G and H } } }
    for x=1...x_max {
        for y=1...y_max {
            for t=1..t_max {
            convolve S[x,y,t] with G and H } } }
}
```

## 3   3-D Wavelet Decomposition on Shared Memory MIMD Computers

### 3.1   A Message Passing Algorithm

When computing the 3-D decomposition in parallel one has to decompose the data and distribute it among the processor elements (PE) somehow. In contrast to previous work [11] we decompose the data along the time-axis into parallel-epipeds (see Figure 3). The obvious reason is that the dimension of the data is expected to be significantly larger in t (=time) direction as compared to the spatial directions. This results in better parallel scalability for large machines operating on comparatively small data sets (an important property in real-time processing). Also, the amount of data partition boundaries should be minimized [11] since at these boundaries one has to deal with the border problems of the wavelet transform.

In the literature two approaches for the boundary problems have been discussed and compared [16, 14]. During the *data swapping* method (also known as *non-redundant data calculation*) each processor calculates only its own data and exchanges these results with the appropriate neighbour processors in order to get the necessary border data for the next decomposition level. Employing *redundant data calculation* each PE computes also necessary redundant border data in order to avoid additional communication with neighbour PE to obtain this data. Therefore, redundant data calculation requires a larger initial data set and subsequently trades off computation for communication.

In this work we employ the data swapping approach. Figure 3 shows the stages of the algorithm.

(1) The video data is distributed uniformly among the PE.
(2) The PE exchange the necessary border data (light gray) for the next decomposition level with their corresponding neighbours.
(3) The 3-D decomposition is performed on the local data on each PE.

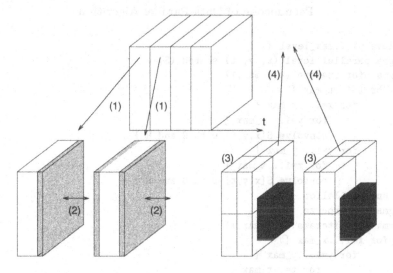

**Fig. 3.** Message passing with data swapping.

(4) All subbands but the approximation subband (dark gray) are collected (there is no more work to do on these data).

(5) Repeat steps 2 - 4 until the desired decomposition depth is reached.

Please note that for the first decomposition level steps 1) and 2) are combined by sending overlapping data blocks to the PE. Given $p$ PE and a convolution filter of length $f$ this algorithm shows a computational complexity of order $O(\frac{8}{7}\frac{xyt3f}{p})$ and a transfer complexity of order $O(xy(\frac{2t}{p} + \frac{4(f-2)}{3}))$.

## 3.2   A Data Parallel Algorithm

Data parallel programming on a shared memory architecture is easily achieved by transforming a sequential algorithm into a parallel one by simply identifying areas which are suitable to be run in parallel i.e. in which no data dependencies exist. Subsequently, local and shared variables need to be declared and parallel compiler directives are inserted.

Since the order of execution is not important and there are no data dependencies among different loop runs (except for the level loop) we may apply the following simple parallelization strategy. We distribute the two outer loops t and x among the PE. Between those two loops the PE need to by synchronized since before the temporal filtering step the spatial filtering has to be completed. Only the indices for the three coordinate axes are declared to be local variables, the data S[x,y,t] are of course declared to be shared.

### Pseudocode of Data-Parallel Algorithm

```
for level=1...max_level {
#pragma parallel local (x, y, t) shared (S)
#pragma pfor iterate (t,t_max,1)
    for t=1..t_max {
        for x=1...x_max {
            for y=1...y_max {
            convolve S[x,y,t] with G and H } }
        for y=1...y_max {
            for x=1...x_max {
            convolve S[x,y,t] with G and H } }
    endparallelfor
#pragma synchronize
#pragma pfor iterate (x,x_max,1)
    for x=1...x_max {
        for y=1...y_max {
            for t=..t_max {
            convolve S[x,y,t] with G and H } }
    endparallelfor
endparallelregion
}
```

## 4   Experimental Results

We conduct experiments on a SGI POWERChallenge GR (at RIST++, Salzburg Univ.) with 20 MIPS R10000 processors and 2.5 GB memory. The size of the video data is $128 \times 128$ pixels in the spatial domain, combined to a 3-D data block consisting of 2048 frames. The PVM version used as default setting is a native PVM developed for the POWERChallenge Array. We use QMF filters with 4 coefficients.

When performing convolutions on 2-D data using a sensible ordering of the computations modern architectures usually produce not too many cache misses due to the large cache size of the available microprocessors. The situation changes dramatically when changing to 3-D data. In our concrete implementation the spatial wavelet transform (the parallelized t loop) is not affected by cache misses since the data is organized frame by frame in the memory. The problem arises when computing the temporal transform (the parallelized x loop) since the data involved in this transform is not close in terms of memory location, in contrary, these values are separated by a data block of the size of one frame (this is why prefetching techniques do not work properly as well). Moreover, since we operate with power-of-two sized data, these in-between datablocks are as well of power-of-two size which means that the data values required for the temporal transform are loaded into the same cache lines each time when required within the transform stage (which is quite a few times) which produces a significant amount of cache misses in the transform.

In the original version of our algorithm we result of an sequential execution time of 187 seconds. This execution time can be improved with a simple trick (since it includes many cache misses): instead of having power-of-two sized data in between the data for the temporal transform we insert a small amount artificial dummy data in between the single frames. With this technique we result in an sequential execution time of 82 seconds. Subsequently, all speedup results (despite of those in Figure 4) refer to this improved sequential algorithm.

The cache misses problems do not only affect sequential performance but reduce as well parallel efficiency significantly. Figure 4 shows speedup values of the message passing implementation comparing the original algorithm with the cache corrected version (note that speedup is computed with respect to the corresponding sequential algorithms, to 187 and 82 seconds, respectively). Although the sequential execution time of the original algorithm is much higher (and we would therefore expect a better efficiency) we observe an almost constant speedup for more than 8 PE for the not corrected algorithm. On the other hand, the optimized algorithm shows increasing speedup across the entire PE range. This on first sight surprising behaviour is caused by the cache misses of the not corrected algorithm which lead to a congestion of the bus if the number of PE gets too high.

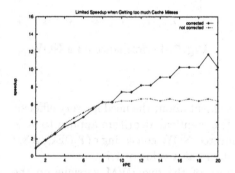 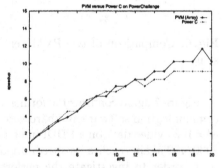

**Fig. 4.** Reduced efficiency due to cache misses

**Fig. 5.** Comparison of message passing and data parallel algorithms

Figure 5 shows a comparison of the message passing (PVM) and data parallel (PowerC) algorithms. Whereas both algorithms perform almost equally up to 12 PE, the message passing algorithm performs better for higher PE numbers. It should be noted that the message passing implementation uses the host-node paradigm, the number of PE gives in the figure describing the number of node processes being spawned. The actual number of PE in use is therefore one higher. Keeping this is mind we have to compare speedup the speedup at positions PE + 1 (PowerC) and PE (PVM). With this interpretation of the plot the results are different - up to 8 PE actually used the data parallel algorithm exhibits better efficiency, whereas for larger PE numbers the message passing algorithm still shows better speedup (e.g. we observe speedup 9 versus speedup 12 for 20 PE actually in use.

In order to highlight the importance of an efficient message passing library for satisfying results with this algorithm, we have used as well the public domain version of PVM available at www.netlib.org (denoted PVM (Standard) in the plots). We observe a dramatic decrease in efficiency across the entire range of PE numbers. The public domain version of PVM does not allow to reach speedup larger than 2 at all. One fact worth noticing is that for initial data distribution it is not possible to send a single message from the host process to each node since the PVM buffer can not handle data set of this size. Therefore the data is cut into smaller pieces and is sent within a loop.

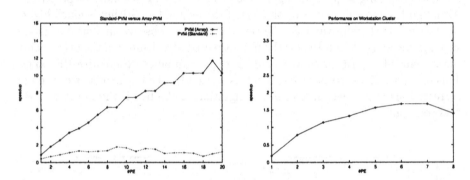

**Fig. 6.** Comparison of two PVM versions    **Fig. 7.** Performance on a NOW

Figure 7 again confirms the former statement about the need for very efficient communication software and hardware. The identical algorithm applied to $64 \times 64 \times 1024$ video data on a FDDI interconnected NOW consisting of 8 DEC/AXP 3000/400 again reaches its speedup maximum at a value of 1.7.

In order to investigate the performance of the two PVM versions on the SGI POWERChallenge in some more detail we measure the effect of sending differently sized messages and of a varying number of messages, respectively.

Figure 8 clearly shows that for both cases (large messages and a high number of messages) the native PVM version dramatically outperforms the public domain version.

Having now sufficient information about the scalability of our algorithms applied to a fixed data size, we finally investigate the effect of varying the size of the video data in the spatial (just imagine the difference between QCIF and SHDTV) as well as in the temporal domain (it might be desirable to use (temporal) smaller blocks in order to keep the coding delay to a minimum in a real-time application).

Figures 9 and 10 show that both varying temporal and spatial dimensions do not change the relation between parallel and sequential execution times (since both curves are nearly parallel).

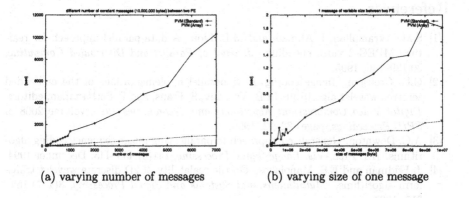

(a) varying number of messages          (b) varying size of one message

**Fig. 8.** Public domain PVM vs. POWERChallenge Array PVM

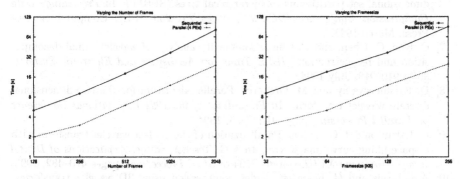

**Fig. 9.** Temporal scalability          **Fig. 10.** Spatial scalability

## 5  Conclusion

In this work we have discussed several aspects of performing 3-D wavelet decomposition on a shared memory MIMD architecture. It has been shown that special attention has to be paid towards cache misses and the right choice of a message passing library. The message passing approach outperforms the data parallel implementation for high PE numbers, whereas we observe the contrary behaviour for small PE numbers.

## Acknowledgements

The first author was partially supported by the Austrian Science Fund FWF, project no. P11045-ÖMA. We want to thank Andreas Pommer for his help in resolving the cache problem on the POWERChallenge.

# References

[1] S.M. Akramullah, I. Ahmad, and M.L. Liou. A data-parallel approach for real-time MPEG-2 video encoding. *Journal of Parallel and Distributed Computing*, 30:129–146, 1995.

[2] C.D. Creusere. Image coding using parallel implementations of the embedded zerotree wavelet algorithm. In B. Vasudev, S. Frans, and P. Sethuraman, editors, *Digital Video Compression: Algorithms and Technologies 1996*, volume 2668 of *SPIE Proceedings*, pages 82–92, 1996.

[3] A.C. Downton. Generalized approach to parallelising image sequence coding algorithms. *IEE Proc.-Vis. Image Signal Processing*, 141(6):438–445, December 1994.

[4] J. Fridman and E.S. Manolakos. On the scalability of 2D discrete wavelet transform algorithms. *Multidimensional Systems and Signal Processing*, 8(1–2):185–217, 1997.

[5] K.H. Goh, J.J. Soraghan, and T.S. Durrani. New 3-D wavelet transform coding algorithm for image sequences. *Electronics Letters*, 29(4):401–402, 1993.

[6] B.J. Kim and W.A. Pearlman. An embedded wavelet video coder using three-dimensional set partitioning in hierarchical trees (SPHIT). In *Proceedings Data Compression Conference (DCC'97)*, pages 251–259. IEEE Computer Society Press, March 1997.

[7] C. Koc, G. Chen, and C. Chui. Complexity analysis of wavelet signal decomposition and reconstruction. *IEEE Trans. on Aereospace and Electronic Systems*, 30(3):910–918, July 1994.

[8] D. Krishnaswamy and M. Orchard. Parallel algorithm for the two-dimensional discrete wavelet transform. In *Proceedings of the 1994 International Conference on Parallel Processing*, pages III:47–54, 1994.

[9] G. Lafruit and J. Cornelius. Parallelization of the 2D fast wavelet transform with a space-filling curve image scan. In A.G. Tescher, editor, *Applications of Digital Image Processing XVIII*, volume 2564 of *SPIE Proceedings*, pages 470–482, 1995.

[10] A.S. Lewis and G. Knowles. Video compression using 3D wavelet transforms. *Electronics Letters*, 26(6):396–398, 1990.

[11] H. Nicolas, A. Basso, E. Reusens, and M. Schutz. Parallel implementations of image sequence coding algorithms on the CRAY T3D. Technical Report Supercomputing Review 6, EPFL Lausanne, 1994.

[12] J.N. Patel, A.A. Khokhar, and L.H. Jamieson. Scalability of 2-D wavelet transform algorithms: analytical and experimental results on coarse-grain parallel computers. In *Proceedings of the 1996 IEEE Workshop on VLSI Signal Processing*, pages 376–385, 1996.

[13] K. Shen, G.W. Cook, L.H. Jamieson, and E.J. Delp. An overview of parallel processing approaches to image and video compression. In M. Rabbani, editor, *Image and Video Compression*, volume 2186 of *SPIE Proceedings*, pages 197–208, 1994.

[14] S. Sullivan. Vector and parallel implementations of the wavelet transform. Technical report, Center for Supercomputing Research and Development, University of Illinois, Urbana, 1991.

[15] D. Taubman and A. Zakhor. Multirate 3-D subband coding of video. *IEEE Transactions on Image Processing*, 5(3):572–588, September 1993.

[16] M-L. Woo. Parallel discrete wavelet transform on the Paragon MIMD machine. In R.S. Schreiber et al., editor, *Proceedings of the seventh SIAM conference on parallel processing for scientific computing*, pages 3–8, 1995.

# On the Parallel Implementation of the Fast Wavelet Packet Transform on MIMD Distributed Memory Environments

Stefania Corsaro[3], Luisa D'Amore[2,3], and Almerico Murli[1,3]

[1] University of Naples "Federico II"
[2] Second University of Naples, Caserta, Italy
[3] Center for Research on Parallel Computing and Supercomputers (CPS), CNR
Complesso Monte S.Angelo, via Cintia, 80126 Naples, Italy
{corsaro, damore, murli}@matna2.dma.unina.it

**Abstract.** This work describes the design and implementation issues of the *Fast Wavelet Packet Transform* (FWPT) 1D and 2D on a parallel distributed memory multiprocessors. In particular, we describe two different approaches in the development of a parallel implementation of the FWPT of a matrix $A$.

In section 2 we introduce some notations and definitions, in section 3 we describe the computational environments, and, in section 4, we discuss the parallel implementation of the bidimensional FWPT. In section 4.3 we finally show a numerical experiment.

## 1 Introduction

Wavelets have generated a tremendous interest in both theoretical and applied areas, especially over the past few years. In particular, the *Fast Wavelet Transform* (FWT) is being a very powerful computational tool in Image Processing. Most applications of wavelets deal with compression, transmission, synthesis and reconstruction of signals in one or more dimensions, moreover wavelets have been applied in the numerical solution of Partial Differential and Integral Equations[4], in the Approximation and Interpolation of data[6]. At the present time several routines and mathematical software packages that perform the computation of the FWT and the FWPT (1D anf 2D) are available, but there are only few parallel implementations, especially for the FWPT (see for example [1] and [8]). Most of real applications (for instance Image Processing in astrophysics and in medical imaging), requires either a big amount of data or needs solutions in a suitable "turnaround" time, so the only way to effectively solve them is the use of the advanced architectures resources. In particular, our interest focuses on Image Restoration problems described by Integral Equations of the first kind, which require the solution of ill-conditioned linear systems. In this case, a least square solution is computed by using the Preconditioned Conjugate Gradient (PCG). Moreover, to compute a reasonable solution, we should use a regularization technique to smooth out the noise which perturbs the data: the computational kernel of this problem is the bidimensional FWPT.

P. Zinterhof, M. Vajteršic, A. Uhl (Eds.): ACPC'99, LNCS 1557, pp. 357–366, 1999.
© Springer-Verlag Berlin Heidelberg 1999

## 2    Preliminaries

One of the main features of wavelets is "localization": wavelets allow to study a signal with finite energy, that is a function $f(t)$ belonging to $L^2(R^+)$, localizing both the time interval and the frequency one. This property is particularly clear looking at the wavelet functions in the context of a Multiresolution Analysis (MRA): projecting $f(t)$ onto a space of a MRA allows to obtain information about it, depending on the *resolution* of the space. The mapping that leads from the $m$-th level resolution to the $(m-1)$-th level retaining the information that is lost in this process is the FWT and, more generally, the FWPT[2,9].

Given two sequences $(h_k)_{k \in Z}$ and $(g_k)_{k \in Z}$, the *filters* of the wavelets, and the vector $c^m = (c_n^m)_{n \in Z}$ of the coefficients of the projection of $f(t)$ onto the $m$-th resolution subspace of the MRA, let us define the FWT operator.

**Definition 1.** *The FWT operator, W, is defined as follows:*

$$W : c^m \in l^2(Z) \longrightarrow (c^{m-1}, d^{m-1}) \in l^2(Z) \times l^2(Z)$$

*where $Z$ is the set of integers, $l_2(Z) = \{(c_k)_{k \in Z} \quad c_k \in C, \quad \sum_k |c_k|^2 < \infty\}$, and*

$$\begin{cases} c_n^{m-1} = \sum_{k \in Z} h_{k-2n} c_k^m \\ d_n^{m-1} = \sum_{k \in Z} g_{k-2n} c_k^m \end{cases} \tag{1}$$

□

where $h$ and $g$ are the *low-pass* and the *high-pass filters* respectively.
In matrix form, if $L = (\tilde{h}_{i,j} = h_{j-2i})$ is the low-pass operator and $H = (\tilde{g}_{i,j} = g_{j-2i})$ is the high-pass operator, relations (1) can be written as:

$$\begin{pmatrix} c^{m-1} \\ d^{m-1} \end{pmatrix} = \begin{pmatrix} L \\ H \end{pmatrix} \cdot c^m \iff \begin{cases} c^{m-1} = Lc^m \\ d^{m-1} = Hc^m \end{cases}$$

The transformation of the vector $c^m$ using the filters $h_k$ retains the information about the low frequencies, while the filters $g_k$ "detect" the high frequencies: so the vector $d^{m-1}$ contains the *details*, that is, the information that is lost passing from the resolution $m$ to the resolution $m-1$.
From a computational point of view, it is worth emphasizing that, if $l$ is the length of the two sequences $h_k$ and $g_k$, then the number of floating-point operations required for the computation of the FWT of a vector of length $N$ is $O(lN)$[9].

Figure 2 shows that the vector $d^{m-1}$ is not transformed after the first step: if we use *Wavelet Packets* we have a further subdivision of the details and, from a computational point of view, more efficient parallel algorithms.

**Definition 2.** *The FWPT operator, WP, is defined as follows:*

$$WP : c^m \in l^2(Z) \longrightarrow (c_0^{m-s}, ..., c_{2^s-1}^{m-s}) \in l^2(Z)^s$$

*where s is the number of transform steps and*

$$(c_l^{m-j})_n = \begin{cases} \sum_{k \in Z} h_{k-2n}(c_{l/2}^{m-j+1})_k & \text{if } l \text{ is even} \\ \sum_{k \in Z} g_{k-2n}(c_{\frac{l-1}{2}}^{m-j+1})_k & \text{if } l \text{ is odd} \end{cases} \tag{2}$$

*with $j = 1, s$ and $l = 0, 2^j - 1$.* □

In matrix form, if we define $H_0 := L$, $H_1 := H$ and:

$$F_i := \prod_{j=1}^{s} H_{\epsilon_j} \qquad i = 0, 2^s - 1$$

where $\epsilon_j$ is the $j - th$ binary digit of $i$, then relations (3) can be written, for $j = s$, as:

$$\begin{pmatrix} c_0^{m-s} \\ c_1^{m-s} \\ \vdots \\ c_{2^s-1}^{m-s} \end{pmatrix} = \begin{pmatrix} F_0 \\ F_1 \\ \vdots \\ F_{2^s-1} \end{pmatrix} \cdot c^m$$

We define the FWPT of a matrix $A$ as well.

**Definition 3.** *If*

$$Q_s := \prod_{i=0}^{s-1} diag_{2^i} \begin{pmatrix} L \\ H \end{pmatrix}$$

*where $diag_{2^i}$ is the block diagonal matrix with $2^i$ diagonal blocks, then the FWPT in s steps of a matrix $A$ is defined as $A^s = Q_s A Q_s^T$* □

In particular, we will consider *compressible* matrices[2], since they allow to obtain sparse representations of the operators.

**Definition 4.** *A square matrix $A = (a_{i,j})$ of dimension $2^m$ is said to be compressible if two constants $M$ and $C$ exist such that $a_{i,j} = 0$ if $i = j$, and:*

$$|a_{i,j}| \le \frac{2^m C}{|i-j|}$$

$$\left| \sum_{k=0}^{M} (-1)^{M-k} \binom{M}{k} a_{i+k,j} \right| + \left| \sum_{k=0}^{M} (-1)^{M-k} \binom{M}{k} a_{i,j+k} \right| \le \frac{2^m C}{|i-j|^{M+1}}$$

*otherwise.* □

In figure 1 an example of compressible matrix is shown.

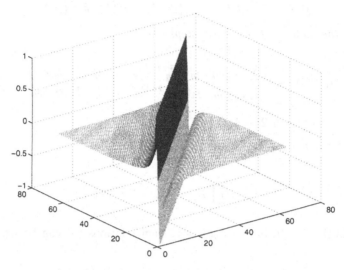

**Fig. 1.** $A(i,j) = \frac{1}{i-j}$ if $i \neq j$, $A(i,j) = 0$ otherwise.

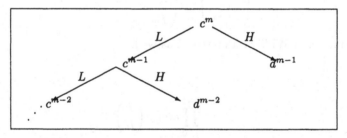

**Fig. 2.** representation of the "cascade structure" of the FWT operator

## 3   The Computational Environment

The parallel implementation is based on the Single Program Multiple Data programming model, that is each processor executes the same algorithm on different data.

Let *nprocs* be the number of processors and $P_r, P_c$ two integers such that $nprocs \leq P_r \cdot P_c$; we map the processors onto a logical bidimensional mesh with $P_r$ rows and $P_c$ columns and, if $0 \leq i < nprocs$, we denote by $P_i$ the processor whose id number is equal to $i$. More precisely, this map can be defined as follows:

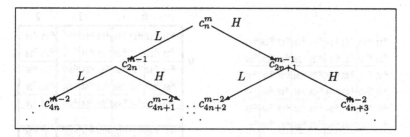

**Fig. 3.** representation of the cascade structure of the FWPT operator

$$F : i \in \{0, 1, , ..., nprocs - 1\} \longrightarrow (r_i, c_i) \in \{0, 1, ..., P_r - 1\} \times \{0, 1, ..., P_c - 1\}$$

where $r_i = nprocs/i$ and $c_i = mod(nprocs, i)$.
If $P_r = 1$ or $P_c = 1$, we have a ring interconnection among the processors.

One of the main difficulties about the development of parallel implementations concerning applications of FWPT is the choice of an appropriate data distribution strategy that ensures that the meaningful elements of the compressed matrices, mainly located in a square block whose dimension depends on the number $s$ of FWPT steps, are distributed over all the processors, in order to guarantee a good work load balancing. Here we refer to a theoretical model of parallel computer such as the one used in the Scalapack Library[5], the high-performance library designed to solve linear problems on distributed memory multiprocessors. According to Scalapack conventions, we use a *block cyclic distribution* of the compressed matrix $A_B^s$, so that each processor has a "dense" block of $A_B^s$. The block cyclic data distribution is parameterized by the four numbers $P_r, P_c, r, c$, where $P_r \times P_c$ is the process template and $r \times c$ is the block size. The generical element of global indexes $(m, n)$ of a matrix $A$ is stored in the position $(i, j)$ of the block $(b, d)$ in the processor $(p, q)$, where

$$(p, q) = \left( \left\lfloor \frac{m}{r} \right\rfloor \mathrm{mod} P_r, \left\lfloor \frac{n}{c} \right\rfloor \mathrm{mod} P_c \right)$$

$$(b, d) = \left( \left\lfloor \frac{\lfloor \frac{m}{r} \rfloor}{P_r} \right\rfloor, \left\lfloor \frac{\lfloor \frac{n}{c} \rfloor}{P_c} \right\rfloor \right) \tag{3}$$

$$(i, j) = (m \,\mathrm{mod}\, r, n \,\mathrm{mod}\, c)$$

In the following section we describe the main idea of the parallel implementation of the FWPT of a matrix $A$; two different strategies of parallelization are discussed, the former based on a *domain decomposition* strategy, the latter on the *parallelization of the sequence of floating-point operations*.

$$
\begin{array}{|cc|cc|cc|cc|c|}
\hline
a_{11} & a_{12} & a_{13} & a_{14} & a_{15} & a_{16} & a_{17} & a_{18} & a_{19} \\
a_{21} & a_{22} & a_{23} & a_{24} & a_{25} & a_{26} & a_{27} & a_{28} & a_{29} \\
\hline
a_{31} & a_{32} & a_{33} & a_{34} & a_{35} & a_{36} & a_{37} & a_{38} & a_{39} \\
a_{41} & a_{42} & a_{43} & a_{44} & a_{45} & a_{46} & a_{47} & a_{48} & a_{49} \\
\hline
a_{51} & a_{52} & a_{53} & a_{54} & a_{55} & a_{56} & a_{57} & a_{58} & a_{59} \\
a_{61} & a_{62} & a_{63} & a_{64} & a_{65} & a_{66} & a_{67} & a_{68} & a_{69} \\
\hline
a_{71} & a_{72} & a_{73} & a_{74} & a_{75} & a_{76} & a_{77} & a_{78} & a_{79} \\
a_{81} & a_{82} & a_{83} & a_{84} & a_{85} & a_{86} & a_{87} & a_{88} & a_{89} \\
\hline
a_{91} & a_{92} & a_{93} & a_{94} & a_{95} & a_{96} & a_{97} & a_{98} & a_{99} \\
\hline
\end{array}
$$

Right — the matrix mapped onto a 2 × 3 process grid:

|   | **0** | | **1** | | **2** | |
|---|---|---|---|---|---|---|
| **0** | $a_{11}\ a_{12}$ | $a_{17}\ a_{18}$ | $a_{13}\ a_{14}$ | $a_{19}$ | $a_{15}\ a_{16}$ | |
|  | $a_{21}\ a_{22}$ | $a_{27}\ a_{28}$ | $a_{23}\ a_{24}$ | $a_{29}$ | $a_{25}\ a_{26}$ | |
|  | $a_{51}\ a_{52}$ | $a_{57}\ a_{58}$ | $a_{53}\ a_{54}$ | $a_{59}$ | $a_{55}\ a_{56}$ | |
|  | $a_{61}\ a_{62}$ | $a_{67}\ a_{68}$ | $a_{63}\ a_{64}$ | $a_{69}$ | $a_{65}\ a_{66}$ | |
|  | $a_{91}\ a_{92}$ | $a_{97}\ a_{98}$ | $a_{93}\ a_{94}$ | $a_{99}$ | $a_{95}\ a_{96}$ | |
| **1** | $a_{31}\ a_{32}$ | $a_{37}\ a_{38}$ | $a_{33}\ a_{34}$ | $a_{39}$ | $a_{35}\ a_{36}$ | |
|  | $a_{41}\ a_{42}$ | $a_{47}\ a_{48}$ | $a_{43}\ a_{44}$ | $a_{49}$ | $a_{45}\ a_{46}$ | |
|  | $a_{71}\ a_{72}$ | $a_{77}\ a_{78}$ | $a_{73}\ a_{74}$ | $a_{79}$ | $a_{75}\ a_{76}$ | |
|  | $a_{81}\ a_{82}$ | $a_{87}\ a_{88}$ | $a_{83}\ a_{84}$ | $a_{89}$ | $a_{85}\ a_{86}$ | |

**Fig. 4.** On the left, a 9 × 9 matrix partitioned in 2 × 2 blocks; on the rigth, the matrix is mapped onto a 2 × 3 process grid

## 4   The FWPT Parallel Implementation

We describe two different strategies of parallelization of step 1 and 2 of the algorithm described in figure 5.

### 4.1   First Strategy

Let $A$ be a compressible square matrix of dimension $N = 2^m$ and $nprocs$ the number of processors. The matrix $A$ is distributed in *column-block* fashion, that is, if $MB = (N + nprocs - 1)/nprocs$ and $lp = (N + MB - 1)/MB - 1$, then the processors whose identification number belongs to the set $\{0, 1, ..., lp - 1\}$ have $MB$ columns; the processor whose identification number is $lp$ has $MB$ columns if $num = mod(N, MB) = 0$, it has $num$ columns otherwise.

If $0 \le i < lp$, let us denote by $A^i = A(:, i \cdot MB : i \cdot MB + (MB - 1))$ the block formed by the columns $i \cdot MB, ..., i \cdot MB + (MB - 1)$ or, if $i = lp$,

$$
A^i = A(:, i \cdot MB : i \cdot MB + (MB - 1)) \quad \text{if} \quad num = 0
$$
$$
A^i = A(:, i \cdot MB : i \cdot MB + (num - 1)) \quad \text{if} \quad num \ne 0
$$

is the block formed by the columns $i \cdot MB, ..., i \cdot MB + (MB - 1)$ if $num = 0$, or the one formed by the columns $i \cdot MB, ..., i \cdot MB + (num - 1)$ if $num \ne 0$; if $P_i$ is the processor whose id number is equal to $i$, then $P_i$ has the block $A^i$. This distribution can be derived from (3) by setting $P_r = nprocs, P_c = 1, r = c = \lceil N/nprocs \rceil$.

Let us look at figure 6: steps 1. and 3. do not require communication among the processors, while step 2. does.

Each processor exchanges with each other a block of dimension $k = N^2/nprocs^2$, so the total amount of data exchanged by each processor over $(nprocs - 1)$ steps of communication is $O(k \cdot (nprocs - 1))$.

## 4.2 Second Strategy

Let $A$ be a compressible square matrix of dimension $N = 2^m$ and $nprocs = 2^d$ the number of processors. The matrix $A$ is distributed in *row-block* fashion, that is, if $n = 2^{m-d}$, then $A_i = A(n \cdot i : n \cdot (i+1) - 1, :)$ is distributed to the processor with id number $i$, $0 \le i < nprocs$.

In figure 7 the second strategy is represented: only the first step, that is the computation of the FWPT of the columns of $A$, requires communication among the processors, since the columns are distributed and, therefore, each processor has a part of the vectors to be transformed.

If $s$ is the number of FWPT steps, then for each $0 \le k < s$ we compute $2^k$ FWT of sequences of length $2^{m-k}$: since these sequences could be distributed, we divide the processors in groups so that each sequence corresponds to a group, and only processors belonging to the same group must communicate.

At each step $k$, where $0 \le k \le \min\{s - 1, d - 1\}$, each processor exchanges four blocks, two of dimension $nN$ and two of dimension $(2M - 2)N$, over four steps of communication, where $M$ is the number of vanishing moments of the wavelets, and it is always $M << N$.

---

**Procedure** FWPT2D$(A, M, N, h, s)$
% $A$: matrix to be transformed
% $M$: number of vanishing moments of the wavelets
% $N$: dimension of $A$
% $h$: Ingrid Daubechies low-pass filters
% $s$: number of transform steps
**begin**
% **step 1:** FWPT in $s$ steps of the columns of $A \Longleftrightarrow \bar{A} = Q_s A$
    **for** $i = 0, N - 1$ **do**
        *call* FWPT1D$(A^i, M, N, h, s)$
    **endfor**
% **step 2:** FWPT in $s$ steps of the rows of $\bar{A} \Longleftrightarrow A^s = \bar{A}Q_s^T$
    **for** $i = 0, N - 1$ **do**
        *call* FWPT1D$(\bar{A}_i, M, N, h, s)$
    **endfor**
**end**

---

**Fig. 5.** FWPT in $s$ steps of a matrix $A$

---

**Procedure FWPT2D** $(A, M, N, h, s, ...)$
**begin**
    **step 1:** { *each processor* $P_i, 0 \leq i < nprocs$ *computes* $\bar{A}^i$ }
    **step 2:** { *global transposition of* $\bar{A}$ }
    **step 3:** { *each processor* $P_i$ *computes* $(A^s)^i$ }
**end**

---

**Fig. 6.** parallel bidimensional FWPT: first strategy

---

**Procedure FWPT2D** $(A, M, N, h, s, ...)$
**begin**
    **step 1:** { *each* $P_i, 0 \leq i < nprocs$ *computes the FWPT of the columns of* $A_i$ }
    **step 2:** { *each* $P_i$ *computes* $(A^s)_i$ }
**end**

---

**Fig. 7.** parallel bidimensional FWPT: second strategy

## 4.3  Numerical Experiments

In this section we describe only one of some numerical experiments we carried out. The experiment has been performed on the IBM SP, with 12 nodes Thin Power 2 connected via a switch, available at the Center for Research on Parallel Computing and Supercomputers, CNR, Naples.

The routines have been written in C, and the message passing software used is BLACS (Basic Linear Algebra Communication Subprograms) which provides ease of use and portability in parallel linear algebra applications.

The global transpose of the matrix $\bar{A}$ has been computed using the f01crfp routine as described in [3].

*Example 1.* Let $A$ be a square matrix of dimension $N = 2^{11}$, $s = 2$ the number of FWPT steps we want to perform and $nprocs = 2^2$ the number of processors. If we use the first strategy then about $2^{19}$ elements over 3 steps of communication are exchanged, while using the second one the total amount of data exchanged is about $2^{22}$ elements over 8 steps of communication.  □

*Example 2.* Let $N = 2^{18}$, $s = 2$ and $nprocs = 2^7$. If we use the first strategy then about $2^{29}$ elements over 127 steps of communication are exchanged, while using the second one the total amount of data exchanged is about $2^{30}$ elements over 8 steps of communication.  □

The amount of data exchanged at each step of communication is smaller if we use the first strategy, and the communication does not depend on the number of transform steps. On the other side, the number of steps of communication grows "linearly" with the number of processors, while it is always $\leq 4 \cdot log_2 (nprocs)$ if we use the second strategy. So, we expect the first algorithm to perform better than the second one with a few processors and with a large number of FWPT steps.

In figures 8 and 9 the efficiency $E_p = T_1(N)/(p \cdot T_p(N))$ of the two methods discussed are shown.

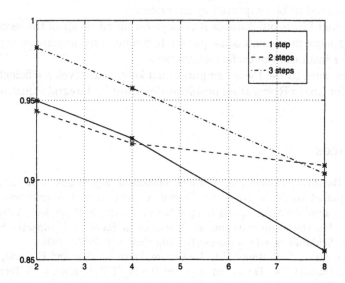

**Fig. 8.** efficiency of the first strategy. The dimension of the matrix is $N = 2048$ and the number of vanishing moments is 7

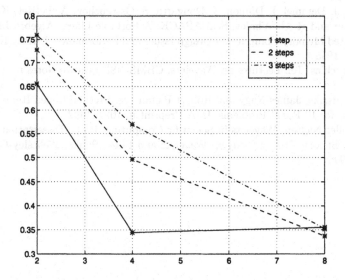

**Fig. 9.** efficiency of the second strategy. The dimension of the matrix is $N = 2048$ and the number of vanishing moments is 7

# 5  Conclusions

In this work we have described two different strategies of parallelization of the bidimensional FWPT, and a comparison between them: further experiments, expecially on massively parallel architectures, have to be carried out, since we mostly compared the two strategies from a theoretical point of view and our statements need to be supported by experience.

In the next future our purpose is to improve the efficiency of the second strategy taking more in account some parameters related to the hardware platform, such as, for instance, the cache memory size.

We besides want to use these computational kernels to develop efficient preconditioners for Image Restoration problems described by Integral Equations of the first kind.

# References

1. Laura Bacchelli Montefusco, Parallel numerical algorithms with orthonormal wavelet packet bases, in Wavelets: Theory Algorithms, and Applications, Charles K. Chui, Laura Montefusco, and Luigia Puccio, Academic Press, Inc., 1994
2. Beylkin, On the Representation of Operators in Bases of Compactly Supported Wavelets, Schlumberger-Doll Research, Ridgefield, CT 06877, 1991
3. L. Carracciuolo, I. de Bono, M.L. De Cesare, D. di Serafino and F. Perla, Development of a Parallel Two-Dimensional Mixed-Radix FFT Routine, CPS Tech. Rep. n. TR-97-8, 1997
4. Philippe Charton, Valerie Perrier, Produits rapides matrice-vecteur en bases d'ondelettes: application à la résolution numérique d'èquations aux dérivés partielles, draft, 1995
5. J.Choi, J. Demmel, I. Dhillon, J. Dongarra, S. Ostrouchov, A. Petitet, K. Stanley, D. Walker and R. C. Whaley, ScaLAPACK: A Portable Linear Algebra Library for Distributed Memory Computers - Design Issues and Performance, LAPACK Working Note 95, 1995
6. I. Daubechies, Ten Lectures on Wavelets, CBMS-NSF Lecture Notes nr. 61, SIAM, 1992
7. Martin Hanke, James Nagy and Robert Plemmons, Preconditioned Iterative Regularization for Ill-Posed Problems, IMA Preprint n. 1024, 1992
8. Ole Moller Nielsen, Wavelets in Scientific Computing, Ph.D. Dissertation Thesis
9. Gilbert Strang, Truong Nguyen, Wavelets and Filter Banks, Wellesley-Cambridge Press, 1996

# Algorithms and Programming Paradigms for 2-D Wavelet Packet Decomposition on Multicomputers and Multiprocessors

Manfred Feil and Andreas Uhl

RIST++ & Department of Computer Science and System Analysis
University of Salzburg, AUSTRIA
{mfeil,uhl}@cosy.sbg.ac.at

**Abstract.** In this work we describe and analyze algorithms for 2-D wavelet packet decomposition for multicomputers and multiprocessors. In the case of multicomputers we especially focus on the question of handling of the border data among the processing elements. For multiprocessors we discuss several optimizations of data parallel algorithms and finally we compare the results obtained on a multiprocessor employing the message passing and data parallelism paradigm, respectively.

## 1 Introduction

Wavelet packets [9] represent a generalization of the method of multiresolution decomposition and comprise the entire family of subband coded (tree) decompositions. Whereas in the wavelet case the decomposition is applied recursively to the coarse scale approximations (leading to the well known (pyramidal) wavelet decomposition tree), in the wavelet packet decomposition the recursive procedure is applied to all the coarse scale approximations and detail signals, which leads to a complete wavelet packet tree (i.e. binary tree and quadtree in the 1D and 2D case, respectively) and more flexibility in frequency resolution.

Recently, wavelet packet based compression methods have been developed [11, 3] which outperform the most advanced wavelet coders (e.g. SPHIT [5]) significantly for textured images in terms of rate-distortion performance. Therefore, wavelet packet decomposition currently attracts much attention and a thorough examination of parallelization possibilities seems to be desirable.

Whereas a significant amount of work has been already done concerning parallel algorithms for the fast wavelet transform, only few papers have been devoted to parallel wavelet packet decomposition and its specific features and demands (e.g. approaches for performing the best basis algorithm and the irregular decomposition into such a basis on parallel MIMD [2, 7, 8] and SIMD [1] architectures, application of parallel wavelet packet decomposition in numerics [4]).

P. Zinterhof, M. Vajteršic, A. Uhl (Eds.): ACPC'99, LNCS 1557, pp. 367–376, 1999.
© Springer-Verlag Berlin Heidelberg 1999

# 2    Multicomputer Algorithms (Employing the Message Passing Paradigm)

In this section we discuss parallel algorithms for the two-dimensional periodic wavelet packet transform based on the *data swapping* and the *redundant data calculation* techniques. The proposed algorithms assume that the size of processing element (PE) memory is sufficient to keep the wavelet packet coefficients of all levels. For machines or applications which do not meet this condition a specific technique has been derived for parallel decomposition into any given basis [8].

In this and the following sections we use a notation introduced in [2]. Suppose we have a two-dimensional signal $S = (s_{k,l})$ with size $0 \leq k < 2^{Xmax}$ and $0 \leq l < 2^{Ymax}$. Further we have $4^{jp}$ PEs, on which we want to compute the two-dimensional wavelet packet decomposition of signal $S$.

## 2.1    Data Swapping Method

To provide the necessary border data to each PE the following algorithm uses the so-called *data swapping* method (also known as *non-redundant data calculation*): each PE computes only non-redundant data and exchanges these results with the appropriate neighbour PE in order to get the necessary data for the next calculation step (i.e. the next decomposition level). This classical approach of data distribution has been realized for 1-D and 2-D FWT [10, 6].

We begin our calculation with the initial subband $^0C^0_{k,l} = s_{k,l}$ and define every other subband recursively by the formula:

$$^{4f_j+d}C^{j+1}_{k_{j+1},l_{j+1}} = \sum_{n_j,m_j} F(d,f_j)_{n_j-2k_{j+1},m_j-2l_{j+1}} * {}^{f_j}C^j_{n_j,m_j}$$

with the indices in the range:

$$0 \leq j < jmax = min(Xmax, Ymax) \ldots \text{last calculated level}$$
$$0 \leq f_j < 4^j \ldots \text{index of subband in use}$$
$$0 \leq d < 4 \ldots \text{index of filter in use}$$
$$0 \leq n,m < filter\_length \ldots \text{indices of filter coefficient}$$
$$0 \leq k_{j+1} < 2^{Xmax-(j+1)} \ldots \text{coordinate of data-point in x-direction}$$
$$0 \leq l_{j+1} < 2^{Ymax-(j+1)} \ldots \text{coordinate of data-point in y-direction}$$

From a one-dimensional wavelet $\Psi$ (defined by its filter $g_n$) and its smoothing function $\Phi$ (defined by its filter $h_n$) we obtain four two-dimensional filters by tensor products: $Filter(0)_{n,m} = h_n * h_m$, $Filter(1)_{n,m} = h_n * g_m$, $Filter(2)_{n,m} = g_n * h_m$, and $Filter(3)_{n,m} = g_n * g_m$. Simplifying, we may say that convolving the coefficients of a subband $C^j$ at level $j$ with 4 different filters $Filter(0)\ldots Filter(3)$ we obtain 4 subbands $C^{j+1}$ at level $j+1$.

Fig.1 shows the data arrangement for levels $j = 0\ldots 2$: at level $j$ we get $4^j$ subbands, each with $2^{Xmax-j}$ by $2^{Ymax-j}$ coefficients.

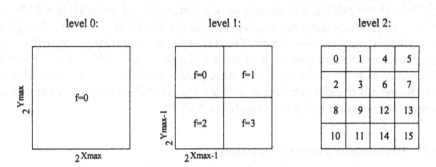

**Fig. 1.** Data arrangement and labeling for a 2D wavelet packet decomposition

To do the composition in parallel, we will separate each level into $4^{jp}$ parts of equal size $2^{Xmax-jp}$ by $2^{Ymax-jp}$. Each PE always stores the same part of the data field so we can identify the PE label by the data it contains at scale $jp$. Fig.2 shows a repartition onto 4 PE $(jp = 1)$ from level $j = 0$ to 2.

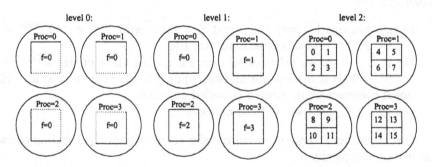

**Fig. 2.** Repartition of the wavelet packets onto 4 PE

After cutting up the field as shown above, we will use the decomposition formula in two different ways upon whether $j$ is smaller or greater than $jp$.

First we consider the case when $j < jp$: a subband at level $j$ and frequency $f$ is not assigned to a single PE but is shared by the PE $p = 4^{jp-j} * f$ to $p = 4^{jp-j} * (f + 1) - 1$. Therefore in the initialisation step those $4^{jp-j}$ PE will exchange their data in order to have the entire shared subband on each of them. Then, in the second step, they will calculate their own part of the packet they share at level $j + 1$.

In the second case (when $j \geq jp$) each PE has $4^{j-jp}$ subbands and holds also the four children of each of them. Thus, no communication between PE is needed and we will use the sequential algorithm on the subset of subbands each PE carries at level $j$.

For $0 < j \leq jp$ we have a transfer complexity of order $O(2 * jp * (jp + 1) * 2^{Xmax} * 2^{Ymax}/4^{jp})$ and a decomposition complexity of order $O(jp * 2^{Xmax} * 2^{Ymax}/4^{jp})$, where the constant depends on the length of the filters. For $jp < j \leq jmax$ ($jmax = min(Xmax, Ymax)$) there is no transfer complexity and the decomposition complexity is of order $O((jmax - jp) * 2^{Xmax} * 2^{Ymax}/4^{jp})$. Thus, if $N = 4^{jmax}$ is the number of data points and $NP = 4^{jp}$ is the number of PE the overall transfer complexity is $O(2 * [\log_4(NP)]^2 * N/NP)$ and the overall computational complexity is $O(\log_4(N) * N/NP)$.

## 2.2  Redundant Data Calculation

A second strategy of data distribution is the *redundant data calculation* approach [10, 6]: in the initialisation step we do not only provide a partial block $2^{Xmax-jp} * 2^{Ymax-jp}$ of the original signal to a PE but broadcast the whole data set of size $2^{Xmax} * 2^{Ymax}$ to all PE (which limits this approach to moderately sized data sets). Subsequently, in each calculation step each PE computes also redundant data in order to avoid additional communication with neighbour PE to obtain this data.

We proceed from the algorithm in the last section and extend it with some minor changes:

- In the initialization step the signal $S = (s_{k,l})$ with $0 \leq k < 2^{Xmax}$ and $0 \leq l < 2^{Ymax}$ is broadcasted onto each PE.
- For $0 < j \leq jp$ we calculate in each PE a data block of size $2^{Xmax-j} * 2^{Ymax-j}$. This is the amount of data required in the next step, so no additional communication substep is needed anymore.
- For $jp < j \leq jmax$ we continue as above.

This algorithm shows an overall transfer complexity of $O(1)$ and an overall calculation complexity of

$$O\left(2^{Xmax} * 2^{Ymax} * \left(\sum_{j=1}^{jp}\frac{1}{4^j} + \sum_{j=jp+1}^{jmax}\frac{1}{4^{jp}}\right)\right)$$

$$= O\left(N * \frac{NP - 4 + 3\log_4(N/NP)}{3NP}\right)$$

## 2.3  Experimental Results

We carried out our experiments on several distributed memory architectures having different relations concerning communication/computation:

- Parsytec GCel 1024 (Transputer T805 nodes at 30 MHz)
- Meiko/Quadrics CS-2HA (SuperSPARC nodes at 50MHz)
- Parsytec GC/PP-64 (two Motorola PowerPC 601 at 80MHz per node)

As test image we employ a $1024 \times 1024$ pixel version of the *Lena*-image (except for a $256 \times 256$ version on the GCel 1024 due to memory constraints) and perform a complete decomposition (i.e. 10 levels) with Daubechies $W20$ filters using the data swapping and the redundant data calculation variant of the wavelet packet algorithm. For comparison purposes we additionally execute the fast wavelet transform (fwt) using the data swapping technique which has communication in *every* step, but also reduces the size of data computations are performed on to a quarter (since only one subband is processed further in the next decomposition level). We use native versions of the PVM message passing library on every machine except for the POWERChallenge where the public domain version available at www.netlib.org is employed.

The results shown in Fig.3 give the speedup of the different parallel algorithms with respect to the corresponding execution times of the sequential ones and verify their behaviour predicted by the complexity examinations:

- all results on multicomputers show a better performance of the data swapping approach for more than 4 PE.
- the significance of the superiority of the data swapping approach is architecture dependent. Whereas this effect is very clear for the GCel 1024, it is less pronounced on the Meiko and hardly noticeable on the GC/PP. The reason is that the PE speed of the architectures is very different, i.e. it is increasing significantly in the order given before. On the other hand, the communication cost stays constant within an order of magnitude for all machines. For this reason the redundant data approach becomes competitive on architectures with higher PE speed.
- the *fwt* implementation shows a worse performance as compared to both wavelet packet implementations since the proportion communication/computation is much worse (in fact it is getting worser with each decomposition level). We may say that this approach of parallelizing the fast wavelet transform is not very suitable for message passing implementations (a simple improvement could be achieved if only few levels would be executed in parallel).

Figure 3.d shows the performance of both algorithms on a POWERChallenge GR using PVM. In accordance to the trend found before we notice that the redundant data calculation approach now outperforms the data swapping algorithm. The PE speed of this architecture (MIPS R10000, 195MHz) is much higher as compared to those of the multicomputers discussed. Therefore, it is more efficient to perform some extra computations instead of invoking communication procedures (although this communication takes place via shared memory) which explains the effect.

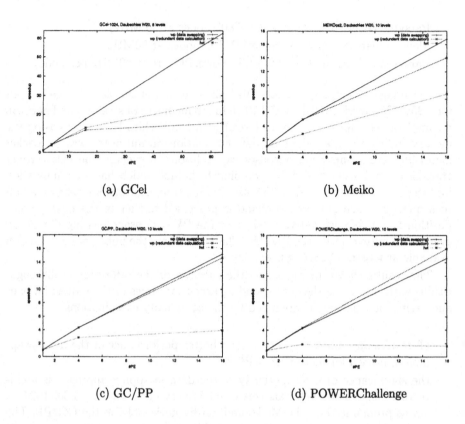

(a) GCel                                    (b) Meiko

(c) GC/PP                                 (d) POWERChallenge

**Fig. 3.** Comparison of data swapping and redundant data calculation on multi-computers and a multiprocessor

# 3    Multiprocessor Algorithms (Employing the Data Parallel Paradigm)

## 3.1    A Straightforward Approach (wp1)

The nice feature of data parallel programming on a shared memory architecture is that a sequential algorithm may be transformed into a parallel one by simply identifying areas which are suitable to be run in parallel i.e. in which no data dependencies exist. Subsequently, local and shared variables need to be declared and parallel compiler directives are inserted.

We start with a sequential pseudo code of the wavelet packet algorithm, which in a very simplified form is as follows:

```
loop level=1...max_level
  loop subband_nr=1...subbands(level)
    loop xy=1...xrange(level)*yrange(level)
      coefficient(subband_nr,xy)=0
```

```
    loop filter_nr=1...filter_length
        coefficient=coefficient+data*filter
end all loops
```

We try now to identify loops in which single loop runs can be run concurrently. We have to pay attention to that:

- there must not be any data-dependency between two single runs (such as in `loop level`).
- there must be a sufficient large number of runs, otherwise the parallelization overhead would be too large (such as in `loop subband_nr`).
- each run should consist of a sufficient large number of statements otherwise the same problem as in the previous point arises (`loop filter_nr` would not be a good choice).

These requirements in mind we choose `loop xy` for parallelization in algorithm *wp1*. The only local variables are `xy`, `filter_nr` and `filter`, all other variables are shared among all PE.

For a parallel implementation of the fast wavelet transform we can use exactly the same strategy, i.e. parallelizing `loop xy`.

## 3.2  An Adaptive Parallelization Strategy (*wp2*)

We notice in the sequential pseudo code that the parameters `subbands`, `xrange` and `yrange` depend on the variable `level` i.e. in each consecutive level the number of parallel loop runs in *wp1* is reduced by half up to `level=max_level` where only one parallel run will be done. Therefore we expect to find for *wp1* a reasonable speedup only in the first decomposition levels because in the last levels the parallelization overhead will exaggerate.

To improve this situation we derive a threshold value *level_th* from which on it is advisable to parallelize the outer `loop subband_nr`. From the theoretical point of view this should be at `loop level`= *level_th* where the number of outer loop runs exceeds the number of inner ones i.e. where:

$$packets > xrange * yrange$$
$$4^{level} > 4^{jmax-level}$$
$$level > \frac{jmax}{2}$$

For the threshold value we therefore get *level_th* $= \lceil \frac{jmax}{2} \rceil$, and implementation *wp2* works as follows:

- for *level* < *level_th*: do `loop xy` in parallel (as in *wp1*).
- for *level* ≥ *level_th*: do `loop packet_nr` in parallel.

## 3.3    Experimental Results

Practical experiments were conducted on the POWERChallenge GR using again the 1024 * 1024 pixel version of the *Lena*-image. Figure 4 clearly shows the superiority of the adaptive parallelization approach. Moreover we notice the excellent performance of the parallel fast wavelet transform (which is due to less access to shared variables).

In Fig.5 we investigate the behaviour of the algorithms over the full range of computed levels (1 ... 10).

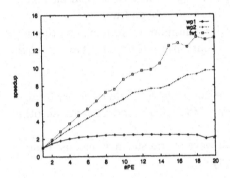
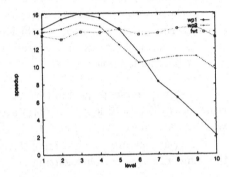

**Fig. 4.** Comparison of *wp1, wp2* and *fwt* with Daubechies W20

**Fig. 5.** Speedup development over the full range of computed levels

We see that the proportion between parallel and sequential loop runs is of prime importance for the performance of all implementations:

- *fwt*: Parameter `xrange*yrange` is always larger than `subbands=1`, therefore this algorithm shows a constant good speedup over the whole `level`-range in Fig.5.
- *wp1*: If *level* > 5 the amount of computation inside the parallelized loops gets too small and reduces the speedup achieved in the levels 1...5 (see Fig.5).
- *wp2*: *level_th* = 5 is the theoretical threshold for changing the parallelization strategy. For $1 \leq level \leq 5$: `xrange*yrange>packets` and the inner loop is parallelized as in *fwt* and *wp1*; for $6 \leq level \leq 10$ the outer loop is parallelized, and the speedup should be kept over the full range of levels.

We find in this parallel region more accesses to shared variables which implies more communication overhead and reduces the expected speedup (Fig.5: speedup decreases for $4 \leq level \leq 10$). This is also due to the fact that we have considered the number of loop runs as our criterion, but have ignored the changing amount of computation inside the loop runs.

To investigate this effect in more detail we performed further tests with different numbers of PE and threshold values. Fig.6 shows results for runs on 16 and 20 PEs with threshold values *level_th* = 1 ... 7:

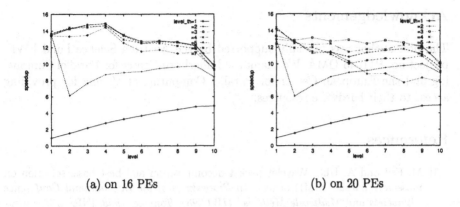

(a) on 16 PEs                    (b) on 20 PEs

**Fig. 6.** *wp2* with different parameter sets

We see that the optimal speedup values are achieved at *level_th* = 3 (this is also true for smaller PE numbers). Simplifying, we may say that the change in parallelization strategy should be placed at the level, where *#subbands* >> *#PE* for the first time.

### 3.4 Comparison between Message Passing and Data Parallelism on a Multiprocessor

Figure 3.d shows speedup 16 using 16 PE for the message passing algorithm. This value is not entirely fair since due to the host-node programming used 16 node PE and 1 host PE have been employed. The correct value for comparison with data parallel programming is speedup 15 for 16 PE. In Figure 4 we observe a speedup of 8.8 and the optimal threshold choice (Figure 6.a) leads to speedup 10.

Though simple to use and to implement the data parallel approach leads to significant lower efficiency as compared to the message passing algorithm. Nevertheless, it should be pointed out that the message passing algorithm in the form discussed here is restricted to PE numbers which are a power of 4 – for PE numbers not satisfying this condition the corresponding algorithm is of lower efficiency [8] and might therefore loose its lead. This issue is subject to further investigations.

## 4  Conclusion

In this work we have shown that the data swapping approach is the algorithm of choice for 2-D wavelet packet decomposition as long as the speed of the PE remains moderate. For architectures with high speed PE the redundant data calculation approach is superior. Moreover we have demonstrated that on a multiprocessor even an especially tuned data parallel algorithm does not reach at all the performance of a message passing algorithm.

## Acknowledgements

The first author was partially supported by the Austrian Science Fund FWF, project no. P11045-ÖMA. We thank the Paderborn Center for Parallel Computing and the European Center for Parallel Computing at Vienna for providing access to their hardware resources.

## References

[1] M. Feil and A. Uhl. Wavelet packet decomposition and best basis selection on massively parallel SIMD arrays. In *Proceedings of the International Conference "Wavelets and Multiscale Methods" (IWC'98), Tangier, 1998.* INRIA, Rocquencourt, April 1998. 4 pages.

[2] E. Goirand, M.V. Wickerhauser, and M. Farge. A parallel two-dimensional wavelet packet transform and some applications in computing and compression analysis. In R. Motard and B. Joseph, editors, *Applications of Wavelet Transforms in Chemical Engineering*, pages 275–319. Kluwer Academic Publishers Group, 1995.

[3] F.G. Meyer, A.Z. Averbuch, J.O. Strömberg, and R.R. Coifman. Fast wavelet packet image compression. In *Proceedings Data Compression Conference (DCC'98)*, page 563. IEEE Computer Society Press, March 1998.

[4] L. Bacchelli Montefusco. Parallel numerical algorithms with orthonormal wavelet packet bases. In C.K. Chui, L. Montefusco, and L. Puccio, editors, *Wavelets: Theory, Algorithms and Applications*, pages 459–494. Academic Press, San Diego, 1994.

[5] A. Said and W.A. Pearlman. A new, fast, and efficient image codec based on set partitioning in hierarchical trees. *IEEE Transactions on Circuits and Systems for Video Technology*, 6(3):243–249, 1996.

[6] S. Sullivan. Vector and parallel implementations of the wavelet transform. Technical report, Center for Supercomputing Research and Development, University of Illinois, Urbana, 1991.

[7] A. Uhl. Adapted wavelet analysis an moderate parallel distributed memory MIMD architectures. In A. Ferreira and J. Rolim, editors, *Parallel Algorithms for Irregulary Structured Problems*, volume 980 of *Lecture Notes in Computer Science*, pages 275–284. Springer, 1995.

[8] A. Uhl. Wavelet packet best basis selection on moderate parallel MIMD architectures. *Parallel Computing*, 22(1):149–158, 1996.

[9] M.V. Wickerhauser. *Adapted wavelet analysis from theory to software*. A.K. Peters, Wellesley, Mass., 1994.

[10] M-L. Woo. Parallel discrete wavelet transform on the Paragon MIMD machine. In R.S. Schreiber et al., editor, *Proceedings of the seventh SIAM conference on parallel processing for scientific computing*, pages 3–8, 1995.

[11] Z. Xiong, K. Ramchandran, and M.T. Orchard. Wavelet packet image coding using space-frequency quantization. *IEEE Transactions on Image Processing*, 7(6):892–898, June 1998.

# Real-Time Layered Video Compression Using SIMD Computation

Morten Vadskær Jensen and Brian Nielsen

Aalborg University
Department of Computer Science
Fredrik Bajersvej 7E
DK-9220 Aalborg, Denmark
{mvj | bnielsen}@cs.auc.dk

**Abstract.** We present the design and implementation of a high performance software layered video codec, designed for deployment in bandwidth heterogeneous networks. The codec facilitates layered spatial and SNR (signal-to-noise ratio) coding for bit-rate adaption to a wide range of receiver capabilities. The codec uses a wavelet subband decomposition for spatial layering and a discrete cosine transform combined with repeated quantization for SNR layering.

Through the use of the Visual Instruction Set on SUN's UltraSPARC platform we demonstrate how SIMD parallel image processing enables layered real-time software encoding and decoding. The codec partitions our $384 \times 320 \times 24$-bit test video stream into 21 layers at a speed of 39 frames per second and reconstructed at 28 frames per second. The Visual Instruction Set accelerated encoder stages are about 3-4 times as fast as an optimized C version. We find that this speedup is well worth the extra implementation effort.

## 1 Introduction

Smart technology for multicasting live digital video across wide area networks is necessary for future applications like video conferencing, distance learning, and tele-commuting. The Internet Multicast Backbone (MBone) [1] is already popular and allows people from anywhere on the planet to exchange modest quality video and audio signals. However, a fundamental problem with nearly all large computer networks is that network capacity (bandwidth) varies extremely from one network segment to another. This variation is the source of a serious video multicast problem. All users wish to participate in the video conference with the highest quality video their connection capacity (varying between 128kbps for ISDN dialup connections to 100 Mbits LAN connections) and host computational resources allow. High capacity receivers are capable of processing good quality video with bit-rates of 4Mbps or more, but this outperforms the low capacity receivers by more than an order of magnitude. Even a modest bit rate may overflood low capacity receivers.

Conventional video compression techniques code the video signal to a fixed target bit rate, and is then multicasted to all receivers. A low target bit rate

P. Zinterhof, M. Vajteršic, A. Uhl (Eds.): ACPC'99, LNCS 1557, pp. 377–387, 1999.

is typically chosen to enable as many receivers as possible to participate. However, the resulting quality is unacceptable for the high capacity receivers. Alternatively, the video signal can be encoded repeatedly to a multitude of data rates. This is, however, very inefficient: Firstly, it involves extra computational overhead from compressing the same frame repeatedly. Secondly, it introduces bandwidth redundancy as each high bit-rate video stream contains all the information also included in the lower bit-rate ones. This translates to inefficient bandwidth utilization from the root of the the multi-cast tree.

We propose *layered coding* coupled with multi-casting, where the video stream is coded and compressed into several distinct layers of significance as shown in Figure 1. In principle, each layer is transmitted on its own multi-cast channel. The receivers may subscribe to these channels according to their capabilities and preferences. The more layers received, the higher video quality, but also higher bandwidth and processing power requirements. The most significant layer constitutes the *base layer* and the following layers *enhancement layers*.

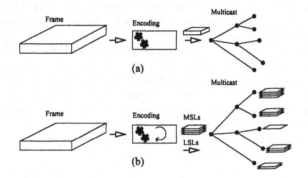

**Fig. 1.** Conventional single bit-rate video coding (a) vs. layered coding (b).

Live video requires that video frames are encoded and decoded in real-time. This is even more challenging for layered coding than for conventional coding because layer construction and reassembly requires use of additional image filter functions and repeated processing of image data, and hence requires more CPU-processing. We believe that it is important that this coding is possible in real-time on modern general purpose processors without dedicated external codec hardware—partly because no current hardware unit has the functionality we advertise for, but more importantly, because applications in the near future will integrate video as a normal data type and manipulate it in application dependent manner. This require significant flexibility. Fortunately, most modern CPU-architectures have been extended with SIMD instructions to speed up digital signal processing. Examples include VIS in Sun's UltraSPARC, MMX in Intel's Pentium(II), MAX-2 in Hewlett-Packards PA-RISC, MDMX in Silicon Graphics' MIPS, MVI in Digital's Alpha, and, recently, Altivec in Motorola's PowerPC CPUs.

In this paper we show a high performance implementation of a layered codec capable of constructing a large set of layers from reasonably sized test-video in real-time. We demonstrate how SIMD parallelism and careful considerations of superscalar processing can speedup image processing significantly. Our encoder implementation exists in both an optimized C-version and in a version almost exclusively using SUN microsystem's Visual Instruction Set (VIS) available on the SUN UltraSPARC platform. The decoder only exists in a VIS-accelerated version.

The remainder of the paper is structured as follows. Section 2 presents our layered coding scheme and our codec model which incorporates it. Section 3 presents the implementation of an instance of our codec model along with our performance optimization efforts. Section 4 evaluates the coding scheme through a series of measurements. Finally, section 5 discusses related work, and section 6 concludes.

## 2   The Codec Model

Our codec combines two independent layering facilities: spatial layering and signal-noise-ratio (SNR) layering. Spatial layering controls the resolution of the image—subscription to more spatial layers means higher image resolution. SNR layering controls the number of bits used to represent image pixels—the more bits the higher pixel precision, and thus a more sharp and detailed image.

The wavelet filter performs a subband decomposition of the image and decomposes it into one low (L) and one high frequency (H) subband per dimension. The effect of subband decomposition on the Lena-image is depicted in Figure 2. The LL subband is a subsampled and filtered version of the original image. The three subbands, HL, LH, and HH contain the high frequency information necessary to reconstruct the horizontal, vertical, and diagonal resolution, respectively. Therefore, the HL, LH, and HH layers act as refinement (or enhancement) layers for the LL layer. Further, the LL subband can itself be subject to further subband decompositions, or can be subjected to the familiar block-based DCT coding scheme known from other codecs for further psycho-visual enhanced compression. We use a 2D-version of the 4-tap 1-3-3-1 spline wavelet [5, p. 136], because it achieves *good pixel value approximation* and because it is highly summetric and consequently can be implemented very efficiently.

SNR layering divides the incoming coefficients into several levels of significance, whereby the lower levels include the most significant information, resulting in a coarse image representation. This is then progressively refined with the number of layers. SNR layering is a flexible and computationally efficient way of reducing the video stream bit-rate, and allows fine-grain control of image quality at different layers. In our codec SNR-layering is achieved through a DCT-transform combined with repeated uniform quantization.

Figure 3 shows our codec model. Encoding is performed by the following components: The *Wavelet Transformer* transforms YUV images into LL, HL, LH, and HH layers corresponding to four spatial layers. The enhancement layers

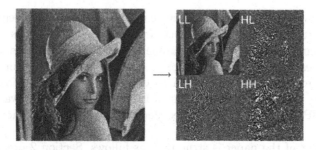

**Fig. 2.** Wavelet transformation of the Lena image into LL, HL, LH, HH layers. The LL layer is a downsampled version of the original image and the HL, LH, and HH layers contain the information required to restore horizontal, vertical, and diagonal resolution, respectively.

HL, LH, HH are sent directly to the quantizer, while the LL layer is passed on to the DCT stage. The *DCT transformer* performs the Discrete Cosine Transform on the LL image from above. The resulting coefficient blocks are passed on to the quantization stage. The *Quantizer* splits incoming coefficients into several layers of significance. The base layer contains a coarse version of the coefficients, and each layer adds precision to coefficient resolution. The *Coder* takes the quantized coefficients from each separate layer and Huffman compresses them to produce a single bit-stream for each layer.

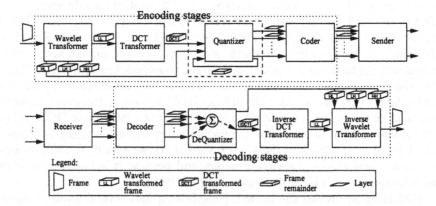

**Fig. 3.** Our codec model. In the encoding stages, the wavelet transformer sends the LL layer to the DCT stage before all layers go through the quantizer and on to transmission. The quantizer reuses coefficient remainders, as depicted by the arrow reentering the quantizer.

The decoding stages essentially perform the inverse operations. The *Decoder* uncompresses the received Huffman codes. The *Dequantizer* reconstructs coefficients based on the decoded layers; the dequantizer works both as coefficient reconstructor and as layer merger. Like the encoder, there are two different de-quantizers, one for DCT coefficients, which are passed on to the IDCT stage, and one for wavelet coefficients which are passed directly on to the wavelet reconstruction stage. The *Inverse DCT Transformer* performs IDCT on incoming DCT coefficient blocks, which reconstructs the lowest resolution spatial layer, LL. The LL layer is used in the *Inverse Wavelet Transformer*, which performs wavelet reconstruction.

The *Sender* and *Receiver* modules, which are outside the actual codec model, must support multi-casting for efficient bandwidth utilization. The layers are transmitted to different multi-cast groups, so receivers can adjust their reception rate by joining and leaving multi-cast groups. Further details on this design can be found in [4].

Our design does not offer temporal layering besides simply dropping frames. The most effective temporal compression schemes use motion compensation, but this does not in it self add any layering, only a lower bit rate. Our goal is to provide high bandwidth versatility. Moreover, it is untrivial to obtain scalability from a motion compensated stream, because frames are interdependent. That is, both the desired frame and the frames it depends on must be received. We have therefore decided to postpone temporal motion compensated layering to future work.

# 3 Implementation

The visual instruction set found on SUN's UltraSPARC CPUs [10] is capable of processing $8 \times 8$ bit, $4 \times 16$ bit, or $2 \times 32$ bit partitioned data elements in parallel. In addition to the usual SIMD multiplication and addition instructions VIS contains various special instructions dedicated to video compression and manipulation of 2-3 dimensional data. Furthermore, the UltraSPARC CPU has two pipelines and is therefore able to execute pairs of VIS instructions in parallel [7].

The VIS instructions are available to the application programmer through a set of macros which can be used from C-programs with a reasonable amount og effort, although it must be realized that even with these C-macros, programming VIS is essentially at the assembler level. The programmer is, however, alleviated from certain aspects of register allocation and instruction scheduling.

Our implementation is optimized at the architecture level as well as at the algorithmic level. The architecture level optimizations fall in three categories: 1) using the VIS to achieve SIMD parallel computation of 4 data elements per instruction, 2) using super scalar processing to execute two (independent) instructions in parallel per clock cycle, and 3) reducing memory access by keeping constants and input data in registers, and by using the 64-bit load/store capabilities. These optimization principles are applied in all stages of the codec. Below we exemplify these on the layer decomposition stage.

The the decomposition filter is applied by multiplying each pixel in a $4 \times 4$ input block with the corresponding element in the filter coefficient matrix. The 16 results are then summed into a single value and divided by 16 to produce an "average", which is the final output. This is done for each layer type. The filter dictates a two-pixel overlap between the current and the next input block, which therefore becomes the image data starting two pixels to the right relative to the current block. The implementation of the decomposition routine operates on 8 × 4 pixel blocks at a time, see Figure 4, producing 4 32-bit outputs; one for each of the LL, HL, LH, HH layers. It uses 64-bit loads to load input data, and since all data fits in registers, it requires only 4 64-bit loads and 4 32-bit stores pr. pixel block. The routine spills overlapping pixels from one 8 × 4 block to the next, so horizontal traversal causes no redundant memory accesses. Vertically there is a 2 pixel overlap, dictated by the filter, so each pair of vertical lines is read twice.

By utilizing the VIS fmul8x16 instruction, which multiplies four 16-bit values with four 8 bit values producing four 16 bit results, the four pixels in row 1 in block 1 can be multiplied with row 1 of the filter coefficients in parallel. Similarly, row 2 in group 1 can be SIMD-multiplied with row 2 of the filter coefficients. Moreover, both of these multiplications can execute in parallel: The super scalar processing in the UltraSPARC CPU allows execution of two independent VIS instructions in parallel per clock cycle. Two instructions are independent if the destination of one instruction is different from the source of the next instruction, as is indeed the case here.

The UltraSPARC CPU is incapable of out-of-order execution, so instructions must be carefully scheduled, either manually or by the compiler, to fully exploit instruction independence. However, we found that the compiler supplied with the platform did not do a satisfactory job on this point and we have therefore manually organized the VIS-code to pair independent instructions, and thereby maximize the benefit from super scalar processing. Thus, SIMD parallelism reduces the number of calculations by three quarters, and super scalar parallelism further halves this number.

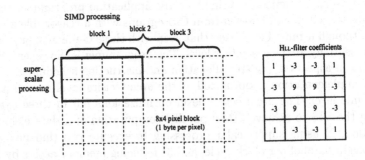

**Fig. 4.** Utilizing SIMD and super scalar processing. SIMD instructions allow 4 elements in the same row to be multiplied with the corresponding filter coefficient row in parallel. Super scalar processing enables two rows in the same block to be computed in parallel. Thus the 8 enclosed pixels are processed in parallel.

At the algorithmic level we exploit the fact that the filter coefficients for the four filters are highly symmetric. This permits computation of the LH-block after the LL-block has been computed through simple sums and differences of data already cached in CPU-registers, that is, without the need for multiplication of a new set of filter coefficients. Similarly, the HH-block can be derived from the HL-block. Therefore, all four types of data blocks are constructed from a single scan of each YUV picture components (Y is the luminance component, and U, V are chrominance components).

With these optimizations, the implementation uses only 0.505 memory accesses per pixel for creating all 4 wavelet layers, and a maximum of 0.344 memory accesses per pixel per reconstructed layer.

## 4    Performance Measurements

The input is a test video stream with 174 frames of $384 \times 320$ pixels at 24-bit colour, digitized at 18 frames per second. The video stream is a typical "talking head" sequence, with one person talking and two persons entering and leaving the image as background movement. The CPU-usage was measured on a Sun Ultra Creator-1 machine with one 167MHz UltraSPARC CPU and 128MB RAM. The test video stream is stored in YUV format, so the colour space conversion stage is not included in the encoding CPU-usage measurements. Likewise the colour space conversion and display times for the decoder are not included as they combined take a nearly constant 13ms.

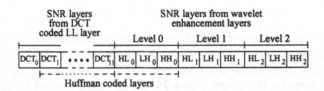

**Fig. 5.** The layer definitions used for testing along with their significance ordering. $DCT_0$-$DCT_{11}$ are layers constructed from the DCT coded LL layer. $HL_0$, $LH_0$, and $HH_0$ contain the most significant bits from the corresponding wavelet enhancement layers. Levels 1 and 2 are refinement layers to these.

Our codec configuration uses one level of subband decomposition, and has a total of 21 layers as seen in Figure 5. Layer 0 is the most significant, down to layer 21 as the least significant. The first 12 layers, $DCT_0$ - $DCT_{11}$, are DCT coded layers constructed from the wavelet LL layer. The remaining 9 layers, $HL_0$ - $HL_2$, $LH_0$ - $LH_2$, $HH_0$ - $HH_2$, are constructed from the corresponding wavelet enhancement layers, distributed with 3 SNR layers from each. The 3 most significant SNR layers, level 0, consist of the upper 4 bits of each coefficient from each of the wavelet layers. Layers from levels 1 and 2 consist of 2 additional bits per coefficient per layer. All three levels thus add $4 + 2 + 2 = 8$ bits of precision. The most significant layers, i.e. those on level 0, are Huffman coded. The layers

384 Morten Vadskær Jensen and Brian Nielsen

on levels 1 and 2 are sent verbatim because their contents are very random and Huffman coding would add very little, if any, compression.

The resulting accumulated bitrates are evenly distributed between 3.5kbits with one layer and 777.3kbits with all 21 layers as shown in Figure 6. The lowest number of layers necessary to produce clearly recognizable persons in the test-movie was found to be 3 ($DCT_0$–$DCT_2$), corresponding to an average frame size of 16.4kbits. This corresponds to a factor 47.4 in difference between lowest capacity receiver and highest capacity receiver—a significant variation. Sample images can be found on the World Wide Web  http://www.cs.auc.dk/~bnielsen/codec/

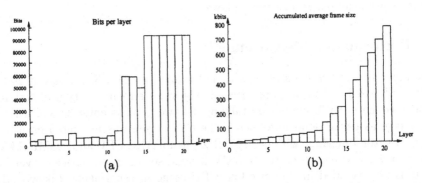

**Fig. 6.** Bandwidth distribution on layers on the 21-layer codec. (a) Average size of the individual layers. (b) Accumulated frame size versus number of added layers.

The CPU-usage measurements show average encoding times (for both the C- and VIS-version), average decoding times (for the VIS-version), and identifies the cost distribution between all codec-stages. The results are summarized in Table 1. On average, the encoder uses 25.4ms to construct and compress all 21 layers. This enables the encoder to process 39 frames per second, which is more than fast enough for real-time software encoding. Similarly, the total average decoding time is 34.81ms or 28fps. An additional 13ms is required for color space conversion and display drawing, but this still allows real-time decoding and display of the test video stream. Further, in most real-life applications, one would rarely reconstruct all 21 layers, meaning even faster decoding times.

Decoding appears more expensive than encoding, which is unusual. However, the reason for this is that the decoder is designed to be very flexible with respect to which frames it decodes, and in what order it does so. It permits seperate decoding of each layer which require a separate iteration across the image per layer. This makes it easier to change the set of layered subscribed to dynamically, e.g., to reduce congestion. Also decoding can begin as soon as a layer is received. Finally note that two of the stages are not VIS-accellerated.

The SIMD implementation provides a significant performance improvement. For comparison, an otherwise identical, efficient and compiler optimized non-VIS accelerated C-implementation of the encoder is capable of encoding only

**Table 1.** Average encoding and decoding time. The Wavelet Transform produces all 4 wavelet layers, the DCT transforms the LL layer, and the Quant & Huff (DCT/wavelet) stages quantize and code the 9 output wavelet enhancement layers, and the 12 DCT SNR layers, respectively. The DeQuant & UnHuff (DCT) stage decodes all 12 SNR layers. DeQuant & UnHuff(wavelet) decodes the 9 wavelet enhancement layers. Wavelet reconstruction upscales the LL image, and adds resolution by reconstructing the HL, LH, and HH layers.

| Encoding time (ms) | | | Decoding time (ms) | |
|---|---|---|---|---|
| | VIS | C | | VIS |
| Wavelet Transform | 7.89 | 28.15 | DeQnt&UnHuff(DCT)[‡] | 4.58 |
| Quant & Huff (wavelet)[‡] | 7.62 | 9.11 | IDCT[†] | 2.13 |
| DCT[†] | 2.51 | 8.51 | DeQnt&UnHuff(wavelet)[‡] | 14.94 |
| Quant & Huff (DCT) | 7.38 | 22.01 | Wavelet reconstruction | 13.36 |
| Total | 25.40 | 67.78 | | 35.01 |

[†] The C and VIS versions of DCT and IDCT functions are from SUN's MediaLib graphics library[9].

[‡] Only the C version exists of the (de)quantization and huffman (de)coding stages.

14.8 fps. A more detailed inspection of the individual stages reveals that VIS provides speedups in the range 3-4 for these particular algorithms: The C-version of wavelet transformation takes 28.15ms on average as opposed to the 7.89ms needed by the VIS-version. This yields a speedup factor of 3.6. The speed of quantization and Huffman coding of the DCT coefficients is increased by a factor 3. Also, we measure a speedup of similar magnitude, factor 3.4, for SUN's MediaLib [9] implementation of DCT. The overall effect of VIS acceleration is that real-time coding becomes possible—with time to spare for network communication, display updates and other application processing.

## 5 Related Work

Layered video coding and related network design issues are active and recent research areas. Related work on layered video codecs exist, notably [3], [6]. The codec in [6] resembles our codec in that they use wavelets for subband decomposition and DCT coding for the LL wavelet layer. Their design does not like ours allow for several levels of subband decomposition. They include temporal layering in the codec using conditional block replenishment. The authors stress the need for error resilient coding mechanisms for error prone networks such as the Internet. Their receiver-driven layered multicast (RLM) scheme, where receivers adjust their reception rate by joining and leaving multi-cast groups, was developed to work in environments like the MBone [1] using IP multi-casting. Although suggestions for implementation optimizations are presented in the paper, they present very little information about run-time performance.

MPEG-2 was the first standard to incorporate a form of layering, called *scalability*, as part of the standard [2]. But MPEG-2 is intended for higher bit-rate video streams, and therefore only allows for three enhancements to the base layer; one from each of scalability category: spatial, SNR, and temporal scalability. Also, no MPEG-2 implementation exists that includes the scalabilities. A combination of MPEG and wavelet based spatial layering for very high bandwidth video is proposed in [11]. The video is repeatedly downsampled until the resolution reaches the resolution of the common intermediate format (cif). The cif-sized $LL^n$-layer is then further processed by an MPEG based codec. Their proposal also offer hierarchical motion compensation of the high frequency subbands, but not temporal scalability.

Compression performance of our codec may be improved by using other methods than Huffman compression. One of the most efficient methods is the embedded zerotree wavelet coding [8], but it is most effective when using several levels of decomposition.

## 6   Conclusions

This paper addresses the problem of efficiently distributing a video stream to a number of receivers on bandwidth heterogeneous networks. We propose layered coding and multi-cast distribution of the video. We design a proprietary video codec incorporating wavelet filtering for spatial layering and repeated quantization for SNR layering. We contribute with a high performance implementation using the SIMD capabilities of our platform. Our measurements show that the codec is capable of real-time software encoding a test movie at 39fps and decoding at 28fps. We found that the Visual Instruction Set accelerated stages in the codec was 3-4 times as fast as an otherwise identical, efficient C-implementation; we therefore find SIMD acceleration to be worth the extra implementation effort.

As this work was carried out as part of a master's thesis, our experience suggest that the new media instructions can be successfully applied by programmers without years of signal processing experience. Indeed, in several cases, the implementation was in fact simplified by using SIMD, although it requires a different line of thought. Since SIMD is integrated into virtually all modern CPU-architectures, we think that developers should consider using it, given the possible speedups. Unfortunately, the SIMD engines are not compatible between CPU-architectures.

## References

[1] Hans Eriksson. MBONE: The Multicast Backbone. *Communications of the ACM*, 37(8), 1994.
[2] Barry G. Haskel, Atul Puri, and Arun N. Netravali. *Digital Video: An Introduction to MPEG-2*. Chapman and Hall, 1997.
[3] Klaus Illgner and Frank Müller. Spatially Scalable Video Compression Employing Resolution Pyramids. *IEEE Journal on Selected Areas in Communications*, 15(9):1688–1703, December 1997.
[4] Morten Vadskær Jensen and Brian Nielsen. Design and Implementation of an Efficient, Layered Video Codec. Technical Report R-98-5008, Aalborg University, Dept. of Computer Science, 9220 Aalborg SØ, Denmark, August 1998.

[5] Martin Vetterli and Jelena Kovačević. *Wavelets and Subband Coding*. Prentice Hall, 1995.
[6] Steven McCanne, Martin Vetterli, and Van Jacobson. Low-Complexity Video Coding for Receiver-Driven Multicast. *IEEE Journal on Selected Areas in Communications*, 15(6):983–1001, August 1997.
[7] Sun Microsystems. UltraSPARC and New-Media Support, 1995. WPR-95-028.
[8] Jerome M. Shapiro. Embedded Image Coding Using Zerotrees of Wavelet Coefficients. *IEEE Trans. on Signal Processing*, 41(12):3445–3462, December 1993.
[9] Sun Microsystems. *mediaLib Users Guide*, June 1997. http://www.sun.com/microelectronics/vis/mlib_guide.pdf.
[10] Marc Tremblay, J. Michael O'Connor, V. Narayanan, and Liang He. VIS Speeds New Media Processing. *IEEE Micro*, 16(4):10–20, August 1996.
[11] Qi Wang and Mohammed Ghanbari. Scalable Coding of Very High Resolution Video Using the Virtual Zerotree. *IEEE Transactions on Circuits and Systems for Video Technology*, 7:719–727, October 1997.

# Parallelisation of a Satellite Signal Processing Code - Strategies and Tools

Ian Glendinning

VCPC
European Centre for Parallel Computing at Vienna
Liechtensteinstraße 22, A-1090 Vienna, Austria
ian@vcpc.univie.ac.at

**Abstract.** This paper presents strategies and tools that have been used in work to parallelise a satellite signal processing code. The Magellan mission to map the surface of Venus using synthetic aperture radar (SAR) is briefly described, the Magellan SAR processor code is outlined, and a parallelisation strategy is presented. The code's large size and limited documentation made the use of program analysis tools essential to implement this strategy. Three tools, FORESYS, IDA and FORGExplorer, are compared, and the use of FORGExplorer to perform code analysis is described in detail. The techniques presented are of general applicability to the parallelisation of codes in other application areas.

## 1 Introduction

VCPC (European Centre for Parallel Computing at Vienna) is collaborating with the ICG (Institute for Computer Graphics) in Graz, Austria, and JPL (Jet Propulsion Laboratory) in Pasadena, California, USA, to parallelize a program which performs image analysis on the data collected by the Magellan spacecraft on its mission to Venus. Due to the code's complexity, and the limited amount of documentation available for it, the use of program analysis tools has proved to be essential, and this paper reports on tools and techniques that have been employed, which are also relevant to parallelisation in other application areas.

## 2 The Magellan Mission

Magellan was carried into Earth orbit in May 1989 by space shuttle Atlantis. Released from the shuttle's cargo bay, it was propelled by a booster engine toward Venus, where it arrived in August 1990. Magellan used a sophisticated imaging radar to pierce the cloud cover enshrouding the planet Venus and map its surface. During its 243-day primary mission, referred to as Cycle 1, the spacecraft mapped over 80 percent of the planet with its high-resolution Synthetic Aperture Radar (SAR). By the time it completed its third 243-day period mapping the planet in September 1992, Magellan had captured detailed maps of 98 percent of the planet's surface.

P. Zinterhof, M. Vajteršic, A. Uhl (Eds.): ACPC'99, LNCS 1557, pp. 388–397, 1999.
© Springer-Verlag Berlin Heidelberg 1999

Magellan orbited Venus approximately once every 3.5 hours, passing over the poles, and mapped the surface of the planet in thin north-south strips about 25 km wide and 16000 km long [2], with range and azimuth resolution of 88 and 120 metres respectively [3]. These strips were nicknamed noodles, and some 4000 of them were mapped during the whole mission, 1800 per cycle. The radar system operated in burst mode, sending out trains of pulses and listening for echoes in between the pulses, and each noodle contained data from around 5000 bursts. The radar gathered data while looking perpendicular to the direction of motion (Fig. 1), and measured the strength of reflected signals, as well as how long each signal took to make the round trip, and changes in the signal frequency. This enabled the Magellan SAR processor, running on computers back on earth, to calculate range and azimuth coordinates from the raw data, and produce high-resolution two-dimensional images of the planet's surface.

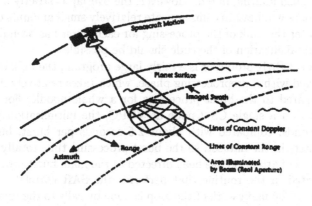

**Fig. 1.** Magellan SAR geometry

# 3   The Magellan SAR Processor

The Magellan SAR Processor is a program that implements the digital signal processing operations needed to convert the raw signal data to image form representing normalised backscatter return from the surface of Venus. The program operates on burst data and produces single-look image framelets, which are then superimposed to form multi-look image data orbit strips. Each burst is processed using an FFT range correlation for range compression, followed by a corner turn and azimuth compression using a "deramp-FFT" algorithm [1]. Side-lobe control, range-walk and phase shift compensation is also done during range compression. Azimuth processing includes side-lobe control and interpolation in azimuth spacings. Finally, geometric, radiometric and multi-look processing are performed to produce image framelets in a (oblique) sinusoidal projection. A utility program called mosaic can be used to assemble the framelets into an image of a noodle for viewing.

# 4    Parallelisation Strategy

From the outset it was decided to perform the parallelisation using the message-passing style of programming, using the portable MPI message passing interface [6]. The 128 node QSW CS-2 machine at VCPC is being used as a development platform, together with the Argonne National Lab / Mississippi State University implementation of MPI for the CS-2.

It was quickly realised that radar bursts represent a natural unit of parallelism in the program, since each one can be analysed independently, except for the final stage, called 'look buildup', which merges results from overlapping observations in neighbouring bursts. However, the overlap is strictly local, so only a few neighbours will have to communicate relatively small amounts of data with each other, after the bulk of the processing for each burst has been done, and so an efficient parallelization of the code should be possible

Although the SAR processor is a fairly large program, the bulk of the source code is involved with a preprocessing phase, which takes relatively little time to execute compared to the burst processing for a whole noodle. For the sample dataset tested on a single CS-2 node at VCPC, the initialisation phase takes about 59 seconds, compared with about 2.5 seconds per burst, but there are over 4000 bursts in the dataset, so the burst processing time totally dominates. Thus, the focus of attention has been directed at the main burst processing loop, which is located in the routine that performs the SAR correlator processing, process_corr. The main work of the loop is done by calls to the routines shown in Fig. 2, whose parameters are omitted for brevity.

```
call corr_pp      ! Fill processing parameter (pp) common block
call pp_keyvar    ! Copy key pp values to correlator common block
r_nbytes = cread  ! Read raw burst data
call decode_sar   ! Decompress raw intensity values
call range_comp   ! Perform range compression
call corner_turn  ! Perform a non-symmetric matrix transpose
call az_comp      ! Perform the Azimuth compression
call radio_comp   ! Perform the radiometric compensation
call geo_rect     ! Perform a geometric rectification
call multi_look   ! Generate multi-look image
```

**Fig. 2.** Routines in the main burst processing loop in process_corr

The final routine multi_look is the only one that does not operate independently on each burst, and is the one that writes the image to disk. The basic strategy is to re-code this loop so that each iteration can be executed as a task on a separate processor. In practice, a pool of processors, each with code to execute a single burst, would have burst data distributed to them by a master process. The resulting image data could be collected by the same master process, or by another process, but for now let's assume that there is a single master controller. In order to be able to re-code the loop like this, it is necessary to understand its data dependences. In particular it is necessary to identify:

1. Variables which are set before the loop, and read inside it, as this implies communication of initial values from the master to worker processes.
2. Variables which are set inside the loop and read after it, which implies communication of values from the workers back to the master.
3. I/O, which must be properly sequenced. File handles need special treatment, as you can't write to a handle opened on another processor.
4. Variables whose values are read within the loop, before later being updated, as their values depend on an assignment from the previous iteration, and imply communication between worker processes. This represents an anti-dependence from the first statement to the second.

Not only variables which are local to the subroutine containing the loop must be considered, but also any variables in common blocks that are used either by the routine itself, or any other routines that it calls, directly or indirectly.

Other work [5] has used a finer-grained approach, parallelising the range and azimuth compression code, but for the Magellan SAR processor that would give a limited maximum possible speedup, due to the residual sequential code. For example, it is estimated that if the range_comp, corner_turn and az_comp routines were parallelised with 100% efficiency, a maximum speedup of only 3.0 could be obtained.

The source code for the SAR processor consists of approximately 125 thousand lines of Fortran, divided among approximately 450 subroutines, together with approximately 6500 lines of C, and although focusing on the main burst processing loop means that only around 5500 lines (plus included common block declarations) need to be analysed in depth, this is still a substantial body of code, so manual analysis would be extremely laborious, and support from parallelisation tools is highly desirable.

# 5  Evaluation of Parallelisation Tools

## 5.1  FORESYS

FORESYS is a commercial tool marketed by SIMULOG. It is a 'Fortran engineering system', and has a range of features that help improve scientific software. Initially version 1.4.1 was used, and later version 1.5 when it became available. The features of the tool that were evaluated were code browsing, program analysis, and program restructuring.

**Code Browsing** FORESYS has a graphical user interface which simplifies code browsing, for example by being able to open a new window displaying the source code for a routine, simply by highlighting its name in another window and selecting a menu option. Although conceptually simple, this was found to be quite useful in practice.

**Program Analysis** The feature of FORESYS that seemed potentially most useful for this work was its program analysis component, called PARTITA, which was able to graphically display dependences (Fig. 3). However, in practice the display became cluttered with information about types of dependence which weren't relevant to the coarse-grained task parallelism that had been selected for the Magellan code. There was also no obvious way to identify dependences between variable references inside the main loop and those outside it.

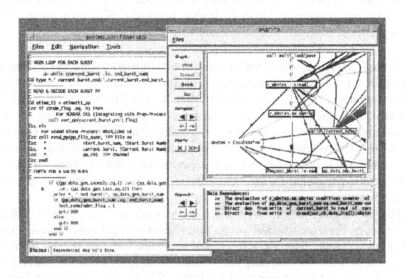

Fig. 3. Program analysis with FORESYS

**Program Restructuring** The tool was found to have an impressively robust front end, which was able to handle the non-standard Fortran extensions used in this large code, almost without complaint. It was also able to transform the program into equivalent code in standard Fortran 77 (it can also produce Fortran 90, but that was not relevant in this case). This rather straightforward sounding feature turned out to be its most useful one in the context of this work, as neither IDA nor FORGExplorer could have been used without it. Neither of them had a front end robust enough to handle the non-standard Fortran code of the Magellan SAR processor, which had to be cleaned up first using FORESYS.

## 5.2  IDA

IDA (Inter-procedural Dependency Analyser), a public domain tool available from the University of Southampton and VCPC [4]. It provides interprocedural information about Fortran programs, such as:

1. Call graphs: the calling relationships between program units
2. Traces of variables: where and how a variable is used throughout the program
3. Common block partitioning and usage: how common blocks are partitioned into variables in each program unit, and how those variables are used in that unit and its descendents
4. Procedure references and argument associations: the location and actual arguments of every call to a particular procedure

IDA was designed for speed and simplicity of operation rather than sophistication. It provides a text command interface, not a graphical one, and it only performs code analysis, not code transformation. However, it is simple to use and its analysis is fast, and its variable trace feature proved to be a useful starting point for investigating some of the dependences within the Magellan SAR code, which were later investigated more fully using FORGExplorer. IDA actually provides much of the functionality offered by FORGExplorer, but the latter tool's more sophisticated user interface made it much easier to use in practice.

## 5.3  FORGExplorer

FORGExplorer is a commercial tool marketed by Applied Parallel Research. It has a Motif GUI, and presents a global, interprocedural view of a program. It can perform searches and variable traces, and features an interactive, interprocedural Distributed Memory Parallelizer (DMP). Initially version 2.1 of the tool was used, and later version 2.2 when it became available. The front end of version 2.1 did not accept all of the non-standard Fortran features in the code, and so FORESYS was used to produce a cleaned up version of the code which FORGExplorer could process. The features of the tool that were evaluated were code browsing, the DMP option, and the global analysis views.

**Code Browsing** Similar facilities to those of FORESYS are provided for code browsing, and they are similarly useful, though there are slightly more options. The Call Subchain display, which represents the call graph as a list of routine names, indented according to their call level, was found to be particularly useful.

**Distributed Memory Parallelizer** The idea of an automatic parallelizer was obviously attractive, and so some time was invested in 'profiling' the routines that were unknown to the tool, that is the ones for which it does not have source code, such as the C routines in the Magellan code. The types of their arguments must be specified, and whether they are read from or written to. Unfortunately, having profiled the unknown routines, it was discovered that the program was too complex for FORGExplorer to handle, and it looped indefinitely. It may have been possible to get further with the DMP using semi-automatic parallelisation options, but it was decided to be less ambitious, and to investigate the manual inter-procedural analysis features of the basic FORGExplorer tool instead.

**Global Analysis Views** Manual analysis of the code using FORGExplorer's global analysis views proved to be a much more fruitful approach than the DMP option, and in particular the variable tracing and common block usage displays were found to be extremely helpful for identifying dependences. The variable tracing display (Fig. 4) lists all program lines where a variable is referenced, following the path that items passed through common or subprogram arguments take, and the common block usage display shows a grid of common blocks versus subprograms. At the intersecting cells, the usage of the common block by each routine is summarised, according to whether variables are set and/or read.

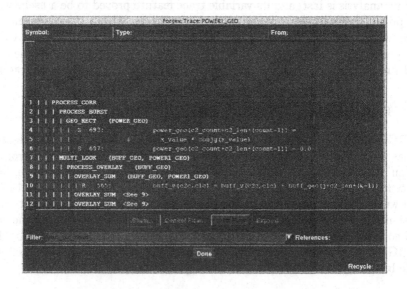

Fig. 4. The FORGExplorer Trace window

# 6    Program Analysis and Parallelisation

Having evaluated the tools, FORGExplorer was the clear winner in terms of its program analysis features, and so it was applied to the analysis of the Magellan SAR code, to implement the parallelisation strategy.

## 6.1   Isolating the Slave Process Code

The first step was to identify precisely which code from the burst loop (Fig. 2) would be run as a slave process, and to localise it in a separate subroutine, giving it a clean interface to the rest of the code, exclusively through common blocks. It was decided to keep the cread() statement in the master process, together with the two routines called before it, as that represented a particularly clean break in the loop. The remaining routines in the loop were moved into a new subroutine, called process_burst. The new routine inherited all of the common block and local data declarations from process_corr, and in addition, all of the local variables were declared to be in a new common block, so that values set in one routine could be read by the other.

## 6.2   Using FORGExplorer's 'Reference Node'

The next step was to restrict the area of interest to the process_burst routine, and routines called by it, by setting FORGExplorer's 'Reference Node' to it. The 'Common Blocks' display shows that the whole program uses 415 routines, 91 of which reference data in common, spread over a total of 61 common blocks. Setting the reference node to process_burst reduced the area of interest to 15 routines, referencing 14 common blocks. Potentially that means $15 \times 14 = 210$ common block references, but FORGExplorer showed that in fact there were only 40, and so had succeeded in reducing the size of the problem considerably.

## 6.3   A Simplified Parallelisation Strategy

Although the complexity of the program analysis had now been much reduced, the dependencies within the multi_look code were still fairly complex, and in order to obtain an initial parallel version of the code as quickly as possible, it was decided to first construct a simpler parallel version of the code than originally planned, by moving the call to multi_look() out of process_burst, into its calling routine process_corr. In this scheme, the maximum amount of parallelisation is limited by the sequential code remaining in multi_look, but the analysis is significantly simplified, as just 8 routines need to be considered in process_burst, and there are a total of just 14 references to 8 common blocks.

## 6.4   Dependence Analysis

The dependences between the master and slave processes were then analysed, so that the communications necessary between them in the parallelised code could be determined. This was done using the common block displays for both process_burst and process_corr, together with the variable tracing display.

The 'Common Blocks' display allows a display of each individual common block to be opened, which shows a table of variables in the common block versus the routines that access them, and for each variable reference it it indicated whether it is read and/or set. Variables in the common blocks that were referenced within process_burst fell into three categories, according to whether they contained variables that were:

1. Read but not set within process_burst
2. Set within process_burst and read afterwards in process_corr
3. Set and then read within process_burst, but not used in process_corr

Those in the first category were the easiest to identify, since it was enough that all their references were reads, but for those in the other two categories, it was necessary to check the read and write usage of the individual variables in the common blocks, since different variables set in a common block may or may not be read later. This was done using the variable tracing display, which also revealed accesses to common block variables that were passed to routines as arguments, although the common block was not declared in the routine, in which case the references were not indicated in the common blocks display. The variable tracing feature was also used to check for anti-dependences within the burst loop (a variable read followed by a set), but none were found. Fig. 5 summarises the results of the dependence analysis, where the numbers in the grid cells correspond to the categories of reading and writing defined above.

| | ant_weight | az_comp | cfft | geo_rect | init_fft | process_burst | radio_comp | range_comp |
|---|---|---|---|---|---|---|---|---|
| /buffer/ | | | | | | 2,3 | | |
| /fft_aux/ | | | 3 | | 3 | | | |
| /key_var/ | 1 | 1 | | 1 | | 1 | 1,3 | 1 |
| /overlay_c12/ | | | | 2,3 | | | | |
| /pc_loc/ | | | | | | 1 | | |
| /po_a/ | | | | 1 | | | | |
| /s_weight/ | | | | 1 | | | | |
| /test_burst_no/ | | | | | | | 1 | |

**Fig. 5.** Results of dependence analysis - Common blocks vs. routines

The communication of variables between master and slave processes can now easily be deduced. Variables in category 1 must be sent to the slave when it starts to execute, those in the category 2 must be sent back to the master when the slave has finished its processing, and those in category 3 require no action.

## 6.5 Parallelisation

Although the slave code was isolated from the rest at an early stage in the work, the whole program was kept sequential for as long as possible, so that it could be compiled and tested after each modification, to make sure that it still behaved as before, and so that the global analysis features of FORGEexplorer could continue to be used on the whole code. Eventually, having performed the dependence analysis, the code was split into two executables, one for the slave process, containing the code for the process_burst routine, and one for the master process containing the rest of the code. Further analysis showed that the large arrays in the common block /BUFFER/ could be split between the master and the slave, which reduced the size of both executables considerably, which is important as they are close to the limit of available memory on the CS-2. The insertion of the MPI communication calls is currently being performed.

## 7 Conclusion

The parallelisation strategy that has been adopted for the Magellan SAR processor is to re-code the burst processing loop so that part of each iteration can be executed as a task on a separate processor. Due to the complexity of the code, program analysis tools were needed to help implement this strategy. The FORESYS, IDA and FORGExplorer tools were evaluated regarding their suitability for the task, and it was found that the most effective approach was to use FORGExplorer's variable trace and common block usage facilities, in conjunction with FORESYS as a preprocessor to clean up the code. The data dependences for a slightly simplified parallelisation strategy were analysed, and work has begun to implement the corresponding parallel program. FORGExplorer is a general tool, and the techniques that have been described are also applicable to the parallelisation of codes in other application areas.

## Acknowledgements

This work has been supported by the Austrian Fonds zur Förderung der Wissenschaftlichen Forschung (FWF) through FSP project S7001, "Theory and Applications of Digital Image Processing and Pattern Recognition".

The author wishes to thank Scott Hensley for discussions about the Magellan SAR code, Ivan Wolton for advice on using FORGExplorer, and Rainer Kalliany and Alois Goller for much help and advice.

## References

1. Curlander, J. C., McDonough, R. N.: Synthetic Aperture Radar: Systems and Signal Processing. Wiley Interscience (1991)
2. JPL: The Magellan Venus Explorer's Guide. NASA; Jet Propulsion Laboratory, California Institute of Technology, Pasadena, CA, JPL Publication 90-24 (1990)
3. Leberl, F., Maurice, K., Thomas, J., Kober, W.: Radargrammetric Measurements from the Initial Magellan Coverage of Planet Venus. Photogrammetric Engineering & Remote Sensing, Vol. 57 No. 12 (1991) 1561-1570
4. Merlin, J. H., Reeve, J. S.: IDA - An aid to the parallelisation of Fortran codes. Technical report, Department of Electronics and Computer Science, University of Southampton (1995)
5. Miller C., Payne, D. G., Phung, T. N., Siegel, H., Williams, R.: Parallel Processing of Spaceborne Imaging Radar Data. Proceedings of Supercomputing '95, IEEE Computer Society Press, San Diego, CA (1995)
6. Message Passing Interface Forum: MPI: A message-passing interface standard. International Journal of Supercomputer Applications 8(3-4) (1994)

# MMIPPS - A Software Package for Multitemporal and Multispectral Image Processing on Parallel Systems

Jürgen Janoth[1], Markus M. Eisl[1], Erwin M. Bakker[2], Rogier van Sterkenburg[2],
Roberta Borgia[3], Salvo Sabina[3], and Fabio Volpe[4]

[1] GEOSPACE, Jakob Haringerstr. 1, A-5020 Salzburg, Austria
{juergen, eisl}@geospace.co.at
[2] Leiden University, Niels Bohrweg 1, P.O.Box 9512, NL-2300 RA Leiden,
The Netherlands
erwin@wi.leidenuniv.nl
[3] Intecs Sistemi, Via Livia Gereschi, 32, I-56127 Pisa, Italy
{salvo, mips}@pisa.intecs.it
[4] Italeco S.p.A., Via Carlo Pesenti, 109, I-00156 Roma, Italy
Italeco@mclink.it

**Abstract.** This paper presents the parallel image processing package MMIPPS (multitemporal and multispectral image processing on parallel systems) developed in a project of the partly EC funded Parallel Computing Initiative II. Image classification and rectification are computational intensive image processing routines, capable to benefit from parallelization. This paper reports the development approach and the performance results achieved on standard networks of PC's and workstations. It is shown, that the MMIPPS package provides the expected results and speed-ups for the selected image processing tasks, leading to shorter completion times and a more efficient service.

## 1 Introduction

Computational intensive image processing tasks are typical for various remote sensing applications [1-7]. Commonly processed optical satellite images, eg. of the Landsat Thematic- Mapper series, have an average size of 36 Mbytes per spectral band and up to 7 bands (or more for multitemporal processing). The CPU processing time on such images can depend considerably on the complexity of the service to be provided. Moreover, the response and turn-around times of image processing increase with the image size. In many cases the productivity and efficiency is limited by inadequate computational power available by sequential image processing software packages.

Therefore, the MMIPPS image processing package has been developed to demonstrate the benefits of High Performance Computing and Networking (HPCN) technologies in the domain of Earth Observation by the means of the development of a HPCN image processing software accelerator. Computationally intensive image processing tasks are accelerated by the MMIPPS Image processing package, leading

P. Zinterhof, M. Vajteršic, A. Uhl (Eds.): ACPC'99, LNCS 1557, pp. 398-407, 1999.

to reduced service-time, and/or a better service quality within the same time frame. The package provides a fully scalable environment, which is interoperable with many existing (sequential) image processing packages, thus allowing the optimal exploitation of available computational resources and reaching excellent speed-up figures. In particular, the MMIPPS Image processing package provides following tasks, which can be launched through an user-friendly graphical environment:

- Unsupervised Clustering, for the generation of a set of clusters that are used in the Supervised Image Classification phase.
- Supervised Image Classification, to classify image pixels. Following decision rules are provided: Parallelepiped, Maximum Likelihood, or Minimum Distance.
- Image Rectification, to perform transformations of satellite data either to fit to another image or to a map projection. The selected spatial transforms are: Elmert, Polynomial Interpolation on Ground Control Points, and Delaunay triangulation. The resampling is done with the Nearest Neighbour, Bilinear Interpolation, or Cubic Convolution method.

The integration of the MMIPPS package into the existing image processing environment is guaranteed through the support of most of the import image formats in remote sensing applications such as *Landsat (BSQ), SPOT (BIL), TIFF,* and *RAW.* Moreover, MMIPPS can be launched on both homogeneous and heterogeneous networks of Workstations and/or Personal Computers, taking advantage of the already available processing environment. Currently the package is available on platforms (e.g. PC-, HP-, SUN-, Silicon Graphics workstations, etc.) which are equipped with the Posix Operating System as well as on PCs which are equipped with Windows95. The MMIPPS package is fully configurable; the number and names of the processing elements of the network can easily be specified by the user, and the package automatically selects the appropriate partition size of the data to be processed. The adopted communication model and related run-time environment (i.e. Parallel Virtual Machine) guarantees a high degree of portability. Furthermore, the MMIPPS Image processing package is scalable, both in terms of the number of processing elements within the network and of problem size.

In the following the paper presents a short description of the selected image processing functionality, an overview over the development approach used (section 3), and the main results (section 4), closing with the conclusions.

## 2    The Image Processing Functionality

The MMIPPS Image processing package is designed to boost selected computationally intensive image processing tasks offering a considerable potential for parallelization. Among the standard image processing tasks image rectification and image classification have been identified as the most relevant tasks to be considered in the MMIPPS package.

## 2.1    Classification Functionality

Classification routines are used to infer thematic information from multispectral and/or multitemporal satellite data. Prominent examples in satellite remote sensing are land-use/land-cover classifications. The most widely applied classification routines are unsupervised clustering, parallelepiped, minimum distance, and maximum likelihood. In general, two major classification approaches can be distinguished - firstly the so-called supervised and secondly the unsupervised classification. However, the underlying algorithmical structure of these routines is the same.

The class, to which a pixel of a multispectral image is assigned, is determined by means of prescribed statistical information. This information is used as parameters of a decision function that implements the classification process. The decision function can be quite simple and straightforward, as shows the case of parallelepiped classification, where the decision function is merely a membership test for intervals or more complex as in the case of Maximum Likelihood where a discriminating function bases on statistic probability is used.

Anyway these routines have all the same algorithmical structure:

> **for each** pixel $p$ in the input image
> **do**
>     determine the class $c$ in $[1,...,N_c]$ that
>     minimises/maximises/fullfills: $D(p,c)$
> **od**

where D(p,c) is either a Boolean function (Parallelepiped Classifier), a distance function (Minimum Distance Classifier) or a discriminating function (Maximum Likelihood Classifier).

## 2.2    Unsupervised Classification Functionality

In unsupervised classifications the clusters are automatically computed. The basis of the unsupervised classification process is an iterative process that, based on an initial guess of class-characteristics, determines the centers of representative classes. During each iteration a sub-sample image is classified using the minimum distance classification routines which leads to new class statistics. The algorithm scheme can be directly derived from the above one including it inside a loop ending when either a prefixed number of iterations or a statistic convergence is reached.

## 2.3    Rectification Functionality

Rectification routines are used to resample a satellite image either to another image or to a map projection. Basically, the resampling process consists of two operations (i) a spatial transformation, which defines the "re-arrangement" of pixels on the image

plane, and (ii) a gray level interpolation, which deals with the assignment of gray levels to pixels in the spatially transformed image. The spatial transformation algorithms implemented by the Rectification module are:
- Elmerts Method,
- Delaunay Triangulation,
- Polynomial Transformation.

During the spatial transformation one of the following resampling methods is used: nearest neighbor, bilinear interpolation, and cubic convolution.

# 3    Development Approach

The requirement analysis has shown that the development of a complete HPCN Image Processing Software Accelerator from scratch was more efficient than the re-use of the code of the available sequential packages [8,9]. The software package was designed to act as an "accelerator" for a subset of the CPU intensive standard services related to Unsupervised Clustering, Supervised Image Classification, and Image Rectification. Furthermore, it was designed and developed to allow the interoperability of the MMIPPS Accelerator with the existing (sequential) image processing packages: standard image file formats such as TIFF and Raw and more satellite specific but widely used formats such as Landsat TM and Spot can be directly processed and produced as output by the MMIPPS accelerator. The input and output data are compatible with other image processing software packages and no post processing is required because of the introduction of parallelism. The following sub-sections provide some details about the technology and parallelization approach used and the results, mainly the speed-up factors, obtained.

## 3.1    Technology and Parallelization Approach

In order to guarantee a high degree of portability and to allow the use of cheaper platforms (e.g. PC), the MMIPPS Accelerator has been implemented in ANSI C and Tcl/Tk [10-12]. The parallel version of the image processing algorithms has been implemented using the Parallel Virtual Machine (PVM) as message passing programming paradigm. PVM was chosen because of its ease in programming and high portability. Furthermore it allows the use of heterogeneous network of PCs and workstations [13]. A top-down design methodology was adopted during the design phase of the parallel algorithms; in particular, the following aspects were evaluated:
- the identification of the programming functions that could be executed in parallel - each of these functions was encapsulated in a separate programming module,
- the definition of distribution of the data among modules,
- for each algorithm class (i.e. unsupervised clustering, supervised classification, and rectification) the assessment of the dependency of the theoretical speed-ups on parallelization parameters such as the image block size,

– the computation of the complexity of the algorithms' parallel version - the comparative analysis with the corresponding sequential algorithms imposed the re-design of some parallel algorithms,
– the design of the MMIPPS Accelerator using a structured design method.

The remote sensing algorithms evaluated exhibited (i) a limited amount of *control parallelism* (i.e. the simultaneous execution of different instructions streams) but (ii) a high amount of *data parallelism* (i.e the possibility to assign data elements to various processors, each of which performs identical computations on its data) [8]. In fact an accurate analysis showed that the image processing operations could be executed independently, and therefore in parallel, on each pixel. It followed that the parallelism grain could be as fine as necessary: the available resources (i.e. number of processing elements, their relative speed, amount of local memory, speed of the network and work-load of processors and network) determine the optimum level of parallelism.

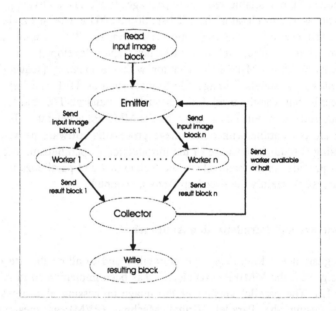

**Fig. 1.** Diagram of the skeleton architecture of the MMIPPS parallel algorithms.

Fig. 1 depicts the skeleton architecture of all the parallel algorithms composing the MMIPPS accelerator [14]. An Emitter – Worker – Collector approach was followed for managing dynamic work-load balancing and process failures; in particular:

• the *Emitter* is responsible for activating the other components, partitioning the input data image and sending each input data partition to an idle Worker;
• each *Worker* is in charge of processing its input data partition in accordance with the specific algorithm implemented (e.g. a type of worker for each image processing algorithm) and sending the results (which can be either processed image blocks or statistical data, depending on the algorithm ) to the Collector;

- the *Collector* component is responsible for collecting the results of processing carried out by each Worker, and producing the result. This means that it rebuilds the complete output classified or rectified image starting from the received processed blocks, or, in the case of Unsupervised Clustering, it produces global statistics from the partial ones.

The MMIPPS package runs on a network of platforms (e.g. PC-, HP-, SUN-, Silicon Graphics workstations, etc.) equipped with *Unix* operating systems or a network of Personal Computer equipped with *Windows 95*.

## 4    Results

The system validation has shown that the MMIPPS modules work correctly and that the quality of the results is high. The qualitative validation was largely based on a comparative assessment with results from comparable sequential software packages. In order to guarantee a systematic approach each MMIPPS feature was tested and validated individually. Furthermore, within each feature (rectification, unsupervised clustering, classification) all relevant parameter settings were evaluated separately.

Performance tests have been carried out under various network environment configurations to demonstrate the efficiency of the MMIPPS package. In order to base the tests on a reproducible approach, quantitative measures such as Absolute Running Time, Efficiency, Speed-up and Overhead have been selected to evaluate the computational power of MMIPPS. Such measures are defined as follows:
- the **Absolute Running Time** $T$ is the global time spent for the execution of a function F.
- the **Speed-up** S is defined as:

$$S = T_s/T_p \tag{1}$$

where $T_s$ is the running time of the sequential implementation of F and $T_p$ is the running time of its parallel implementation. The speed-up shows the speed gain of the parallel implementation of a function with respect to the sequential one.
- the **Efficiency** $E$ is defined as:

$$E = S/N \tag{2}$$

where $N$ is the number of processors used to execute the parallel version of a function F. The efficiency expresses the utilization degree of the processors, $0<E\le 1$.
- The **Overhead** $O$ is defined as:

$$O = T_p - T_{id} \tag{3}$$

where $T_{id} = T_s / N$ represents the lower bound for the running time that could be reached using $N$ processors (ideal speed-up). The overhead expresses the time spent executing the operations required to support the parallel implementation of a function F, such as communications, processes scheduling etc.

Most of the performance tests have been performed in a realistic commercial working environment. System loading and network traffic varied significantly, so the tests have been performed repeatedly. The average time is as result.

## 4.1 Unsupervised Classification

Performance tests of the unsupervised clustering routine have been performed on a network of 9 SUN SLC and SUN ELC machines. The results showed that the speed-up obtained is close to the ideal one when equivalent workers are used (Fig. 2). The tests also demonstrated that the performance was improved significantly by the introduction of a dynamic load balancing model. The efficiency values shown in Fig. 3 indicate an optimal use of the available resources by the parallel algorithm.

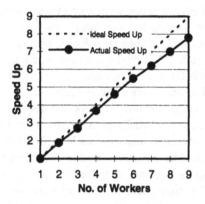 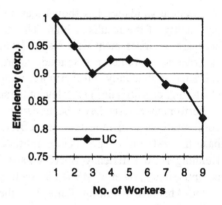

**Fig. 2.** Unsupervised clustering algorithm, measured speed-up on a homogeneous network of Sun 4 Solaris Workstations Input: 3-band image, 41520 sub image pixels, 64 classes.

**Fig. 3.** Unsupervised clustering algorithm efficiency.

## 4.2 Supervised Classification

The performance tests of the classification routines have been based on several statistics files with a varying number of classes. The tests have been performed using the possible parameter constellations for the provided methods.

Due to differences in the complexity of the algorithms, the computing times vary significantly. The minimum distance method takes about two times, the maximum likelihood classification even 20 times the computing time of the parallelepiped method. Other parameter settings (ranking, distance measure, band weight, etc) have only limited influence on the performance of the routines.

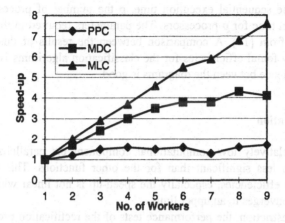

**Fig. 4.** Supervised classification algorithm: measured speed-up on a homogeneous of a network of Silicon Graphics Workstations, Input: 4-band image, 4MB per band, 64 classes.

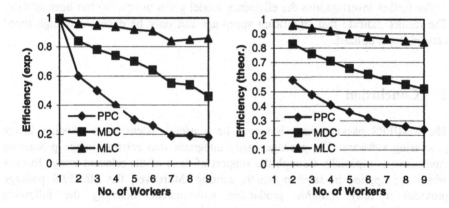

**Fig. 5.** Comparison between the measured (left hand) and the theoretical (right hand) efficiency of the parallelization (PPC parallelepiped classification, MLD maximum likelihood classification, MDC minimum distance classification).

The large number of processing steps per pixel leads to good speed-ups for the parallelized image classification module. The less sophisticated classification routine PPC lacks this complexity, and hence is expected to give good speed-ups only on very special parallel computers with fast communication networks.

A simple performance model [15] has been applied to calculate theoretical efficiencies for the parallelization. In this model, which uses the number of processors and the communication time only, the efficiency E is given by the equation

$$E = \frac{T_s}{T_s + pT_c(p)} \qquad (4)$$

where $T_s$ is the sequential execution time, $p$ the number of processors and $T_c$ the communication time for $p$ processors. The parameters required in this equation have been derived from [15]. A comparison between the results of this model and the experimentally found efficiencies for the classification algorithms is given in Fig.5., the correspondence between the diagrams is good.

## 4.3    Rectification

Due to the relatively low computational component the parallelization effect for rectification is less significant than for the other functions. This results in lower speed-ups and efficiencies, especially the speed-up is not linear with the number of workers but converges to an upper limit.

With this restriction, the performance tests of the rectification procedures showed the expected overall behavior. The influence of the resampling method decreases with the order of the rectification algorithm as the computation time consumed for the calculation of the positions increases whereas the computation time consumed for the interpolation remains constant.

For further investigations the efficiency model given in equ. (4) has been applied. The results indicate that significant speed-ups can only be expected if high-speed networking is utilized.

# 5    Conclusion

The MMIPPS package has shown to be a useful extension to existing image processing software. The package easily integrates into existing image processing environments especially through the support of most of the relevant image formats which are applied in satellite remote sensing. Moreover, the MMIPPS package provides a fully *scalable* production environment, boosting the following computationally intensive image processing tasks:

- *Unsupervised Clustering*, for the generation of a set signatures of clusters that are used in the Supervised Image Classification phase.
- *Supervised Image Classification*, to classify image pixels by using one of the following decision rules: Parallelepiped, Minimum Distance, or Maximum Likelihood.
- *Image Rectification*, to perform map projections in different cartographic systems. The available spatial transforms are Elmert, Polynomial Interpolation on Ground Control Points, and Delaunay Triangulation. The resampling is done with the Nearest Neighbour, Bilinear Interpolation or Cubic Convolution method.

Performance tests of the MMIPPS package both in homogeneous and heterogeneous environments have shown the expected effects - the speed-up scales with the number of workers and the scaling with respect to the amount of data to be processed is nearly linear. Especially in the case of heterogeneous and loaded environment, MMIPPS proved to be robust and well capable to compensate for bottlenecks of smaller available CPU or network capacities. The performance on homogeneous unloaded networks is good and is considerably improved for heterogeneous and loaded networks through the introduction of a dynamic load balancing.

## Acknowledgements

This package has been developed during the ESPRIT sub-project No. 20970 project partly financed by the European Commission, DG II, in the framework of the PCI-II (parallel computing initiative II) project No. 21037. Further information on the project can be found under http://www.pisa.intecs.it/projects/MMIPPS.

## References

1. Schowengerdt R.A., Techniques for image processing and classification in remote sensing, Academic Press, Orlando (1983)
2. Niblack W., An Introduction to Digital Image Processing, Prentice-Hall International, London (1986)
3. Castleman K.R., Digital Image Processing, Prentice-Hall International, London (1996)
4. Schuermann, Pattern Classification - A Unified view of statistical and neural approaches, John Wiley & Sons, Inc., New York (1996)
5. Tou J.T., Gonzalez R.C., Pattern Recognition Principles, Addison-Wesley Publishing Company, Inc., Reading (1974)
6. Wolberg, Digital Image Warping, IEEE Computer Society Press Monograph, Los Alamitos, California (1990)
7. Pitas I., Digital Image Processing Algorithms, Prentice Hall International, London (1993)
8. Kumar V. et al., An introduction to Parallel Programming - Design and Analysis of Algorithms, The Benjamin/Cummings Publishing Company (1994)
9. Wilson, Practical Parallel Programming, The MIT Press, Cambridge (1995)
10. Pratt, (Programmer's Imaging Kernel System - the ISO image processing API) Piks Foundation C Programmer's Guide, Manning Publications Co., Greenwich (1995)
11. Dongarra B., Tourancheau (Eds.), Environments and Tools for Parallel Scientific Computing, North-Holland, Amsterdam (1993)
12. Danelutto, Di Meglio, Orlando, Pelagatti, Vanneschi, A methodology for the development and the support of massively parallel programs, Future Generation Computer Systems, North Holland Volume 8 Numbers 1-3, (1992)
13. Geist A. et al., PVM: Parallel Virtual Machine - A User's Guide and Tutorial for Nteworked Parallel Computing, The MIT Press, Cambridge (1994)
14. Bakker E., Parallel Image Processing, in Proceedings of DAS Symposium, in print
15. Langendoen K., Hofman R., Bal H., Challenging Applications on Fast Networks, Technical Report, Dept of Mathematics and Computer Science, Vrije Universiteit, Amsterdam (1997)

# Parallel Matching of
# Synthetic Aperture Radar Images

Alois Goller

Institute for Computer Graphics and Vision (ICG),
Graz University of Technology,
Münzgrabenstraße 11, A-8010 Graz, Austria
goller@icg.tu-graz.ac.at

**Abstract.** This paper presents an MPI-parallelized version of a matching algorithm for synthetic aperture radar (SAR) images. First the necessity for automated matching and the chosen algorithm are outlined briefly, and the algorithmic structure and the applied parallelization strategy are then presented and discussed. Performance results on a Cray T3D and a PowerChallenge from SGI are given, showing unexpected differences between these architectures. Performance analysis indicates a bottleneck in data I/O.

## 1  Introduction

Producing an atlas of planet Venus based on data sensed by the Magellan probe is challenging due to the amount of data and the algorithms needed. More than 400 Gbyte of raw data must be processed, partially with specialized SAR algorithms that are very compute-intensive.

NASA created the so-called Magellan Stereo Toolkit (MST) [1] as a collection of sequential algorithms which an individual data user could employ to process small subsets of the Magellan images. The most frequently used algorithms are *image matching, resampling and gridding, shape-from-shading* and *perspective rendering*. Matching and shape-from-shading are candidates for parallelization since they are the most time consuming algorithms. However, shape-from-shading already has been parallelized [4].

Initial results (e.g. Leberl [6]) demonstrated that matching SAR stereo pairs is a successful strategy to obtain a Digital Elevation Model (DEM). This is especially valuable for regions that cannot be explored otherwise, which holds for the whole planet Venus. Hence, matching is the central part for processing stereo pairs and thus accurately overlapping basic Magellan data stripes.

Parallel and therefore faster matching allows more compact production cycles and consequently better quality, since a nearly interactive usage of matching will allow accompanying quality control and an online re-adjustment of matching parameters.

P. Zinterhof, M. Vajteršic, A. Uhl (Eds.): ACPC'99, LNCS 1557, pp. 408–416, 1999.

# 2  SAR Matching

*Image matching* establishes a set of corresponding points in two overlapping images. Various approaches exist, which for radar images are mainly based on correlation of pixel arrays, yet must cope with the radiometric differences due to illumination differences. Therefore, some algorithms use additional information derived from edge filters and local image statistics [3].

In figure 1 a pair of images is shown which is a typical input to the matching algorithm. Additionally, some of the corresponding points (matchpoints) are numbered and plotted in both images.

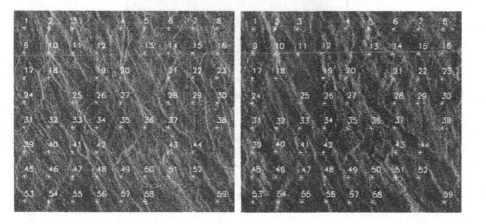

**Fig. 1.** A pair of SAR images with some matchpoints found by Xmatch.

As one can see, the images are very noisy and features have different brightness, thickness and shape. The speckle noise is common to all SAR images and caused by the imaging technique itself. The variation of the features in both radiometric and geometric extent mainly arises because the two images are taken from different look angles. This is necessary for computing a DEM based on matchpoints, but also makes matching more difficult. Effects only present in SAR images like foreshortening, layover and shadow further complicate the matching process.

The algorithm we parallelized is called Xmatch and was developed by Scott Hensley at JPL [5]. It is based on template matching. In order not to be irritated by the noise, it first matches at a very coarse grain level and then refines the accuracy of the matchpoints by taking more and more detail information, meaning not performing a low-pass filter before correlating the templates. Although the algorithm becomes very robust due to this hierarchical structure, it requires a lot of memory and computations.

In figure 2 the areas important for matching a single point are illustrated. Since Xmatch is based on template matching, such a template or region of interest

(ROI) is defined around the point in question. A template with the same size is then moved over the second image, any time calculating the cross-correlation of these two templates. Since finding a matchpoint requires surrounding points (ROI) to be available and in memory, a halo region can be defined, in which no matchpoints can be found.

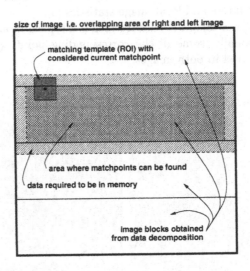

**Fig. 2.** Halo and template required for matching.

## 3    Algorithmic Analysis

Xmatch mainly is written in Fortran 77, but some I/O subroutines and control of the X11 display are encoded in C. The sequential version was developed for and tested on Sun workstations. Portability between SUN and SGI was no problem, but getting the code to run on the Cray T3D was difficult due to different length of integers and different calling conventions between C and Fortran.

The algorithmic structure of Xmatch is shown in figure 3. The easiest way parallelization can be applied is by distributing memory blocks across the processors. This coarse grain parallelization matches quite well to the use of MPI and the manager/worker paradigm, as outlined in section 5.

The next loop, the iteration loop, cannot be parallelized since results from one iteration are used to calculate more accurate points in the next iteration. However, within one iteration, there are 5 double-nested loops which might be of interest.

The best candidate is the "find matches" loop, since 98% of the time is spent there. To parallelize at this level, several arrays must be interchanged between processors and many global sums are to be calculated. This level of parallelism

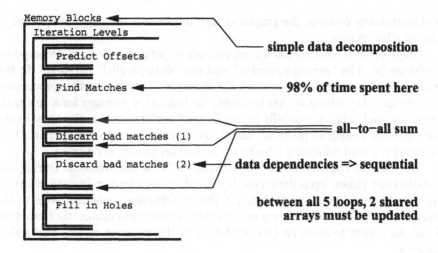

**Fig. 3.** Xmatch algorithmic structure.

is better suited for an implementation using communication directly via shared-memory.

We focus on the possibility for coarse grain parallelization. We expect the load-imbalance to be less problematic than the additional overhead in communicating all the common arrays.

## 4    Computing Platform

Due to an opportunity in connection with the TRACS program we were able to utilize the Cray T3D supercomputer hosted at the Edinburgh Parallel Computing Center (EPCC). It consists of 512 nodes, each containing a DEC-Alpha RISC processor and 64 Mbyte of main memory. Xmatch could only be ported to the Cray T3D by solving of a variety of problems, dominated by non-compatible data types for REAL and INTEGER, missing BYTE (or INTEGER*1) data type on Cray T3D and different interfaces between C and Fortran 77 functions.

The second architecture is a PowerChallenge from SGI containing 20 processors. Each R10000 processor is clocked with 194 MHz and equipped with primary data and instruction cache of 32 Kbytes each, and 2 Mbyte secondary cache (p:I32k+D32k, s:2M). All 20 processors share an 8-way interleaved, global main memory of 2.5 Gbyte.

## 5    Parallelization

We decided to keep the C and Fortran mixture and to apply the Message Passing Interface (MPI [7]) standard library for parallelization. In specific, we used CHIMP on the Cray T3D and MPICH on the PowerChallenge. For performance and portability reasons, the graphical user interface is no longer supported in the parallel version.

There exist two widely known approaches to parallelize domain-decomposed problems [8]: The "manager/worker" and the "data-parallel" paradigm. In the manager/worker model, one process, the manager, controls the other processors, the workers. Any time a worker becomes idle, it asks the manager for a new task. Dynamic load balancing usually produces good performance results here. In the data-parallel model, all nodes are equal, each processing a predefined part. This rather static load balancing scheme works well on equally sized tasks.

We tried to combine these two techniques for several reasons: During the initialization phase, input data must be moved to every worker. Instead of letting the manager read in the whole files and then communicate selected pieces to the workers, the workers themselves read in the necessary pieces directly from disk. Thus, we expect to avoid an I/O bottleneck at the manager, as also outlined in figure 4.

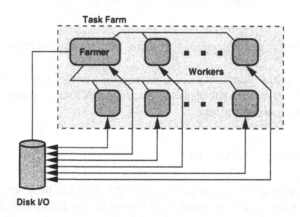

**Fig. 4.** I/O communication scheme: Every worker accesses the disk.

Except at the very beginning and at the end, the manager also computes matchpoints as the workers do. This helps to fully utilize all processors, since otherwise the manager would be idle during that time. At the beginning, the master only reads meta-information, computes controlling directives and distributes both to all workers before actually starting to work. Finally, the master sorts the matches that have been reported to him during the previous stage and writes the output file.

Furthermore, partitioning the input images into larger pieces causes less overlapping regions (see figure 2) and therefore less overhead due to parallelization. We expect a short startup time, minimal internal data communication and high utilization of processors. This combination also appeared to be an easy programming and communications model.

# 6  Results

There are two widely used figures to illustrate a computer's performance: Speed-up $S$ and efficiency $\varepsilon$. Aiming to maximize speed under the constraint of maximal utilization, Ghosal [2] defines efficacy $\eta$ as benefit $B$ ($\sim S(p)$) divided by cost $C$ ($\sim 1/\varepsilon(p)$), leading to:

$$\eta := \frac{B}{C} = \frac{S(p)}{\frac{1}{\varepsilon(p)}} = \frac{S(p)^2}{p}.$$

Additionally we use the scaleup, plotting the time one processor needs for a fixed piece of work over the number of processors in total. This requires that the size of the input images is doubled whenever the number of processors doubles. Constant time will be expected for all experiments.

## 6.1  Cray T3D

Performance of the Cray T3D is shown in figure 5. The left plot summarizes the speedup ($S$), efficiency ($\varepsilon$) and efficacy ($\eta$) for a pair of images each 1K × 1K pixels large. Because of the coarse grained parallelization, with this data set not more than 8 processors can be utilized. However, efficacy already has its peak at 4 processors, pointing to other problems than load-balancing.

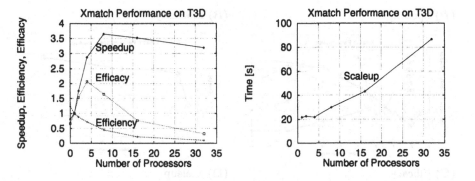

**Fig. 5.** Performance on Cray T3D.

To get an impression of scalability, we kept the amount of work to be performed at one processor constant and measured the scaleup as illustrated in the right plot of figure 5. Each processor has to match 250K pixels in this configuration. This equals to an 1K × 1K image for 4 processors and to a 4K × 2K image for 32 processors. Such an analysis bypasses the load balancing problem and inefficiencies caused by the parallelization method, and shows bottlenecks of the system.

The ideal horizontal line only can be seen for less than 4 processors. We found disk I/O bandwidth to be the bottleneck. All files are stored on a single disk and this disk is interconnected to the T3D by an ordinary Ethernet 10 Mbit/s line. However, within the TRACS program, there were no means to put data onto a faster disk.

## 6.2   SGI PowerChallenge

The experiments on the PowerChallenge were more extensive. All data were located on a disk of the machine itself and thus connected by a fast SCSI bus. In figure 6 several plots are collected. Input image size doubles from line to line. As one can see, only the first three images are too small to utilize all processors due to the coarse load balancing strategy.

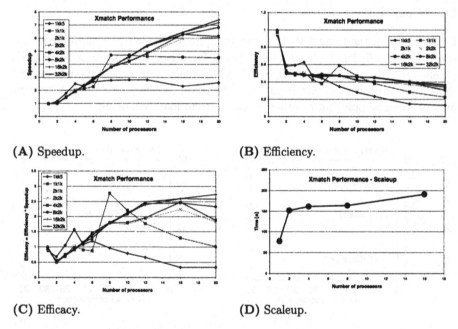

(A) Speedup.                              (B) Efficiency.

(C) Efficacy.                             (D) Scaleup.

**Fig. 6.** Performance on SGI PowerChallenge.

The scaleup plot (D) shows good performance over the whole range of used processors. Consequently, no bottlenecks regarding data I/O occur on this architecture. As noticeable on all other plots, too, the time one processor requires is about half of the time when invoking two or more processors.

We found that this is caused by two reasons: First, concurrent access to the same file while reading from different positions is in fact a degradation in performance. Simple I/O experiences have shown that it is best to do all data

I/O via the manager, although it then obviously becomes the bottleneck. We hope that MPI-2 will provide a solution therefore.

Secondly, MPI internal problems occurred because many messages are sent to the master during the matching phase, extensively filling or even overloading the input buffer of the manager. Additionally, it takes disproportionately long to actually receive and process these messages afterwards, and to write the match-points to disk. It is a good advice to the manager to take a smaller piece of work, and to check for incoming messages more often instead.

# 7  Conclusion and Outlook

This paper shows that Xmatch can be parallelized using MPI combined with a very coarse grained data decomposition and simple load balancing resulting in reasonable performance figures. It further showed that it is important to place the input data very close to the processors. Otherwise, as experienced on the Cray T3D, data I/O appears to be the bottleneck.

The suggested combination of the manager/worker and the data-parallel paradigm did not show the expected advantages. Avoiding the bottleneck when reading input data caused problems hidden within MPI and the operating system. Partially, performance became even worse. With respect to massive parallel systems, utilizing more than some 10 processors while keeping the efficiency above 50% will be very difficult for Xmatch. We also argue that porting Xmatch to MPPs might not be that useful since execution time already can be reduced to acceptable values using fewer processors.

We plan to advance the communications structure and the data I/O part. MPI-2 will be a welcome means therefore. Moreover, it would be also interesting to run Xmatch on clusters of workstations, since they are more common than supercomputers like the Cray T3D. Implementing a shared-memory version incorporating fine-grained parallelism might also be a future task.

## Acknowledgments

This work has been supported by the Austrian Fonds zur Förderung der Wissenschaftlichen Forschung (FWF) through FSP project S7001, "Theory and Applications of Digital Image Processing and Pattern Recognition".

I also wish to thank the Edinburgh Parallel Computing Centre (EPCC) for using the Cray T3D and the Research Institute for Software Technology in Salzburg for allowing to run Xmatch on their 20-node PowerChallenge.

## References

1. Curlander, J., K. Maurice, *Magellan Stereo Toolkit User Manual* (Vexcel Corporation, Boulder, CO, 1992).

2. Ghosal D., G. Serazzi, S. K. Tripathi, The Processor Working Set and Its Use in Scheduling Multiprocessor Systems, *IEEE Transactions on Software Engineering* **17(5)** (May 1991) 443–53.
3. Gelautz, M., G. Jakob, G. Paar, S. Hensley, F. Leberl, Automated Matching Experiments with Different Kinds of SAR Imagery, *Proceedings of IGARSS 96* (Lincoln, NE, 1996) 31–33.
4. Goller, A., M. Gelautz, F. Leberl, Concurrent Radar Image Shape-from-Shading on High-Performance Computers, *Proceedings of ICPR 96* (Vienna, Austria, 1996) D589–93.
5. Hensley, S., S. Shaffer, Automatic DEM Generation Using Magellan Stereo Data, *Proceedings of IGARSS 94* (Pasadena, CA, 1996) 1470–72.
6. Leberl, F. W., J. K. Thomas, K. E. Maurice, Initial Results From the Magellan Stereo-Experiment, *Journal of Geophysical Research* **97(E8)** (1992) 13675–87.
7. Message Passing Interface Forum: MPI: A message-passing interface standard. International Journal of Supercomputer Applications **8(3-4)** (1994) home page at http://www.mcs.anl.gov/mpi/.
8. Zomaya A. Y. H., editor. *Parallel and Distributed Computing Handbook*, chapter 9, Partitioning and Scheduling. McGraw-Hill, 1996.

# Parallel Decomposition of Distance-Hereditary Graphs

Sun-yuan Hsieh

Institute of Information Science, Academia Sinica, Taipei, Taiwan
hsiehsy@iis.sinica.edu.tw

**Abstract.** A distance-Hereditary graph $G$ has a binary tree representation called a decomposition tree. Given a decomposition tree, many graph-theoretical problems can be efficiently solved on $G$ using the binary tree contraction technique. In this paper, we present an algorithm to construct a decomposition tree in $O(\log^2 n)$ time using $O(n+m)$ processors on a CREW PRAM, where $n$ and $m$ are the number of vertices and edges of $G$, respectively.

## 1 Introduction

A graph is *distance-hereditary* [2,13] if the distance stays the same between any of two vertices in every connected induced subgraph containing both (where the *distance* between two vertices is the length of a shortest path connecting them). Distance-hereditary graphs form a subclass of perfect graphs [9,12,13] that are graphs $G$ in which the maximum clique size equals the chromatic number for every induced subgraph of $G$ [11]. Two well-known classes of graphs, trees and cographs, both belong to distance-hereditary graphs. Properties of distance-hereditary graphs are studied by many researchers [2,3,4,7,8,9,10,12,13,14,16,15] which resulted in sequential or parallel algorithms to solve quite a few interesting graph-theoretical problems on this special class of graphs.

Previous works of designing parallel algorithms on distance-hereditary graphs can be found in [8,7,14]. In [14], Hsieh *et. al.* presented algorithms to find a minimum weighted connected dominating set and a minimum weighted Steiner tree of a distance-hereditary graph in $O(\log n)$ time using $O(n + m)$ processors on a CRCW PRAM. A minimum connected $\gamma$-dominating set and a $\gamma$-dominating clique of a distance-hereditary graph can be found in $O(\log n \log \log n)$ time using $O((n + m)/\log \log n)$ processors on a CRCW PRAM [15]. In [16], Hsieh *et. al.* defined a general problem-solving paradigm for subgraph optimization problems. Problems fit into the paradigm include (a) the maximum independent set problem, (b) the maximum clique problem, (c) the vertex connectivity problem, (d) the edge connectivity problem, (e) the dominating set problem, (f) the independent dominating set problem and (g) the connected dominating set problem. They showed that all the above problems can be solved in $O(max(T(n), \log n))$ time using $O(max(P(n), n/\log n))$ processors on distance-hereditary graphs, where $T(n)$ (respectively, $P(n)$) is the time (respectively, pro-

P. Zinterhof, M. Vajteršic, A. Uhl (Eds.): ACPC'99, LNCS 1557, pp. 417–426, 1999.
© Springer-Verlag Berlin Heidelberg 1999

cessor) required to construct a binary tree representation, called a decomposition tree, of a distance-hereditary graph.

In this paper, we construct a decomposition tree in $O(\log^2 n)$ time using $O(n+m)$ processors on a CREW PRAM. Our result directly imply that Problems (a)–(g) belong to the NC class.

## 2   Preliminaries

This paper considers finite, simple and undirected graphs $G = (V, E)$, where $V$ and $E$ are the vertex and edge sets of $G$, respectively. Let $n = |V|$ and $m = |E|$. For graph-theoretic terminologies and notations, see [11].

For a vertex $v \in V$ of a graph $G = (V, E)$, the *neighborhood of* $v$ is $N_G(v) = \{u \in V \mid (u, v) \in E\}$ and the *closed neighborhood of* $v$ is $N_G[v] = N_G(v) \cup \{v\}$. For a subset $X$ of $V$, the *neighborhood of* $X$ is $N_G(X) = \bigcup_{v \in X} N_G(v)$ and the *closed neighborhood of* $X$ is $N_G[X] = \bigcup_{v \in X} N_G[v]$. The subscript $G$ in the notations used in this paper can be omitted when no ambiguity arises. Let $G[X]$ denote the subgraph of $G$ induced by $X \subseteq V$. The *degree* of a vertex $v$ is $deg(v) = |N(v)|$. We say that vertex $u$ is a *pendant vertex* attached to vertex $v$ if $deg(u) = 1$ and $v$ is the vertex adjacent to $u$. Two vertices $u$ and $v$ are called *true* (respectively, *false*) *twins* if $N[u] = N[v]$ (respectively, $N(u) = N(v)$). Given an ordering $\delta = v_1, v_2, \ldots, v_n$ of $V$, we call $\delta$ a *one-vertex-extension ordering* of $G$ if $v_j$ is a pendant vertex attached to $v_i$, $i < j$, or $v_i$ and $v_j$ are twins in $G[\{v_1, v_2, \ldots, v_j\}]$ for $1 \leq j \leq n$. It was shown that a graph is distance-hereditary if and only if it has a one-vertex-extension ordering [2,12].

Let $G = (V, E)$ be a connected distance-hereditary graph and $\delta = (v_1, v_2, \ldots, v_n)$ is a one-vertex-extension ordering of $G$. In [4], Chang *et. al.* define the *one-vertex-extension tree* $\mathcal{E}_G$ *with respect to* $\delta$ as follows. One-vertex-extension tree $\mathcal{E}_G$ is a rooted ordered tree with the root $v_1$ and $V(\mathcal{E}_G) = V$. From $\delta$, $\mathcal{E}_G$ can be generated as follows. For a vertex $v_j$, $j > 1$, which is either a pendant vertex attached to $v_i$ or $v_j$ and $v_i$ are twins in $G[\{v_1, v_2, \ldots, v_j\}]$ for some $i < j$, we let $v_j$ be the rightmost child of $v_i$ in the current tree. We use $[v_i, v_j]$ to denote an edge of $\mathcal{E}_G$. If $v_j$ is a pendant vertex attached to $v_i$ in $G[\{v_1, v_2, \ldots, v_j\}]$, $[v_i, v_j]$ is labeled $P$. If $v_i$ and $v_j$ are true twins in $G[\{v_1, v_2, \ldots, v_j\}]$, $[v_i, v_j]$ is labeled $T$. Otherwise, $[v_i, v_j]$ is labeled $F$. A one-vertex-extension tree of a distance-hereditary graph can be constructed in $O(n + m)$ time.

For a node $v$ in $\mathcal{E}_G$, the *twin set of* $v$, denoted by $S(v)$, is the set of $V(\mathcal{E}_G)$ which are $v$ and the descendants of $v$ that can be reached from $v$ through only $T$ edges or $F$ edges. Suppose $v_{h_1} < v_{h_2} < \cdots < v_{h_j} < v_{h_{j+1}} < \cdots < v_{h_k}$ are children of $v_i$, denoted by $child(v_i)$, in $\mathcal{E}_G$. For an edge $[v_i, v_{h_j}]$ in $\mathcal{E}_G$, let $S_r[v_i, v_{h_j}] = S(v_i) \setminus (\bigcup_{l=1}^{j} S(v_{h_l}))$. Let $\mathcal{E}_G[v_i, v_{h_j}]$ denote the subtrees of $\mathcal{E}_G$ induced by $v_i, v_{h_j}, v_{h_{j+1}}, \ldots, v_{h_k}$ and all descendants of $v_{h_j}, v_{h_{j+1}}, \ldots, v_{h_k}$. For a node $v$ in $\mathcal{E}_G$, let $\mathcal{E}_G[v]$ denote the subtree rooted at $v$.

We say that two disjoint vertex subsets $X$ and $Y$ of $V$ *form a join* in graph $G = (V, E)$ if every vertex of $X$ is connected to every vertex of $Y$.

**Lemma 1.** *[4] Suppose $v_j$ is a child of $v_i$ in $\mathcal{E}_G$. Then,*
*(1) If $[v_i, v_j]$ is labeled $P$ or $T$, then $S(v_j)$ and $S_r[v_i, v_j]$ form a join in $G$.*
*Moreover, for every vertex $v \in V(\mathcal{E}_G[v_j]) \setminus S(v_j)$, $N[v] \subseteq V(\mathcal{E}_G[v_j])$.*
*(2) If $[v_i, v_j]$ is labeled $F$, then every vertex of $V(\mathcal{E}_G[v_j])$ is not adjacent to any*
*vertex of $V(\mathcal{E}_G[v_i, v_j]) \setminus V(\mathcal{E}_G[v_j])$ in $G$.*

Suppose $G = (V, E)$ with $|V| > 1$ is a distance-hereditary graph and $\mathcal{E}_G$
is its one-vertex-extension tree. In the remainder of this section, we assume
each vertex of $G$ is represented by its corresponding one-vertex-extension order.
Note that vertex 1 is the root of $\mathcal{E}_G$ and vertex 2 is the first child of 1. The
vertex set $V$ can be partitioned into four disjoint sets: $V_1 = V(\mathcal{E}_G[2]) \setminus S(2)$,
$V_2 = S(2)$, $V_3 = (V(\mathcal{E}_G[1]) \setminus V(\mathcal{E}_G[2])) \setminus S_r[1, 2]$ and $V_4 = S_r[1, 2]$. By Lemma 1,
no vertex in $V_1$ is adjacent to any one in $V_4$. We can consider $G$ is formed from
$G_1 = G[V(\mathcal{E}_G[1]) \setminus V(\mathcal{E}_G[2])]$ and $G_2 = G[V(\mathcal{E}_G[2])]$ according to the label of
$[1, 2]$. If $[1, 2]$ is labeled $T$ or $P$, then $G$ is formed from $G_1$ and $G_2$ by connecting
every vertex of $S(2)$ to all vertices of $S_r[1, 2]$ by Lemma 1(1). If $[1, 2]$ is labeled $F$,
then $G$ is the union of $G_1$ and $G_2$ by Lemma 1(2) (where by union of two graphs
$G_1 = (V_1, E_1)$ and $G_2 = (V_2, E_2)$ we mean the graph $G = (V_1 \cup V_2, E_1 \cup E_2)$).
Note that a graph consisting of a single vertex $v$ is clearly a distance-hereditary
graph. Chang *et. al.* call it a *primitive distance-hereditary graph with the twin
set* $\{v\}$ [4]. Let $G_1$ and $G_2$ be distance-hereditary graphs with the twin sets $S_1$
and $S_2$, respectively.

**Lemma 2.** *[4] (1) The union of $G_1$ and $G_2$ is a distance-hereditary graph with
the twin set $S_1 \cup S_2$. (2) The graph obtained from $G_1$ and $G_2$ by connecting every
vertex of $S_1$ to all vertices of $S_2$ is a distance-hereditary graph with the twin set
$S_1 \cup S_2$. (3) The graph obtained from $G_1$ and $G_2$ by connecting every vertex of
$S_1$ to all vertices of $S_2$ is a distance-hereditary graph with the twin set $S_1$.*

A distance-hereditary graph $G$ is said to be formed from $G_1$ with the twin
set $S_1$ and $G_2$ with the twin set $S_2$ by the *false twin* (respectively, *true twin*)
operation if $G$ is obtained through (1) (respectively, (2)) of Lemma 2, and by
the *attachment operation* if $G$ is obtained through (3) of Lemma 2.

From the above discussion, a distance-hereditary graph can be represented
by a binary tree form, called a *decomposition tree*, which is defined as follows.

**Definition 1.** (1) The tree consisting of a single vertex labeled $v$ is a decomposition tree of the primitive distance-hereditary graph $G = (\{v\}, \emptyset)$.
(2) Let $\mathcal{D}_1$ and $\mathcal{D}_2$ be the decomposition trees of the distance-hereditary graphs
$G_1$ and $G_2$, respectively.
  (a) If $G$ is formed from $G_1$ and $G_2$ by the true twin operation, then a tree $\mathcal{D}$
    with the root $r$ represented by $\otimes$ and with the roots of $\mathcal{D}_1$ and $\mathcal{D}_2$ being the
    two children of $r$ is a decomposition tree of $G$.
  (b) If $G$ is formed from $G_1$ and $G_2$ by the attachment operation, then a tree
    $\mathcal{D}$ with the root $r$ represented by $\oplus$ and with the roots of $\mathcal{D}_1$ and $\mathcal{D}_2$ being
    the right and left child of $r$, respectively, is a decomposition tree of $G$.
  (c) If $G$ is formed from $G_1$ and $G_2$ by the false twin operation, then a tree $\mathcal{D}$
    with the root $r$ represented by $\odot$ and with the roots of $\mathcal{D}_1$ and $\mathcal{D}_2$ being the
    two children of $r$ is a decomposition tree of $G$.

# 3    Parallel Construction of a Decomposition Tree

In Section 3.1, we construct one-vertex-extension trees for cographs (the subclass of distance-hereditary graphs). In Section 3.2, we show that, by properly partitioning $G$ into several cographs $H$'s, a one-vertex-extension tree of $G$ can be generated by merging all the one-vertex-extension trees of $H$'s. We then present a method to transform a one-vertex-extension tree of $G$ into a decomposition tree in Section 3.3.

## 3.1    Constructing a One-Vertex-Extension Tree for a Cograph

A graph is *cograph* [6] if it is either a vertex, the complement of a cograph, or the union of two cographs. The cograph is also called the $P_4$-free graph which does not contain any induced path of length three [6]. It has been shown that the class of cographs is properly contained in distance-hereditary graphs [12]. A cograph $G$ has a tree representation called *cotree*, denoted by $T_G$, with (a) the leaves of $T_G$ are the vertices of $G$; (b) the internal nodes of $T_G$ are labeled 0 or 1; (c) 0 nodes and 1 nodes alternate along every path starting from the root; (d) two vertices $x$ and $y$ of the cograph are adjacent if and only if the least common ancestor of $x$ and $y$ in $T_G$ is a 1 node.

For a node $v$ in a rooted tree $T$, let $par(v)$ (respectively, $child(v)$) denote the parent (respectively, children) of $v$ in $T$ and let $T[v]$ denote the subtree of $T$ rooted at $v$. Let $leaf(T)$ be the leaves of $T$. For any two vertices $u$ and $v$ in $G$, let $dist(u, v)$ denote the distance between $u$ and $v$.

**Lemma 3.** *Let $u$ and $v$ be two leaves in $T_G$ such that $par(u) = par(v)$. Then, $u$ and $v$ are true (respectively, false) twins if $par(u)$ is a 1 (respectively 0) node.*

Given a cotree $T_G$ with the root $r$, $G$ can be reduced to a single vertex by repeatedly merging two vertices which are twins as follows. Arbitrarily find two leaves $u$ and $v$ of the current tree which have the same parent. Let $par(u) = par(v) = w$. By Lemma 3, $u$ and $v$ are true (respectively, false) twins if $w$ is a 1 (respectively, 0) node. Delete $u$ from the current graph and the current tree. After the deletion, if $v$ is the only child of $w$ in the current tree, $w$ is also deleted from the current tree. Also let $par(w)$ be the new parent of $v$ if $w \neq r$. We repeatedly execute the above procedure until $G$ is reduced to a single vertex. Clearly, a one-vertex-extension ordering of $G$ can be obtained by reversing the reducing process. The above procedure can be efficiently implemented with the following algorithm.

**Algorithm 1**
INPUT: A cograph $G$
OUTPUT: A one-vertex-extension tree $\mathcal{E}_G$

Step 1.  Construct a cotree $T_G$. Let $r$ denote the root of $T_G$.
Step 2.  Order the leaves of $T_G$ from 1 to $|leaf(T_G)| = k$. Let $order(v)$ denote the order associated with $v \in leaf(T_G)$.

**Step 3.** For each node $u \in T_G$, compute the label $L(u)$ as follows. If $u \in leaf(T_G)$, let $L(u) = u$. Otherwise, $u \notin leaf(T_G)$. Find the vertex $v \in leaf(T_G[u])$ such that $order(v) = max\{order(w)| \ w \in leaf(T_G[u])\}$, and let $L(u) = v$.

**Step 4.** For each $v \in V(G)$, find the set $\mathcal{L}_v = \{x \in V(T_G)| \ L(x) = v\}$ and $level(v) = min_{x \in \mathcal{L}_v}\{dist(x, r)\}$.

**Step 5.** Construct $\mathcal{E}_G$ as follows. Let $L(r)$ be the root of $\mathcal{E}_G$. For each non-root node $v \in T_G$, let $par(L(v)) = L(par(v))$ if $L(v) \neq L(par(v))$. Label $(L(v), L(par(v))) \in E(\mathcal{E}_G)$ as $T$ edge (respectively, $F$ edge) if $par(v)$ is a 1 (respectively, 0) node.

**Step 6.** For each non-leaf node $v \in \mathcal{E}_G$ with $child(v) = \{w_1, w_2, \ldots, w_p\}$. Order the children of $v$ with $w_{i_1} < w_{i_2} < \cdots < w_{i_p}$, where $level(w_{i_1}) \leq level(w_{i_2}) \leq \cdots \leq level(w_{i_p})$.

Since a cotree can be constructed in $O(\log^2 n)$ time using $O(n+m)$ processors on a CREW PRAM, the following complexity can be achieved.

**Theorem 1.** *Algorithm 1 correctly constructs a one-vertex-extension tree for a cograph in $O(\log^2 n)$ time using $O(n + m)$ processors on a CREW PRAM.*

## 3.2 Constructing a One-Vertex-Extension Tree for a Distance Hereditary Graph

Throughout this section, $G$ is used to denote a distance-hereditary graph whenever no ambiguity occurs. Given a vertex $u \in V$, the *hanging* of a connected graph $G = (V, E)$ rooted at $u$, denoted by $h_u$, is the collection of sets $L_0(u), L_1(u), \ldots, L_t(u)$ (or simply $L_0, L_1, \ldots, L_t$ when no ambiguity arises), where $t = max_{v \in V} dist(u, v)$ and $L_i(u) = \{v \in V| \ dist(u, v) = i\}$ for $0 \leq i \leq t$. For any vertex $v \in L_i$ and any vertex set $S \subseteq L_i, 1 \leq i \leq t$, let $N'(v) = N(v) \cap L_{i-1}$ and $N'(S) = N(S) \cap L_{i-1}$. Any two vertices $x, y \in L_i$ ($1 \leq i \leq t-1$) are said to be *tied* if $x$ and $y$ have a common neighbor in $L_{i+1}$.

Given a graph $G = (V, E)$, a vertex subset $S$ is *homogeneous* in a graph $G = (V, E)$ if every vertex in $V \setminus S$ is adjacent to either all or none of the vertices of $S$. We call a family of subsets *arboreal* if every two subsets of the family are either disjoint or comparable (by set inclusion). For a hanging $h_u = (L_0, L_1, \ldots, L_t)$, Hammer and Maffray [12] defined an equivalence relation $\equiv_i$ between vertices of $L_i$ by $x \equiv_i y$ means $x$ and $y$ are in the same connected component of $L_i$ or $x$ and $y$ are tied. Let $\equiv_a$ be defined on $V(G)$ by $x \equiv_a y$ means $x \equiv_i y$ for some $i$.

**Lemma 4.** *[2,9,12] Let $h_u$ be the hanging of $G$ rooted at $u$ and let $R_1, R_2, \ldots, R_r$ be the equivalence classes with respect to $h_u$.*

1. *For any two vertices $x$ and $y$ in some $R_i$, $N'(x) = N'(y)$.*
2. *The graph obtained from $G$ by shrinking each $R_j$ into one vertex is a tree rooted at $u$;*
3. *Each $R_j$ induces a cograph;*
4. *The family $\{N'(R_k)| \ N'(R_k) \subseteq R_i\}$, for $1 \leq i \leq r$, is an arboreal family of homogeneous subsets of $G[R_i]$.*

Let $R$ be an equivalence class of $G$ with respect to $h_u$. We call $\Gamma_R = \{ S \subset R |$ there is an equivalence class $R'$ with $N'(R') = S \}$ the *upper neighborhood system in* $R$ and call each $S \in \Gamma_R$, where $S = N'(R')$, the *upper neighborhood of* $R'$. By Lemma 4, $\Gamma_R$ is an arboreal family of homogeneous subsets of $R$. We can define a partial order $\preceq$ between two different sets $S_p$ and $S_q$ in $\Gamma_R$ with $S_p \preceq S_q \Leftrightarrow S_p \subset S_q$. According to the partial order $\preceq$ defined on $\Gamma_R$, let $\mathcal{U}_R = \{ S_i | S_i \not\subseteq S_k,$ for all $S_k \in \Gamma_R$ and $k \neq i \}$, that is, $\mathcal{U}_R$ is the set of those maximal elements of $(\preceq, \Gamma_R)$. We call $\mathcal{U}_R$ the *maximal upper neighborhoods in* $R$. For a set $S$ which is the upper neighborhood of some equivalence class, we can also define $\Gamma_S$ and $\mathcal{U}_S$ similarly. In what follows, the notation $R$ is referred as an equivalence class or an upper-neighborhood of some equivalence class if it is not specified.

**Lemma 5.** *Let* $\mathcal{U}_R = \{ Q_1, Q_2, ..., Q_k \}$ *and let* $x_i$ *be an arbitrary vertex of* $Q_i$, $1 \leq i \leq k$. *The graph* $G[(R \setminus \cup_{i=1}^{k} Q_i) \cup \{ x_1, x_2, \ldots, x_k \}]$ *is a cograph.*

*Proof.* By the property that every induced subgraph of a $P_4$-free graph is still $P_4$-free.      $\square$

Let $T$ be a one-vertex-extension tree of a distance-hereditary graph $G$. Note that $G$ is generated by a one-vertex-extension ordering represented by $T$. In what follows, we say $G$ *is generated by* $T$ for convenience.

**Lemma 6.** *Suppose* $G = (V, E)$ *is a cograph and* $Q_1, Q_2, \ldots, Q_k$ *are the homogeneous sets of* $G$ *such that* $Q_i \cap Q_j = \emptyset$, *where* $1 \leq i, j \leq k$ *and* $i \neq j$. *Let* $\mathcal{E}$ *be a one-vertex-extension tree of* $G[(V \setminus (\cup_{i=1}^{k} Q_i)) \cup \{ x_1, x_2, \ldots, x_k \}]$, *where* $x_i \in Q_i$, *and let* $\mathcal{E}_i$ *be a one-vertex-extension tree of* $G[Q_i]$, *where* $1 \leq i \leq k$. *Then, the tree obtained by identifying the root of* $\mathcal{E}_i$ *with the vertex* $x_i$ *in* $\mathcal{E}$ *for all* $1 \leq i \leq k$ *is a one-vertex-extension tree of* $G$.

*Proof.* The proof is by induction on $k$. The base case of $k = 0$ holds clearly. Suppose now that $k > 0$. By the proof of Lemma 5, the graph $G' = G[(V \setminus Q_1) \cup \{ x_1 \}]$ is a cograph with $k - 1$ homogeneous sets $Q_2, Q_3, \ldots, Q_k$. By the induction hypothesis, a one-vertex-extension tree $\mathcal{E}'$ of $G'$ can be obtained with the desired method. Since $Q_1$ is a homogeneous set, we have (1) $N_{G'}(x_1) = (N_G(y) \setminus Q_1)$ for all $y \in Q_1$ and (2) $E = E(G') \cup E(G[Q_1]) \cup \{ (y, b) | b \in N_{G'}(x_1) \}$. Note that $G'$ (respectively, $G[Q_1]$) is generated by $\mathcal{E}'$ (respectively, $\mathcal{E}_1$). Let $T$ be the resulting tree obtained by identifying the root of $\mathcal{E}_1$ with $x_1$ in $\mathcal{E}'$. By the definition of the one-vertex-extension tree, the graph generated by $T$ ensures that (1) and (2) hold. Moreover, $V(T) = V(G)$. Hence, $T$ is a one-vertex-extension tree of $G$.      $\square$

**Definition 2.** *Let* $R$ *be an equivalence class or an upper-neighborhood of some equivalence class. A one-vertex-extension tree* $\mathcal{E}_{G[R]}$ *is said to be canonical if for each* $Q \in \Gamma_R$, *there exists a subtree which is a one-vertex-extension tree of* $G[Q]$.

**Lemma 7.** *There exists a canonical one-vertex-extension tree of* $G[R]$.

*Proof.* We first construct a tree as follows. Let $\Gamma_R \cup \{R\} = \Gamma'_R$. For each $Q \in \Gamma'_R$, let $\mathcal{U}_Q = \{Q_1, Q_2, \ldots, Q_k\}$. We select an arbitrary vertex $x_i \in (Q_i \setminus \cup_{B \in \mathcal{U}_{Q_i}} B)$ as the *shrunk vertex*. Let $H_Q$ be the subgraph of $G[Q]$ induced by $(Q \setminus \cup_{i=1}^{k} Q_i) \cup \{x_1, x_2, \ldots, x_k\}$. By Lemma 5, $H_Q$ is a cograph. By executing Algorithm 1, a one-vertex-extension tree of $H_R$ and a one-vertex-extension tree of $H_Q$, for each $Q \in \Gamma_R$, whose root is the shrunk vertex of $Q$ can be constructed. Let $\mathcal{E}_{H_X}$, where $X \in \Gamma'_R$, be the constructed one-vertex-extension tree. For each $Y \in \Gamma'_R$, we identify the vertices $y_1, y_2, \ldots, y_q$ in $\mathcal{E}_{H_Y}$ with the root of $\mathcal{E}_{H_{Y_1}}, \mathcal{E}_{H_{Y_2}}, \ldots, \mathcal{E}_{H_{Y_q}}$, respectively, where $\mathcal{U}_Y = \{Y_1, Y_2, \ldots, Y_q\}$ and $y_i$ is the shrunk vertex of $Y_i$.

We next show the resulting tree, denoted by $T$, is a one-vertex-extension tree. The proof is by induction on $|\Gamma'_R|$. The base case of $\Gamma'_R = \{R\}$ is clearly correct. Now we consider $|\Gamma'_R| = t$. Suppose $\mathcal{U}_R = \{U_1, U_2, \ldots, U_p\}$ and $u_i$ is the shrunk vertex of $U_i$. By the induction hypothesis, the canonical one-vertex-extension trees of $G[U_i]$'s can be constructed correctly. By Lemma 4(4), $U_i$ is a homogeneous set of $G[R]$. Moreover, $U_j \cap U_l = \emptyset$, where $1 \le j, l \le p$ and $j \ne l$, according to the definition of $\mathcal{U}_R$. From the construction, $T$ can be viewed as merging $\mathcal{E}_{H_R}$ and the canonical one-vertex-extension trees of $G[U_i]$'s with the method described in Lemma 6. Hence, $T$ is a one-vertex-extension tree of $G[R]$. The property of canonical can be checked easily. □

We now present an algorithm to construct a one-vertex-extension tree of a distance-hereditary graph.

## Algorithm 2
INPUT: A distance-hereditary graph $G$.
OUTPUT: A one-vertex-extension tree $\mathcal{E}_G$.

**Step 1.** Build a hanging $h_u$ and compute the equivalence classes with respect to $h_u$.

**Step 2.** For each equivalence class $R$, compute $\Gamma_R = \{Q_1, Q_2, \ldots, Q_l\}$.

**Step 3.** Generate a canonical one-vertex-extension tree $\mathcal{E}_{G[R]}$ for each equivalence class $R$ with $\Gamma_R = \{Q_1, Q_2, \ldots, Q_l\}$.

**Step 4.** For each equivalence class $R$ with $\Gamma_R = \{Q_1, Q_2, \ldots, Q_l\}$, find the equivalence classes $R_1, R_2, \ldots, R_k$ such that $N'(R_i) \subseteq R$. Let $z_i$ be the root of $\mathcal{E}_{G[R_i]}$. Let $H_i$ be a subtree in $\mathcal{E}_{G[R]}$ which is a one-vertex-extension tree of $G[Q_i]$ and $x_i$ be the root of $H_i$.
CASE 1 $N'(R_i) \subset R$ for all $1 \le i \le k$. Without loss of generality, assume $N'(R_1) = N'(R_2) = \cdots = N'(R_{i_1}) = Q_1$, $N'(R_{i_1+1}) = N'(R_{i_1+2}) = \cdots = N'(R_{i_2}) = Q_2, \ldots,$ and $N'(R_{i_{l-1}+1}) = N'(R_{i_{l-1}+2}) = \cdots = N'(R_{i_l}) = Q_l$, where $i_l = k$. Let $i_0 = 0$. For each $1 \le j \le l$, we additionally add $z_{i_{j-1}+1}, z_{i_{j-1}+2}, \ldots, z_{i_j}$ to be $i_j - i_{j-1}$ children of $x_j$ in $\mathcal{E}_{G[R]}$ such that (a) the children of $x_j$ are $c_1, c_2, \ldots, c_t, z_{i_{j-1}+1}, z_{i_{j-1}+2}, \ldots, z_{i_j}, c_{t+1}, c_{t+2}, \ldots, c_w$, (b) the subtree in $\mathcal{E}_{G[R]}$ which is induced by $\{x_j, z_{i_{j-1}+1}, z_{i_{j-1}+2}, \ldots, z_{i_j}\}$ and the descendents of $z_{i_{j-1}+1}, z_{i_{j-1}+2}, \ldots, z_{i_j}$ is $H_j$.
Label each $[x_j, z_q]$ as $P$, $i_{j-1} + 1 \le q \le i_j$.

CASE 2 There exists an equivalence class $R_j$ such that $N'(R_j) = R$. Without loss of generality, assume $N'(R_1) = N'(R_2) = \cdots = N'(R_{i_0}) = R$, $N'(R_{i_0+1}) = N'(R_{i_0+2}) = \cdots = N'(R_{i_1}) = Q_1, \ldots$, and $N'(R_{i_{l-1}+1}) = N'(R_{i_{l-1}+2}) = \cdots = N'(R_{i_l}) = Q_l$, where $i_l = k$. Execute the following two operations: (i) Additionally add $z_1, z_2, \ldots, z_{i_0}$ to be the first $i_0 (\neq 0)$ children of the root $z$ of $\mathcal{E}_{G[R]}$, and label $[z, z_1], [z, z_2], \ldots, [z, z_{i_0}]$ as $P$, (ii) For each $1 \leq j \leq l$, we additionally add $z_{i_{j-1}+1}, z_{i_{j-1}+2}, \ldots, z_{i_j}$ to be $i_j - i_{j-1}$ children of $x_j$ in $\mathcal{E}_{G[R]}$ such that (a) the children of $x_j$ are $c_1, c_2, \ldots, c_t, z_{i_{j-1}+1}, z_{i_{j-1}+2}, \ldots, z_{i_j}, c_{t+1}, c_{t+2}, \ldots, c_w$, (b) the subtree in $\mathcal{E}_{G[R]}$ which is induced by $\{x_j, z_{i_{j-1}+1}, z_{i_{j-1}+2}, \ldots, z_{i_j}\}$ and the descendents of $z_{i_{j-1}+1}, z_{i_{j-1}+2}, \ldots, z_{i_j}$ is $H_j$.
Label each $[x_j, z_q]$ as $P$, $i_{j-1} + 1 \leq q \leq i_j$.

From Lemma 4(2), shrinking each equivalence class with respect to the hanging $h_u$ forms a tree, denoted by $T_{h_u}$. For each equivalence class $R$, let $\nu_R$ denote the node representing $R$ in $T_{h_u}$. We define $\psi(R) = \{Q \mid \nu_Q \in V(T_{h_u}[\nu_R])\}$. Let $\psi'(R) = \bigcup_{X \in \psi(R)} X$. In the following, we show the correctness of Algorithm 2.

**Lemma 8.** *Suppose $R$ is an equivalence class with respect to a given hanging. Algorithm 2 correctly constructs a one-vertex-extension tree of $G[\psi'(R)]$.*

*Proof.* The proof is by induction on $|\psi(R)|$. The base case of $\psi(R) = \{R\}$ clearly holds. Suppose now that $|\psi(R)| = t > 1$. Let $R_1, R_2, \ldots, R_k$ be the equivalence classes such that $N'(R_i) \subseteq R$. Let $\Gamma_R = \{Q_1, Q_2, \ldots, Q_l\}$. From Step 3, a canonical one-vertex-extension tree $\mathcal{E}_{G[R]}$ of $G[R]$ can be constructed. Note that $|\psi(R_i)| < t$ for all $1 \leq i \leq k$. By the induction hypothesis, the one-vertex-extension trees $\mathcal{E}_{G[\psi'(R_i)]}$'s of $G[\psi'(R_i)]$'s can be correctly constructed. From Algorithm 2, the construction of the desired tree, denoted by $\mathcal{E}_{G[\psi'(R)]}$, can be viewed as merging $\mathcal{E}_{G[R]}$ and $\mathcal{E}_{G[\psi'(R_i)]}$'s. Let $R_i$ be the equivalence classes with $N'(R_i) \subseteq R$, $1 \leq i \leq k$. By executing Step 4, the graph generated by $\mathcal{E}_{G[\Phi'(R)]}$ is obtained from $G[\Phi'(R)]$ (generated by $\mathcal{E}_{G[\Phi'(R)]}$) and $G[\Phi'(R_i)]$'s (generated by $\mathcal{E}_{G[\Phi'(R_i)]}$'s) by letting $R_i$ and $N'(R_i)$ to form a join. According to the structure characterization described in Lemma 4, the graph generated by the tree is surely $G[\psi'(R)]$. □

Therefore, we have the following result.

**Theorem 2.** *Algorithm 2 correctly constructs a one-vertex-extension tree of a distance-hereditary graph in $O(\log^2 n)$ time using $O(n + m)$ processors on a CREW PRAM.*

*Proof.* By Lemma 8, Algorithm 2 constructs correctly a one-vertex-extension tree of $G[\Phi'(u)] = G$, where $u$ is the root of the given hanging. We now analyze the complexity of Algorithm 2. Step 1 can be done in $O(\log^2 n)$ time using $O(n + m)$ processors on a CREW PRAM [8]. Step 2 can be done in $O(\log n)$ time using $O(n + m)$ processors on an EREW PRAM [14]. Based on the method described in previous section and the proof of Lemma 7, Step 3 can be done in $O(\log^2 n)$ time using $O(n + m)$ processors on a CREW PRAM. Step 4 can be done in $O(1)$ time using $O(n)$ processors. Thus the result holds. □

### 3.3 Transforming a One-Vertex-Extension Tree into a Decomposition Tree

Suppose $G = (V, E)$ with $|V| > 1$ is a distance-hereditary graph and $\mathcal{E}_G$ is its one-vertex-extension tree. In this section, we assume each vertex of $G$ is represented by the corresponding one-vertex-extension order. Note that vertex 1 is the root of $\mathcal{E}_G$ and vertex 2 is the first child of 1. There are two subgraphs $G_1$ and $G_2$ such that $G$ is formed from $G_1$ and $G_2$ by the true twin or the false twin or the attachment operation. We can consider $G_1 = G[V(\mathcal{E}_G[1]) \setminus V(\mathcal{E}_G[2])]$ with the twin set $S_r[1,2]$ and $G_2 = G[V(\mathcal{E}_G[2])]$ with the twin set $S(2)$. The above observation provides a recursive method to transform a one-vertex-extension tree into a decomposition tree: Given a one-vertex-extension tree $\mathcal{E}$ with the root $x$, let $y$ denote the first child of $x$. If $[x, y]$ is a $T$ edge, create a $\otimes$ node as the root of a decomposition tree $\mathcal{D}_{G[V(\mathcal{E}[x])]}$. If $[x, y]$ is a $P$ edge, create a $\oplus$ node as the root. Otherwise, create a $\odot$ node as the root. Recursively construct $\mathcal{D}_{G[V(\mathcal{E}[y])]}$ (respectively, $\mathcal{D}_{G[V(\mathcal{E}[x]) \setminus V(\mathcal{E}[y])]}$) from transforming $\mathcal{E}[y]$ (respectively, the tree obtained by deleting $\mathcal{E}[y]$ from $\mathcal{E}$). Let the roots of $\mathcal{D}_{G[V(\mathcal{E}[y])]}$ and $\mathcal{D}_{G[V(\mathcal{E}[x]) \setminus V(\mathcal{E}[y])]}$ be the left child and right child of the root of $\mathcal{D}_{G[V(\mathcal{E}[x])]}$, respectively. Based on the above observation, the following algorithm is presented.

**Algorithm 3**
INPUT: A one-vertex-extension tree $\mathcal{E}_G$.
OUTPUT: A decomposition tree $\mathcal{D}_G$.
**Step 1.** For each vertex $v$ in $\mathcal{E}_G$, let $num(v)$ be the one-vertex-extension order associated with $v$. For each edge $e = [par(v), v]$ in $\mathcal{E}_G$, let $num(e) = num(v)$.
**Step 2.** For each edge $e$ in $\mathcal{E}_G$, create an internal node $\nu_e$ ($\otimes$ or $\oplus$ or $\odot$) for $\mathcal{D}_G$ depending on the type of $e$.
**Step 3.** For each node $\nu_e$, where $e = [par(v), v]$, created in Step 2, execute the following operations. (a) If $par(v) \in V(\mathcal{E}_G)$ contains no child $w$ satisfying $num([par(v), w]) > num(e)$, then create a node representing $par(v)$ to be the right child of $\nu_e$. Otherwise, find the edge $e'$ such that $num(e) < num(e')$ and no other edge $y$ satisfying $num(e) < num(y) < num(e')$. Let the node created for $e'$ be the right child of $\nu_e$. (b) If $v \in V(\mathcal{E}_G)$ is a leaf, create a node representing $v$ to be the left child of $\nu_e$. Otherwise, find the edge $e' = (v, z)$ such that $num(z) = min\{num(x)| \ x \in child(v)\}$. Let the node created for $e'$ be the left child of $\nu_e$.

**Theorem 3.** *Algorithm 3 correctly transforms a one-vertex-extension tree into a decomposition tree in $O(1)$ time using $O(n)$ processors.*

## 4  Conclusion

In this paper, we have represented a distance-hereditary graph with a binary tree form in $O(\log^2)$ time using $O(n + m)$ processors on a CREW PRAM. The bottleneck of our algorithm is the cotree construction. If such a tree can be optimally constructed, the parallel complexity of our algorithm can be further reduced to $O(\log n)$ time using $O(n / \log n)$ processors.

# References

1. K. Abrahamson, N. Dadoun, D. G. Kirkpatrick, and T. Przytycka, A simple parallel tree contraction algorithm. *Journal of Algorithms.*, 10, pp. 287-302, 1989.
2. H. J. Bandelt and H. M. Mulder. Distance-hereditary graphs. *Journal of Combinatorial Theory Series B*, 41(1):182-208, Augest 1989.
3. A. Brandstädt and F. F. Dragan, A linear time algorithm for connected $\gamma$-domination and Steiner tree on distance-hereditary graphs, *Networks*, 31:177-182, 1998.
4. M. S. Chang, S. Y. Hsieh, and G. H. Chen. Dynamic Programming on Distance-Hereditary Graphs. Proceedings of 7th International Symposium on Algorithms and Computation, ISAAC97, to appear.
5. R. Cole, "Parallel merge sort," *SIAM Journal on Computing*, 17(4):770-785, 1988.
6. D. G. Corneil, H. Lerchs, and L. S. Burlingham, Complement reducible graphs. *Discrete Applied Mathematics.*, 3, pp. 163-174, 1981.
7. E. Dahlhaus, "Optimal (parallel) algorithms for the all-to-all vertices distance problem for certain graph classes," Lecture notes in computer science 726, pp. 60-69.
8. E. Dahlhaus. Efficient parallel recognition algorithms of cographs and distance-hereditary graphs. *Discrete applied mathematics*, 57(1):29-44, February 1995.
9. A. D'atri and M. Moscarini. Distance-hereditary graphs, steiner trees, and connected domination. *SIAM Journal on Computing*, 17(3):521-538, June, 1988.
10. F. F. Dragan, Dominating cliques in distance-hereditary graphs, *Algorithm Theory-SWAT'94 "4th Scandinavian Workshop on Algorithm Theory, LNCS 824, Springer, Berlin*, pp. 370-381, 1994.
11. M. C. Golumbic. *Algorithmic graph theory and perfect graphs*, Academic press, New York, 1980.
12. P. L. Hammer and F. Maffray. Complete separable graphs. *Discrete applied mathematics*, 27(1):85-99, May 1990.
13. E. Howorka. A characterization of distance-hereditary graphs. *Quarterly Journal of Mathematics (Oxford)*, 28(2):417-420. 1977.
14. S.-y. Hsieh, C. W. Ho, T.-s. Hsu, M. T. Ko, and G. H. Chen. Efficient parallel algorithms on distance-hereditary graphs. *Parallel Processing Letters*, to appear. A preliminary version of this paper is in *Proceedings of the International Conference on Parallel Processing*, pp. 20–23, 1997.
15. S.-y. Hsieh, C. W. Ho, T.-s. Hsu, M. T. Ko, and G. H. Chen. A new simple tree contraction scheme and its application on distance-hereditary graphs. *Proceedings of Irregular'98*, Springer-Verlag, to appear.
16. S.-y. Hsieh, C. W. Ho, T.-s. Hsu, M. T. Ko, and G. H. Chen. Characterization of efficiently solvable problems on distance-hereditary graphs. *Proceedings of ISAAC'98*, Springer-Verlag, to appear.
17. J. Ja'Ja'. *An Introduction to Parallel Algorithms*. Addison Wesley, 1992.

# Asynchronous Parallel Construction of Recursive Tree Hierarchies

Dirk Bartz and Wolfgang Straßer

WSI/GRIS,
University of Tübingen,
Auf der Morgenstelle 10/C9,
D72076 Tübingen, Germany
{bartz,strasser}@gris.uni-tuebingen.de
http://www.gris.uni-tuebingen.de

**Abstract.** Multi-resolution methods are widely used in scientific visualization, image processing, and computer graphics. While many applications only require an one-time construction of these data-structures
which can be done in a pre-process, this pre-process can take a significant amount of time. Considering large datasets, this time consumption
can range from several minutes up to several hours, especially if this pre-
process is frequently needed. Furthermore, numerous new applications,
such as visibility queries, arise which often need a dynamic reconstruction of a scene database.
In this paper, we address several problems of the construction or reconstruction of recursive tree hierarchies in parallel. In particular, we focus
on parallel dynamic memory allocation and the associated synchronization overhead.
**Keywords**: Parallel hierarchies, recursive tree structures, octrees, memory synchronization, shared memory, thread model.

## 1 Introduction

In computer graphics, hierarchical methods are widely used to reduce the complexity of common problems. Specifically, multi-resolution methods are used
to reduce the polygon count of large models [4], to reduce the light interaction
between different parts of a scene [6], and so forth. Among the most popular spatial multi-resolution representations are recursive tree structures like quadtrees
and octrees [13], and binary-space-partitioning trees (BSP-trees) [3]. Unfortunately, the construction or reconstruction of these representations is very costly.
Therefore, we proposed to perform this process in parallel [1] by decoupling the
recursive parent/children relationship of the tree elements. However, dynamic
memory allocation is limiting the scalability of our algorithm by introducing
additional synchronization overhead. In this paper, we discuss three approaches
to remove this bottleneck on three different MIMD architectures.

P. Zinterhof, M. Vajteršic, A. Uhl (Eds.): ACPC'99, LNCS 1557, pp. 427–436, 1999.
© Springer-Verlag Berlin Heidelberg 1999

Our paper is organized as follows: We first outline the background and the related work of our paper. Next, we briefly describe the basic algorithm for the parallel and balanced tree (re-)construction and isosurface extraction. In Section 4, we discuss three approaches for the reduction of the overhead during the dynamic allocation of memory. Finally, we draw a conclusion and give perspectives to future work.

## 2    Background and Related Work

Tree-based spatial subdivision schemes are widely used in scientific visualization, image processing, and computer graphics. Of special importance are schemes like quadtrees - which represent a regular two-dimensional subdivision - and BSP-trees - which represent an irregular binary subdivision of arbitrary dimensions. Although our results are generally valid for all recursive tree structures, in this paper we only focus on the three-dimensional counterpart of quadtrees, the octrees.

An octree is a hierarchical spatial data-structure to represent three-dimensional volumetric data at different levels of details [13]. Starting with the superblock - representing the whole dataset - each octant (an octree block) is subdivided into eight child blocks. Each of these child blocks has half the size of the parent in each dimension (Fig. 1). This subdivision is performed until the lowest level is reached, where each block represents eight volumetric sample values (voxels). These bottom level blocks are called cells.

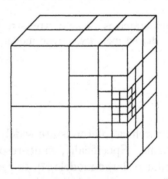

**Fig. 1.** Octree.

Due to the subdivision, the size of each octant is a power of two. Unfortunately, datasets usually do not have a size of this scheme. Therefore, some octants are "empty" - they do not intersect with the dataset, according to the alignment of the dataset within the octree. In order to save space, we use a minimal octree. The minimal octree approach enumerates only the octants - and their children - that are not empty.

Octrees are used in several applications to provide a multi-resolution representation. Laur and Hanrahan presented an octree-based scheme for hierarchical splatting [10]. Splats of different size and shape are used, according to the standard deviation of the color values of the different octree blocks. Grosso et al. presented a parallel implementation of this algorithm [5]. In their approach, a static parallelization is used, where up to eight threads are processing up to eight children blocks of the superblock. Greene et al. use an octree and an image pyramid for visibility queries in large polygonal environments [4]. Shekkar et al. use an octree representation of a volumetric dataset to generate a block-oriented polygon reduction scheme of its isosurface [14]. Levoy presented an approach to accelerate ray casting by using octrees [11], where coherent (non-contributing) data can be skipped rapidly.

As an example application of our algorithm, an isosurface is extracted from a volumetric dataset which represents a surface with the same scalar value (isovalue) throughout this dataset. We are using a parallel implementation of the approach by Wilhelms and van Geldern [16]. By storing the minimum and maximum values of the voxels at each block of the octree, the blocks which do not contain the isovalue in their minimum/maximum interval, can be skipped rapidly. After selecting all these contributing voxels of these blocks, the isosurface is generated.

The construction of a multi-resolution representation can be very time consuming. This is especially true if we consider large datasets. If this construction is needed frequently, the parallelization of this process quickly becomes worthwhile. Furthermore, several applications require fast reconstructions of octrees (or other recursive tree structures). Changes of the color table or the transfer functions in volume rendering applications require a reconstruction of the used hierarchical representations [5, 11]. In occlusion culling applications, moving objects cause a partial reconstruction of the scene representation [15].

# 3   Parallel Construction of Tree Hierarchies

In general, recursive tree structures are constructed in two stages; a split-down of a parent into several children, and a push-up of the results of the children back to the parent, i.e. the standard deviation, or - in our case - the minimum and maximum voxel values.

The parallelization of a recursive split-down is a rather simple task. Depending on the workload and the available processors, a subtree could be assigned to a thread. Usually, the second stage causes difficulties for a balanced parallelization. Due to their recursive relationship, we need to maintain the parent/child information. On the other hand, a balanced parallelization requires a decoupling of the structure. A simple distributed top-down subdivision, as suggested for the first stage, only provides the top-down information; every parent knows its children. For a push-up, we also need the bottom-up information - i.e. which block is the parent of the current block. In our approach, we solve this problem by combining a central workload splitting job queue and our new asynchronous push-up [1].

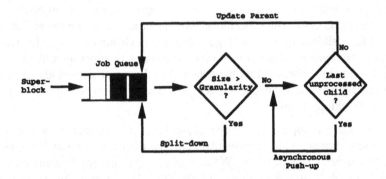

**Fig. 2.** General design of recursive tree construction.

In Figure 2, we outline the general design of our algorithm. After initially adding the superblock of the octree[1] to the empty job queue, the algorithm starts to read the first job from the queue. If the size of this job, the block size of the octant, exceeds a certain granularity value, this block is splitted into its children which are inserted into the queue. Thereafter, a new job is read from the queue. If the block size is below the granularity value, the processing thread proceeds sequentially with the octant and the associated subtree. This differentiation is necessary in order to guarantee a balance between the queue and synchronization overhead, and the parallelization benefits.

After processing the octant, we update its parent. Subsequently, we check if this octant was the last child of its parent which had not completed its computation. If it is the last uncompleted child, the processing thread continues processing the parent block and the flow of control has returned to the parent. Otherwise, the thread simply gets a new job from the queue. We call this semantic an asynchronous push-up (apu).

Figure 3 outlines the flow of control of the asynchronous push-up. Thread $t0$ is splitting the parent $j$ into $m$ children blocks, where thread $t1$ is processing child $j, 0$, thread $t2$ is processing child $j, i$, and thread $t0$ is processing child $j, m - 1$. After the completion of child $j, 0$ and child $j, m - 1$, threads $t0$ and $t1$ get a new job from the job queue. Thread $t2$ processes the last uncompleted child of parent $j$. Therefore, after completion of child $j, i$, thread $t2$ performs the asynchronous push-up and continues with parent $j$.

In [1], we show that all potential bottlenecks added by our algorithm - the mutex protected job queue access and the mutex protected asynchronous push-

---
[1] As already mentioned earlier, we focus in this paper on octrees.

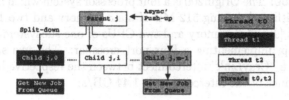

Fig. 3. Asynchronous Push-Up.

up - turned out to be of no significance. Furthermore, at no point during the computation do the threads stall because of an empty job queue. Until the final phase of the octree construction, the queue is always sufficiently filled, even with a large number of threads. However, the construction process introduces heavy memory allocation which limits the scalability of the algorithm. We will discuss this problem in detail in Section 4.

**Parallel Isosurface Extraction**

Following Wilhelms and van Geldern [16], we store the minimum and maximum isovalues of the voxels at all levels-of-detail. Therefore, we can rapidly decide if a subtree contains contributing cells[2], thus limiting the number of voxels examined by the Marching Cubes algorithm [12].

After assigning the contributing cells to the threads using a round-robin scheme, the threads are generating the polygons which represent the isosurface within the volumetric dataset. Due to page limitation, we refer to [12] for more details on the Marching Cubes algorithm, and to [1] for more details on the measurements of the parallel Marching Cubes implementation.

# 4   Reducing Synchronization Overhead of Parallel Memory Allocation

In [1], we discussed the presented algorithm on two different memory architectures; an NUMA-architecture[3] (Convex SPP 1600) and on an UMA-architecture[4] (SGI Challenge). In this paper, we base our discussion on the pthread implementation on three different memory architectures; two NUMA systems (SGI Onyx2, SGI Origin200), and one UMA system (SGI Challenge). All systems show different synchronization behavior, due to their architectural differences.

---

[2]  A cell is called contributing if the isosurface passes through the cell.
[3]  Non-Uniform-Memory-Access
[4]  Uniform-Memory-Access

**SGI Origin200:** The Origin200 is a four processor system which is split into two subsystems with each having 512 MB of main memory and two 180 MHz Mips R10000 CPUs [9]. The memory and two CPUs of one subsystem are connected via a hub chip, implementing a four port crossbar. The two subsystems are connected via a "CrayLink" interconnect between the respective hub chips. The peak performance of the interconnect is 1.44 GB/s.

**SGI Onyx2:** The Onyx2 used for our measurements has 10 195 MHz R10000 CPUs. These CPUs are organized on two processor modules, with a total of five node boards [8]. The first processor module contains four node boards, while the second only contains one node board. Each node board contains up to 512 MB main memory and two CPUs. The CPUs, the memory, and the connection to other parts of the Onyx2 are interlinked via a hub chip, implementing a four port crossbar. Two node boards are connected via a six port crossbar, thus interlinking both node boards via a router to the interconnection fabric of the processor modules. This interconnection fabric interlinks both processor modules using a hypercube topology. Similar to the Origin200, the peak performance of the interconnect is 1.44 GB/s.

**SGI Challenge:** The SGI Challenge is an UMA architecture system. 3 GB main memory is connected with 16 195 MHz R10000 CPUs via a global bus running at 1.2 GB/s [7].

### 4.1   Memory Allocation Strategies

Three different methods for the parallel allocation of memory are examined. Two of these methods produce very different results on the three different architectures. For the measurements, we used two different volume datasets of different origin (Table 1). Dataset A represents a velocity field generated by a computational fluid dynamics (CFD) simulation, where two sides of a fluid filled cavity are heated differently. Dataset B is a MRI scan of a human head with special focus on the ventricle system of the human brain. On both datasets, our algorithms showed the same behavior. Therefore, we only look in detail at dataset B and present only the general results of dataset A.

**Standard Memory Allocation** This technique uses the memory allocation functions of the standard library (stdlib). The thread-safe versions of these functions are using a global locking mechanism to guarantee mutual exclusion, usually denoted as a "big lock" [2].

Closer examination of the memory allocation using malloc/calloc shows that this mechanism introduced a significant synchronization overhead (Fig. 4); approximately 95% of the time spend for memory allocation is used only for synchronization. While synchronization is scaling on the SGI Challenge down to

| Dataset/Size | Total number of cells in octree | Contributing cells | #triangles |
|---|---|---|---|
| A: Cavity dataset<br>191x191x191 | 6,968K<br>100% | 2,204K<br>6.6% | 4,960K |
| B: Ventricle system<br>258x258x212 | 14,112K<br>100% | 1,781K<br>12.6% | 1,523K |

**Table 1.** Datasets

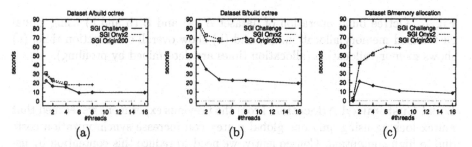

(a)                                    (b)                                    (c)

**Fig. 4.** Octree build operation of dataset A (a) and B (b and c) using standard memory allocation. (a) and (b) show overall construction time, (c) shows memory allocation (allocation times are determined by profiling).

a the constant overhead, memory allocation on the SGI NUMA-architecture machines (Origin200 and Onyx2) deteriorates severely.

**Global Customized Allocation** The previous experiment showed that the standard thread-save memory allocation functions introduced an expensive memory-locking mechanism. However, the mutexes used in the experiments suggested that the standard mutex locking mechanism is a faster, and therefore cheaper synchronization mechanism. Consequently, we introduced an alternative memory allocation method. We pre-allocate a huge chunk of memory before entering the parallel region of our code. Later, we assign blocks of this memory to the octants using a customized data-structure, similar to an array of octant arrays. The actual assigning action is protected by a data-structure local mutex. However, only one structure is used for the whole construction process.

Using this method, we obtained good scaling on the SGI Origin200 architecture (Fig. 5). Memory synchronization in particular scaled down to a fraction of the original amount. The SGI Onyx2 architecture showed a different picture. While the four CPU Origin200 only needs one additional crossbar hop to the CPUs on the other subsystem, access to all other CPUs of the Onyx2 requires up to two hops via the interconnection fabric, thus increasing the synchronization overhead. On the SGI Challenge, increasing memory requests of the threads increase the costs of synchronization. In contrast to memory locking using the standard library functions, global mutex locking does not scale. This probably results from increasing contention of the growing number of threads.

434     Dirk Bartz and Wolfgang Straßer

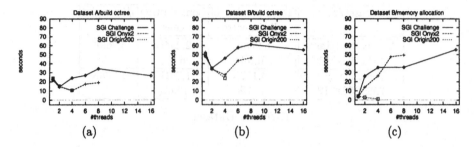

**Fig. 5.** Octree build operation of dataset A (a) and B (b and c) using global customized memory allocation. (a) and (b) show overall construction time, (c) shows memory allocation (allocation times are determined by profiling).

**Local Customized Allocation** From the previous experiment we learned that mutex locking using only one global mutex can increase synchronization costs due to high contention. Consequently, we need to reduce this contention by using multiple locks. Due to the facts that all systems are shared-memory systems and that memory access through the interconnect always scaled nicely despite the interconnection technology (bus or crossbar), this approach uses the previous pre-allocating data-structure for each thread. Therefore, a specific memory locking is not necessary.

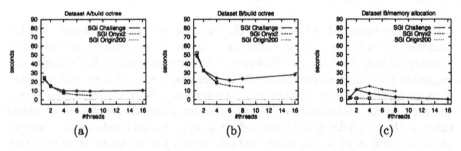

**Fig. 6.** Octree build operation of dataset A (a) and B (b and c)using local customized memory allocation. (a) and (b) show overall construction time, (c) shows memory allocation (allocation times are determined by profiling).

Figure 6 shows the results of this approach. Memory allocation time (including the synchronization overhead) could be reduced to a fraction of the previous amounts on all three systems. Furthermore, memory allocation scales throughout all CPUs, resulting in a balanced parallelization of the complete construction process.

# 5 Conclusion and Future Work

In this paper, we presented an algorithm for the parallel construction or recon-struction of recursive tree structures. Although this algorithm was discussed only for octrees, it is also suited to quadtrees, BSP-trees, or other recursive tree structures.

A main bottleneck of this algorithm is the global memory locking mecha-nism. Consequently, we presented three different approaches for solving this problem. The final approach using a local customized allocation scheme solved the problem on three different architectures and produced a scaling scheme for the construction of tree hierarchies.

However, while using an increasing number of threads, the job queue access might introduce a more significant overhead. Therefore, future work will focus on a distributed job queue.

Another focus for future work will be support for fast visibility queries in large dynamic polygonal datasets, which require a fast reconstruction of the hierarchy.

While most CFD datasets are based on a curvilinear grid, our current ap-plications are only based on rectilinear grids. However, for octrees, only the topology needs to be rectilinear, but not the geometry. Therefore, we can apply our scheme to curvilinear grids as well, which is another field of future work.

## Acknowledgments

Numerous people helped us preparing this paper. Especially, we would like to thank Roberto Grosso and Thomas Ertl of the Computer Graphics Group at the University of Erlangen-Nürnburg for the collaboration on previous work. We like to thank Arie Kaufman of the Center of Visual Computing at the SUNY Stony Brook for using the SGI Challenge at Stony Brook, Stephan Braun and Heinrich Bülthoff for support and using of the SGI Onyx2 at the Max-Planck-Institute for Biological Cybernetics. The cavity dataset is courtesy of the Institute for Fluid Dynamics of the University of Erlangen-Nürnberg. Last but not least, we thank Michael Meißner and Michael Doggett for proof-reading.

This work was supported by the MedWis program of the German Federal Ministry of Education, Science, Research, and Technology.

## References

[1] D. Bartz, R. Grosso, T. Ertl, and W. Straßer. Parallel construction and isosurface extraction of recursive tree structures. In *Proc. of WSCG'98*, volume III, 1998.
[2] D. Butenhof. *Programming with POSIX Threads*. Addison-Wesley, Reading, 1997.

**Fig. 7.** Vortex breakdown of a fluid which is injected in another fluid.

[3] H. Fuchs, Z. Kedem, and B. Naylor. On visible surface generation by a priori tree structures. In *Proc. of ACM SIGGRAPH*, pages 124–133, 1980.

[4] N. Greene, M. Kass, and G. Miller. Hierarchical Z-buffer visibility. In *Proc. of ACM SIGGRAPH*, pages 231–238, 1993.

[5] R. Grosso, T. Ertl, and R. Klier. A load-balancing scheme for parallelizing hierarchical splatting on a MPP system with non-uniform memory access architecture. In *Proc. of High Performance Computing for Computer Graphics and Visualization*, pages 125–134, 1995.

[6] P. Hanrahan, D. Salzman, and L. Aupperle. A rapid hierarchical radiosity algorithm. In *Proc. of ACM SIGGRAPH'93*, pages 197–206, 1993.

[7] Silicon Graphics Inc. Power Challenge. Technical report, Silicon Graphics Inc., Mountain View, 1994.

[8] Silicon Graphics Inc. Onyx2 Reality and Onyx2 InfiniteReality. Technical report, Silicon Graphics Inc., Mountain View, 1997.

[9] J. Laudon and D. Lenoski. System overview of the SGI Origin 200/2000 product line. Technical report, Silicon Graphics Inc., Mountain View, 1997.

[10] D. Laur and P. Hanrahan. Hierarchical splatting: A progressive refinement algorithm for volume rendering. In *Proc. of ACM SIGGRAPH'91*, pages 285–288, 1991.

[11] M. Levoy. Efficient ray tracing of volume data. *ACM Transactions on Graphics*, 9(3):245–261, 1990.

[12] W. Lorensen and H. Cline. Marching Cubes: A high resolution 3D surface construction algorithm. In *Proc. of ACM SIGGRAPH*, pages 163–169, 1987.

[13] H. Samet. *The Design and Analysis of Spatial Data Structures*. Addison-Wesley, Reading, 1994.

[14] R. Shekkar, W. Fayyad, R. Yagel, and J. Frederick. Octree-based decimation of marching cubes surface. In *Proc. of IEEE Visualization*, pages 287–294, 1996.

[15] O. Sudarsky and C. Gotsman. Output-sensitive visibility algorithms for dynamic scenes with applications to virtual reality. In *Proc. of Eurographics'96*, pages 249–258, 1996.

[16] J. Wilhelms and A. van Geldern. Octrees for faster isosurface generation. *ACM Transaction on Graphics*, 11(3):201–227, 1992.

# The Locality Property in Topological Irregular Graph Hierarchies*

Helmut Kofler, Ernst J. Haunschmid, Wilfried N. Gansterer, and
Christoph W. Ueberhuber

Institute for Applied and Numerical Mathematics,
Technical University of Vienna,
Wiedner Hauptstraße 8-10/115, A-1040 Vienna, Austria
{hkofler,ernst,ganst}@aurora.tuwien.ac.at,
christof@uranus.tuwien.ac.a

**Abstract.** Graph contraction is applied in many areas of computer science, for instance, as a subprocess in parallel graph partitioning. Parallel graph partitioning is usually implemented as a poly-algorithm intended to speed up the solving of systems of linear equations. Image analysis is another field of application for graph contraction. There regular and irregular image hierarchies are built by coarsening images.
In this paper a general structure of (multilevel) graph contraction is given. The graphs of these coarsening processes are given a topological structure which allows to use concepts like the neighborhood, the interior and the boundary of sets in a well-defined manner. It is shown in this paper that the various coarsenings used in practice are continuous and therefore local processes. This fact enables the efficient parallelization of these algorithms. This paper also demonstrates that the efficient parallel implementations which already exist for multilevel partitioning algorithms can easily be applied to general image hierarchies.

## 1 Introduction

*Graph contraction (graph coarsening)* is a method used in various fields of scientific computing to reduce the complexity of a graph structure while preserving important information. Graph contraction occurs, for instance, in *parallel graph partitioning* algorithms, which are used for the efficient solution of sparse linear systems (Karypis, Kumar [6]). In image processing graph contraction is used to construct *image hierarchies* (Kropatsch [12]). Instead of geometric operations on images (like Euclidean and affine operations or projections) topological operations can be used. These operations which maintain the neighborhood structures of elements of matrices or irregular graphs (like pixels or objects in digital images) preserve their topological structures and therefore enable the development of efficient algorithms and storage schemes.

* This work was supported by the Austrian Science Fund (*Österreichischer Fonds zur Förderung der wissenschaftlichen Forschung*).

P. Zinterhof, M. Vajteršic, A. Uhl (Eds.): ACPC'99, LNCS 1557, pp. 437–449, 1999.

In Section 2 a general coarsening process is formulated. Instead of using the entire information of a graph $G_o = G(V, E)$, a coarser graph $G_n$ is constructed which represents some global structure of the original graph $G_o$. The regular contraction of a grid, i.e., the merging of vertices in a locally uniform way, preserves a regular graph structure. Contractions of heterogeneous type are irregular contractions (see Figs. 1 and 2). For a particular application it has to be decided how to choose the contraction condition of each coarsening step.

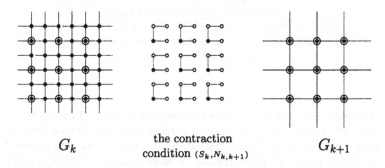

$$G_k \qquad \text{the contraction} \qquad G_{k+1}$$
$$\text{condition } (S_k, N_{k,k+1})$$

**Fig. 1.** A regular contraction $C = C_{k,k+1}$ of a regular structure.

In Section 3 the hierarchy of reduced irregular graphs $G_o, G_1, \ldots, G_n$, which is the result of a successive coarsening process, is given a topological neighborhood structure, i.e., the neighborhood $U(v)$ of each vertex $v \in V(G)$ is defined explicitly.

A *topological graph structure* $(G, \tau_G)$, called $\tau_G$ graph, is constructed in Section 3.2. This graph structure is compatible with the concept of adjacency and, moreover, it is the optimal discrete counterpart to a *Hausdorff space* (the Euclidean topology on the continuum), a fact, which can be demonstrated by using a special convergence process (Kofler [9]).

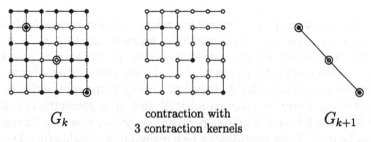

$$G_k \qquad \text{contraction with} \qquad G_{k+1}$$
$$\text{3 contraction kernels}$$

**Fig. 2.** An irregular contraction $C_{k,k+1}$ on a regular structure.

The topological graph structure enables to use concepts like neighborhood, interior and boundary of sets, and connectedness in a well-defined setting. In Section 3.3 the general contraction process for irregular graphs is given a topological structure. It is shown why the coarsening method which is applied to graph partitioning algorithms and to image hierarchies is a continuous and therefore a local process.

Continuity of *graph contraction*, which implies that the respective algorithm has a *locality of reference property*, can be strictly formulated only in an underlying discrete topological structure. Its locality of reference property is the reason why the coarsening algorithm can be efficiently parallelized. Furthermore, general basic topological considerations are helpful, for instance, in deciding whether it is better to use 4-grid or 8-grid structures. These considerations are sketched in a survey about topologies on graph structures in Section 3.1.

In Section 4 the general result about coarsening, i.e., the preservation of its topological structure, is applied to graph partitioning algorithms and to image hierarchies. It is shown that existing *matrix multilevel partitioning algorithms* can be slightly modified to make them applicable to *image hierarchies*. This means that *dual graph contraction* can be implemented using the efficient parallel algorithms implemented in METIS [6], PARTY [14], WGPP [4] and others.

## 2   The General Graph Coarsening Process

A graph coarsening process constitutes a hierarchical structure of irregular graphs. Regular contractions of an *digital image* are named *image pyramids* (see Fig. 1). In addition to a graph contraction also the *dual graph* and its contraction can be discovered, the term for both together is the *dual graph contraction*.

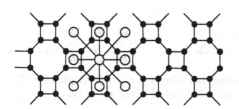 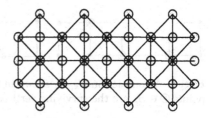

Fig. 3. Example: The dual graph of a 3-adjacency graph structure.

**Graph Coarsening.** Let $G_0(V_0, E_0) := G(V, E)$ be the input graph, for instance, an $n \times m$ - matrix, an $n$-grid, a regular graph or an irregular graph. A *graph contraction process* resp. *graph coarsening process* with $N - 1$ steps is defined by the contractions

$$C_{k,k+1} : G_k(V_k, E_k) \to G_{k+1}(V_{k+1}, E_{k+1}), \qquad k = 0, \ldots, N-1, \quad N \in \mathbb{N},$$

where the *decimation parameters* $(S_k, N_{k,k+1})$ (i.e., the contraction condition) give the conditions of the contraction operator $C_{k,k+1}$. The set $S_k = V_{k+1} \subseteq V_k$ specifies the *surviving vertices* while the set $N_{k,k+1} \subseteq E_k$ defines the edges which are contracted. A specification of the *contraction condition* is given below.

**Dual Graph Contraction.** The *dual graph contraction* consists of the contraction of a planar *input graph* and of the *dual* of the input graph $\overline{G}_o(\overline{V}_o, \overline{E}_o)$ (for instance, see Fig. 3). The contraction of the dual graph is denoted by $\overline{C}_{k,k+1}$ : $\overline{G}_k \rightarrow \overline{G}_{k+1}$. The hierarchical structure of irregular graphs $G_k(V_k, E_k)$ and $\overline{G}_k(\overline{V}_k, \overline{E}_k)$, $k = 0, \ldots, N$, forms two *irregular pyramids*. The whole structure is called the *dual graph contraction* (Kropatsch [12]).

**Composition of Contractions.** The contraction $C_{m,n}$ denotes a composition of contractions from level $m$ to level $n$:

$$C_{m,n} := C_{m,m+1} \circ C_{m+1,m+2} \circ \cdots \circ C_{n-1,n}, \quad m, n+1 \in \{0, 1, \ldots, N-1\} .$$

**Contraction Condition.** The contraction condition $(S_k, N_{k,k+1})$ of the contraction $C_{k,k+1}$ is specified. The graph $G_k$ is partitioned into $|S_k|$ path-connected sub-graphs. Each sub-graph, called *contraction kernel*, has exactly one surviving vertex in his structure. The set $N_{k,k+1}$ definitely adjoins the number of *non-surviving vertices* to each surviving vertex and decides how the vertices $V_k - S_k$ are merged with the surviving vertices $S_k$. An example of a *regular contraction* and an *irregular contraction* is shown in Figs. 1 and 2.

The contraction of vertices automatically builds loops and multiple edges (double edges). In *image hierarchies* these special types of edges are omitted if they appear (see Fig. 4). In *graph partitioning algorithms* multiple edges are merged into one and counted by weights.

**Fig. 4.** A contraction $C_{k,k+1}$ shown in 2 steps, handling a *loop* and a *double edge*. The points $\circledcirc \in S_k$ are the surviving vertices, the bold marked edges are in $N_{k,k+1}$.

Applications of the contraction process in *multilevel graph partitioning processes* and general *image pyramids* are discussed in Section 4.

# 3 The Topological Structure of Graphs and Coarsening Processes

## 3.1 A Short Survey of Topologies on Graph Structures

A topological structure is induced by the following canonical (adjacency) neighborhood definition. Two vertices $v, w \in V(G)$ are *neighbors* if $(v, w) \in E(G)$. Therefore the neighborhood $U(v)$ of a vertex $v \in V(G)$ consists of all vertices which are adjacent to $v$. The vertex $v$ is *per definitionem* element of the set $U(v)$ itself, which means that the vertex $v$ can be interpreted as a neighbor of itself.

Due to the lack of the intersection property, in this case the topological structure produces no topology, except the *discrete topology*. Every vertex $v \in V(G)$ forms a neighborhood $U(v) = \{v\}$ itself. The *discrete topology* on the set $V(G)$ is the result. But the *discrete topology* is not supportive to describe topological entities like connectedness—nothing is connected in this case—or convergence—every sequence is convergent. The above interpretation of neighborhood is not sufficient, for instance, in the case of *directed graphs*, where the neighborhood property is not symmetric. A vertex $v$ is a neighbor of $w$ but $w$ is not a neighbor of $v$. Therefore a comprehensive theory helps to find suitable topological characterizations to overcome the appearing problems.

There is no homogeneous topology on an $n \times m$ matrix resp. $\mathbf{Z}^2$, except the discrete topology, which follows from the following result: *any local finite $T_1$-topology is discrete*. But there exist non-homogeneous $(T_0 \, minus \, T_1)$ topologies like *cellular models* (Alexandroff, Hopf [2], Khalimsky, Kopperman, Meyer [8], Kovalevsky [11]), the *star-topology* (Ahronovitz, Aubert, Fiorio [1]) or the n-dimensional *Marcus-Wyse topology* (Wyse, Marcus [17]). The topological space $(\mathbf{Z}^2, \tau_{MW})$ of Marcus-Wyse is not homogeneous, but the property that *path-connectedness* is equivalent to *topological connectedness* holds in this environment. This is not true for 6 or 8-adjacency (Ptak, Kofler, Kropatsch [15], Wyse, Marcus [17]). Non-homogeneous topologies of the Marcus-Wyse kind on irregular graphs, discussed in Kofler [10], can be based on *irregular graph contractions* as well.

In this case a topological graph structure is preferred $(G, \tau_4)$ which is near the *semi-topological approach* (Latecki [13]), which fulfills the homogeneity property and the topological connectedness property is equivalent to path-connectedness of the underlying adjacency structure. Therefore it is compatible to the concept of adjacency and at the same time it is very close to the *Hausdorff space*, i.e., the Euclidean topology on the continuum, which is demonstrated in a convergence process in Kofler [10].

## 3.2  The Topological Structure $\tau_4$ on $\mathbf{Z}^2$

The basic *structure elements* of the topological graph structure $(G, \tau_G)$ are the neighborhoods $U(v) := \{w \in V; (v, w) \in E(G)\} \cup \{v\}$ of a vertex $v \in V(G)$ called *star sets*. This neighborhood structure of the open sets induced by *star sets* is distributed into classes using an equivalence relation $\sim_G$.

The structure is defined on the *regular n-grid* in the plane with $n = 4$, i.e., $\mathbf{Z}^2$ with the 4-adjacency condition. Therefore the *star set* $U_4(x)$ of each point $x \in \mathbf{Z}^2$ looks like Fig. 5 (i).

**The Topological Structure $(\mathbf{Z}^2, \tau_4)$.** The topological neighborhood structure $\tau_4$ on $\mathbf{Z}^2$ is defined by an *equivalence relation* $\sim_4$, $\tau_4 := \mathcal{P}(\mathbf{Z}^2)/\sim_4$. Two elements $O_1, O_2$ are in the same equivalence class, this means $O_1 \sim_4 O_2$, if there exists a set $O_r = \cup\{U_4(x); \text{ for some } x \in \mathbf{Z}^2\}$, $O_r \subseteq O_1, O_2$, where $O_1 - O_r$ and $O_2 - O_r$ have no neighborhoods $U_4(.)$ inside. The elements of the class $O \in \tau_4$ and its representative $O_r$ can be identified and are called open sets.

442     Helmut Kofler et al.

The single elements $\{x\}$, $x \in \mathbb{Z}^2$, are called *open points*. *Single open points* are not neighborhoods of themselves. Therefore, an intersection of any sets which result in a set of single *open points*, i.e., there is no neighborhood structure $U_4(.)$ in this intersection, will be identified with the empty set $\emptyset$.

The set $U_4(x)$ forms a *neighborhood base* of each $x \in \mathbb{Z}^2$, called the *structure element*. The collection of this neighborhoods $U_4(x)$, $x \in \mathbb{Z}^2$ constitute the *base of the topological structure*.

With this construction on the open sets the 4-neighborhood adjacency structure is placed into the foreground and the structure $(\mathbb{Z}^2, \tau_4)$ looks like a topological space:

(1) The empty set $\emptyset$ and the whole set $\mathbb{Z}^2$ are in $\tau_4$.
(2) The union of any family of $\tau_4$ is again in the set $\tau_4$.
(3) The finite intersection property holds, regarding the characterization above:
    If $U(x), U(y) \in \tau$ are two 4-neighborhoods where $y$ is a neighbor of $x$, i.e., $y \in U(x)$, then $U(x) \cap U(y) \stackrel{\sim 4}{=} \emptyset$.

**Properties.** The topological structure $(\mathbb{Z}^2, \tau_4)$ has a $T_2$ structure[1] and the *symmetry property* holds: if $y \in U_4(x)$ then there exists a neighborhood of $y$ such that $x \in U_4(y)$.

The *cardinality* of *the neighborhood base* $\chi$ of any point $x \in \mathbb{Z}^2$ is $\chi(x) = 1$. Therefore the *character* of the space is $\chi(\mathbb{Z}^2) := \sup\{\chi(x), x \in \mathbb{Z}^2\} = 1$. Compared with that situation the character of the continuum with the topology induced by the 1-norm is $\chi(\mathbb{R}^2) = \infty$.

The topological structure $(\mathbb{Z}^2, \tau_4)$ with these structure elements $U_4(x)$ fulfills the *homogeneity property*. A couple of single open points are like finite sets in measure theory with zero-measure. They exist, but are not important for the neighborhood structure of the space.

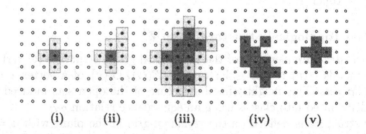

(i)        (ii)        (iii)        (iv)        (v)

**Fig. 5.** Types of open resp. closed sets in $\mathbb{Z}^2$: (i) the structure element of the topological structure, the interior is marked in dark, (ii) an open set identical to the first set respective the relation $\sim_4$, (iii) a typical open set, (iv) a closed set with an isolated point, two interior points, (v) the closure of the structure element.

---

[1] The topological characterizations like $T_2$-space, *relative topology*, *base of a topology* and others are used synonymously in this wider topological content.

**Closed Sets.** A *closed set* is the complement of an open set. The sets $\emptyset$, $\mathbb{Z}^2$ are open and closed simultaneously by definition.

The *open sets* are formally constructed to give the connection to the classical theory, but in the digital case it is better to work with *closed sets*. With a complement argument every closed set looks like an open set. These two types of sets can be distinguished by using a measure (*see below*). Moreover, the whole axiomatic formulation of the topological structure can be given by the closed sets instead of open in a dualistic manner.

**Interior and Boundary of a Set, Isolated Points, and Erosion Points.**
The *interior* or *kernel* of an arbitrary set $D \subseteq \mathbb{Z}^2$ (see Fig. 5 (iii)) is defined as

$$interior(D) := \{x, U_4(x) \subseteq D\} \ .$$

The $interior(D)$ is a subset of the *open interior*, called $interior_O(D)$ which describes the largest union of open subsets in $D$. The *closure* of the set $D$ is the smallest closed set $C \subseteq \mathbb{Z}^2$ with $D \subseteq C$. The *boundary* or *frontier* of the set $D$ is the closed set $boundary(D) := closure(D) - interior(D)$. Therefore

$$interior(D) \subset interior_O(D) \subseteq D \subseteq closure(D)$$

for $D \subseteq \mathbb{Z}^2$. Other equivalent characterizations can be substituted with these definitions, for instance, the following lemma characterizes the boundary as well: If $x \in boundary(D)$ then in its neighborhood $U_4(x) - \{x\}$ there exist elements $x_1 \in interior(D)$ and $x_2 \in complement(closure(D))$.

An *isolated* point $x \in D$ has an open neighborhood $(U_4(x) - \{x\}) \cap D = \emptyset$. The boundary of a set $D$ often enclose *erosion points*. These points are expressed by taking $D$ minus the closure of the $interior_O$ of the set $D$ (see Fig. 5 (iv)). A set $D$ is called *regular*, if $D = closure(interior_O(D))$. The *morphological operators* of *erosion* and *dilation* can be introduced easily by $\tau_4$ at this place, *opening* and *closing* respectively (Eckhardt, Hundt [3], Heijmans [5]).

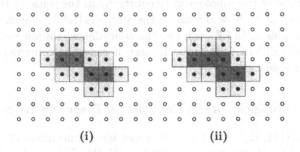

(i)                          (ii)

**Fig. 6.** (i) A non-connected and (ii) a connected open set, the interior points are marked in dark.

**Connected Sets.** A closed set $D \subseteq \mathbb{Z}^2$ is *connected*, if there is no partition of $D$ into two disjoint sets $O_{rel1}, O_{rel2}$, such that $O_{rel1} \cup O_{rel2} = D$, for $O_{reli} := D \cap interior(O_i)$, $O_i \in \tau_4$ (very similar to relatively open sets). In the case a set $D$ is closed path-connectedness is equivalent to topological connectedness, which is an immediate result of the investigations of Latecki [13]. An open set is connected, if it's closed *interior* is connected. This definition of open sets is described in Fig. 6.

**Description of Open Sets by a Measure to Distinguish between Open and Closed.** Open neighborhoods $U_4(x)$ consist of a *nucleus* which is the crude point itself and the four points, the *boundary*, forming the neighborhood structure. The *boundary* of an open set can be measured by $m(x) = 1/2^N$, $N \in \mathbb{N}$ fix, and the *interior* by $m(x) = 1$. The closed set $D$ has measure $m(x) = 1$ for all $x \in D$. The measure $m(x)$ is zero if $x$ is not element of the set. The measure is represented by different grey levels in Fig. 5. The intersection of two sets $D_1, D_2$ is defined in *fuzzy set* manner:

$$D_1 \cap_m D_2 := \{(x, m(x)); x \in \mathbb{Z}^2, m(x) = \min\{m_{D_1}(x), m_{D_2}(x)\}\} .$$

The union of two sets $D_1 \cup D_2$ is defined by the max-*function*. If the grid width decreases, i.e., $N \to \infty$, the measure of the boundary points becomes smaller. If the limit of the measure of the boundary points of open sets is zero the situation corresponds to the real case (Kofler [9]).

### 3.3  The Topological Graph Structure $(G, \tau_G)$

By analogy with the topological structure on a 4-grid this structure can easily be given to all regular graphs ($n$-grids) as well as to *irregular graphs* $G$ using an embedding of a general graph into a 4-grid and taking the *relative topology* (Kofler [9]).

A general graph $G$ can be directly equipped with $\tau_G$, which is called $\tau_G$ graph. The *star neighborhood sets* must be constructed by the adjacency condition of the graph $G$. This basic set of open neighborhoods of each vertex $v \in V(G)$ constitutes the base of the topological structure $\tau_G$ in the sense of the topological characterization of $\tau_4$. Now the size of the *star sets* locally differs, but the properties are the same as in $(\mathbb{Z}^2, \tau_4)$, i.e., the equivalence relation $\sim_G$ now depends on these locally different *star sets*. To keep it easy, the construction was given in the regular case.

### 3.4  The Topological Graph Contraction Preserves Its Structure

The graphs $G_k(V_k, E_k)$, $k = 0, \ldots, N$, given by a contraction can be equipped with the topological structure $\tau_G$ named as $(G_k(V_k, E_k), \tau_k)$, $k = 0, \ldots, N$. The same procedure can be applied to dual graphs $\overline{G}_k(\overline{V}_k, \overline{E}_k)$, $k = 0, \ldots, N$.
**Theorem.** The contraction $C_{k,k+1}$, $k \in \{0, \ldots, N-1\}$, $N \in \mathbb{N}$,

$$C_{k,k+1} : (G_k(V_k, E_k), \tau_k) \to (G_{k+1}(V_{k+1}, E_{k+1}), \tau_{k+1})$$

(resp. its dual contraction $\overline{C}_{k,k+1}$) is a *surjective* mapping. The mapping is also a *continuous* and *open* function with respect to the topological structure $\tau_G$.

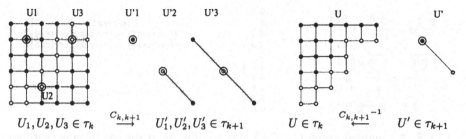

$$U_1, U_2, U_3 \in \tau_k \xrightarrow{C_{k,k+1}} U_1', U_2', U_3' \in \tau_{k+1} \qquad U \in \tau_k \xleftarrow{C_{k,k+1}^{-1}} U' \in \tau_{k+1}$$

**Fig. 7.** Examples of $C_{k,k+1}(U)$ and $C_{k,k+1}^{-1}(U')$ of basic *open sets* $U, U'$.

**Structure of the Proof.** Let $O \in \tau_k$ and $O' \in \tau_{k+1}$ be open sets. *Continuous* means: for any $O' \in \tau_{k+1}$ there exists a set $O \in \tau_k$ that the *inverse image* $\{v \in G_k; C_{k,k+1}(v) \in O'\} = O$. *Open* means: for any $O \in \tau_k$ there exists a set $O' \in \tau_{k+1}$ that $C_{k,k+1}(O) = O'$. I.e., the neighborhoods in each scale have their parts (resp. the *empty set* $\emptyset \in \tau_k, \tau_{k+1}$) in higher or lower scales. Open sets are unions of *structure elements* resp. the equivalence relation $\sim_G$. Therefore the properties *continuous* and *open* are demonstrated by the structure elements.

The idea of the proof is given graphically in Fig. 7 using the contraction example of Fig. 2 with the structure elements $U_4(.)$. The set $C_{k,k+1}(U_1) = U_1'$ consists of one open single point which is equal to $\emptyset$ resp. $\sim_4$, as well as the set $U_2'$. The set $U_3'$ consists of 3 nodes, which form a complete neighborhood in $\tau_{k+1}$. The set $U'$ is a basic neighborhood of the black node and its *inverse image* leads to the set $C_{k,k+1}^{-1}(U') = U$ which is an open set in $\tau_k$.

Investigating all different cases give the result: regular and irregular contractions (in particular image hierarchies) preserve their topological structure.

**Remark.** A graph contraction $C_{k,k+1}$ is not an injective function, therefore the mapping cannot be homeomorph either. But the special condition of the contraction kernels $(S_k, N_{k,k+1})$ guarantees continuity. If the contraction kernels are not path-connected subgraphs of $G_k$, which means contractions on different locations occur, the contracted structure $G_{k+1}$ would be topologically different to $G_k$. A simple *counterexample* can verify the importance of the special contraction condition, which forces the contraction process to be a "local process".

## 4    Parallel Coarsening of Matrices and Irregular Graphs

In *graph partitioning algorithms* to solve systems of linear equations, the basic graph $G(V, E)$ represents a sparse symmetric matrix by interpreting it as an adjacency matrix. A digital image with pixel structure is completed to a graph structure, mostly 4- or 8-adjacency. In general, the given graph is represented in an *adjacency list data structure*. In topological terms this means that the graph is stored in star-neighborhoods $U(v)$, $v \in V$ (Karypis, Kumar [6]). (Another often used format is the *compressed row storage format* CRS.) The graph may also be provided with weights $w(v)$ and $w((v_1, v_2))$, for each vertex $v \in V$, and for each edge $(v_1, v_2) \in E$ (see Fig. 8).

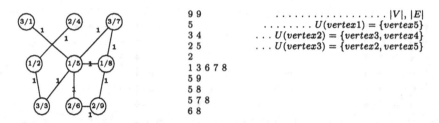

Fig. 8. A weighted graph $G_0(V, E)$ and an often used storage scheme (without weights). $w(v)/i$ denotes the weight $w$ of each vertex $v \in V$ and $i$ gives its number.

## 4.1    The Contraction Kernels in Parallel Partitioning Algorithms

The coarsening for irregular graphs in *parallel partitioning algorithms* generally uses a contraction kernel named *heavy-edge matching* (HEM) (Karypis, Kumar [7]): In a graph contraction step the set $S \subseteq V$ of *surviving vertices* is gathered in random order, with the condition that only pairs of adjacent vertices are merged, i.e., *random matching* (RM). Further, in HEM the vertex of an edge with the highest possible weight (if the vertex is not used by another merging) in the star neighborhood $U(v)$, $v \in S$, is merged. A contraction step is named *maximal matching* if no pairs are left to contract (see Fig. 9). In the package WGPP (Gupta [4]) *heavy-triangle matching* (HTM) *three vertices* in one star set are merged together. *Heaviest-edge matching* searches the heaviest edges in the graph for merging the incident vertices one by one. There exist slightly modified versions to improve the contraction types, and one new version—the HLM—is proposed below. Often a combination of various methods succeeds.

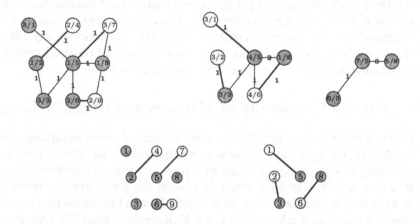

Fig. 9. An example of a contraction $C_{0,2} = C_{0,1} \circ C_{1,2}$ of the graph $G_0$ in 2 steps with *maximal* RM in the *first row*. The contraction condition of each *random matching* is explicitly represented in the *second row*.

In general the notion of neighborhood is interpreted as "nearness". But neighborhoods can also depend on non-spatial features like grey levels or color, stored by weights. The *weight-function* $w : V(G) \to \mathbb{R}^n$, $n \in \mathbb{N}$, indicates an *intensity level* at any vertex $v \in V$. Neighborhoods can be defined using this intensity

$$U_w(v) := \{w^{-1}(r); \|w(v) - r\| \leq c, r \in \mathbb{R}^n\} \cap U(v), \quad c \in \mathbb{R}.$$

**Heavy-Leveled-Edge-Weight Matching (HLM).** Instead of merging 2 or 3 vertices (HEM, HTM), all vertices of a star set $u \in U(v)$, $v \in V$, are taken if the weight $w(u)$ is larger than a given threshold $l \in \mathbb{N}$. This means that the topological neighborhood sets of the graph are defined as $U_l(v) := \{w^{-1}(r); r \geq l, r \in \mathbb{N}\} \cap U(v)$ with level $l \in \mathbb{N}$. The threshold can be adapted in each contraction step during the algorithm.

## 4.2   Locality of Coarsening in Graph Partitioning Algorithms

It was shown in Section 3.4 that coarsening in general is a local process. All the coarsening methods in *graph partitioning algorithms* are identical with the conditions of a contraction process which is easily seen in Fig. 9. Two consecutive contractions with an RM kernel are shown which can be comprehended to one process (see Fig. 10). The weights of the vertices resp. edges count the non-surviving vertices resp. multiple edges. In the case of RM and HEM *multilevel contraction kernels* have the characteristics of paths with one surviving vertex.

For instance, the method of *minimum degree* (MD) of a *fill-in reducing ordering algorithm* for symmetric sparse matrices is no such coarsening process, but also a local process. It is easy to prove that it is a continuous process with respect to the topological structure $\tau_G$: Introducing a distance $d$, where $d(v_1, v_2) = 1$ if $v_1$ is adjacent to $v_2$, $v_1, v_2 \in V(G)$, it can be seen that vertices of the neighborhood $U_2(v) := \{w; d(v, w) \leq 2\}$, $v \in V$, are mapped into the star set $U_1(v) = \{w; d(v, w) \leq 1\} \cap U(v)$ of $v \in V$, i.e., the neighborhoods of vertices merge to smaller neighborhoods.

Using distributed memory parallel computers in parallel processing, the basic graph is subdivided to $l$ processors $i = 1, \ldots, l$. The multilevel $k$-way partitioning algorithm relies extensively on using a *coloring of vertices* by Luby's algorithm also in the coarsening stage (Karypis, Kumar [6]).

## 4.3   Parallel Algorithms in Image Hierarchies

Regular pyramids and irregular image hierarchies are multilevel contractions of images. The contraction $C_{k,k+1} : G_k \to G_{k+1}$ of an image fulfills the underlying contraction condition (Section 2) which forces the contraction process to preserve its topology in the regular and irregular case. Examples of *image contractions* can be found, for instance, in Kropatsch [12].

Image contraction is of the same type as coarsening in parallel partitioning algorithms. The contraction kernel of parallel partitioning generally is a *heavy-edge matching* HEM. Within image analysis various contractions can occur. The

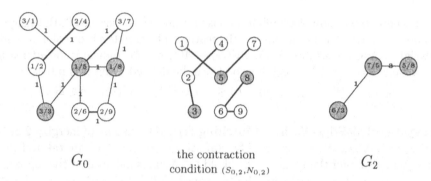

$$G_0 \qquad \text{the contraction} \atop \text{condition } (S_{0,2}, N_{0,2}) \qquad G_2$$

**Fig. 10.** The contraction $C_{0,2}$ of the graph $G_0$ in 1 step.

contraction, i.e., the segmentation, depends on the given context of an image. The type of the image is most important for the choice of the contraction-mapping. There is a need to find such types of contractions.

If contraction kernels can be used optionally in those packages, i.e., the contraction of a star set neighborhood or neighborhoods with *arbitrary conditions* are possible, the existing packages of parallel graph partitioning could be applied easily in images processing. This would be a great advantage for managing large data sets of images.

For instance, if a certain type of digital image is given, and the background of the image is set to zero, a sparse matrix is the result. In that case the packages of graph partition algorithms can be applied efficiently to pattern recognition tasks.

## Conclusion

In this paper a general contraction process of irregular graphs has been defined. Using topological techniques it has been demonstrated that this contraction process (with its special requirements on the *contraction kernels*) is a continuous and consequently a *local process*. Locality is an important property which enables parallel implementations.

Coarsening has been used to reduce the complexity of several parallel partitioning algorithms while preserving the quality of the solutions obtained. These multilevel methods are implemented in, for instance, in METIS, CHACO and WGPP. In this paper the most important contraction type, heavy-edge matching HEM, has been investigated and slightly modified.

Other applications of coarsening are regular pyramids and irregular image hierarchies. It has been shown that image hierarchies preserve their topological structure. Further, coarsening in parallel partitioning algorithms and in image contraction is of the same type. Therefore existing packages for efficient parallel partitioning can easily be applied to image pyramids for image segmentation.

# References

1. Ahronovitz, E., Aubert, J.-P., Fiorio, Chr.: The Star-Topology: A Topology for Image Analysis. 5th International Workshop, DGCI'95 (1995)
2. Alexandroff, P., Hopf, H.: Topologie, Erster Band. Springer-Verlag, Berlin (1935)
3. Eckhardt, U., Hundt, E.: Topological Approach to Mathematical Morphology. preprint (1997)
4. Gupta, A.: Fast and Effective Algorithms for Graph Partitioning and Sparse-Matrix Ordering. IBM, Journal of Research & Development (1997)
5. Heijmans, H.: Morphological Image Operators. Academic Press (1994)
6. Karypis, G., Kumar, V.: A Coarse-Grain Parallel Formulation of Multilevel k-Way Partitioning Algorithm. Proceedings of the 8th SIAM conf. on Parallel Processing for Scientific Computing (1997)
7. Karypis, G., Kumar, V.: METIS, a Software Package for Partitioning Unstructured Graphs, Partitioning Meshes, and Computing Fill-Reducing Orderings of Sparse Matrices, version 3.0.3. METIS, Minnesota (1997)
8. Khalimsky, E., Kopperman, R., Meyer, P.R.: Computer Graphics and Connected Topologies on Finite Ordered Sets. Topology Appl. 36 (1980)
9. Kofler, H.: Irregular Graph Hierarchies Equipped with a Topological Structure. 14th Intern. Conference ICPR'98, Brisbane, Australia (1998)
10. Kofler, H.: The Topological Consistence of Path Connectedness in Regular and Irregular Structures. 7th Intern. Workshop, SSPR'89, Sydney (1998)
11. Kovalevsky, V.A.: Finite Topology as Applied to Image Analysis. Computer Vision, Graphics and Image Processing 46 (1989)
12. Kropatsch, W.G.: Building Irregular Pyramids by Dual Graph Contraction. IEE Proceedings Vis. Image Signal Process., Vol. 142 (1995)
13. Latecki, L.: Digitale und Allgemeine Topologie in der Bildhaften Wissensrepräsentation. Ph.D.-Thesis, Hamburg (1992)
14. Preis, R., Diekmann, R.: The PARTY Partitioning-Library. User Guide-Version 1.1. Univ. Paderborn, Germany (1996)
15. Ptak, P., Kofler, H., Kropatsch, W.: Digital Topologies Revisited. 7th International Workshop, DGCI'97, Montpellier, France, Springer series (1997)
16. Ueberhuber, C.W.: Numerical Computation 1 and 2. Methods, Software, and Analysis. Springer-Verlag, Heidelberg (1997)
17. Wyse, F., Marcus, D. et al.: Solution to Problem 5712. Am. Math. Monthly 77 (1970)

# Geometry-Aided Rectilinear Partitioning of Unstructured Meshes

Rainer Koppler

Department of Computer Graphics and Parallel Processing (GUP)
Johannes Kepler University, A-4040 Linz, Austria/Europe
koppler@gup.uni-linz.ac.at

**Abstract.** This paper introduces an infrastructure for parallel mesh computations running on distributed-memory computers. The infrastructure consists of the mesh partitioning algorithm *GARP* and the domain-specific communication library *GRAPHlib*. Unlike existing algorithms, GARP exploits geometrical properties of the mesh shape in order to produce shape-adequate rectilinear partitions. The structure of such partitions is exploited by GRAPHlib using an optimized message ordering strategy. We describe the concepts behind GARP and GRAPHlib and show that for meshes with particular shapes our infrastructure provides better utilization of the parallel computer than solutions using existing partitioning algorithms and communication libraries.

## 1 Introduction

Unstructured meshes are used in a large number of scientific codes, for example, in computational fluid dynamics and structural analysis codes. Nowadays computations with large meshes are increasingly being performed on parallel computers because of size considerations and computational requirements. Mesh computations have been predominantly performed on distributed-memory computers because such systems offer advantages over their shared-memory counterparts concerning scalability.

With distributed-memory computers the user is faced with the non-trivial problem of *data distribution.* Meshes must be partitioned across the local memories of the processors in such a way that a good load balancing is achieved and interprocessor communication is kept minimal. With virtual-shared-memory machines data distribution is also relevant because a good distribution promotes data locality and so improves overall performance.

The quality of a mesh distribution depends on the suitability of a distribution algorithm for the given mesh, the problem to be solved, and the parallel computer. Thus, data distribution should not be a fully automatic process. Users can choose from a variety of algorithms, which show different weaknesses and strengths with repect to each of the above factors.

We observed that today's data distribution algorithms are too general with respect to the *mesh shape*, that is, they neglect geometrical shape properties in order to produce good distributions. In many cases fast algorithms could be

P. Zinterhof, M. Vajteršic, A. Uhl (Eds.): ACPC'99, LNCS 1557, pp. 450–459, 1999.
© Springer-Verlag Berlin Heidelberg 1999

employed in order to produce rectilinear partitions, which allow lossless mapping to various processor topologies and contention-free communication during local exchange operations. Furthermore, the advantages of rectilinear partitions are not exploited by today's communication libraries, for example, for optimization of exchange operations.

This paper introduces an infrastructure for parallel mesh computations on distributed-memory computers that use meshes with particular shapes. The fast data distribution algorithm *GARP* exploits symmetry and circle-shaped contours of the mesh shape in order to produce rectilinear partitions with good quality. The communication library *GRAPHlib* exploits the grid structure of rectilinear partitions using optimized message scheduling algorithms that minimize the number of exchanged messages and exclude network conflicts. We give an overview of GARP and GRAPHlib and report on first evaluations using a parallel mesh code, which we have run on the nCUBE 2 and the Intel iPSC/860.

# 2  Parallel Mesh Computations

Today there exists a variety of parallel schemes for the solution of problems on unstructured meshes, which can be classified into *explicit* and *implicit*, just as their sequential equivalents [4]. In explicit schemes mesh entities are visited in loops and entity values are updated with respect to values of adjacent entities. Implicit schemes operate on a global matrix formed from the mesh and solve a system of linear equations. Since implicit schemes require matrix distribution instead of mesh distribution, they are beyond the scope of this paper. Mesh distribution is also required for substructuring-based computations, which divide a problem into several subproblems and an interface problem.

The mesh distribution problem is usually divided into two NP-complete subproblems: *partitioning* and *mapping*. During partitioning the mesh is divided into several parts such that the parts are balanced and the boundaries between adjacent parts show minimal lengths. The result is called a partition, the respective graph is called *task interaction graph* (TIG) [12]. During mapping the partition is mapped to the parallel computer such that adjacent parts are assigned to adjacent processors in the interconnection network. Due to NP-completeness of partitioning and mapping, heuristic algorithms are employed, which usually imply a trade-off between the quality of a distribution and its computing time.

Partitioning algorithms can be classified into *graph-based* and *coordinate-based* algorithms. Graph-based algorithms consider the topology of mesh entities. Thus, they are generally applicable. Well-known examples are *Greedy* [3], *Recursive Graph Bisection* [13], and *Recursive Spectral Bisection* [11]. Coordinate-based algorithms can be used whenever coordinates of mesh entities are available. They are fast and can exploit geometrical properties of the mesh. Popular examples are *Orthogonal Recursive Bisection* [1] and *Recursive Inertial Bisection* (RIB) [10]. Today graph-based algorithms are used predominantly due to their generality. On the other hand, generality implies considerably longer run times than with coordinate-based algorithms and causes neglection of the mesh shape, which often leads to irregular TIGs.

Mapping of mesh partitions to parallel computers is relevant if the TIG of the partition is *irregular*. It to the problem of embedding the TIG into the computer's processor communication graph, which has been investigated intensively in the literature (see [2] for an overview). On the other hand, *regular* TIGs allow lossless embedding into popular topologies, for example, grids or hypercubes. Practical importance of mapping has been considered differently in the literature. Notable impact has been observed on computers with a large cost differential between nearest-neighbour communication and global communication, in particular, when many processors are used [5,15].

Besides mesh distribution, efficient implementation of communication is important with distributed-memory computers. Parallel mesh codes resolve data dependencies between submeshes using *local pairwise exchange*. In order to minimize communication overhead send and receive operations must be scheduled such that network conflicts are avoided. Since senders and receivers are defined by the mesh distribution, optimization of scheduling is difficult with irregular TIGs. Machine-dependent heuristics have been used for this purpose [14]. On the other hand, regular TIGs allow ordering of messages such that network conflicts are excluded and the number of required send and receive operations is minimized [12].

# 3    The Data Distribution Algorithm GARP

Partitions that exhibit regular TIGs are called *rectilinear* [9]. The previous discussion makes obvious that such partitions show several advantages over partitions showing irregular TIGs. Thus, a partitioning algorithm should primarily aim at rectilinear partitions, whenever it is appropriate, and then optimize known criteria such as load balance and minimum boundary lengths. Clearly, there are many cases where a rectilinear partition cannot be the optimal one, for example, if the mesh exhibits a highly inconvex shape or strongly varying vertex density. However, there are also many cases where the optimal partition of a mesh is a rectilinear one that fits the shape of the mesh. Throughout this paper we use the term *shape-adequate* for such partitions.

We observed that today's partitioning algorithms generally fail to produce shape-adequate partitions for particular meshes. Figure 1 shows partitions of two triangular meshes. The topology of triangles is highly irregular but the mesh shapes let rectilinear partitioning appear appropriate.

Figure 1b shows that the coordinate-based RIB algorithm produces rectilinear partitions even with a higher number of parts. On the contrary, graph-based algorithms cause loss of TIG regularity already with 16 processors. On the other hand, while RIB exploits the rectangular shape of the first mesh, it fails to produce a rectilinear partition with 32 parts for a circle-shaped mesh, as shown in Fig. 1c. If partitioning is based on polar coordinates instead of cartesian coordinates, the rectilinear partition shown in Fig. 1d can be obtained easily.

Partitioning of circular shapes based on polar coordinates was introduced by Chrisochoides [2]. We evaluated all four partitions on the nCUBE 2 using a parallel Laplace solver. It showed that the shape-adequate partitions perform slightly better than their irregular counterparts.

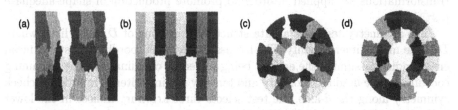

**Fig. 1.** Partitions: (a) graph-based, (c) RIB, (b,d) shape-adequate

Our experiments led to the conclusion that only coordinate-based partitioners provide a fast and easy way to producing shape-adequate rectilinear partitions. Furthermore, such a partitioner must the able to detect geometric characteristics of the mesh shape. Our algorithm GARP (Geometry-Aided Rectilinear Partitioner) is based on the previous considerations. During practical evaluation of a variety of mesh partitions we obtained two promising characteristics: *symmetry* and *contour shape*. Exploitation of symmetry shows advantages with small part numbers. Exploitation of contour shapes promotes shape adequacy especially with high part numbers.

Exploitation of symmetry is motivated by the fact that the bisection of a mesh along the axis of affinity yields perfect load balance if there are no mesh entities lying on this axis. Furthermore we observed that such a bisection often produces the minimum number of cut edges, i. e., the number of edges that connect mesh entities belonging to different parts. Symmetry can be exploited frequently because many real-world objects used in scientific computing exhibit symmetry, for example, car-bodies, airplanes, space shuttles, or ship hulls.

Exploitation of contour shapes has shown useful in Fig. 1. Since also a number of 3D meshes benefit from coordinate transformation, for example, cylinder liners (see Fig. 4ab) and sphere surfaces, we extended Chrisochoides' technique in order to apply transformation into cylindrical and spherical coordinates as well.

GARP employs fast heuristics in order to detect the above characteristics automatically. In the rest of this section we outline GARP's heuristic geometry tests and its rectilinear partitioning strategy.

### 3.1 Geometry Tests

Before partitioning, coordinates of the mesh entities are subjected to two geometry tests. The *symmetry test* checks if the mesh exhibits symmetry along one or more coordinate axes. The *contour test* attempts to classify the shape of the mesh

454    Rainer Koppler

contour. In three dimensions, GARP recognizes object shapes through combination of contour shapes that are detected in normal projections. For example, it assumes a cylinder if two projections exhibit the same rectangular contour and the third projection shows a circular contour. A sphere is assumed if all projections are circles. Depending on the detected shape appropriate coordinate transformations are applied in order to promote production of shape-adequate partitions.

The symmetry test uses a data structure consisting of $D$ index lists, where $D$ is the mesh dimensionality. The list associated with coordinate $d$ stores mesh entity indices assuming the entities being sorted by coordinate $d$. The remaining coordinates are used as secondary and tertiary sorting criteria. In order to check symmetry along the $d$-axis, the test selects three arbitrary indices in the lower half of the $d$-th list, which yields three entities $v_1$, $v_2$, and $v_3$. After this, it determines the three assumed mirror entities $w_1$, $w_2$, $w_3$ by mirroring the indices into the upper half of the list. Finally, the mid-points $m_i$ between each entity and its mirror are computed, as shown in Fig. 2a. Symmetry is assumed if the three mid-points are collinear. In three dimensions, the test selects four indices and checks if the four mid-points are coplanar.

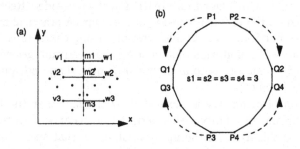

**Fig. 2.** (a) symmetry test, (b) contour test

The contour test is always applied to a two-dimensional domain. It receives two coordinate axes, where it treats the first one as the horizontal direction and the second one as the vertical direction.

The first phase, called *boundary tracking*, searches for boundary vertices. It starts at the top and at the bottom of the domain and proceeds towards the middle. The top-left, top-right, bottom-left, and bottom-right entity are chosen as starting points, which are labelled $P_1$, $P_2$, $P_3$, and $P_4$, respectively. The search directions for each starting point are depicted in Fig 2b. For example, the search starting at $P_1$ proceeds in the left and lower direction. Whenever a new leftmost entity $P$ is encountered, the distance between $P$ and the current leftmost entity, which is $P_1$ at start, contributes to the perimeter approximation $U$.

The second phase, called *shape classification*, evaluates $U$ and the number of search steps $s_i$ in order to detect a convex boundary shape. For example, the condition required for detecting a circle is $U \approx (x_{Q_2} - x_{Q_1}) \cdot \pi \wedge \forall i\; s_i > 0$.

Since both tests work with discretized data, their reliability depends on the granularity of the discretizations. Consequently, each test accepts an additional parameter $\epsilon$ that holds a test-specific tolerance value.

## 3.2 Partitioning Strategy

GARP divides a set of mesh entities into $N$ parts using $\log_2 N$ bisections along the coordinate axes. Thus, it requires the number of parts $N$ to be a power of two. By default, the resulting TIG is a grid showing the same dimensionality as the mesh.

For each mesh dimension an appropriate number of bisections must be determined. A dimension that has been bisected $n$ times exhibits $2^n$ sections with probably different widths. The next dimension to be bisected is the dimension with the largest section width. The cartesian product of the sections over all dimensions yields the partition.

By default, GARP gives symmetry precedence over section widths, such that the first bisections are performed along symmetric dimensions. For reasons stated above this approach has often shown profitable with low part numbers. However, this behaviour can be disabled as well.

After each bisection GARP checks the current partition because sole consideration of section extents may cause poor load balance, for example, with inconvex mesh shapes or high bisection numbers. If the partition has empty parts or its load balance falls beyond a user-specified value, the last bisection is undone and GARP proceeds with another dimension. Furthermore, the last bisection becomes locked such that no more bisections are performed along it. If all dimensions are locked before the required number of bisections has been reached, GARP considers the mesh inappropriate for rectilinear partitioning and aborts.

# 4   The Communication Library GRAPHlib

GRAPHlib is a MPI-based library for distributed-memory computers that supports development of parallel graph-oriented applications, for example, explicit mesh computations. It supports applications where the main loop sweeps over the edges of a graph. GRAPHlib hides the details of message passing from the programmer using *Gather* and *Scatter* abstractions [5].

Our main motivation for the development of GRAPHlib was the lack of a compact and simple testbed for the evaluation of mesh distribution using optimized and unoptimized message scheduling. In particular, GRAPHlib should fully exploit the advantages of rectilinear partitions discussed in section 2. Whenever a parallel application is run with a rectilinear mesh partition, the library performs lossless mapping of the regular TIG to the processor topology and exchange operations employ an optimized message ordering strategy. In all other cases partitions are directly mapped and messages are exchanged in arbitrary order.

Message ordering in GRAPHlib is based on a strategy developed by Sadayappan and Ercal [12]. The strategy exploits the presence of directions with rectilinear partitions, for example, the presence of left, right, lower, and upper neighbour processors. Figure 3 shows a submesh consisting of nine four-sided elements, where nodes enclosed by frames are shared with neighbour processors. The strategy orders messages such that network conflicts are excluded and direct data exchange with diagonal neighbours is avoided. For example, data to be exchanged with the upper right neighbour are first sent to the right neighbour, which forwards them to its upper neighbour.

Sadayappan and Ercal's strategy has also limitations. First, it does not perform "true" data exchange with diagonal neighbours but data accumulation using an associative operator. Secondly, it has been tailored to meshes consisting of four-sided elements and requires generalization in order to be applicable with unstructured meshes. Finally, the order of send and receive operations can be changed in order to achieve better utilization of the interconnection network. GRAPHlib's strategy overcomes all these limitations. Our optimized message ordering is shown in Fig. 3 on the right. Moreover, we extended the strategy for use with three-dimensional partitions, where communication with up to 26 neighbours is performed using just six communication steps.

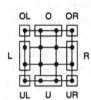

OL    O    OR

L                    R

UL    U    UR

1. Copy data for OL, L, and UL to buffers. Send OL, L, and UL to L.
   Copy data for OR, R, and UR to buffers. Send OR, R, and UR to R.

2. Receive OL, L, and UL from L. Consume L.
   Receive OR, R, and UR from R. Consume R.

3. Copy data for U to buffer U. Send UL, U, and UR to U.
   Copy data for O to buffer O. Send OL, O, and OR to O.

4. Receive UL, U, and UR from U. Consume UL, U, and UR.
   Receive OL, O, and OR from O. Consume OL, O, and OR.

**Fig. 3.** Optimized message ordering for 2D rectilinear partitions

## 5    Experiences

In our experiments we compared a number of partitions produced by GARP with corresponding results produced by RIB, which is the best coordinate-based partitioner available at the moment. We did not consider graph-based algorithms because we expected RIB to yield the best results for the used meshes.

We also evaluated the parallel gains of GARP and GRAPHlib using an Euler solver kernel, which is based on explicit first-order scheme [8]. For each partition we took the total execution time of the solver's node program after 100 iterations. Each iteration performs just a few computations and thus is dominated by communication.

The run times were taken on the nCUBE 2 and the iPSC/860. Since both computers are known to show a small cost differential between global and local communication, we did not expect considerable gains through the use of GARP and GRAPHlib. However, notable gains are to be expected on computers where global communication is expensive compared to nearest-neighbour communication. The numbers given in the sequel refer to the nCUBE 2. On the iPSC/860 we observed the same trends as with the nCUBE 2.

Figure 4ab shows partitions of a mesh that describes one half of a cylinder liner, which can be found in automotive industry simulations. GARP detects symmetry along the X and the Z axis, classifies the XY projection as a semiannulus, and performs transformation into cylindrical coordinates. GARP's partitions are optimal in terms of submesh surface-volume ratio and fit the mesh shape well. The results of GARP and RIB coincide with up to four parts, then RIB's partitions become slightly irregular. Using less than 16 processors GRAPHlib's optimized message ordering does not show effect whereas using 16 processors our infrastructure improves parallel run time by 3,5 percent.

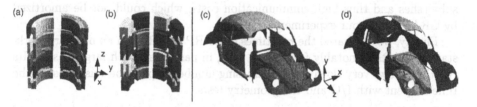

**Fig. 4.** *liner:* (a) RIB, (b) GARP. *beetle:* (c) RIB, (d) GARP

Figure 4cd shows another mesh that comes from the real world. It indicates that exploitation of symmetry with small part numbers is advantageous. GARP detects symmetry along the X axis. The initial bisection along X causes perfect load balance because there are no vertices lying on the axis of affinity. As mentioned in section 3.2, this requires precedence given to symmetry over section widths. Otherwise, the initial bisection along Z produces a suboptimal partition, which causes run time loss of 3,7 percent.

RIB does not exploit symmetry here and thus yields irregular partitions. Although they are slightly better optimized with respect to load balance and cut edges, GARP's results exhibit shorter run times with up to four processors and achieve a gain of 4,6 percent. With eight processors RIB's partition is better by 1,6 percent, due to the high variance of cut edges among the submeshes of GARP's partition.

It must be noted that all improvements, on which we have reported so far, were obtained for two-dimensional TIGs. In order to evaluate GRAPHlib with three-dimensional TIGs we used a volume mesh that describes a cubic solid and consists of 4096 vertices. Since finding the optimal partition of such a mesh is trivial, we could concentrate on the impact of mapping and message scheduling. It showed that with eight processors arbitrary mapping and scheduling is sufficient whereas with 64 processors GRAPHlib's optimizations improve parallel run time by 10 percent.

On the other hand, our experiments also revealed weaknesses of GARP. When a mesh shows a highly irregular shape or strongly varying vertex density, rectilinear partitions, primarily those with many parts, may suffer from critical load imbalance. Hence, some rectilinear partitioners are coupled with a load balancing heuristic [12]. The RIB partitions shown in this paper were also produced by an implementation that employs the Kernighan-Lin heuristic [7] for boundary optimization. We observed that GRAPHlib's communication optimizations can amortize bad partition quality up to a specific number of processors, otherwise RIB's partitions perform better.

Furthermore, rectilinear partitioning is not appropriate for meshes with offset extremities, for example, airplanes with wide wing spans. If more than one bisection is carried out along the wings, further bisections along the body may lead to partitions with empty parts, which are not accepted by GARP. Hence, all bisections except one are carried out along the body. This causes elongated submeshes and thus high communication costs, which could not be amortized by GRAPHlib in our experiments.

Finally, we compared the run times of GARP and RIB on an ordinary workstation. GARP is notably faster than RIB, in particular, with large meshes. Its run time grows very slowly with increasing number of parts because most of the time is spent with I/O and the geometry tests.

## 6   Related Work

Rectilinear partitioners make up a small part in the mass of existing partitioning algorithms. Development was primarily motivated by the impact of communication with some parallel computers. Moreover, rectilinear partitions have been regarded due to simplified data organization with grid-shaped meshes.

Sadayappan and Ercal developed a graph-based approach named *1D/2D Strip Partitioning* [12]. Their algorithm groups mesh vertices into parts using a technique called *levelization*. Hinz introduced an iterative coordinate-based partitioner based on so-called *N4 nets* [6]. His algorithm divides a continuous two-dimensional data space into rectangular regions and distorts the regions to general quadrilaterals in order to achieve good load balance. Nicol gave a comprehensive treatment of rectilinear partitioning based on *workload matrices* [9]. He developed a fast algorithm for optimal 1D partitioning and an iterative heuristic for two dimensions. He also proved that optimal rectilinear partitioning in three dimensions is NP-complete.

## 7   Conclusion

We introduced an infrastructure for parallel mesh computations that includes the mesh partitioning algorithm GARP and the communication library GRAPHlib.

GARP yields best results for meshes with a largely uniform vertex density and without offset extremities. It works well with a small number of parts if the mesh exhibits symmetry and as well with a high number of parts if the mesh shape can be classified. A drawback of GARP is poor load balancing with meshes that show highly varying vertex density. Hence, coupling with a load balancing heuristic is highly recommended for practical use.

First measurements using GARP and GRAPHlib on computers with fast communication showed modest improvement against Recursive Inertial Bisection and arbitrary message scheduling. We observed that GRAPHlib's optimized scheduling strategy achieves most gain with 3D volume meshes. On the whole, our infrastructure demonstrates the importance of shape-adequate partitioning for maximum parallel efficiency, in particular, with large meshes.

In the future we are going to couple GARP with a parallel load balancing heuristic and to study the impact of message ordering on parallel computers with more restrictive processor topologies, for example, mesh-connected computers.

# References

1. Berger, M., Bokhari, S.: A partitioning strategy for non-uniform problems on multiprocessors. IEEE Transactions on Computers **36** (1987) 570–580
2. Chrisochoides, N.: On the mapping of PDE computations to distributed memory machines. Ph.D. thesis, Purdue University (1992)
3. Farhat, C.: A simple and efficient automatic FEM domain decomposer. Computers and Structures **28** (1988) 579–602
4. Farhat, C., Lesoinne, M.: Automatic Partititioning of Unstructured Meshes for the Parallel Solution of Problems in Computational Mechanics. International Journal for Numerical Methods in Engineering **36** (1993) 745–764
5. Hammond, S.: Mapping Unstructured Grid Computations to Massively Parallel Computers. Ph.D. thesis, Renesselaer Polytechnic Institute (1992)
6. Hinz, D.: A run-time load balancing strategy for highly parallel systems. Acta Informatica **29** (1992) 63–94
7. Kernighan, B., Lin, S.: An efficient heuristic procedure for partitioning graphs. Bell System Technical Journal **49** (1970) 291–307
8. Mavriplis, D.J.: Three-Dimensional Multigrid for the Euler Equations. AIAA Paper 91-1549CP, American Institute of Aeronautics and Astronautics (1991) 824–831
9. Nicol, D.: Rectilinear Partitioning of Irregular Data Parallel Computations. Journal of Parallel and Distributed Computing **23** (1994) 119–134
10. Nour-Omid, B., Raefsky, A., Lyzenga, G.: Solving finite element equations on concurrent computers. In Noor, A.K. (ed.): Parallel computations and their impact on mechanics. American Society on Mechanical Engineering (1986) 209–227
11. Pothen, A., Simon, H., Liou, K.-P.: Partitioning sparse matrices with eigenvectors of graphs. SIAM Journal of Matrix Analysis and Applications **11** (1990) 430–452
12. Sadayappan, P., Ercal, F.: Nearest Neighbor Mapping of Finite Element Graphs onto Processor Meshes. IEEE Transactions on Computers **36** (1987) 1408–1424
13. Simon, H.: Partitioning of unstructured problems for parallel processing. Computing Systems in Engineering **2** (1991) 135–148
14. Wang, J.-C.: Load Balancing and Communication Support for Irregular Problems. Ph. D. thesis, Syracuse University (1993)
15. Walshaw, C., Cross, M., Everett, M., Johnson, S., McManus, K.: Partitioning and Mapping of Unstructured Meshes to Parallel Machine Topologies. In Ferreira, A., Rolim, J. (eds.): Parallel Algorithms for Irregular Structures Problems. Springer LNCS 980 (1995) 121–126

# Reducing Cache Conflicts by a Parametrized Memory Mapping

Daniela Genius[1] and Jörn Eisenbiegler[2]

[1] Institut für Programmstrukturen und Datenorganisation, Universität Karlsruhe
Zirkel 2, 76128 Karlsruhe, Germany
genius@ipd.info.uni-karlsruhe.de, Fax: (+49) 721 30 0 47
[2] Forschungszentrum Informatik (FZI) an der Universität Karlsruhe
Haid-und-Neu Straße 10-14, 76131 Karlsruhe, Germany
eisen@fzi.de, Fax: (+49) 721 9654 603

**Abstract.** Algorithms which access memory regularly are typical for scientific computing, image processing and multimedia. Cache conflicts are often responsible for performance degradation, but can be avoided by an adequate placement of data in memory. The huge search space for such compile time placements is systematically reduced until we arrive at a class of very simple mappings, well known from data distribution onto processors in parallel computing. The choice of parameters is then guided by a cost function which reflects the tradeoff between additional instruction overhead and reduced miss penalty. We show by experiment that when keeping the overhead low, a considerable speedup can be achieved.

**Keywords:** cache optimization, memory mapping, data distribution

## 1 Introduction

Algorithms that cause a lot of memory accesses in nested loops are typical for the areas of scientific computing and multimedia applications. Lately, the popularity of the latter has increased enormously. In great parts, namely in the areas of image processing and compression, they can be treated by the method presented here. In high performance computing, a trend towards running applications on workstation clusters emerges due to lower hardware cost [13]. We focus on workstations with instruction level parallel (ILP) architectures. Fully associative caches are expensive and thus rarely employed. While vector computers such as CRAY often have no cache at all, first level caches in workstations typically implement a set associativity of two (INTEL PENTIUM) or a direct mapping (DEC ALPHA). Caches with a such low set associativity suffer from conflict misses. When data is accessed regularly, conflicts or even cache thrashing are responsible for a severe performance degradation [9]. If high performance is the issue, the problem of making good use of the cache mostly remains with the programmer, for whom conflicts in particular are difficult to predict. We show that this can be accomplished by the compiler for typical benchmarks.

P. Zinterhof, M. Vajteršic, A. Uhl (Eds.): ACPC'99, LNCS 1557, pp. 460–469, 1999.
© Springer-Verlag Berlin Heidelberg 1999

Low set associativity allows to represent cache lines by corresponding memory areas. First we propose a general representation of memory mappings. We then reduce the large search space of general mappings step by step until we arrive at a compiler-controlled data distribution technique for SIMD programs [4] which we adapt for caches. Preliminary measurements show that a great number of conflict misses can thus be avoided.

In section 2, related approaches to the conflict problem are presented. Basic terminology and some prerequisites can be found in section 3. Starting with the description of a very general memory mapping, section 4 develops the block-cyclic memory mapping step by step; section 5 provides a general cost function for selecting the most suitable mapping. Results of preliminary measurements are shown in section 6. In the final section, an overview is given on directions of further work.

# 2 Related Work

There are two main ways to change the temporal-spatial structure of memory accesses: modification of the source code, usually by loop transformations, and changes to the memory mapping [1]. Tiling in general has a large impact on performance, particularly through better utilization of cache capacity. The problem of conflict misses has also been tackled by tiling [7, 3]. The former suggest square tiles and provide a cost measure that considers also cross interference. The latter suggest a loop transformation for rectangular tiling, handling conflicts between different arrays by estimation only. Tiling is however limited to tile-able algorithms, and it suffers from irregular performance. Changing the distribution of data onto memory promises a much smoother performance curve [10]. Also, some access reordering can not be achieved by loop transformations.

*Example 1.* Given the following C fragments with the usual row wise mapping,

```
for(i=0;i<n;i++){        for(j=0;j<n;j++){        for(i=0;i<n;i++){
  for(j=0;j<n;j++){        for(i=0;i<n;i++){        for(j=0;j<n;j++){
    A[i][j]=B[j][i];}        A[i][j]=B[j][i];}        A[i][j]=B[i][j];}
}                        }                        }
```

loop interchange e.g. as shown in the middle will improve reuse for $B$ while destroying reuse for $A$. Transposing $B$ as shown on the right allows row wise access for both arrays.

There are still few memory mapping based approaches due to the following reason: the tradeoff between reduced miss penalty and increased indexing cost has to be taken into account, otherwise run time improvements will be small. A basic technique, *padding*, originates from avoiding bank conflicts on vector machines with banked memory [2]. Unused data is inserted into relevant data thus changing the temporal-spatial structure. Rivera and Tseng lately provided a comprehensive experimental study [11]. Both propose to insert padding also between different arrays. Panda, Nicolau et.al. [10] apply padding after having determined the tile size by the method of [3]. Again, conflicts between different arrays are not sufficiently dealt with. Pad sizes are tried out successively, which is hardly feasible in the case of several arrays because pad sizes depend on each other. Run time is only slightly improved because the additional cost of more complex index operation is not considered. We describe a more flexible approach aiming at both conflict and run time reduction: block-cyclic mapping allows "pads" of different sizes with varying distances. Our approach is also related to classical techniques like *skewing* [8].

## 3   Basic Notions

Whenever a value requested in a calculation is present neither in registers nor in the cache, a *cache miss* occurs. *Compulsory misses* occur when filling up an empty cache. *Capacity misses* happen when the data does not fit fully into the cache. *Conflict misses* are due to the competition of memory locations for the same cache line. *Full associativity* allows free choice of candidates for replacement (e.g. LRU, FIFO, see [6]), thus no conflicts can occur. In a *set associative* cache, this policy is applied to smaller units, called *sets*. As the design of fully associative caches is complex and expensive, mostly a low set associativity or a direct mapping is chosen in practice. By *self-interference* we denote the fact that

a) LL7

```
for(k=1; k<SIZE; k++){
    x[k] = u[k]+r*(z[k]+r*y[k])
        +t*(u[k+6]+r*(u[k+5]
        +r*u[k+4])+t*(u[k+3]
        +r*(u[k+2]+r*u[k+1]))));
}
```

b) Matrix Multiply

```
for (i=0; i<SIZE; i++){
    for (j=0; j<SIZE; j++){
        for (k=0; k<SIZE; k++){
            c[i][j]+=a[i][k]*b[j][k];
        }
    }
}
```

c) Filter

```
for ( k=1 ; k<SIZE ; k++ ){
    for ( j=1 ; j<SIZE ; j++ ){
        b[i][j]=0.25*
            (a[i-1][j]+a[i][j-1]
            +a[i+1][j]+a[i][j+1]);
    }
}
```

d) FFT

```
le=2^n; windex=1; wptrind=0;
for(l=0;l<n;l++){
    le=le/2;
    for(j=0;j<le;j++){
        wpr=wreal[wptrind];
        wpi=wimag[wptrind];
        for(i=j;i<2^n;i+=2*le){
            tmpr=sigreal[i];
            sigreal[i]+=sigreal[i+le];
            tmpi=sigimag[i];
            sigimag[i]+=sigimag[i+le];
            tr=tmpr-sigreal[i+le];
            ti=tmpi-sigimag[i+le];
            sigreal[i+le]=tr*wpr-ti*wpi;
            sigimag[i+le]=tr*wpi-ti*wpr;
        }
        wptrind+=windex;
    }
    windex=windex*2;
}
```

**Fig. 1.** Benchmarks: a) matrix multiply b) Livermore loop 7 c) filter d) FFT

there are conflicts caused by accesses to the same array. *Cross-interference* occurs when the same is the case between different arrays. Conflicts lead to *cache thrashing* when in turn cache lines are evicted that will shortly be reloaded. *Temporal reuse* of data in the cache occurs when the same data item is accessed several times, *spatial reuse* occurs when accessing data in the same cache line.

Consider some typical algorithms from scientific computing given in figure 1. Example 1, matrix multiply (figure 1.a), does not contain any uniform references. Loop index $k$ occurs in the first as well as in the second argument of the index tuple, causing array $A$ to be accessed row major while $B$ is accessed column major. The Livermore kernel (1.b) exhibits interesting opportunities for our optimization: four arrays are accessed uniformly, values of $U$ are heavily reused. A typical application taken from image processing is *filtering* (1.c), where values from former iterations will have been evicted before they can be reused. Fast Fourier Transform (FFT, figure 1.d) suffers from the same problem from the (n-1)th iteration on, as is pointed out in [10]. The additional problem with FFT is that problem sizes are always a power of two, which is pathological wrt. cache conflicts. The examples contain "near-perfect" loop nests; no branches are allowed, however operations outside the innermost loop are admissible. The array indices are generated by an *affine mapping* of the loop counter vector.

It is also typical for scientific programs that data items are small and of constant size (e.g. `float` and `double`). Innermost loops are primarily considered, as the greatest effects can be achieved here. Reuse wrt. outer loops is more unlikely unless the cache size is large compared to the problem size. It is assumed that array sizes are known to the compiler. We concentrate on first-level data caches; unless loops are unrolled extensively, it is legitimate to leave the instruction cache behaviour out of the focus. In order to isolate the effects of conflict

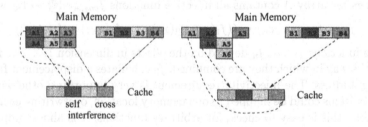

**Fig. 2.** Conflict-Avoiding Memory Mapping: General Principle

misses, we assume that transformations for reducing capacity misses have been applied (such as transposition or tiling [14]). Memory is virtually subdivided into cache lines, see section 3. Technically however, attention must be payed that any memory space delivered by routines such as `malloc` is aligned to element size only, whereas we need alignment to cache line boundaries. It is then possible to relocate entire cache lines and blocks of cache lines in memory without running the risk of destroying spatial locality — which is a phenomenon inside one cache line. We can now concentrate exclusively on conflicts.

# 4   Deriving a Memory Mapping

*Naive Memory Mapping* Conventionally, compilers map arrays to memory in the following way. Let there be $m$ arrays. Consider one $n$-dimensional array. The $x_l$ denote the offsets in direction of one of the dimensions, $size_k$ the number of data items in dimension $k$. Remember that C maps arrays row major, FORTRAN column major. Consider a C-style mapping as function $f_{C,A} : \mathbb{N}_0^n \to \mathbb{N}_0$ of the array offsets for one $n$-dimensional array $A$:

$$f_{C,A}(x_1, \ldots, x_n) = x_1 * size_2 * \ldots * size_n + x_2 * size_3 * \ldots * size_n + \ldots + x_n.$$

Such mappings pay no heed to the structure of accesses, thus running a high risk of self- and cross-interference. Both situations are depicted on the left of figure 2: in iteration $i$, $A$ suffers from self-interference in $A1$ and $A4$, $A$ and $B$ interfere due to $A6$ and $B4$. On the right, a conflict-avoiding mapping is shown.

*General Memory Mapping* Let us start over by assuming that an element can be placed according to any injective function into memory addresses. One could argue that a *table* can be employed to directly determine the memory mapping. This is problematic because in this case, table size depends on the matrix size. Also, accesses to a table kept in memory may cause additional cache misses. Our goal is to find a method where complexity only depends on the number of physical cache lines.

*Position-Dependent Memory Mapping* We restrict to functions that depend only on the *position* of an element inside an $n$-dimensional array. This enables us to take loop indexes into consideration. A *family* of position-dependent memory mappings for array $A$ contains all injective functions $f_{pos,A} : \mathbb{N}_0^n \to \mathbb{N}_0$ where

$$f_{pos,A}(x_1, \ldots, x_n) = f(f_1(x_1), f_2(x_2), \ldots, f_n(x_n)) + f_0(start).$$

The functions $f_1, \ldots, f_n$ depend on the offsets in dimension $1, \ldots, n$. $f$ determines the way in which they are combined, $f_0$ calculates a displacement from the starting address. The injectivity requirement is very important; otherwise, several data items could be mapped to one memory location, overwriting each other. In a table, this is easy to check; for arbitrary functions, it is almost impossible. Moreover, the complexity of finding suitable functions among those described by the above mapping is unmanageable in practice. Let us now restrict to a subset of position-dependent mappings that promises good results in practice.

*Innermost Memory Mapping* As in a preliminary step data was reordered in order to have elements in direction of $x_n$ accessed consecutively by the innermost loop, it seems obvious to maintain the standard mapping, letting the new functions affect the offset only in direction $x_n$. $f_{inner,A} : \mathbb{N}_0^n \to \mathbb{N}_0$, where

$$f_{inner,A}(x_1, \ldots, x_n) = x_1 * size_2 * \ldots * size_n + x_2 * size_3 * \ldots * size_n + \ldots + map(x_n).$$

Innermost mapping is still leaving open a variety of choices for *map*. It might also be difficult to check injectivity. In the following, we propose a set of simple mapping functions where injectivity can be determined by a modulo calculation.

*Block-Cyclic Memory Mapping* Let us further restrict the choice of mapping functions among those affecting only the last dimension to "block-cyclic" functions. Such mappings have been used in the field of data distribution onto processors in parallel computation. Eisenbiegler showed that for our class of applications, block size and cycle length can be computed at compile time [4]. Let

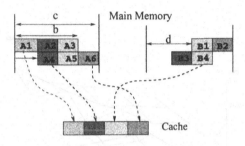

**Fig. 3.** Block-cyclic Mapping

us now demonstrate how the idea can be transferred to cache-conscious compile time memory mapping. For "block-cyclic" mapping functions $f_{bcd,A} : \mathbb{N}_0^n \to \mathbb{N}_0$:

$$f_{bcd,A}(x_1, \ldots, x_n) = x_1 * size_2 * \ldots * size_n + \ldots + (x_n \text{ div } b) * c + x_n \bmod b + d$$

only three parameters can be varied: block size, cycle distance and displacement. Block size $b$ denotes the number of data items relocated together, usually a multiple of the cache line size $l$. Cycle distance $c$ is an integer multiple of $b$, where $c = 1$ means a consecutive mapping; for $c > 1$, the gap between the blocks increases. The principle is depicted in figure 3. Starting address displacement $d$, shown only for array $B$, represents the distance of the array's starting address wrt. a reference address. Injectivity can now be assured by simple modulo calculation. For one array, parameters $b$, $c$ and $d$ have to be chosen in order to preserve the one-to-one mapping. For a pair of offsets in the last dimension $x_i, x_j : ((x_i \text{ div } b) * c + x_i \bmod b + d) \neq ((x_j \text{ div } b) * c + x_j \bmod b + d)$. Different arrays map to different memory areas.[1]

Concerning compile time complexity, the function is fixed, only its parameters can be varied. For $k$ cache lines we have at most $k^{|\{b,c,d\}|} = k^3$ choices for one array and $k^{3*m}$ alternatives for $m$ arrays with the additional restrictions of injectivity and $c = a*b, a \in \mathbb{N}$ considerably reducing the search space. Note that we have achieved our goal that complexity is dependent on the number of cache lines instead of on the array size.

---

[1] When intermixing arrays for reducing cross interference, the additional injectivity condition can be guaranteed easily in some cases. If for two arrays $A, B$ block sizes are equal, $c_A = a_A * b, c_B = a_B * b$, then for $d_B = d_A + d', d' \geq b$ the remaining condition is $x * (a_A - a_b) \neq d'$.

*Example 2.* Figure 4 depicts part of the search space for a $300 * 300$ matrix multiplication. For presentation reasons, we vary only parameters $c_B$ and $c_C$ $(1, \ldots, 5)$ with $m = 3$, $d_A, d_B$ and $d_C$ are multiples of $k$ (256 on a DEC ALPHA) and $c_A = b_A = b_B = b_C = 1$ fixed, $c_C = 1$. The minimal number of misses can be achieved for $c_B = 2$ and $c_C = 1$.

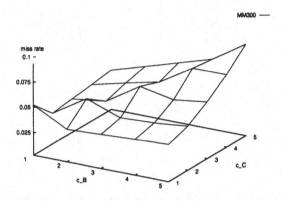

**Fig. 4.** Search Space: Variation of $c_B$ and $c_C$ for Matrix Size 300

In the example, we measured resulting miss rates of the entire program at run time. To select a good admissible parameter combination at compile time, a compiler would naively have to evaluate all admissible combinations of $b,c$ and $d$ for all three arrays, totalling $256^{3*3} = 2^{21}$. This is still unrealistic at compile time, even if injectivity and restrictions to $c$ are checked beforehand. We supply a *cost* function in the following section to guide more efficient search heuristics.

## 5   Cost Model

To ensure that a conflict minimal memory mapping is also optimal wrt. run time, only the relevant costs of the proposed mappings in the innermost loop bodies $b$ have to be compared against each other: Mapping 1 is preferred over mapping 2, if

$$\sum_{op_1 \in b_1} C_{ind}(op_1) + miss_1 * C_{miss} < \sum_{op_2 \in b_2} C_{ind}(op_2) + miss_2 * C_{miss}.$$

Both $C_{ind}$ and $C_{miss}$ are highly architecture dependent. We use the DEC ALPHA 21064 as a reference architecture. The costs per index operation $C_{ind}(op)$ can hardly be approximated for modern multiple-issue processors. For simplicity we assume single-issue, (static) latencies and one level of data cache. Integer multiply needs 8 cycles while other integer operations take 2 cycles. The ALPHA architecture offers no integer divide, only its floating point equivalent, which is extremely expensive (30 cycles single precision). Division by an integer constant function can be replaced by a multiplication followed by a right shift. The load latency on a cache miss is assumed to be 100 cycles. $C_{miss}$ is dependent on the memory hierarchy *latency* which is system dependent and difficult to determine. Miss rates *miss* can be estimated at compile time for the loop nest in question, e.g. by the techniques described in [9] or [7].

*Example 3.* Indexing into a two dimensional array normally costs 10 cycles. Revisiting example 2, assume an estimated miss rate of 30% for the loop nest.[2] Optimal block-cyclic indexing yields only 10% cache misses at cost of $2 * 8 + 3 * 2 + 2 * 30 = 82$ cycles if naively implemented, $4 * 8 + 4 * 2 = 40$ using shifts and $2*8+2*2 = 20$ in example 2 because division by $b = 1$ can be optimized away. The comparison of the total costs, $20 + 100*0.1 = 30 < 10 + 100*0.3 = 40 < 50 < 92$, indicates that the new mapping pays off if the miss rate can be significantly reduced *and* indexing functions are carefully chosen.

# 6   Experimental Results

Our intention is to measure both run times and cache misses for entire programs on workstations. The examples in figure 1 show just interesting loops of such C programs. The DEC ALPHA 21064 memory hierarchy is well documented, and presents a typical example of a direct mapped cache architecture with separate data and instruction caches. The analysis tool ATOM [12] allows specifying analyses on C programs by instrumenting binaries.

Figure 5 shows a comparison of miss rates and run times between our method (GE98) without intermixing, [7] (LRW91) and [3] (CoMK95) for selected sizes of the matrix example. It turns out that our method is not affected by matrix sizes of a power of two which are pathologic cases for some tiling methods. Generally, the significant reduction of cache misses exceeds the overhead for the more expensive indexing functions. Table 1 shows the miss rate and run time results for the rest of our set of benchmarks. The Livermore kernel and FFT are one dimensional and thus not covered by the tiling approaches. The column labeled "none" shows the non cache optimized case, "thrash" means cache thrashing, the worst case. Thrashing is a major problem for such one dimensional arrays, so miss rates are significantly reduced by our method. Our explanations why run time improvements fall short behind reduced miss rates are the following: Firstly, cache misses have an effect on run time only if the floating point pipeline is not completely filled and suffers a stall; otherwise, the actions overlap. Secondly, when arrays are mapped to memory less densely, there is also the danger that paging activity is increased. Additional care should be taken when selecting parameters $b$ and $c$. It often suffices to adapt only $d$ to obtain good results.

---

[2] Estimation considers only the loop nest in question while measurements totalize misses over the entire program, accounting for lower measured miss rates.

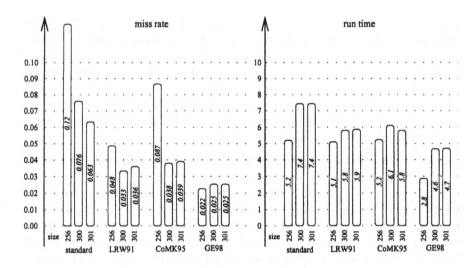

**Fig. 5.** Results for Matrix Multiplication

# 7   Conclusions and Future Work

By exploiting the mapping properties of low-associative caches, a new approach
to compiler-controlled cache optimization is presented. We give a general description of the class of possible memory mappings and then restrict the search
space in several steps. It has turned out that for run time reduction, it is crucial
to take conflicts as well as instruction overhead into account. We are currently

**Table 1.** Conflict Miss and Run Time Reduction: Benchmarks

| problem | N | miss rate % | | | run time sec | | |
|---------|---|------|-----|--------|------|-----|--------|
| | | none | GE | thrash | none | GE | thrash |
| LL 7 | $10^5$ | 1.535 | 0.776 | 7.03 | 0.057 | 0.057 | 0.066 |
| filter | 1000 | 3.11 | 2.61 | 3.12 | 0.212 | 0.189 | 0.212 |
| fft | 1024 | 3.61 | 3.58 | 8.47 | 0.065 | 0.064 | 0.067 |

integrating our method into the prototype compiler described in [4]. By providing a cost-based configuration heuristics, this will help us cope with the problem
of evaluating runtime reordering vs. compile time replication between different
loop nests. A coloring-directed method [5] with a more exact conflict prevention
(at higher instruction overhead) can also be expressed in the general framework.
The two methods will have to be compared against each other. We expect further
improvement from the combination with tiling.

   Experimentation will have to cover other processor types, such as multimedia and DSP processors. We are examining a number of image processing and

multimedia algorithms, namely parts of MPEG. Our processor model delivers a rather coarse approximation. Apart from latencies, interaction with (cache-aware) instruction scheduling has to be considered; by this, we hope to further narrow the gap between miss rate and run time improvement.

## References

[1] D. F. Bacon, S. L. Graham, and O. J. Sharp. Compiler transformations for high-performance computing. *ACM Computing Surveys*, 26(4):345–420, Dec. 1994.
[2] G. J. Burnett and J. Coffman, E.G. A study of interleaved memory systems. In *Proc., AFIPS 1970 Spring Jt. Computer Conf.*, volume 36, pages 467–474, Montvale, NJ, 1970. AFIPS Press.
[3] S. Coleman and K. S. McKinley. Tile size selection using cache organization and data layout. In D. W. Wall, editor, *ACM SIGPLAN '95 Conference on Programming Language Design and Implementation (PLDI)*, volume 30(6) of *ACM SIGPLAN Notices*, pages 279–290, New York, NY, USA, June 1995. ACM Press.
[4] J. Eisenbiegler. Optimization of SIMD programs with redundant computations. In *Proceedings of Europar '98*, volume 1470 of *LNCS*, pages 456–462. Springer, 1998.
[5] D. Genius. Handling cross interferences by cyclic cache line coloring. In *Proceedings of the 1998 Parallel Architectures and Compilation Techniques Conference (PACT'98)*, pages 112–117, Paris, France, October 14-16 1998. IEEE.
[6] J. L. Hennessy and D. A. Patterson. *Computer Architecture - A Quantitative Approach*. Morgan Kaufman, 2nd edition, 1996.
[7] M. S. Lam, E. E. Rothberg, and M. E. Wolf. The cache performance and optimizations of blocked algorithms. In *Proceedings of the Fourth International Conference on Architectural Support for Programming Languages and Operating Systems*, pages 63–74, Santa Clara, California, Apr. 8–11, 1991.
[8] D. H. Lawrie. Access and alignment of data in an array processor. *IEEE Trans. on Computers*, C-24(12):1145–1155, Dec. 1975.
[9] K. S. McKinley and O. Temam. A quantitative analysis of loop nest locality. In *Seventh International Conference on Architectural Support for Programming Languages and Operating Systems*, pages 94–104, Cambridge, Massachusetts, 1–5 Oct. 1996. ACM Press.
[10] P. R. Panda, H. Nakamura, N. D. Dutt, and A. Nicolau. Improving cache performance through tiling and data alignment. In *IRREGULAR 1997*, pages 167–185. Springer LNCS 1253, 1997.
[11] G. Rivera and C.-W. Tseng. Data transformations for eliminating conflict misses. In *Proceedings of the 1998 ACM SIGPLAN Conference on Programming Language Design and Implementation*, Montreal, Canada, june 1998.
[12] A. Srivastava and A. Eustace. ATOM: A system for building customized program analysis tools. In *Proceedings of the ACM SIGPLAN '94 Conference on Programming Language Design and Implementation*, pages 196–205, June 1994.
[13] T. M. Warschko, J. M. Blum, and W. F. Tichy. The parastation project: Using workstations as building blocks for parallel computing. In *Proceedings of PDPTA*, pages 9–11, Sunnyvale, California, USA, 1996.
[14] M. E. Wolf and M. S. Lam. A data locality optimizing algorithm. *SIGPLAN Notices*, 26(6):30–44, jun 1991. *Proceedings of the ACM SIGPLAN '91 Conference on Programming Language Design and Implementation*.

# Optimizing I/O for Irregular Applications on Distributed-Memory Machines *

Jesús Carretero[1], Jaechun No[2], and Alok Choudhary[2]

[1] Arquitectura y Tecnología de Sistemas Informáticos
Universidad Politécnica de Madrid, Spain
[2] Dept. of Electrical Engineering and Computer Science
Northwestern University, USA

**Abstract.** In this paper we present the design, implementation and evaluation of a runtime system based on collective I/O techniques for irregular applications. Its main goal is to provide parallel collective I/O, by having all processors participate in the I/O simultaneously, and making the mapping of the I/O requests simpler. Using such a technique, the input/output of the irregular applications can be greatly simplified by always maintaining global files canonically ordered, thus avoiding the utilization of multiple files and the associated sorting/merging steps. The run-time library has been optimized by applying in-memory compression mechanisms to the collective I/O operations. We also present the results of several evaluation experiments obtained by running a *particle in cell* application on an Intel Paragon machine. Those results demonstrate that significantly high-performance for I/O can be obtained by using our library.

## 1 Introduction

Parallel computers are being increasingly used to solve large computationally and data-intensive applications, such as large-scale computations in physics, chemistry, and other sciences. Large number of those applications are "irregular" applications, where accesses to data are performed through one or more level of indirections. Sparse matrix computations, particle codes, and many CFD applications exhibit these characteristics [4]. Figure 1(a) illustrates an example of an irregular program in which $x$ and $y$ are *data arrays* and $a(i)$, $b(i)$, $c(j)$, and $d(j)$ which are used to index data arrays are called *indirection arrays* This problem can be abstracted into the one shown in Figure 1(b), where we assume that, on each node, there is an indirection array that describes the location of the corresponding data element in a global array. Since indirection arrays are read in at runtime, the compilers cannot analyze the data elements used on a given loop, making it impossible to determine the I/O requirements at compile-time, and generating a huge amount of fine grained I/O operations, most of them including seeking the data element pointed by the indirection array on some file

---

* This work was supported in part by NSF Young Investigator Award CCR-9357840 and NSF CCR-9509143.

P. Zinterhof, M. Vajteršic, A. Uhl (Eds.): ACPC'99, LNCS 1557, pp. 470–479, 1999.
© Springer-Verlag Berlin Heidelberg 1999

or database. To circumvent those I/O requirements, ranging from 100s of MBs to TBs, two solutions are typically used: serial I/O, with only one processor executing I/O operation [9], and storing each processor-local data to a local data file per processor [5]. Both models have inherent problems associated. Serial I/O is a bottleneck for the application performance. The local data file method must rely on some sequential preprocessing to create the local files corresponding to the global mesh, and on some postprocessing to recombine the data files into the single data file of the mesh (see figure 2.) As the number of processors scales and the number of grid points being considered moves to the million-to-billion range, sequential re-combination of the MBytes/GBytes of data into the anticipated terabytes of data simply becomes infeasible.

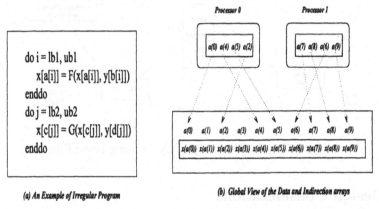

(a) An Example of Irregular Program

(b) Global View of the Data and Indirection arrays

**Fig. 1.** View of Data and Indirection Arrays in an Irregular Program

In this paper we present the design and implementation of a high-performance runtime system which can support all of the types of I/O listed above on large-scale systems. In particular, the system enables accessing irregular data sets; that is, data sets that are accessed via indirections. Our thesis is that the I/O performance of irregular applications can be greatly enhanced by using collective I/O methods on a global, canonically ordered, data file. By always maintaining the global data in some canonical order, we can avoid the explosive number of different file formats, eliminate sequential pre and postprocessing, enable restarting with different number of processors, and share data among different platforms and processor configurations. The rest of this paper is organized as follows. In section 2, we present the design considerations for parallel I/O library to support irregular computations. In section 3, we present the design of our "Collective I/O" library and its influence in the analysis cycle. Section 4 shows some the performance results obtained by executing a *particle in cell* application on a parallel machine. Finally, in section 5 we present a summary and conclusions.

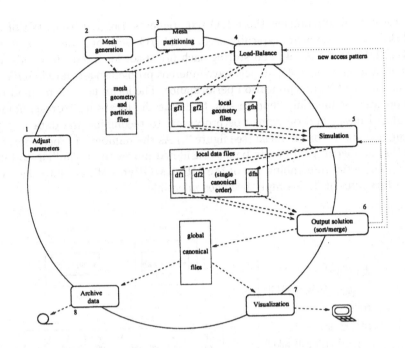

**Fig. 2.** I/O in the Analysis Cycle using Local Files

## 2   Design Considerations

To maintain a compatible data format, the original *particle in cell* application shown in this paper, was parallelized using a file per processor, generated by preprocessing a canonically ordered data set (see figure 2.) To visualize the simulation results, those files must be sorted/merged to get a canonically ordered global data set. However, the *particle-in-cell* application changes the access pattern after executing some iterations of the simulation. As a result, the preprocessing step must be executed again to recreate the local files with the new required geometry data. Moreover, for restarting, the former solution depends on the number of processors. Thus, several configuration files must be maintained to know the load-balance patterns to be used for runs on various number of processors. With the existing solutions [3], fine-grained I/O operations and multiple file management create high contention on the I/O system as the amount of data generated by an unstructured grid calculation increases to GBytes or even TBytes. Our library has three main goals: optimizing the I/O bandwidth utilization, reducing the storage space required by the application, and avoiding the sorting/merging step of the local files in the analysis cycle. To optimize the I/O bandwidth, the design of our library is based on the two-phase *collective I/O* strategies [8]. The basic idea behind two-phase collective I/O is to reorder, at runtime, the access patterns seen by the I/O system such that large number of small and disjoint I/O requests are converted into small number and large contiguous requests. To reduce the storage space, in-memory compression [6] has been incorporated as an option in our collective I/O library. Lately, to avoid the sorting/merging step of the local files, the I/O in our library is executed on *global files canonically ordered.* Those files don't need reordering, as it is done on the collective I/O. Thus, they can be use by any other utility, such a visualization tool, without any postprocessing to sort/merge the files.

## 3   Collective I/O Operations

The design of the collective I/O run-time library relies on three basic functions: schedule construction, reading/writing data from files, and redistributing/collecting data into/from appropriated locations of each processor. In a write operation, collecting precedes the file writing step.

### 3.1   I/O Schedule Construction

Schedule describes the communication and I/O patterns required for each node participating in the I/O operation. When an indirection array is used to reference the data array, it must be scanned to consider each element of the array individually to determine its place in the global canonical representation as well as its destination for communication.

Several factors affect the schedule construction in particular, and overall I/O library design and performance in general. These include the following. 1) *Chunk Size,* which is the amount of buffer space available to the runtime library for the I/O operations. 2) *Number of processors involved in I/O.* The following briefly describes the steps involved in computing schedule information.

- Based on the chunk size, each processor is assigned a data domain for which it is responsible for reading or writing. For example, if there are four processors and total 16 elements to be read/written with a buffer space of two elements on each processor, the chunk size(total) is 8 elements. Processor 0 will be responsible for elements 0,1 and 8,9, processor 1 will be responsible for elements 2,3 and 10,11, and so on. For a chunk, each processor computes its part to read or write data while balancing I/O workload. Next, with each index value in its local memory, processor first decides from which chunk the appropriate data must be accessed and then determines which processor is responsible for reading the data from or writing the data to the chunk.
- Index values in the local memory are rearranged into the *reordered-indirection array* based on the order of destinations to receive them. Therefore, we can communicate consecutive elements between processors(communication coalescing).

The I/O schedule construction algorithm is inherently parallel. Each processor sorts its portion of the indirection array without any dependency. A small amount of communication is needed to communicate to other processors the number of index values to be read or written from/to its access domain.

Many irregular application codes, such as particle-in-cell methods, change the access pattern after a series of iterations. Note that once it is constructed, the schedule information can be used repeatedly during the series of iterations where the access pattern does not change, and thereby amortizing its cost. However, the schedule information must be recomputed every time the access pattern changes.

## 3.2  Collective I/O Operations

A processor involved in the computation is also responsible for reading data from files or writing data into files. Let $D$ bytes be the total size of data and $P$ be the number of processors, each processor then reads a chunk of $D/P$ bytes of data from the file and distributes it among processors based on schedule information. In case of writing, each processor collects $D/P$ bytes of data from other processors, compresses it, and then writes it to the file. Integrating compression in the run-time library allows to reduce the storage space, and the execution time, needed by the application. By performing I/O this way, the workload can be evenly balanced across processors. An *index* has been included into the library to store the parameters of each compressed chunk. Every time an application creates a data array, an *index* is associated with the array. When the array is written the index is also written to an *index file*, which is usually very small. Every time an application reads an array, the index file is also read and stored in memory.

To present the algorithm for the collective write operation, let $C$ be the number of data chunks, $S_{indx}(i, p_j)$ and $S_{data}(i, p_j)$ be the portion of the indirection and data arrays to be sent to $p_j$ in data chunk $i$, and $R_{indx}(i, p_j)$ and $R_{data}(i, p_j)$ be the portion of the indirection and data arrays to be received from $p_j$ in data chunk $i$. Then, $send(o, d, data)$ sends $data$ from processor $o$ to $d$, and $recv(d, o, data)$ receives $data$ from processor $o$ to $d$. $write(fid, data, offset)$ writes $data$ into the position $offset$ of file $fid$. For describing the compression flow of the algorithm, let $S_{offset}(i)$ and $R_{offset}(i)$ be the file offset to be sent/received to/from others after compression, and $S_{localindex}(i)$ and $R_{localindex}(i)$ be the size of the local index of data chunk $i$ to be sent/received to/from other processors to generate global index file. Also, let $asend$ and $arecv$ be the non-blocking versions of the $send$ and $recv$ functions defined before, $compress\_chunk(i)$ be the compression function used in the library, and $set\_chunk\_index(i, chunk\_index)$ be the function storing the index data of chunk $i$ into the specified $chunk\_index$. Figure 3 shows the steps involved in the collective write operation with compression. In *step 3*, each processor exchanges the indirection and data values in each chunk with other processors by taking exclusive-or operation. Each processor writes data to its own data domain in each data chunk in *step 4* if compression is not used. *Steps 5 - 8* in the same Figure show the additional steps when compression is used. In *step 5*, while receiving the offset, each processor compresses its data chunk. After compression, the offset is actually received, aligned to the next file system block, and propagated. Once all the chunk features are known, the chunk index is build, and the chunk data are written to the data file in *step 6*. All steps described above are repeated until the entire chunks assigned to each processor on the schedule phase are written to the file. *Step 7* and *step 8* show how the global index is gathered on a single processor (root) and then written to the index file. Non-blocking I/O and communication is used in the implementation of the algorithm to optimize its cost. Moreover, data is packed in memory before sending it to another processor. Both mechanisms reduces the impact of the personalized communication on the scalability of the collective I/O operation, as shown in the performance results of section 4.

| | |
|---|---|
| step1 | *if* $(io\_access\_pattern\_changed)$ *then* |
| | $\quad compute\_irregular\_io\_sched(IS, OS)$, *where OSisthe* |
| | $\quad$ *ordered indirection array* |
| step2 | $\forall i,\ 0 \leq i \leq C-1\ \{$ |
| step3 | $\quad \forall j,\ 0 \leq j \leq P-1\ \{$ |
| | $\quad\quad p_d = compute\_destination\_processor(p_j), 0 \leq d \leq P-1$ |
| | $\quad\quad send(p_j, p_d, S_{indx}(i, p_d))$ |
| | $\quad\quad recv(p_j, p_d, R_{indx}(i, p_d))$ |
| | $\quad\quad send(p_j, p_d, S_{data}(i, p_d))$ |
| | $\quad\quad recv(p_j, p_d, R_{data}(i, p_d))$ |
| | $\quad \}$ |
| step4 | *if* $\neg(compression\_used)$ *then* $write(data\_file, i, offset(i))$ |
| | $else\{$ |
| step5 | $\quad arecv(me, me-1, R_{offset}(i))$ |
| | $\quad c_i = compress\_chunk(i)$ |
| | $\quad wait(me, me-1, R_{offset}(i))$ |
| | $\quad offset(i+1) = offset(i) + block\_aligned(length(c_i))$ |
| | $\quad asend(me, me+1, S_{offset}(i+1))$ |
| | $\quad set\_chunk\_index(i, localindex(me))$ |
| step6 | $\quad write(data\_file, c_i, offset(i))$ |
| | $\quad \}$ |
| | $\}$ |
| step7 | *if* $(compression\_used)$ *then* $\{$ |
| | $\quad send(me, root, S_{localindex}(me))$ |
| | $\quad if\ (me = root),\ \forall j, 0 \leq j \leq P-1$ |
| | $\quad\quad recv(me, j, R_{localindex}(j))$ |
| | $\quad\quad build(global\_index, R_{localindex}(j))$ |
| step8 | *if* $(me = root)$, $write(indexfile, global\_index)$ |
| | $\}$ |

**Fig. 3.** Algorithm for Collective Write Operation with Compression

Figure 4 shows the I/O in the analysis cycle when our run-time library is used. The local geometry and data files can be eliminated, because our library provides collective I/O to global files canonically ordered. Thus, the sorting/merging step is also avoided. Moreover, the schedule utility of our library allows to recompute the local geometry information in-memory, balancing the load at the same time. This utility, along with the global files, do the application restarting independent of the number of compute nodes used. The archival space needed is also reduced by using the run-time library in-memory compression facilities. As shown in the figure with the modified analysis cycle, if the access patterns changes after a series of iterations, a new schedule operation is needed to accomplish load balancing and indexes ordering. However, in our model, the output files themselves are not affected by the reordering as they are available to all the compute nodes. In the currently used cycle, new local files should be distributed.

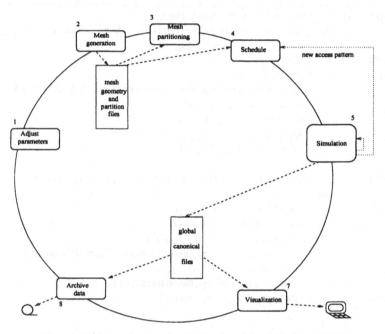

**Fig. 4.** I/O in the Analysis Cycle with Collective I/O

## 4   Performance Evaluation

The test application for our I/O work has been a *particle in cell* code used to study the behavior of a gas by simulating the motion of a large number of particles [1]. Each particle has velocity and position components, plus other physical variables, that are modified with a specified period of time as the molecules are concurrently followed through collisions and boundary interactions. The simulation lays out a Cartesian grid over the domain, associating each particle with a cell of the grid. The application was parallelized by distributing the cells, and their particles, across processors. Periodically, cells must be remapped to processors to adjust load balance. The load distribution changes with time because of particles changing cells every time-step, requiring the regeneration of the in-direction arrays [7]. Figure 5 shows an irregular loop from the application and the movement of particles across cells.

We carried our experiments on an Intel PARAGON machine located at Cal-tech. It has 512 compute nodes and 16 I/O nodes. Each I/O node has a 4 GB Seagate disk attached. The filesystem used is Intel PFS striped across the I/O nodes with a stripe unit of 64 KB. The parameters used in the experiments were chosen as follows: particle number varying from 4 millions to 38 millions, prob-lem size varying from 128 MBytes to 1024 MBytes, and number of processors varying from 32 to 256. The number of particles per processor is 9x128x128, thus having a fixed I/O buffer size per processor of 4 MBytes. The number of time steps executed were 1000, with the domain and access pattern repartitioned ev-ery 40 time steps, and the output data written every 10 iterations for restarting and visualization purposes. The total I/O size on each experiment was equal to $num\_processors * 4MBytes * 100$, thus varying between 12,800 MBytes and 102,400 MBytes. The collective I/O library with and without the compression optimizations was used in the experiments. The *lzrw3* compression algorithm was used, because it showed the best results in former evaluations [2].

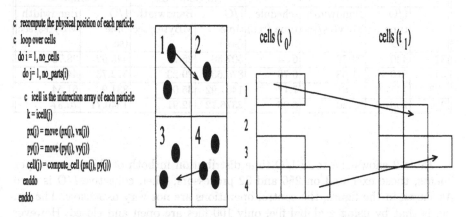

**Fig. 5.** DSMC Application Code Fragment and Example

Table 1 shows the I/O time and bandwidth for the original application and the application using the collective I/O library with and without compression. The total I/O time shown includes I/O control operations, like *open* and *close*, and I/O data related operations, like *write* and *compression*. The bandwidth was obtained dividing the total I/O size by the total I/O time. The compression ratio was 8 : 1 for all the experiments, when the compression optimizations were applied. Several conclusions can be extracted from the results shown in the table. First, the I/O schedule operation is affected more by the I/O buffer size than for the number of processors. The schedule operation is relevant in the analysis cycle because it must be executed each time the access pattern is changed. Thus, it was important to measure the execution time of this operation to understand its effect on the scalability of the application. As can be seen in table 1, the execution time of the schedule operation is very small compared with the total I/O time. Thus, it can be concluded that the parallel algorithm applied in the schedule provides a good execution time, thus not compromising the scalability of the application. Second, the I/O time is almost linear to the size of the file written. Third, for large problems, using in-memory compression is always beneficial. Comparing the original application with our library, it can be concluded that the behavior is very similar or slightly better for our solution in performance terms when compression is not used. Moreover, the results are much better for our library when compression is included. As, our library allows to simplify the I/O cycle of the parallel application, it can be concluded that it is well worth using it to solve the I/O problem in irregular applications.

**Table 1.** Evaluation of DSMC Application with Scaled Load

| Pro-cessors | Original | | | Collective I/O | | | |
|---|---|---|---|---|---|---|---|
| | | | | No Compress | | Compress | |
| | I/O time (sec) | Bandwidth MBytes/sec | schedule time (sec) | I/O time (sec) | Bandwidth MBytes/sec | I/O time (sec) | Bandwidth MBytes/sec |
| 32 | 430 | 29.76 | 6.34 | 507.56 | 25.21 | 445.92 | 28.70 |
| 64 | 810.9 | 31.53 | 6.78 | 875.63 | 29.23 | 572.73 | 44.69 |
| 128 | 1477.6 | 34.61 | 7.17 | 1461.92 | 35.02 | 831.9 | 61.54 |
| 256 | 2681.3 | 38.20 | 6.96 | 2376.72 | 42.91 | 1340.71 | 76.37 |

Table 2 shows the total I/O time distribution in both of the former experiments, those executed on 256 and 32 processors, when collective I/O is used. As shown in the figure, the control operations are not very expensive. The reason is that by using a global file, only 100 files are open and closed. However, the in-memory compression operation became percentually less expensive as the number of processors was increased. That's the behavior expected, because the compression operations are completely parallel, thus being their overhead mostly related with the I/O buffer size and not with the number of processors. In addition, a 8 : 1 compression ratio reduces the storage space required for the 256 processors from 100 GBytes to 12.5 GBytes.

## 5   Conclusions

The results obtained from testing the *particle in cell* application confirmed former behavior of our runtime library for large applications, and the scalability of the schedule operation, which is fully parallel. Comparing the original application with our library, it can be concluded that the behavior is very similar or

**Table 2.** Time Distribution for I/O Operations in the Collective I/O Solution

| Processors | 256 | | 32 | |
|---|---|---|---|---|
| Operation type | Total time (sec) | Percent I/O time | Total time (sec) | Percent I/O time |
| open | 144.88 | 10.81 | 25.34 | 5.68 |
| close | 2.89 | 0.22 | 3.21 | 0.73 |
| irr_write | 1077.69 | 80.38 | 307.06 | 68.93 |
| compression | 115.25 | 8.59 | 109.87 | 24.66 |

slightly better for our solution in performance terms when compression is not used. Moreover, the results are much better for our library when compression is included. As, our library allows to simplify the I/O cycle of the parallel application, it can be concluded that it is well worth using it to solve the I/O problem in irregular applications.

# References

1. G.A. Bird. *Molecular Gas Dynamics and the Direct Simulation of Gas Flows.* Oxford University Press, 1994.
2. J. Carretero, J. No, S. Park, A. Choudhary, and P. Chen. Compassion: a parallel i/o runtime system including chunking and compression for irregular applications. In *Proceedings of the International Conference on High-Performance Computing and Networking 1998*, Amsterdam, Holland, April 1998.
3. M. Christon, D. Crawford, E. Hertel, J. Peery, and A. Robinson. Asci red - experiences and lessons learned with a massively parallel teraflop supercomputer. In *Proceedings of the Supercomputer 1997*, Mannheim, Germany, June 1997.
4. James T. Poole. Preliminary survey of I/O intensive applications. Technical Report CCSF-38, Scalable I/O Initiative, Caltech Concurrent Supercomputing Facilities, Caltech, 1994.
5. L. A. Schoof and V. R. Yarberry. Exodusii: A finite element data model. Technical Report SAND94-2137, Sandia National Laboratories., December 1994.
6. K. E. Seamons and M. Winslett. A data management approach for handling large compressed arrays in high performance computing. In *Proceedings of the Fifth Symposium on the Frontiers of Massively Parallel Computation*, pages 119–128, February 1995.
7. Shamik D. Sharma, Ravi Ponnusamy, Bongki Moon, Yuan shin Hwang, Raja Das, and Joel Saltz. Run-time and compile-time support for adaptive irregular problems. In *Supercomputing 1994*. IEEE Press, November 1994.
8. Rajeev Thakur, Alok Choudhary, Rajesh Bordawekar, Sachin More, and Sivaramakrishna Kuditipudi. Passion: Optimized I/O for parallel applications. *IEEE Computer*, 29(6):70–78, June 1996.
9. Rajeev Thakur, Ewing Lusk, and William Gropp. I/O characterization of a portable astrophysics application on the IBM SP and Intel Paragon. Technical Report MCS-P534-0895, Argonne National Laboratory, August 1995. Revised October 1995.

# Cellular Multiprocessor Arrays with Adaptive Resource Utilization*

Hans-Joachim Stolberg, Martin Ohmacht, and Peter Pirsch

Laboratorium für Informationstechnologie
Universität Hannover, D-30167 Hannover, GERMANY
{stolberg,ohmacht,pirsch}@mst.uni-hannover.de

**Abstract.** A fully scalable, cellular multiprocessor architecture is proposed that is able to dynamically adapt its processing resources to varying demands of signal processing applications. This ability is achieved by migration of tasks between processor cells at run-time such as to avoid cell overload. Several dynamic migration strategies are investigated, and simulation results are provided for different load cases. The results indicate a potential performance gain from dynamic task migration on signal processing applications. By employing a rule-based learning system for an adaptive combination of migration strategies, the migration benefits become independent from the particular application characteristics.

## 1 Introduction

With the tremendous progress in semiconductor technology, an ever-increasing amount of transistor functions can be integrated on a single chip [1]. The dramatic improvements in transistor density are starting to allow a monolithic implementation of processing arrays, translating the large amount of available transistor functions into a high parallel processing capability. Both the increasing dominance of chip-internal wiring on a circuit's delay characteristics as well as the wish for easily scalable processor systems force the use of purely local communication structures, resulting in a cellular system.

As the potential for improved processing capabilities rapidly increases, the computational requirements of algorithms likewise continue to grow, consistently demanding more and more processing power. This trend is driven particularly by the area of signal processing involving large amounts of data to be processed continuously, frequently under real-time constraints. Signal processing algorithms exhibit special processing characteristics that have to be utilized for efficient execution: First, they involve a high amount of parallelism as large sets of data samples are to be processed concurrently in an identical manner. Second, periodic computation patterns can be identified as the same subtasks are repetitively executed on continuously incoming data streams. These properties are well demonstrated, for example, by current algorithms in the area of image, video, or

---

* This work was supported by the Deutsche Forschungsgemeinschaft (DFG) under contract number PI 169/6.

P. Zinterhof, M. Vajteršic, A. Uhl (Eds.): ACPC'99, LNCS 1557, pp. 480–489, 1999.

multimedia processing [2]. However, while selected subtasks of these algorithms used to be good candidates for highly efficient dedicated implementations due to their high regularity, compound processing schemes of higher complexity are evolving that increasingly comprise more irregular tasks with data-dependent computation flow.

In order to respond to the demands of complex processing schemes, a dynamic adaptation capability of processing resources is desired. In a cellular processor array, dynamic task migration [3] can be employed in order to change the allocation of individual tasks to processor cells at run-time. In case of a periodic workload, information from preceding periods can be used to guide the run-time decision on a task transfer. With dynamic task migration, tasks have to be re-scheduled at run-time, taking into account all migrations throughout the system. Data-driven execution of tasks, as employed in data flow processors [4], already provides a self-scheduling execution model that requires only task-specific information and therefore is well suited for distributed computing with local communication. As another advantage of the data flow execution model, task migration within the cellular array is effectively supported as only local communication paths need to be updated after transfer.

In this paper, dynamic task migration strategies in cellular multiprocessor arrays are examined aiming to increase the performance on periodic workloads with dynamically varying computational requirements. Section 2 introduces the architecture of cellular multiprocessor arrays employing data flow processing. In Section 3, strategies for dynamic task migration are discussed, and simulation results are presented for different workloads. The application of a learning system for an adaptive combination of migration strategies is examined in Section 4. Section 5 concludes the paper.

# 2   System Architecture

The processing elements (PEs) of the monolithic multiprocessor proposed here are arranged in a two-dimensional array, as shown in Fig. 1, with each cell connected to its 8-neighborhood. All PEs are assumed to have identical processing capabilities, resulting in a homogeneous architecture.

For parallel processing, a program has to be decomposed into basic units, called tasks. The resulting task graph—consisting of tasks as nodes and inter-task dependencies as directed edges—has to be mapped onto the array, distributing the tasks among the PEs. The granularity of an individual task may range from a single operation (arithmetic, logical, control, I/O) up to a sequence of several operations. The smaller the grain size, the larger may be the amount of parallelism to be exploited, but so is the communication overhead. Within this paper, a single operation has been assumed to require four clock cycles for completion.

A task graph and its possible mapping to the processing array is shown in Fig. 2. In order to enable communicating tasks even to be allocated to PEs that are not directly connected, pseudo router tasks (PRTs) have been introduced which maintain communication by passing on operands from tasks to their successors. The PRTs are created during the initial mapping process as well as dynamically during task migration. The communication cost in the cellular array model is determined by the number of PRTs located between subsequent tasks: Each additional PRT is assumed to add one clock cycle to the communication latency. Communication within the local neighborhood, on the other hand, incurs no additional delay; this assumption can be justified by the locality of interconnects.

The described concept of program execution directly corresponds to the data flow principle (e.g., [4]). Data flow processors can be divided into static [5] and dynamic [6] approaches. The central matching unit required in the dynamic data flow architecture conflicts with the aim of merely local communication, whereas the static approach allows a distributed execution model with result destinations of each operation known a priori. Therefore, in our approach, a static data flow processor model employing the feedback method [5] has been used. Based on this architectural model, dynamic task migration strategies are explored in the following.

## 3   Dynamic Task Migration

The goal of dynamic task migration is to obtain an allocation of tasks to processor cells such as to minimize overall computation time even under varying computational demands. In contrast to dynamic task scheduling or migration problems generally considered in the literature (e.g., [3][7]), the execution sequence of the tasks to be balanced here is not arbitrary, and their weights may differ considerably. Therefore, simple load balancing methods merely considering the number of tasks on each processing node are not sufficient here. A more demanding scenario also allowing for inter-task communication and execution precedence constraints has been considered, e.g., in [8].

### 3.1   Migration Strategies

Several factors may affect the total execution time of a data-flow program representation on a cellular multiprocessor array: First, if some cells are overloaded

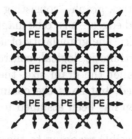

Fig. 1. Cellular processor array.

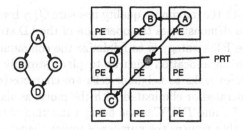

**Fig. 2.** Task graph and its mapping.

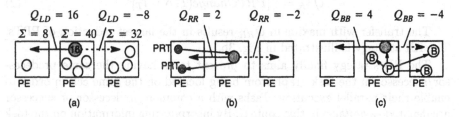

**Fig. 3.** a) Load diffusion (LD); b) Router reduction (RR); c) Brother balancing (BB).

compared to others in the array, execution time increases. Second, if operators to be executed in direct sequence are placed on distant locations in the array, the resulting communication delay may stretch the overall execution schedule. And third, if operators that could be executed in parallel are placed on the same cell, execution time increases due to enforced serialization (assuming individual cells to have no further parallel processing capabilities).

In the scope of this paper, three basic task migration strategies have been devised, each targeting one of the aspects mentioned above. They are denoted accordingly as *Load Diffusion (LD)*, *Router Reduction (RR)*, and *Brother Balancing (BB)*. It is important to understand that all these strategies are purely cell-based and operate in a distributed manner without guidance of a global unit; only local information exchange between direct neighbors once a period is permitted. The cell-based data collected each period consists of the *cell invocation* (number of tasks executed on a cell), *cell activity* (number of clock cycles a cell has been active), and the *task weight* (cycle count of a task) for each task on the cell.

The LD strategy has the goal to equally balance load throughout the array. Each possible transfer of a task to a neighbor cell is characterized according to its load balancing effect as expressed by Eq. 1 (with cell activity $A$, neighbor activity $A_N$, and task weight $W_T$):

$$Q_{LD} = max(A, A_N) - max(A - W_T, A_N + W_T) \qquad (1)$$

The transfer with the maximum quality measure $Q_{LD}$ best possibly balances the load. Figure 3a demonstrates the operation of the LD strategy.

The goal of the RR strategy is to minimize the communication cost among tasks by achieving an allocation such as to place strongly linked tasks close together. Transfers of tasks to neighbor cells are characterized by the number of PRTs being generated or eliminated with the move, as shown in Eq. 2 (with task invoke count $I_T$, and $PRTChange$ being a function of task $T$ and transfer destination cell $N$ that returns the number of router changes when moving $T$ to $N$):

$$Q_{RR} = -[PRTChange(T, N) \cdot I_T] \tag{2}$$

The transfer with maximum $Q_{RR}$ results in the largest reduction of PRTs. The RR strategy is illustrated in Fig. 3b.

The BB strategy finally aims to prevent tasks having the same predecessor/successor in the task graph from being located on the same cell in order to enable their parallel execution. Tasks with a common predecessor or successor are denoted as *brothers* in this context. By incorporating information on the task graph topology, the brothers of the tasks on a cell are identified via their common predecessors/successor, and each task transfer to a neighbor cell is characterized by the change in the number of brothers per cell within the local neighborhood, as specified by Eq. 3 (with current cell $C$, neighbor cell $N$; the task weight $W_T$ converts the brother count into a clock cycle count):

$$Q_{BB} = [Brothers(C) - Brothers(N)] \cdot W_T \tag{3}$$

The task transfer with maximum $Q_{BB}$ best possibly distributes task brothers within the local neighborhood. Figure 3c shows an example of the BB strategy (P = predecessor, B = brother).

In order to incorporate all three aspects discussed above into the transfer decision, the quality measures of all strategies are combined analytically with weight factors according to Eq. 4. The transfer with the highest overall quality measure $Q$ will finally be committed. The vector $[w_{ld}, w_{rr}, w_{bb}]$ comprising the weight factors is referred to as *strategy vector* in the following.

$$Q = \begin{bmatrix} w_{ld}, w_{rr}, w_{bb} \end{bmatrix} \cdot \begin{bmatrix} Q_{LD} \\ Q_{RR} \\ Q_{BB} \end{bmatrix} \tag{4}$$

## 3.2  Synthetic Load Models

The effect of dynamic task migration as introduced above has been examined by simulation of program execution on a cellular array. Within this paper, two synthetic load cases emulating different algorithmic properties have been selected as a test bench:

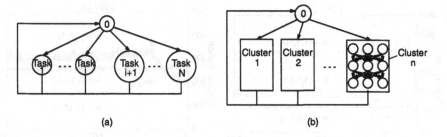

(a)                                                          (b)

**Fig. 4.** Synthetic load structures: a) L1; b) L2.

**Load L1.** This load is characterized by high task parallelism with no communication among individual tasks. Two groups of tasks have been created, having small weight and large weight, respectively. A perfect task migration strategy would have to balance the different task weights most uniformly throughout the array. Figure 4a depicts the task graph of load L1 illustrating the two groups of tasks having different weights.

**Load L2.** This load exhibits a clustered task graph structure with close communication among tasks within the same cluster, but few inter-cluster communication. The corresponding task graph is shown in Fig. 4b. A perfect task migration strategy would be able to separate the different clusters and to group tasks belonging to the same cluster close together; this would minimize communication paths while avoiding overload on individual cells.

These loads, although simply structured, represent structural elements to be found in many real applications; a good behavior in these cases therefore promises good results for a wide range of algorithms.

### 3.3   Simulation Results

In order to evaluate the performance of the dynamic task migration scheme, the execution of the loads L1 and L2 has been simulated on a $7 \times 7$-cell array. Figure 5 shows the simulation results obtained for different strategy vector combinations on the loads L1 and L2. Each diagram shows the number of clock cycles per period during 1000 simulated periods. In both cases, simulation started from a random distribution of tasks over the array.

The three strategy vectors examined here each emphasize one of the basic migration strategies: Vector $[10, 1, 1]$ performs load diffusion in the first place, vector $[1, 10, 1]$ emphasizes brother balancing, and vector $[1, 1, 10]$ concentrates on router reduction; in all cases, however, the other strategies are also active, albeit with a smaller influence.

As can be seen from the simulation results, the strategy $[1, 1, 10]$—dominated by router reduction—does not lead to satisfactory results for any of the loads. While it has a clearly negative effect on load L1, it however appears superior to

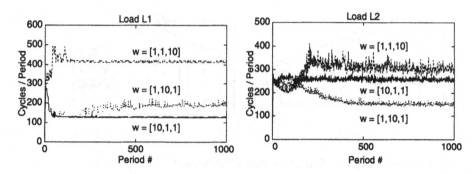

**Fig. 5.** Dynamic task migration applied on synthetic loads L1 and L2 using different strategy vectors.

other strategies in the first phase of simulation on load L2. This observation can be explained by the strong intra-cluster communication present in L2, which is the performance-limiting factor in the initial phase after random distribution. The lack of extensive communication in L1 makes the strategy $[1, 1, 10]$ entirely ineffective for this case.

Strategy $[10, 1, 1]$, in contrast, leads to the best result in the case of load L1. This is due to the load balancing effect which is able to arrange the groups of tasks having different weights such as to avoid overload. This strategy applied to load L2, on the other hand, does not improve performance compared to the initial value at random distribution; this is because all tasks in L2 have been generated with identical weight.

With the third strategy, $[1, 10, 1]$, the effect is just opposite: It delivers the best results on load L2, while the performance of L1 is clearly inferior compared to the second strategy, $[10, 1, 1]$. In the case of L1, the strategy does not take into account the weights of individual tasks as it just balances the number of brothers. Exactly this effect, on the other hand, favorably influences the performance of L2, as particularly the relations between tasks are regarded in a transfer decision.

One conclusion to be drawn from the results presented here is that the proposed dynamic task migration scheme is able to enhance the performance of periodic workloads with different characteristics, but it is sensitive to the particular choice of the strategy vector. Apparently, the strategy vector has to be adapted specifically towards the requirements of a target application. In terms of an autonomous performance improvement regardless of a specific application, or a dynamic change of processing characteristics at run-time, this result is not yet satisfactory. Therefore, the employment of a learning system as a means for an unattended adjustment of strategy parameters is examined in the next section.

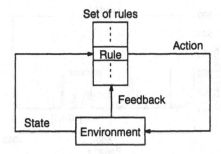

**Fig. 6.** Rule-based learning system.

# 4   Learning-Based Task Migration

The employment of a learning algorithm for adjusting the strategy vector promises to achieve performance improvements independent of the workload characteristics. The application of learning on the problem of task migration has already been a subject of research (e.g., [9][10]).

In our case of learning strategy vectors, we use a rule-based learning classifier system [11] for the task of finding a strategy combination adapted to the load requirements. The general structure of a learning classifier system is depicted in Fig. 6. It consists of a set of rules, each comprising a condition and an action part. The condition part activates a specific rule depending on the state of the environment, whereas the action part specifies a certain measure to be taken in order to improve the system. The success of the rule applied is indicated by the feedback received from the environment. A feedback-based rating of the rules determines their likelihood of being selected again in the future.

In our system, the rules comprise specific combinations of strategies with different weights, i.e., different strategy vectors. System feedback is provided by changes in period durations, reflecting improvement of the system behavior. In case of a positive feedback, the strategy vector being active increases its strength and thus its likelihood of being selected again in the future. The different strategy vectors contained in the rule set have to be defined either a priori or may evolve under the application of genetic operators.

The results of the learning-based adaptation of strategy vectors in comparison to the best static vector for each load—$[10, 1, 1]$ for L1; $[1, 10, 1]$ for L2—are shown in Fig. 7. In this simulation, the rule set consisted of 17 different strategy vectors. In case of load L1, the results from the learning system are not as stable as the best static strategy, but the same lowest level of clock cycles can be reached, particularly as the number of simulated periods increases. In the beginning, the results are dominated by an exploration phase where inferior strategies have to be tested before being discarded for future use. In the further course of operation, more favorable strategies evolve as the preferred choice due to their increase in strength caused by positive feedback.

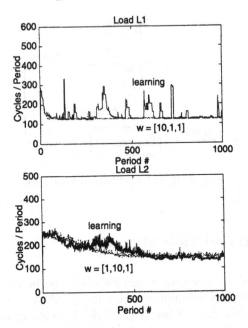

**Fig. 7.** Dynamic task migration applied on synthetic loads L1 and L2 employing the learning system.

The simulation results of load L2 show a similar behavior. After an initial phase of exploration, the number of clock cycles settles in the same range as has been achieved with the best static strategy vector. Moreover, in this case, the performance of the learning system is able to even exceed the performance of the static strategy as the simulation continues. This effect demonstrates the ability of the learning system to autonomously detect favorable strategy vectors that may be difficult to find out a priori.

In the simulations presented here, each cell has been provided with the same strategy vectors from the learning system. As a consequence, the resulting period duration has been related directly to the chosen strategy vector. On the other hand, due to the strictly cellular structure of the array without any global control unit, each cell has to contain its own local rule-based learning system for dynamic strategy selection. Such an arrangement involves some challenges: Due to the different strategies possibly chosen within each cell, the global change of the period duration cannot be related unambiguously to a specific strategy combination. Therefore, some form of local measurement has to be introduced complementing the global feedback figure. This can be achieved by performing an estimate on the possible effect of each strategy combination to be applied, using this estimate to differentiate the global feedback accordingly. Our current research focuses on distributing the learning process in order to comply with the locality principle of the cellular processor array.

# 5  Conclusion

The adaptive resource utilization capability of cellular multiprocessor arrays employing dynamic task migration has been investigated with the aim of increasing the performance of periodic tasks at run-time. A task migration scheme composed of three base strategies has been introduced which allows to adjust its migration effect via a strategy vector. Simulation results with different strategy vectors indicate performance gains due to dynamic task migration, but show a dependence on special algorithm characteristics. A learning classifier system has been employed for unattended adaptation of strategy vectors independent of the particular load. The results indicate that monolithic, cellular multiprocessor arrays employing dynamic task migration constitute a promising architectural direction for signal processing applications.

# References

[1] National technology roadmap. Semiconductor Industry Association (SIA), San Jose, 1997.

[2] R. Chellappa, B. Girod, D. C. Munson Jr., M. Tekalp, and M. Vetterli. The past, present, and future of image and multidimensional signal processing. *IEEE Signal Processing Magazine*, 15(2):21–58, March 1998.

[3] M. H. Willebeck-LeMair and A. P. Reeves. Strategies for dynamic load balancing on highly parallel computers. *IEEE Transactions on Parallel and Distributed Systems*, 4(9):979–993, September 1993.

[4] D. Ghosal and L. N. Bhuyan. Performance evaluation of a dataflow architecture. *IEEE Transactions on Computers*, 39(5):615–627, May 1990.

[5] J. B. Dennis, G.-R. Gao, and K. W. Todd. Modelling the weather with a data flow supercomputer. *IEEE Transactions on Computers*, 33(7):592–603, July 1984.

[6] Arvind and R. S. Nikhil. Executing a program on the MIT tagged-token dataflow architecture. *IEEE Transactions on Computers*, 39(3):300–318, March 1990.

[7] M.-Y. Wu. On runtime parallel scheduling for processor load balancing. *IEEE Transactions on Parallel and Distributed Systems*, 8(2):173–186, February 1997.

[8] J. Watts, M. Rieffel, and S. Taylor. Practical dynamic load balancing for irregular problems. In *Parallel Algorithms for Irregularly Structured Problems: IRREGULAR'96 Proceedings*, volume 1117 of *LNCS*, pages 219–306. Springer Verlag, 1996.

[9] P. Mehra. *Automated Learning of Load-Balancing Strategies for a Distributed Computer System*. PhD thesis, University of Illinois at Urbana-Champaign, 1993.

[10] A. Schaerf, Y. Shoham, and M. Tennenholtz. Adaptive load balancing: A study in multi-agent learning. *Journal of Artificial Intelligence Research*, 2:475–500, May 1995.

[11] John H. Holland. *Adaptation in Natural and Artificial Systems*. University of Michigan Press, 1975.

# NOPE: A Nondeterministic Program Evaluator

Dieter Kranzlmüller and Jens Volkert

GUP Linz, Johannes Kepler University Linz,
Altenbergerstr. 69, A-4040 Linz, Austria/Europe
kranzlmueller@gup.uni-linz.ac.at,
http://www.gup.uni-linz.ac.at:8001/

**Abstract.** Nondeterminism in parallel programs can lead to different results in successive executions even if the same input is supplied. In order to allow debugging of such programs, some kind of replay technique is required. During an initial record phase a program's execution is monitored and information about occurring events is stored in trace files. During subsequent replay steps the traces are used to reproduce an equivalent execution. The problem is that a trace describes one particular execution and therefore limits the user's analysis abilities to this case. Other execution paths can only be analyzed if corresponding program runs can be monitored. This problem is addressed by the nondeterministic program evaluator NOPE, which extends traditional replay to automatically generate other possible execution paths. The idea is to perform combinatorial event manipulation of racing messages on an initial trace to enforce different event orders during replay. If each permutation is tested, different execution paths with previously unknown results and hidden errors may be revealed.

## 1 Introduction

Debugging is the activity of the software-lifecycle that is concerned with detection of computational bugs and run-time failures. Its level of success determines the reliability of the implementation and is therefore critical for the program's quality. Furthermore, the degree of productivity depends on how fast bugs can be tracked down and removed. This can be very tedious and difficult, especially if parallel programs are concerned. The set of errors in parallel programs consists of all errors possible in sequential programs and additionally of errors arising due to several concurrently executing and cooperating tasks (e.g. deadlocks [16]).

A traditional approach to error detection is *cyclic debugging*. The program is executed again and again with the same input, allowing the user to gain more information about program states and intermediate results with a debugger. The basic idea is as follows [17]: determine the results $y$ of the arbitrary function $f$ with a given and valid input $x$. If the input/output pair $\langle x|y \rangle$ does not represents a correct computation of $f$, the error is located somewhere in function $f$. For that reason, $f$ is split up into subfunctions $f_1, f_2,... f_n$ and repeated executions of $f$ are used to determine the correct states between these subfunctions by analyzing intermediate results. In sequential programs this can be achieved with

P. Zinterhof, M. Vajteršic, A. Uhl (Eds.): ACPC'99, LNCS 1557, pp. 490–499, 1999.

breakpoints and single-stepping under control of a debugger. Of course, the input has to be the same and all interactions with the external environment must be reproduced.

In parallel programs the situation is more difficult, because subfunctions of $f$ are executed concurrently and their interactions are strongly influenced by the computer system. At places of communication and synchronization different orders of subtasks may occur due to differences in the state of the hardware, e.g. variations in processor speeds, scheduling decisions, or cache contents [7]. Unfortunately several of these environmental parameters cannot be controlled easily and may lead to different results of the same program, even if the same input data are supplied. As a consequence of this "irreproducibility effect" [16] it is not ad hoc possible to apply the debugging cycle to such nondeterministic parallel programs without special preparations.

A possible solution to this problem is provided by record and replay techniques [9]. During the record phase a program's execution is observed by a monitor, which stores significant information about occurring events in trace files. The data of these traces can be used to perform trace-driven replay of the program under the constraint of providing an equivalent execution [10]. Examples of record and replay techniques are described in [9], [14], and [15].

However, there are still two problems with all these solutions. Firstly, the recorded program run is not necessarily the same as if recording is turned off. This is called the probe effect [3], which means that instrumenting and observing a program always introduces some kind of perturbation. This perturbation not only influences the time of occurrence of each event, but also the order of events in nondeterministic programs [11]. Hence, the user has to be aware that even the monitor with the smallest overhead might influence the program to the extent of event reordering and variations in program results.

The second problem is that the observed results represent not necessarily the only valid execution of a program [1]. For example, in programs computing nondeterministic scenarios of reality (e.g. chaos models, crash simulations,...) several valid solutions may exist. Additionally different execution paths of the program might compute the same results. However, without a suitable steering mechanism it may not be possible to find all prospective solutions, even if a large number of program runs are initiated.

This paper presents an approach to support users experiencing such problems during analysis of nondeterministic message-passing parallel programs. It describes the implementation of the prototype tool *NOPE*, a *NOndeterministic Program Evaluator*. Firstly, *NOPE* records one representative execution of a program while simultaneously trying to meet the requirements for low monitor overhead and small trace size. Secondly, an initial replay step performs an equivalent execution and generates the complete traces. These traces are necessary to evaluate possible execution paths of the program, which are enforced during subsequent re-executions of artificially manipulated replay steps. Of course it is possible that during replay additional nondeterministic behavior is revealed, requiring to perform the evaluation of possible executions iteratively. The results produced with *NOPE* are all possible computations of the program for the provided input.

The paper is organized as follows. The next section describes the parallel computing environment for *NOPE*, which are general message-passing systems. Due to some characteristic restrictions the record and replay method may be simplified. As a result it is easily implemented while still providing a reasonable overhead, which is presented in section 3. Afterwards the basic idea for artificially generating executions of a nondeterministic program that have not been previously traced is introduced. Finally, section 5 describes the evaluation steps with *NOPE* on a small example. After a conclusion some ideas for future improvements are summarized.

## 2   Nondeterminism in Message-Passing Programs

Parallel computation provides two possibilities for implementing the communication and synchronization of concurrently executing processes, shared memory and message-passing. Our technique focuses on message-passing because of some useful properties. Compared with the shared memory model message transmission usually takes place less frequently due to relatively high cost of communication [9]. Additionally, nondeterministic behavior is inserted into a program only through certain statements or function calls. In fact, the potential for non-repeatable behavior arises only where processes interact [9].

In message-passing programs process interaction takes place through functions for sending and receiving messages. Basic operations are point-to-point communication with send and receive, and similar functions are provided by most message-passing libraries. For example, a function for blocking receive, where the process is blocked until a message arrives, is similar in the standard Message-Passing Interface MPI (MPI_Recv [12]), in the Parallel Virtual Machine PVM (pvm_recv [4]), and in vendor-specific libraries like nCUBE 2 (nread [13]). As the main characteristics of these functions are the same for each of these systems, we can define a universally valid primitive for a blocking receive operation:

```
receive(buffer, source, tag,...)
```

The identifier buffer defines the address of the incoming message, whereas source defines the origin of the message. The parameter tag is used to provide some means of classifying the contents of the message. Of course, additional parameters necessary for correct operation of message transmission or for use of additional functionality are available depending on the message-passing systems.

With the receive primitive defined as above a message can only be received if it matches the source and tag values specified by the receive operation. Additionally, the receiver may specify a wild card for source (e.g. ANY_SOURCE) and/or for tag (e.g. ANY_TAG), indicating that any source and/or tag are acceptable. However, due to the possible use of the wild card, nondeterministic behavior may be introduced at receives. At such wild card receives two or more messages might race and the order of their arrival may alter subsequent computations [14].

There are several sources that influence two racing messages. Most important is the invokation time of the send operation, which can vary due to different amount of work, branches and loops in the code. Some more hidden causes are variations in processor speeds, load imbalances, and scheduling decisions, as well as conflicts on the communication network.

Although there are manifold reasons for different message orders to take place, there is only one group of statements - the message receive operations - where nondeterministic behavior may be observable. Furthermore, if programmers wouldn't use wild card receives, program behavior would be deterministic. Yet, there are some reasons for programmer to introduce nondeterminism, for example programmer's laziness, improving performance, or modelling reality.

The first reason is a typical source for errors. If calculating the sender's process number requires too much effort, programmers often tend to use wild card receives which introduces unintended message races. The second reason is rather important, because paradigms like first-come first-served usually mean significant performance improvements. Another important reason is delivered by reality itself, which can often be nondeterministic or chaotic and the same behavior has to be included in computational models. Other reasons are imaginable but each requires a record and replay technique for cyclic debugging.

## 3  Record and Replay Technique

The goal of record and replay techniques is to provide an equivalent execution [10] during successive program runs. To achieve this goal significant information about occurring program events is stored in traces during the record phase, which are used as constraints for the program execution during the replay phases. In the case of message-passing programs, we identify two classes of interesting events, point-to-point communication events (send, receive) and system call return values

The difference between these events is the reproducibility of their parameter values. While the parameters of communication events are easily recomputed during replay, return values from system calls may probably differ between program runs. For example, using the output of the random number generator for calculations clearly influences the program results. Additionally if random numbers are used as destination addresses of send operations, different arrivals at the corresponding receives may be produced. Thus, for system calls the return values have to be stored, while for communication events only ordering information is required during replay.

Ordering information of communication events is usually generated with some kind of clock counters which number the occurrences of events. For example, [15] uses Lamport clocks [8] to obtain a partial order, while [14] applies vector clocks to perform on-the-fly race detection. The drawback is that calculating and updating these clock counters introduces additional monitoring overhead

494    Dieter Kranzlmüller and Jens Volkert

and memory requirement. The advantage is that instead of tracing every event, only events critical for message races have to be stored.

In our approach no clock counters are needed and only a subset of all communication events are traced during the initial record phase. This is achieved by sticking to a useful characteristic of the message-passing interfaces mentioned above. Each of them guarantees that messages in point-to-point communication are non-overtaking (see [12], section 3.5, "Semantics of point-to-point communication"): If a sender sends two messages in succession to the same destination, and both match the same receive, then this operation cannot receive the second message if the first one is still pending. This guarantees that message-passing code is deterministic if the wild card is not used in receives.

Besides that, nondeterministic programs follow a deterministic order of events on each process up to the point where the first wild card receive occurs. At that point several possibilities exist due to possible message races. During record phase the first message to arrive will be accepted. If the replay phase guarantees the same message to be accepted instead of any available message, an equivalent execution is performed. Therefore during record phase an identifier for the incoming message has to be stored. Since overtaking of messages is not possible, this identifier is just the return value of parameter sender, which indicates the address of the sending process. Thus, a record and replay mechanism based on simple point-to-point communication can apply the following simplifications:

- Send events do not have to be traced, because they will be in the same order every time.
- Receive events must be traced only if the parameter sender is a wild card.
- It is sufficient to store the id of the sender. The message tag can be neglected.
- Clock counters are not needed due to implicit message order based on the non-overtaking rule.

These simplifications are the basis for our record and replay technique used in *NOPE*. The operational steps are as follows:

1. Instrument only receive events and system calls [1].
2. During the initial record phase generate initial traces by storing
   - sender id as accepted by receive events which supply wild cards, and
   - return values of system calls.
3. After the initial record phase subsequent replay phases can be used to generate complete traces.

The advantages of this method are that monitoring only occurs at places where nondeterminism may influence the program (wild card receives, system calls). For example, no monitor overhead is generated at send events. Besides that the monitor performs only few and simply activities. Firstly, at receives the parameter sender has to be checked. If its value is no wild card, nothing happens.

---

[1] At present we only instrument calls to the random number generator, which should be sufficient to show the functionality of the system.

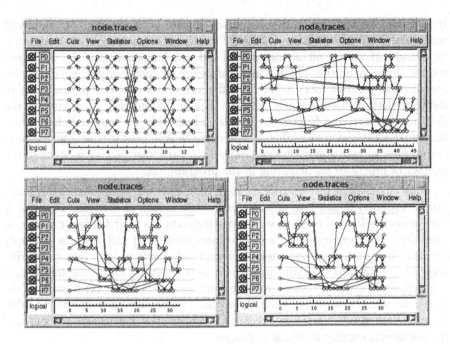

**Fig. 1.** Example event-graphs of a nondeterministic program's execution

Otherwise the monitor has to store the sender id of the incoming message. By using one trace file per process this means that only one integer value has to be saved. Secondly, there is no need to update and store any clock counters, nor to attach any counters to the transmitted messages.

The third step is needed for program analysis. The partial traces of the initial record phase are only necessary to perform a correct replay of the program. Therefore all additional data required for program analysis has to be generated during subsequent replay phases. However, since the initial trace guarantees that there is no perturbation in event order, exhaustive analysis can be performed.

## 4  Event Manipulation

After complete traces have been generated during replay, these traces can serve as the input for any analysis tool. In our case the tool *ATEMPT* (*A Tool for Event ManiPulaTion* [6]) visualizes the traces as an event graph, as for example in figure 1. Various possibilities for program analysis are imaginable, limited only by the functionality of the used inspection tool.

Yet, as mentioned before the generated trace describes exactly one execution of the program. It is possible that for the same input data other execution paths may exist and other results may be obtained. For evaluation of these other possibilities, *ATEMPT* includes a race condition detection mechanism.

It searches the traces for all occurrences of wild card receives. At each wild card receive in the execution, *ATEMPT* evaluates all message race candidates. In general, these are all messages that did not happen before the wild card receive (for a definition of the *happened before* relation see [8]), which is also described as *frontier races* in [14]. A difference exists to their approach due to the fact, that overtaking is not possible. This limits our race candidates to the *first* message arriving from each process, which did not happen before. Only these first messages are possibly accepted at the wild card receive. Furthermore the number of race condition candidates at a particular receive is at most the number of processes participating in the program run.

After all message race candidates have been detected, additional replay steps can be initiated in order to uncover previously hidden executions. The idea is that at each wild card receive every race candidate has to arrive first during one execution. The technique is called *event manipulation* [7] and previous attempts operated by graphically exchanging the order of events in the event graph display. The problem is that with many wild card receives a lot of user interaction is necessary in order to test all possible combinations. In fact, the number of directly connected (follow-up) executions, that can be derived from one recorded trace, is limited by $r^n$ where $r$ equals the number of wild card receives and $n$ is the maximum number of racing messages.

Although in *NOPE* there is no more user interaction necessary to combinatorial generate the follow-up executions, there is still a drawback. The resulting number of executions may be rather large, requiring huge amounts of computing power to automatically test all possibilities [1]. Yet there is a distinction between the races in a program dividing the set of candidates into *intended* and *unintended* races. Intended races may be assumed to be harmless on the outcome of the program, which means that these races need not to be investigated. The experienced user can contribute this knowledge to the testing cycle by graphically removing intended races. Similarly, unintended races that are not wanted by the user can be removed from the code immediately. This reduces the number of replay steps and therefore the time necessary to analyze the races to all cases, where the relation between races and results is unknown. Of course, if the intention of a race is unclear, it is better to leave it in the test sequence.

Based on the set of races that have to be investigated, *NOPE* changes the event order by automatic event manipulation and starts a replay for each permutation. The replay forces the communication to take place as specified in the traces until a point of exchange is detected. From that point on the program's execution is running again without control of the replay mechanism, because there is no knowledge about the event order after that point (see [7] for more details about event manipulation and associated trace driven replay).

The artificially generated subsequent replays deliver all possible executions that are *directly* connected to the original program run. However, there remains the possibility of requiring additional replay steps that could not be detected in the first evaluation step. These problems are introduced with branches and loops [5]. If the number of loop iterations or the decisions at branches depend on the results of the message races, possibly new communication patterns are achieved. This also requires that the evaluation of message race candidates has to be repeated for each generated program execution. Additionally some precautions to limit the number of executions have to be included, for example by allowing a maximum number of loop iterations or restricting the number of indirections from the original trace.

Another problem of *NOPE* is the amount of data generated. Each replay generates a trace necessary for the analysis as well as the usual results of the program. To cope with these space requirements some simple compression mechanisms can be applied. The traces of all executions can be converted from complete traces back to partial traces in the same format as the initial trace by removing all additional data. Only the partial traces are needed to permit equivalent replay and are therefore sufficient for later analysis steps. Of course, if the user wants to analyze the program run more carefully, uncompressing partial traces is only possible through actual replay.

Another compression concerns the possible similarity of partial traces and output data. For example, executions with the same output only need to store the differences in the partial traces instead of storing the same output twice and both traces. Furthermore it is possible that portions of the program outputs resemble each other strongly and are therefore rather qualified for most existing compression algorithms.

# 5  Example

After introducing the basic principles and operation of *NOPE*, we will now demonstrate a case study. The program under consideration performs data circulation on a hypercube machine as used in various numerical algorithms [2]. The task is to circulate data items available on all processes so that these data visit each of the processes exactly once. Therefore each task iteratively performs three activities: (a) compute operation on data, (b) send data to a specific process, and (c) receive data from a specific process. If the program is correct, the resulting communication pattern on 8 processes is equal to the upper left event graph in figure 1.

An important point in the implementation of this function is the selection of the communication partners. Since this algorithm matches specifically the hypercube architecture, so that each transmission always occurs along one cube-dimension, communication takes place only between directly connected neighbors. This defines the destination address that has to be used at the send operations. However, while it is clear that sends have to know exactly where to transmit the message, this may not be necessary for the receive operations. It may seem that the cyclic behavior of the three steps (a), (b), and (c) allows to use a wild card receive. In fact, if the cycle time, determined by the amount of computation time between communication events, is long enough the correct pattern may be retrieved all the time. The possible consequences of this "programmer's laziness" is hidden until cycle time is decreased, for example when porting the program onto a faster machine.

With *NOPE* it is possible to detect this erroneous behavior. Starting with either the correct pattern or any other pattern, *NOPE* would generate the remaining patterns. All possible results of this algorithm are produced and can be analyzed for correctness. The event graphs in figure 1 contain examples for several nondeterministic executions of the same code.

The number of executions needed to generate all combinations is limited with $(p^2 - p)^{\log(p)}$, where $(p^2 - p)$ is the number of wild card receives ( $(p-1)$ receives per process) and $\log(p)$ is the number of racing messages (=hypercube dimension). The memory necessary to store the initial (or compressed) traces can be calculated with $(p-1)(p^2-p)^{\log(p)}$ bytes, if we assume that 1 byte is sufficient for storing the sender's address. Although these numbers are by far not as big as they are when considering complete traces, another decrease can be obtained due to the symmetry of the algorithm. For users investigating the event graph of the initial execution, the communication pattern allows to restrict the replays to permutations on one single process. Therefore only the race candidates on one process may be defined as unintended races and the number of executions is lowered to $(p-1)^{\log(p)}$. Furthermore, if the user considers symmetric properties of the algorithm on the (horizontal) time-axes, only $\log(p)^{\log(p)}$ executions remain.

# 6   Conclusions and Future Work

This paper introduced the *NO*ndeterministic *P*rogram *E*valuator *NOPE*, which is a debugging tool for parallel message-passing programs. It allows cyclic debugging of nondeterministic code by providing an equivalent execution during subsequent program runs with its record and replay mechanism. Compared to previous approaches it reduces monitor overhead and trace size, and therefore the problem of the probe effect. This is achieved by taking advantage of the non-overtaking characteristic of messages originating from one process, which is guaranteed by many available message-passing libraries. There is no need to generate and update clock counters and only one value per wild card receive has to be stored in the traces.

Another feature of *NOPE* is its additional support for testing other possible execution paths. Since nondeterministic programs may produce different results in successive program runs, even if the same input data is supplied, more than one execution has to be analyzed during testing and debugging. Through automatic event manipulation and replay, different event orders are forced to take place during subsequent replay steps. If each combination of message arrival is tested, all possible execution paths of the program are revealed. Nevertheless, as demonstrated with the example of section 5, generating all combinations requires many executions and therefore lots of results and traces.

Therefore future work in this project is concerned with solutions for reduction of test cases. Since races always depend on the characteristics of the program, only few cases can be eliminated from the test cycle automatically. For example, the distinction between intended and unintended races can only be made by the user, while symmetric patterns can also be detected by tools. However, there are still some possibilities which we want to investigate, mainly concerning the probability of message arrival. Yet, we believe that our approach already supports the users and improve debugging of nondeterministic programs.

# References

1. Damodaran-Kamal, S.K. and Francioni, J.M., "Testing Races in Parallel Programs with an OtOt Strategy", Proc. 1994 Intl. Symp. on Software Testing and Analysis, Seattle, WA (1994).
2. Dekel, E., Nassimi, D. and Sahni, S., "Parallel Matrix and Graph Algorithms", Siam Journal on Computing, Vol. 10, No. 4, pp. 657-675 (Nov. 1981).
3. Gait, J., "The Probe Effect in Concurrent Programs", IEEE Software - Practise and Experience, Vol. 16(3), pp. 225-233 (March 1986).
4. Geist, A., Beguelin, A., Dongarra, J., Joang, W., Manchek, R. and Sunderam, V., "PVM 3 User's Guide and Reference Manual", Techn. Rep. ORNL/TM-12187, Oak Ridge Natl. Lab., Oak Ridge, TN (May 1994).
5. Helmbold, D.P. and McDowell, C.E. "Race Detection - Ten Years Later", in: Simmons, M.L. et al (Eds.), "Debugging and Performance Tuning for Parallel Computing Systems", IEEE CS Press, Los Alamitos, CA, pp. 101-126 (1996).
6. Kranzlmüller, D., Grabner, S. and Volkert, J., "Event Graph Visualization for Debugging Large Applications", Proc. of SPDT'96, SIGMETRICS Symposium on Parallel and Distributed Tools, Philadelphia, PA, pp. 108-117 (May 1996).
7. Kranzlmüller, D., Grabner, S. and Volkert, J., "Debugging with the MAD Environment", Parallel Computing, Vol. 23, Nos. 1-2, pp. 199-217 (Apr. 1997).
8. Lamport, L., "Time, Clocks, and the Ordering of Events in a Distributed System", Comm. ACM, pp. 558 - 565 (July 1978).
9. LeBlanc, T.J. and Mellor-Crummey, J.M., "Debugging Parallel Programs with Instant Replay", IEEE Trans. Comp., Vol. C-36, No. 4, pp. 471-481 (Apr. 1987).
10. Leu, E., Schiper, A., and Zramdini, A., "Execution Replay on Distributed Memory Architectures", Proc. 2nd IEEE Symp. on Parallel & Distributed Processing, Dallas, TX, pp. 106-112 (Dec. 1990).
11. Malony, A.D. and Reed D.A., "Models for Performance Perturbation Analysis", Proc. ACM/ONR Workshop on Parallel and Distributed Debugging, ACM SIGPLAN Notices, Vol. 26, No. 12, pp. 15-25 (Dec. 1991).
12. Message Passing Interface Forum, "MPI: A Message-Passing Interface Standard - Version 1.1", http://www.mcs.anl.gov/mpi/ (June 1995).
13. nCUBE Corporation, "nCUBE 2 Programmer's Guide" (1992).
14. Netzer, R.H.B. and Miller, B.P., "Optimal Tracing and Replay for Message-Passing Parallel Programs", Supercomputing '92, Minneapolis, MN (Nov. 1992).
15. Ronsse, M.A. and Kranzlmüller, D., "Rolt$^{MP}$ - Replay of Lamport Timestamps for Message Passing Systems", Proc. 6th EUROMICRO Workshop on Parallel and Distributed Processing, Madrid, Spain, pp. 87-93, (Jan. 21-23, 1998).
16. Snelling, D.F. and Hoffmann, G.-R., "A comparative study of libraries for parallel processing", Proc. Intl. Conf. on Vector and Parallel Processors, Computational Science III, Parallel Computing, Vol. 8 (1-3), pp. 255-266 (1988).
17. Wasserman, H. and Blum, M., "Program result-checking: a theory of testing meets a test of theory", Proc. 35th IEEE Symp. Foundations of Computer Science, pp. 382-392 (1994).

# Visual-MCM: Visualising Execution Histories on Multiple Memory Consistency Models

Alba Cristina Melo and Simone Cintra Chagas

University of Brasilia, Campus Universitario - Asa Norte, Brasilia, Brazil

**Abstract.** The behaviour of Distributed Shared Memory Systems is dictated by the memory consistency model. In order to provide a better understanding on the semantics of the memory models, many researchers have proposed formalisms to define them. Even with formal definitions, it is still difficult to say what kind of execution histories can be produced on a particular memory model. In this paper, we propose Visual-MCM, a visualisation tool that shows what operations orderings could lead to user-defined execution histories on different memory models. We also present a prototype of Visual-MCM that analyses execution histories for two different memory consistency models.

## 1 Introduction

Using the shared memory programming paradigm in parallel architectures is quite complex. The main reason for this is the difficulty to make the shared memory behave exactly as if it was a uniprocessor memory. This characteristic is important since it could make uniprocessor programs automatically portable to parallel architectures. Besides, programmers are already used to the uniprocessor shared memory programming model and using a similar model would make parallel programming easier.

In [6], a memory model called Sequential Consistency (SC) was proposed. In this model, all parallel processes are able to observe all shared memory accesses in the same order. Although SC provides a relatively easy programming model, there is a great overhead on coherence operations to guarantee the global order.

To reduce coherence overhead, researchers have proposed to relax some consistency conditions, thus creating new shared memory behaviours that are different from the traditional uniprocessor one. These memory models are called relaxed memory models because they only guarantee that some of the shared memory accesses are seen by all processors on the same order.

Memory consistency models, have not been originally formally defined. The lack of a unique framework where memory models can be formally defined was also observed by other researchers and some work was indeed done in the sense of formal memory model definitions.

However, it is still difficult to say if some undesirable orderings will be produced when using a particular memory model, even for formally-defined memory consistency models. This led sometimes to unexpected behaviours, and some undesirable execution histories have been produced.

P. Zinterhof, M. Vajteršic, A. Uhl (Eds.): ACPC'99, LNCS 1557, pp. 500–509, 1999.

In this article, we present Visual-MCM, a visualisation tool that assists DSM designers in the task of analysing a memory consistency model. This new tool analyses a particular execution history on a memory model and shows if it is valid or not. Basically, we build an execution tree and traverse it, respecting the constraints imposed by the current memory model.

The rest of this paper is organised as follows. Section 2 describes the formalism used to define memory models. Section 3 describes the approach used to decide the validity of an execution history on a chosen memory model. The prototype of Visual- MCM is presented in section 4. Related work in the area of DSM visualisation is presented in section 5. Finally, conclusions are presented in section 6.

# 2   Formalising Memory Consistency Models

To define memory models formally, we use a history-based system model that was already described in [4]. In table 1, we only review some definitions.

**Table 1.** System Model

| System | A finite Set of Processors |
|---|---|
| Processor $p_i$ | Executes operations on M |
| Shared Global Memory M | Contains all memory addresses |
| Local Memory $m_i$ | Caches all memory addresses of M |
| $o_{pi}(x)v$ | Operation executed by $p_i$ on address $x$ with value $v$ |
| Basic types of operations | read(r) and write(w) |
| Synchronisation operations | sync(x);acquire(x);release(x);user-defined functions |
| $o_{pi}(x)v$ issued | Processor $p_i$ executes the instruction $o_{pi}(x)v$ |
| $r_{pi}(x)v$ performed | The value of $x$ returned to $p_i$ cannot be modified |
| $w_{pi}(x)v$ | $C = \sum_{a=0,n-1} w_{pa}(x)v$ |
| $w_{pi}(x)v$ performed to $p_j$ | Value $v$ is written on position $x$ of $m_j$ |
| $w_{pi}(x)v$ performed | $w_{pi}(x)v$ is performed with respect to all processors |
| Local Execution History $H_{pi}$ | Ordered sequence of memory operations issued by $p_i$ |
| Execution History $H$ | $U H_{pi}$ |
| Memory Consistency Model | Order relation on a set of shared memory accesses |

In our definitions, we use the notion of linear sequences. If Q is a history, a linear sequence of Q contains all operations in Q exactly once. A linear sequence is legal if all read operations r(x)v return the value written by the most recent write operation on the same address in the sequence.

In execution histories, we have some operation orderings that are allowed and some orderings that are forbidden . The decision of which orderings are valid is made by the memory consistency model. One execution history is valid on a memory consistency model if it respects the order relation defined by the model.

An order relation that is used in the definition of nearly all memory consistency models proposed until now in the literature is program order.

> *Program-order.* An operation o1 is related to an operation o2 by program-order ($o1 \xrightarrow{po} o2$) if:
> a) both operations are issued by the same processor $p_i$ and o1 immediately precedes o2 in the code of pi or
> b) $\exists o3$ such that $o1 \xrightarrow{po} o3$ and $o3 \xrightarrow{po} o2$.

The program order that we use in our definitions is, thus, a total order on $H_{pi}$.

## 2.1  Sequential Consistency

Sequential Consistency (SC) imposes an order relation on all shared memory accesses, i.e, it imposes a total order on H. Our formal definition of sequential consistency is derived from the definition presented in [1]:

> *Sequentially Consistent History.* A history H is sequentially consistent if there is a legal linear sequence $\xrightarrow{SC}$ of the set of operations on H such that:
> i) $\forall o1, o2$ where $o1 \xrightarrow{po} o2$ then $o1 \xrightarrow{SC} o2$.

In this definition, we define a new order relation ( SC ) where all processors must perceive the same execution order of all shared memory accesses (legal linear sequence of H). In this order, all operations performed by a process must respect the program order (i).

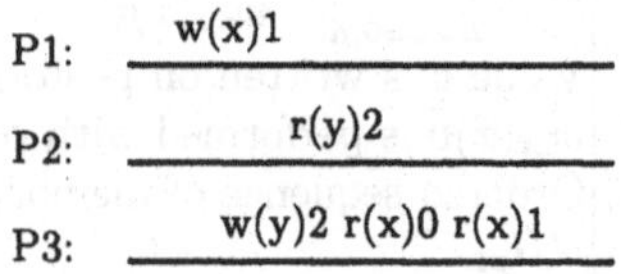

$$P1: \quad \underline{\quad w(x)1 \quad}$$
$$P2: \quad \underline{\quad r(y)2 \quad}$$
$$P3: \quad \underline{\quad w(y)2 \ r(x)0 \ r(x)1 \quad}$$

**Fig. 1.** A Sequentially Consistent Execution History

The execution history shown in figure 1 is sequentially consistent since we are able to derive at least one legal linear sequence of all memory operations where all processors agree. In this history, the program order that must be respected is $w_{p3}(y)2 \xrightarrow{po} r_{p3}(x)0 \xrightarrow{po} r_{p3}(x)1$. The following execution order is a valid execution under SC: $w_{p3}(y)2 \xrightarrow{SC} r_{p2}(y)2 \xrightarrow{SC} r_{p3}(x)0 \xrightarrow{SC} w_{p1}(x)1 \xrightarrow{po} r_{p3}(x)1$. Note, however, that this is not the only legal sequence $\xrightarrow{SC}$ that can be derived.

The sequence $w_{p3}(y)2 \xrightarrow{SC} r_{p3}(x)0 \xrightarrow{SC} w_{p1}(x)1 \xrightarrow{SC} r_{p3}(x)1 \xrightarrow{po} r_{p2}(y)2$. is also a valid execution sequence on SC.

## 2.2 PRAM Consistency

PipelinedRAM Consistency (or PRAM) [7] relaxes some conditions imposed by SC and requires only that writes issued by the same processor are observed by all processors in the order they were issued. In other words, among the operations issued by other processors, only write operations must obey the program order. Before defining PRAM consistency formally, we must define a new execution history:

> *History* $H_{pi+w}$. Let H be a global execution history and $p_i$ be a processor. History $H_{pi+w}$ is the history of writes with respect to processor $p_i$ and it is a sub-history of H that contains all operations issued by processor $p_i$ and all write operations issued by the other processors.

> *PRAM Consistent Execution History.* A history H is PRAM consistent if there is a legal linear sequence $\xrightarrow{PRAM}$ of the set of operations on $H_{pi+w}$ such that:
> i) $\forall o1, o2$ where $o1 \xrightarrow{po} o2$ then $o1 \xrightarrow{PRAM} o2$.

By this definition, we can see that it is no longer necessary for all processors to agree on the order of all shared memory operations. Every processor has its own view ($H_{pi+w}$) of the shared global memory M and the order $\xrightarrow{PRAM}$ is defined for $H_{pi+w}$. In this view, the program order of $H_{pi+w}$ must be respected (i).

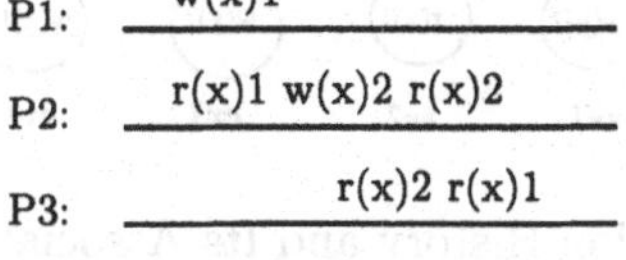

| | |
|---|---|
| P1: | w(x)1 |
| P2: | r(x)1 w(x)2 r(x)2 |
| P3: | r(x)2 r(x)1 |

**Fig. 2.** A PRAM Consistent Execution History

For an execution history to be PRAM consistent, there must be a history $H_{pi+w}$ for every $p_i$ where the program order is respected. Considering the execution history presented in figure 2, the following legal $H_{pi+w}$ can be derived:

$H_{p1+w} : w_{p1}(x)1 \xrightarrow{PRAM} w_{p2}(x)2.$

$H_{p2+w} : w_{p1}(x)1 \xrightarrow{PRAM} r_{p2}(x)1 \xrightarrow{PRAM} w_{p2}(x)2 \xrightarrow{PRAM} r_{p2}(x)2.$

$H_{p3+w} : w_{p2}(x)2 \xrightarrow{PRAM} r_{p3}(x)2 \xrightarrow{PRAM} w_{p1}(x)1 \xrightarrow{PRAM} r_{p3}(x)1.$

As it was possible to derive valid $H_{pi+w}$ for every $p_i$, this execution history is valid on PRAM Consistency. Note, however, that the same history is not a valid history on Sequential Consistency since it is impossible to derive a total order on H.

# 3    Validating Execution Histories

Computing orderings for execution histories is a problem that has been extensively studied in [8] and this problem is shown to be co-NP-hard.

In Visual-MCM, the task of validating a given execution history is decomposed in three main steps. First, an execution tree is constructed for the given history. Second, a memory model is chosen and some of its constraints are extracted. Third, the extracted constraints are used to traverse the execution tree and find out if there are execution paths that are valid on the chosen model.

## 3.1    Step 1: Building the Execution Tree

In order to build a complete execution tree, we must include all possible interleavings of shared memory operations that appear in an execution history. Figure 3 shows an execution history and its associated execution tree.

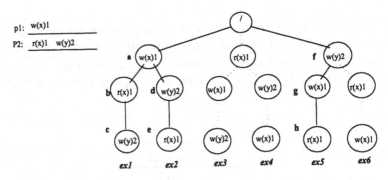

**Fig. 3.** An Execution History and Its Associated Execution Tree

In figure 3, there are six possible execution paths. Each path is an ordered sequence of nodes, from the root to the leaf. For example, the path $ex1$ is the execution order: $w_{p1}(x)1 \longrightarrow r_{p2}(x)1 \longrightarrow w_{p2}(y)2$. As it can be easily seen, there are p! linear execution orders that can be derived, where p is the number of memory operations to be considered. Even for a relatively small number of operations, we observe an explosion of the number of possible paths. Thus, something must be done in order to reduce the size of the execution tree.

Although all sequences in figure 3 are linear, some of them are not well-formed [9]. Sequences ex3, ex4 and ex6 are not well-formed since they read a value before this value has been written. Removing non well-formed sequences from the execution tree will surely reduce its size and that is done with no side-effect, as there is no practical interest in allowing values to be read before they are written. In figure 3, the execution tree built by our system will not contain the dashed paths.

## 3.2 Step 2 - Extracting Constraints from the Memory Consistency Model

For every chosen memory consistency model, there are two decisions that must be taken. These decisions will be based on the formal definitions. As an illustration, we will consider the formal definitions of Sequential Consistency and PRAM Consistency presented in section 2.1 and 2.2.

First, we must identify the set of operations that will be validated. In the case of strong memory models, such as Sequential Consistency, all shared memory operations must be included. In the case of relaxed memory models, only a subset of the memory operations will be verified. In PRAM consistency, $H_{pi+w}$ is this subset.

Second, we must define what order must be respected in this set of operations. In the particular case of Sequential Consistency and PRAM Consistency, the only order to be respected is program order. For the particular history in figure 3, the program order that must be respected is $r_{p2}(x)1 \xrightarrow{po} w_{p2}(y)2$.

## 3.3 Step 3: Traversing the Execution Tree

The last step in validating a history consists to examine all possible paths that belong to the execution tree and verify if they satisfy the constraints imposed by the model. We use a left-to-right in-depth algorithm to traverse the execution tree in search of valid execution paths. When a node is visited, there are two decisions that can be taken. If the operation in the node is valid on the memory consistency model, the in-depth search continues and the next node is examined. If the operation is not valid, the current path is abandoned and the next left-to-right path is examined. If we arrive at one valid leaf, that means that the whole path is valid.

To illustrate this procedure, the search for valid paths on Sequential Consistency will be done on the execution tree presented in figure 3. We start our traversal by examining node a. The operation $w_{p1}(x)1$ is valid since there are no constraints on legality or program order. The next operation in the path is $r_{p2}(x)1$ on node b. The legality is respected since the operation $r_{p2}(x)1$ reads the value written by the most recent write in the sequence. Program order of $p_2$ is also respected: $r_{p2}(x)1$ is the first operation in $p_2$'s program code. Thus, node b is also valid. The next node is node c. Operation $w_{p2}(y)2$ respects program order since it comes after $r_{p2}(x)1$. For this reason, it is also a valid node. As we were able to arrive at one valid leaf, the whole path $w_{p1}(x)1 \longrightarrow r_{p2}(x)1 \longrightarrow w_{p2}(y)2$ is valid on Sequential Consistency. The algorithm continues to examine the tree and chooses node d that contains $w_{p2}(y)2$. This is not a valid node on Sequential Consistency since this order violates the program order of $p_2$. For the same reason, the path starting on node f is not valid. As there is at least one valid execution path, the execution history is valid on SC.

Validating an execution history on PRAM Consistency is much more complex. The main reason for this is that, as it is a relaxed memory model, we must be able to derive valid execution paths for every processor.

## 4    Prototype Implementation

We implemented a prototype of Visual-MCM using the Delphi 3.0 programming environment. The functionality of the prototype is shown in figure 4.

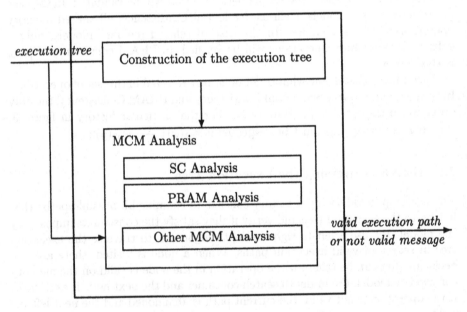

**Fig. 4.** Prototype of Visual-MCM

In our visualisation prototype, there are two main modules: Construction of the Execution Tree and Memory Consistency Model Analysis. The first module receives an execution history as input. This execution history can be selected by the user among predefined execution histories or can be provided by the user himself. The basic function of this module is to construct a tree containing the possible execution paths that could have generated the chosen execution history.

Normal execution trees are generated as explained in section 3.1. Nevertheless, in order to further reduce the number of possible paths, the user can choose to generate an optimised tree. The optimised execution tree will not contain execution paths that do not respect program order. This kind of execution trees cannot be used in memory consistency models that do not impose program order to be entirely respected. If the user asks our visualisation tool to analyse histories on these kind of memory models and an optimised execution tree has been generated, our tool forces the user to regenerate a normal execution tree.

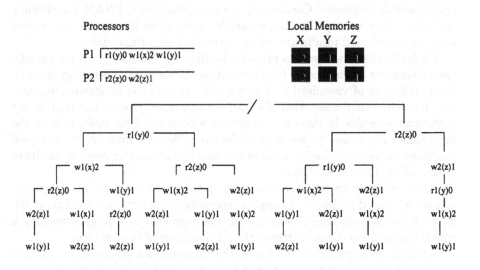

**Fig. 5.** Visualizing Execution Orderings on Visual-MCM

After the tree construction, the user must choose which memory consistency model is to be used in the analysis. By now, we have only implemented Sequential Consistency and PRAM Consistency. The chosen MCM receives the execution history and its associated execution tree and searches the tree for valid paths, as explained in section 3.3.

Our visualisation tool provides both a memory view and an ordering view. The memory view shows the values of each memory position x for every memory Mi. The ordering view shows the execution path that is being examined as well as the valid and invalid execution paths that have already been decided. The analysis of an execution history can be done automatically or in a step-by-step basis.

Figure 5 shows how a particular execution history is analysed on Visual-MCM. Note that the analysis of possible execution paths depends only on the Memory Consistency Model. No particular implementations of memory models are considered.

# 5    Related Work

[5] presents a visualisation tool called StormWatch. This tool is able to analyse the same application on different memory coherence protocols. StormWatch provides trace, source and communication views. The trace view is based on execution histories. Maya is a simulation platform described in [2]. It analyses the behaviour of coherence protocols that implement the following memory consistency models: Sequential Consistency, Causal Consistency, PRAM Consistency and Entry Consistency. The same parallel application is executed on several memory models and results are shown in terms of execution times.

Basically, these two systems provide visualisation of the behaviour of parallel applications on some specific implementations of memory consistency models. While this kind of visualisation is very useful, some of the constraints imposed by a particular implementation of a memory consistency model can lead to bad performance results. In this case, we are not able to tell if the problem is on the implementation or on the memory model itself. Besides, most of the proposed visualisation tools emphasize performance and do not analyse possible results of a parallel application execution.

Analysing feasible event orderings has been previously done in the domain of parallel and distributed debugging, specially to detect race conditions [3]. [8] analysed formally the possible event orderings that can be generated by a parallel program and demonstrated the NP-hardness of this problem.

As far as we know, this is the first work that provides a visualisation tool that can be used to analyse the effects and the potentialities of memory consistency models, in a way that is totally independent from coherence protocols that could implement it. We claim that this separation is necessary and the design of a consistency mechanism should be done in two steps. First, one or more memory consistency models must be chosen. Second, for each chosen memory consistency model, a coherence protocol must be specified. Our visualisation tool will be helpful on the first step of this process.

# 6    Conclusions and Future Work

In this article, we presented a visualisation tool that analyses execution histories and shows graphically what valid execution paths could have led to their production. The analisys of execution histories can be done on a variety of memory consistency models. We claim that visualising the possible interleavings of shared memory accesses is helpful for multiprocessor hardware designers and DSM-system designers in the task of choosing the most appropriate memory model. This can reduce considerably the number of unexpected results that are produced after the DSM- based machine or the DSM-software is released.

As future work, we intend to incorporate other memory models to Visual-MCM. Also, we intend to define a language for specifying memory consistency models based on formal definitions. Using this language, the user can define its own memory models and interactively visualise the interleavings of memory operations that can produced on the newly defined memory consistency model. This facility will be helpful in memory model definition as well as in the comparison of distinct memory consistency models.

# References

1. Ahamad, M. et al.: The Power of Processor Consistency. Technical Report GIT-CC-92/34, GIT (1992) 21 pages
2. Agrawal D., Choy M., Leong H., Singh A.: Evaluating Weak Memories with Maya. In Proceedings of the 8th Workshop on Parallel and Distributed Simulation (1994) 151-155
3. Adve S., Hill M., Miller B., Netzer H.: Detecting Data Races on Weak Memory Systems. In Proceedings of the 18th ISCA (1991) 234-243.
4. Balaniuk A.: Multiple Memory Consistency Models on a SVM Parallel Programming Environment. In Proceedings of the Int. Conf. OPODIS'97 (1997) 249-260.
5. Chilimbi T., Ball T., Eick S., Larus J.: StormWatch: a Tool for Visualizing Memory System Protocols. Supercomputing'95 (1995).
6. Lamport L.: How to Make a Multiprocessor Computer that Correctly Executes Multiprocess Programs. IEEE Transactions on Computers (1979) 690-691.
7. Lipton, R. J. and Sandberg, J. S.: PRAM: A Scalable Shared Memory. Technical Report CS-TR-180-88, Princeton University, (1988).
8. Netzer R., Miller B.: On the Complexity of Event Ordering for Shared-Memory Parallel Program Executions. Technical Report TR-908, University of Wisconsin-Madison, (1990).
9. Heddaya A., Sinha H.: An Overview of Mermera: a System Formalism for Non-Coherent Distributed Parallel Memory. Technical report BU-CS-92-009, Boston University, (1992), 21 pages.

# High Performance Implementation of MPI for Myrinet

Maciej Golebiewski, Markus Baum, and Rolf Hempel

C & C Research Laboratories, NEC Europe Ltd., Sankt Augustin, Germany.
{Golebiewski, Baum, Hempel}@ccrl-nece.technopark.gmd.de

**Abstract.** This paper presents a new implementation of MPI on a cluster of Linux-based, dual-processor PCs interconnected by a Myricom high speed network. A survey of existing software for this hardware configuration resulted in the non-availability of a fully functional, correct and complete MPI library exploiting the full hardware potential. Our library uses MPICH for the high level protocol and FM/HPVM for the basic communications layer. It allows multiple processes and multiple users on the same PC, and passes an extensive test suite, including all test programs from the MPICH distribution, for both C and Fortran. The presented benchmarks, both simple communication kernels and full applications, show good performance. The result is the first high-performance MPI interface which allows regular multi-user service for applications on our PC cluster.

*Keywords:* Workstation clusters; Myrinet; MPI; FM; HPVM.

## 1 Introduction

Continuous price reductions for commodity PC equipment and increasing processor speeds have made PC clusters an attractive low-price alternative to MPP systems for parallel applications. Most of the clusters constructed so far, however, including systems of the Beowulf-class [1], as represented by the Loki cluster, use Fast-Ethernet networks and high-overhead protocols for communication, such as TCP or UDP. This results in a large performance gap between the processing speed of a single PC and the communication between them. On Loki, for example, the communication latency and bandwidth achieved by MPICH 1.1.0 [7] with a switched Fast-Ethernet network are 390 $\mu$s and 8.8 MB/s, respectively [26]. Even using a GigaBit Ethernet would not lead to much improvement because the maximum raw TCP bandwidth under Linux is only 18.405 MB/s, as reported in [6] (latency not given). So, up to now only very coarse-grained parallel applications can use this class of machines efficiently. Recent developments in high-speed networking hardware, however, as, for instance, the Scalable Coherent Interface (SCI) [9] or Myrinet [18], allow the extension of this application domain towards problems with medium-, or even fine-grained, parallelism.

P. Zinterhof, M. Vajteršic, A. Uhl (Eds.): ACPC'99, LNCS 1557, pp. 510–521, 1999.
© Springer-Verlag Berlin Heidelberg 1999

The system installed at our laboratory (as described in Sec. 2) was set up to provide applications programmers with a local and cost-effective parallel program development platform. Most of the relevant codes are parallelized by using the Message-Passing Interface (MPI) [24,16]. A system for our work environment must, therefore, fulfill the following requirements:

1. Reliable, error-free data transmission,
2. support for more than one user at the same time,
3. support for multiple processes per dual-processor PC,
4. a complete and correct implementation of MPI, including the Fortran77 binding, and
5. good communication performance at the MPI level.

We tested all available MPI implementations for Myrinet, but none of them fulfilled all requirements (for details see Section 3). Some of them were optimized for performance, but didn't conform to the MPI standard and didn't pass our test programs. Others did not allow multiple users or check for transmission errors. So, we had to develop our own MPI implementation in order to provide our applications programmers with the high-performance interface they need for their work.

In this paper we present the two versions of the MPI library we have developed during the project (Section 4), followed by results for both kernel and full application benchmarks (Section 5). Finally, we discuss how conflicting design choices, as, for example, optimal ping-pong communication performance versus an improved progress for MPI requests, affect the overall application performance (Section 6).

## 2  Target Platform

Instead of building the cluster from the ground ourselves, we decided rather to base our machine on the previous work by NEC Research Institute (NECI) on the Local Area MultiProcessor (LAMP) [5,20]. With its symmetric multiprocessor (SMP) architecture, the LAMP was well-suited for our intended experiments with clustered shared memory machines, and the Myrinet provided sufficient communication speed.

The cluster installed at our laboratory is one of three LAMPs assembled by NECI (the other two are located at NEC Central Research Laboratory in Tokyo and at their own lab in Princeton). Our machine is a 16 node SMP PC cluster interconnected by the Myrinet network. Dual 8-Port Myrinet-SAN switches, each serving 4 PCs, provide a tetrahedral network topology. Each node has two Pentium Pro 200MHz CPUs installed on a Tyan S1668 motherboard with Intel's 440FX chipset and 512KB of L2 cache per each CPU. The PCs have 128 MB of EDO RAM each, so that the cluster has a total of 32 CPUs and 2 GB of RAM. The nodes are equipped with two network cards: an Ethernet card and a LANai board for accessing the Myrinet network. The LANai cards are basic 32 bit PCI models equipped with only 256 KB of SRAM. The cluster runs under Linux, because this system supports more hardware than any other flavor of UNIX available for PCs. Also, having full UNIX support on all nodes facilitates both the administration and usage of the parallel machine. Initially we were using kernel version 2.0.30 with SMP support, but recently the cluster has been upgraded to the Red Hat 5.1 distribution, so the current kernel version is 2.0.34 (SMP). All machines are also connected to a shared Ethernet segment for the operating system management.

In July 1998 we have added a uniprocessor Pentium II PC (300 MHz, 64 MB RAM) with SCSI disk drives, and use it as a file and compilation server for the cluster. It is not part of the Myrinet network, and connects to the other PCs only via Ethernet. The server runs under the same Linux kernel as the nodes. Moving the management and compile functions from one of the cluster nodes to a machine outside of the Myrinet considerably improved the cluster stability.

## 3 Related Work

Due to space limitations we will describe only projects using Myrinet and supporting Linux. Myricom introduced the Myrinet network in August 1994 [18]. It is a switched network with very low latency (the most recent models of the switches deliver a cut-through latency of about 300 ns) and the very high bandwidth of 1.28+1.28 Gbps on each link. All connections are full duplex. Meanwhile, in a number of projects clusters with Myrinet interconnection networks were built, and the research focused on the development of low level communication libraries optimized to deliver as much raw network performance as possible to the user. Often, those libraries achieved a communication speed similar to the performance of proprietary interconnection networks of MPP systems.

Below we present five software packages we have evaluated on our cluster. Their most significant features are summarized in Tab. 1. The performance figures were obtained by running a ping-pong benchmark on our cluster.

### 3.1 Myricom

Myricom delivers their own "Myrinet API". The software is reliable, supports 16 nodes and multiple users as well as multiple processes per node. However, the best performance we were able to achieve on our cluster was 80 $\mu$s and 20 MB/s for latency and bandwidth, respectively. This is much worse than the raw hardware performance. There is no MPI implementation available for Myrinet API.

### 3.2 BIP

The Basic Interface for Parallelism (BIP) package [21] from Ecole Normale Supérieure de Lyon (France) has the best performance: the latency is only about 4.7 $\mu$s while the bandwith reaches 128 MB/s. The main disadvantage is a very limited functionality: no support is available for multiple users on cluster nor for multiple processes per node. Moreover, only 8 nodes are supported (version 0.9).

**Table 1.** Overview of low-level communication libraries for Myrinet

| project | nodes no. | m. users | m. processes | latency [$\mu s$] | bandwidth [MB/s] |
|---------|-----------|----------|--------------|-------------------|------------------|
| Myricom 3.09 | 16 | yes | yes | 80 | 20 |
| BIP 0.93a | 8 | no | no | 4.7 | 128 |
| VMCC | 16 | no | no | 15 | 83 |
| BDM | 16 | no | no | 22 | 30 |
| HPVM 1.0 | 16 | yes | yes | 8 | 70 |

The library does not provide any flow control, thus it is the user's responsibility to implement mechanisms for avoiding communication deadlocks. BIP provides MPI, but this implementation fails even in simple tests.

### 3.3 VMMC

Virtual Memory Mapped Communication (VMMC) [4] developed at NECI has a latency of 15 $\mu s$ and a bandwith of 83 MB/s. The version we have evaluated supports 16 nodes, but does not allow for multiple users per cluster or multiple processes per node. We experienced frequent deadlocks on our cluster. Meanwhile, NECI has developed a new version which reportedly has much improved stability and supports multiple users and multiple processes. However, the new version requires LANai boards with 1 MB of RAM. MPI is not provided for VMMC.

### 3.4 BDM

The Integrated Concurrent & Distributed Computation Research Laboratory at Mississippi State University has developed the BullDog Myrinet Control Program (BDM) package [11]. This software is reliable and stable, supports 16 nodes and delivers a latency of 22 $\mu s$ and a bandwidth of 30 MB/s. It comes with an MPI implementation which passes most of the MPICH test suite (the network stress test *sendmany* fails), but it does not support multiple users or multiple processes per node.

### 3.5 HPVM

High Performance Virtual Machine (HPVM) [3,13,19] from University of Illinois at Urbana-Champaign supports 16 nodes, simultaneous use of the cluster by multiple users, and up to 2 processes per node. The HPVM distribution contains the low-level Fast Messages (FM) library, v.2.1, together with high-level programming interfaces (MPI, Shmem Put/get and Global Arrays) which are based on FM. The FM library version 2.1 has improved considerably compared to earlier versions. It now provides multi-user and multi-process capabilities and guarantees reliable and error-free transmission. Also, we measured an increase in communication speed to a latency of 8 μs and a maximum bandwidth of about 70 MB/s.

The MPI library that comes with HPVM is based on an old version of MPICH (1.0.8) and does not conform fully to the standard (for example some data types like MPI_Status are declared differently than required by MPI). Many MPICH test suite programs (taken from the MPICH 1.1.0 distribution), and even those for the basic point-to-point tests, fail or deadlock. The Fortran 77 binding is faulty and results in fatal error messages about using invalid communicators.

We discovered a serious problem for applications with unbalanced communication patterns. The MPI library allocates a certain number of slots for unexpected messages, and there is no flow control to throttle the senders when all slots are filled. Instead, this situation which is not uncommon in real applications, causes the whole application to abort. Note that the limit is set by the fixed number of message slots, and not by the memory available on the processor.

## 4   Implementation

Since none of the available MPI implementations fulfilled our requirements, we decided to develop a new MPI library, and base it on FM 2.1 from the HPVM distribution. As explained above, FM 2.1 fulfills all our requirements for a low-level interface, so by using it we saved the substantial effort needed to develop a new low-level library. We used the MPICH software from Argonne National Laboratory for the high-level part, and connected it with FM by our own device driver at the generic channel interface level. Currently we use MPICH 1.1.0, but we plan to migrate to version 1.1.1 soon.

The objective of our work was the development of a device driver, based on FM, with as little additional overhead as possible. We developed two versions of the driver, in order to verify which design is most suitable for our applications and the most promising for further enhancements and optimizations. The two versions differ in performance and functionality:

1. The single-threaded (ST) device driver is optimized for minimum latency and implements only the blocking primitives of the channel interface.
2. The multi-threaded (MT) version implements both blocking and non-blocking primitives, and was developed in order to minimize message delays in applications with bad load balancing.

Both drivers use different protocols for three message length regimes, called *short*, *eager* and *rendezvous*, and switch between protocols at the same message lengths, i.e., at 512 bytes from short to eager, and at 128 KB from eager to rendezvous protocol. Details of these three MPICH standard protocols are described in [8]. In the short protocol the data is sent with the message header. In the eager protocol first a control packet with the message header is sent, followed by the data packet. In the rendezvous protocol the data transfer is deferred until requested by the receiver (by posting a matching receive operation). The rendezvous protocol is less efficient than the eager one, but saves buffer space.

The communication progresses in both versions (as in any program using the FM library) only when a receiver explicitly extracts incoming messages from the network by calling FM_extract(). FM may also decide to call this function implicitly when a send operation is called. A call to FM_extract() passes control to user-defined handler functions. The actual extraction of data from the network is done in a handler associated with a message, by one or more calls to FM_receive(). FM breaks the data into small packets before inserting it into the network, so that they may arrive to the recipient interleaved with packets from other message streams. As a result, a handler might be forced to switch to another message stream during a call to FM_receive(). If the handler tries to receive more data than is actually available, control returns to FM_extract(). If more data arrives, on the next call to FM_extract() the handler execution resumes at the point where it stopped before. This gives handlers a thread-like semantics with clearly defined points where preemption is possible, and the user program must properly handle the situations in which returning handlers leave some data in an inconsistent state.

Sending a message with FM is a three-step process. First, network resources to send the specified number of bytes are allocated by a call to FM_begin_message. Then, one or more calls to FM_send_piece() insert the data to the stream which has been opened previously, followed by a call to FM_end_message(). That function returns only when all message data has been transferred to the receiver's network interface. Thus, if there is enough buffer space at the recipient's LANai board, FM_end_message() will return before the message is extracted by a call to FM_extract(). It is only safe to modify data passed to FM_send_piece after the corresponding call to FM_end_message() returns.

Both versions of our device driver use two handlers: one for start-up and shutdown synchronization of the application, and another one for data transmission. The main difference between both drivers is in the send procedure. Since ST inserts the data into the network directly in the main thread, the sender process blocks unless there is enough buffer space on the receiver's LANai board, or the receiver extracts the message. MT has been created to avoid this blockage: the main thread only appends the data to a request queue, and a dedicated communication thread sends the message. MT also allows for overlapping communication with computations (on multiprocessors), because it is implemented with POSIX threads [2,14] which, under Linux, are scheduled at the kernel level and thus can run concurrently on separate CPUs. As it turned out, however, the overhead caused by having an additional thread, the context switching between threads, the use of synchronization primitives (POSIX mutexes and conditional variables) and by moving the data between the caches of both processors, more than outweighs the benefits. For example, a propagation of a condition between threads on our system costs about $100\,\mu s$, including the locking and unlocking of the associated mutex.

Recently we added to both driver versions the possibility of draining the network periodically (10 ms is the shortest possible period), so that a receiver process involved in a long computation phase does not block the sender. However, this did not improve the performance even for a real application with very high load imbalance.

Both driver versions are fully functional and the MPICH library using any of them passes all the tests provided with the MPICH distribution. ST passes also all the tests for the CLIC library but MT fails on some of them (see Sec. 5).

# 5 Results

In this section we present the results of experiments which we performed to assess the maturity of our MPI implementation, as measured by the requirements given above. In the following, we look at the results from different viewpoints.

## 5.1 Completeness and Correctness

We tested the MPI library, using both our ST and MT device drivers. Both versions passed the whole MPICH test suite. These programs not only check the correct semantics of the MPI functions, but also test the behavior of the implementation under intense network traffic. Additionally, ST passed a much more demanding test suite from the GMD Communications Library (CLIC) for block-structured grids [10,22]. These advanced tests resulted in some errors for the MT version.

## 5.2 Performance Measured with Kernel Benchmarks

The basic performance indicators, latency and bandwidth, have been measured by running the mpptest program from the MPICH distribution in round-trip mode (ping-pong), as well as in head-on mode, where the processes exchange messages using MPI_Sendrecv). The results for the ping-pong are shown in Fig. 1. Figure 2 presents the performance measured in the head-on mode.

The single-threaded version of our device driver shows a very good latency which for zero byte messages exceeds the corresponding figure for the HPVM/MPI library by only about $2\,\mu s$. With more than $100\,\mu s$, the latency is much higher for the multi-threaded version. The reasons for this high latency are the use of mutual exclusion and synchronization primitives from the POSIX threads library, as well as the necessary data structure manipulations (creation of a send request, appending to the queue, retrieving from the queue, and passing the data between threads). If the send thread is scheduled for the same CPU as the main one, the context switch accounts for 8 or $9\,\mu s$. If the send thread executes on another CPU, the data first has to be flushed from cache to memory, and then loaded to the second CPU's cache.

Both our versions and HPVM/MPI exhibit similar bandwidth. The maximum bandwidth reached by our implementations is higher than that obtained with HPVM/MPI. On the other hand, ST performance decreases for very long messages using the rendezvous protocol. MT is slow for shorter messages, but in the rendezvous protocol is the fastest. The decrease in bandwidth seen for all

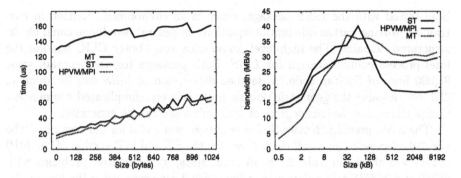

**Fig. 1.** Latency and bandwidth obtained with the ping-pong benchmark

three implementations and very long messages is caused by cache effects, and our implementations seem to suffer more from them than HPVM/MPI. This suggests that there is still some potential for optimization in our device drivers for the rendezvous protocol.

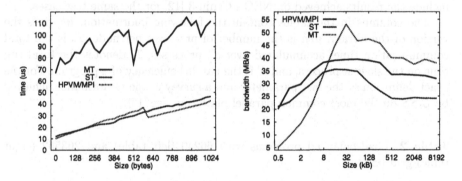

**Fig. 2.** Latency and bandwidth measured in the head-on mode

The sustained performance measured in head-on mode is much higher than that obtained with the ping-pong benchmark. This is the benefit of Myrinet's full-duplex communication channels.

## 5.3   Performance Achieved with Full Applications

While basic test programs can be used to measure some basic performance characteristics of an MPI implementation, they cannot replace the final tests with full application codes, which contain more complex communication patterns and less synchronization between communication partners. Our implementations have

518     Maciej Golebiewski, Markus Baum, and Rolf Hempel

been tested with the $L_iSS$ package, which is an environment, written in Fortran, for solving partial differential equations on general two-dimensional block-
structured domains. The high-level communications library CLIC handles the
inter-process communication for $L_iSS$. Both packages together contain some
85,000 lines of Fortran code. A detailed description of $L_iSS$ can be found in
[23]. By choosing the geometric domain appropriately, complicated communication patterns and various degrees of load imbalance can be generated.

The incompressible Navier-Stokes equations were used for our test runs. The
parallel efficiencies and wall clock times for the ST and MT version of the MPI
library are shown in Tab. 2. In all those tests, load-balanced problems with
66240 and 263552 grid points were solved with 9 processes, using the host-nodes
programming model (a master process and 8 slaves). The tests were repeated
with two slave processes per node to evaluate the performance loss caused by
sharing the LANai boards, and the connection to a Myrinet switch, by two
processes.

Table 3 reports timings and efficiencies for a problem with 14212 grid points
and a great load-imbalance. 30 processes were used for this set of tests, so one or
two application processes were assigned to a single (dual-processor) node. Since
the HPVM/MPI does not have a usable Fortran binding, only our MPI libraries
could be used for the $L_iSS$ tests. For a comparison with an MPP computer, we
include the results achieved on NEC's Cenju-3 [12] for the same test cases.

The columns in the tables contain the following information: $lib$ gives the
version of the library, $p/n$ is the number of processes per node, $T_c$ is the total
communication time, accumulated over all processes, $T_t$ denotes the total run
time on the slowest process, and $E$ is the parallel efficiency of a program. For the
exact definition of the last parameter and a survey of the performance achieved
by $L_iSS$ on the most common parallel platforms, see [15].

**Table 2.** Load-balanced problems with 66240 (left table) and 263552 (right
table) grid points

| lib | p/n | $T_c$ [s] | $T_t$ [s] | $E$ [%] | lib | p/n | $T_c$ [s] | $T_t$ [s] | $E$ [%] |
|---|---|---|---|---|---|---|---|---|---|
| ST | 1 | 35.31 | 62.90 | 92.98 | ST | 1 | 59.24 | 222.88 | 96.68 |
| ST | 2 | 34.43 | 62.97 | 96.16 | ST | 2 | 60.69 | 223.01 | 96.60 |
| MT | 1 | 42.70 | 63.71 | 91.62 | MT | 1 | 75.35 | 222.98 | 95.78 |
| MT | 2 | 145.65 | 77.04 | 76.37 | MT | 2 | 274.73 | 250.20 | 86.27 |
| Cenju-3 | 1 | 399.75 | 164.58 | 69.64 | Cenju-3 | 1 | 836.84 | 519.23 | 79.85 |

Our two implementations perform very well for both the small and big, load-
balanced test cases, with ST always being the better choice. The efficiencies are
well above 90%, except for MT with two processes per node. This is caused by
the increase in CPU load by the additional threads.

For ST we do not observe any significant performance drop by sharing the
network interface between two processes. Apparently, there is not enough data

traffic to saturate either the PCI bus or the network interface. The Cenju-3 performs worst in this test, because its 50 MHz VR4400 RISC CPUs are significantly slower than the 200 MHz PentiumPros of the PC cluster, and the interconnection network of the Cenju-3 is slower than the Myrinet.

**Table 3.** Load-unbalanced problem, 14212 grid points

| lib | $T_c$ [s] | $T_t$ [s] | $E$ [%] |
|---|---|---|---|
| ST | 919.45 | 40.89 | 22.45 |
| MT | 979.72 | 43.50 | 22.33 |
| Cenju-3 | 1690.99 | 74.82 | 22.06 |

The unbalanced test case, because of the signifficant variation of grid points per processor, results in much lower performance than the load-balanced problems. The parallel efficiencies for both ST and MT are only slightly above 22%. This shows that even in test cases for which the MT device driver was designed, it has no advantage over the ST version.

# 6  Conclusions

Both device drivers developed during project have been verified to work correctly for the complete MPICH test suite. The latency and the bandwidth measured with communication kernel benchmarks are close to those achieved by HPVM/MPI. Test with full applications show that the single-threaded version, in spite of its reduced functionality at the device driver level, performs better than the multi-threaded one. This is because MT creates enough overhead to outweigh the advantage of increased functionality.

The work presented in the paper is an ongoing project. We still need to test the libraries' behavior in communication-bound applications. We are also planning to run more standard benchmark suites to learn how our libraries compare to implementations of MPI on MPP systems.

The most important enhancements planned for the nearest future are: improving the performance by implementing an ADI2 device that will replace the current driver for the generic channel device, and developing a multi-device version of the library to use shared memory communications between processes running on the same node. We also plan to migrate to the new MPICH 1.1.1 version, which integrates the MPI-2 [17] parallel I/O functions (the ROMIO library [25]).

The main objective of the project, however, has been reached already: providing our researches with an efficient and reliable platform for their development of complex MPI applications on our cluster, in a multi-user-environment.

# 7  Acknowledgments

We thank our colleague H. Ritzdorf who has evaluated our libraries with the CLIC library tests and the $L_iSS$ package test cases, and provided us with all the performance indicators reported in Sec. 5.3.

# References

1. Beowulf Project at CESDIS,
   *http://cesdis.gsfc.nasa.gov/linux/beowulf/beowulf.html.*
2. Butenhof, D.R.: Programming with POSIX Threads. Addison-Wesley, 1997.
3. Chien, A., Pakin, S., Lauria, M., Buchanan, M., Hane K., Giannini, L., Prusakova, J.: High Performance Virtual Machines (HPVM): Clusters with Supercomputing APIs and Performance. Eighth SIAM Conference on Parallel Processing for Scientific Computing (PP97). March, 1997.
4. Dubnicki, C., Bilas, A., Li, K., Philbin, J.: Design and Implementation of Virtual Memory-Mapped Communication on Myrinet. NECI Technical Report. October, 1996.
5. Edler, J., Gottlieb, A., Philbin, J.: The NECI LAMP: What, Why, and How. Heterogeneous Computing and Multi-Disciplinary Applications: Proceedings of the eighth NEC Research Symposium.
6. GigaBit Ethernet with Linux,
   *http://cesdis.gsfc.nasa.gov/linux/drivers/yellowfin.html.*
7. Gropp, W., Lusk, E.: A High-Performance, Portable Implementation of the MPI Message Passing Interface Standard.
   *http://www.mcs.anl.gov/mpi/mpicharticle/paper.html.*
8. Gropp, W., Lusk, E.: MPICH Working Note: The implementation of the second generation MPICH ADI. ANL.
9. Gustavson, D.: The scalable coherent interface and related standards projects. IEEE Micro, **12**(1), February, 1992.
10. Hempel, R., Ritzdorf, H.: The GMD communications subroutine library for grid-oriented problems. Arbeitspapiere der GMD Nr. **589**, St. Augustin, 1991.
11. Henley G., Doss, N., McMahon, T., Skjellum, A.: BDM: A Multiprotocol Myrinet Control Program and Host Application Programmer Interface. Technical Report, Mississippi State University. May, 1997.
12. Koike, N.: NEC Cenju-3: A Microprocessor-Based Parallel Computer. Proceedings of IPPS. April, 1994.
13. Lauria, M., Pakin, S., Chien, A.: Efficient Layering for High Speed Communication: Fast Messages 2.x. Proceedings of the 7th High Performance Distributed Computing (HPDC7) conference. Chicago Illinois, July, 1998.
14. Leroy, X.: The LinuxThreads library.
   *http://pauillac.inria.fr/~xleroy/linuxthreads/*
15. Linden, J., Lonsdale, G., Ritzdorf, H., Shüller, A.: Scalability aspects of parallel multigrid. FGCS, **10** (1994).
16. Message Passing Interface Forum: MPI: A Message-Passing Interface Standard. June, 1995.
17. Message Passing Interface Forum: MPI-2: Extensions to the Message-Passing Interface. July, 1997.
18. Myrinet Documentation. *http://www.myri.com/scs/documentation.*

19. Pakin, S., Buchanan, M., Connelly, K., Lavery, A., Koenig, G., Giannini, L., Prusakova, J., Herman, G., Chien, A.: HPVM 1.0 User Documentation. August, 1997.
20. Philbin, J., Li, K.: Rationale and Requirements for Low Cost Scalable Servers and Parallel Processors. NECI Technical Note. March, 1995.
21. Prylli, L., Tourancheau, B.: BIP: a new protocol designed for high performance networking on Myrinet. Workshop PC-NOW, IPPS/SPDP98. Orlando, USA, 1998.
22. Ritzdorf, H., Hempel, R.: CLIC - The Communications Library for Industrial Codes Development at GMD-SCAI.
    http://www.gmd.de/SCAI/num/clic/clic.html.
23. Ritzdorf, H., Schüller, A., Steckel, B., Stüben, K.: $L_iSS$ - An environment for the parallel multigrid solution of partial differential equations on general 2D domains. Parallel Computing, **20** (1994).
24. Snir, M., Otto, S.W., Huss-Lederman, S., Walker, D.W., Dongarra, J.: MPI: The Complete Reference. MIT Press, 1996.
25. Thakur, R., Lusk, E., Gropp, W.: Users Guide for ROMIO: A High-Performance, Portable MPI-IO Implementation. ANL Technical Memorandum No. **234**, October, 1997.
26. Warren, M.S., Becker, D.J., Goda, M.P., Salmon, J.K., Sterling, T.: Parallel Supercomputing with Commodity Components. Proceedings of the International Conference on Parallel and Distributed Processing Techniques and Applications (PDPTA'97), 1997.

# Parallel Cluster Computing with IEEE1394-1995

László Böszörményi, Günther Hölzl, and Emanuel Pirker

Institut für Informationstechnologie, Universität Klagenfurt
Universitätsstraße 65–67, A–9020 Klagenfurt
{laszlo@itec,guenther@itec,epirker@edu}.uni-klu.ac.at

**Abstract.** Diverging demands on computer networks, such as high bandwidth, guaranteed quality of service and low latency lead to growing heterogeneity. IEEE1394-1995 is a standardized low-cost high-performance serial-bus-system with both isochronous and asynchronous operation. It might be an interesting candidate for all-round local and system area networks, providing a good compromise in fulfilling the above demands for low costs. Beside providing some technical background we show the possibilities and advantages of building parallel clusters on top of IEEE1394-1995. The main advantage is that minimal speed-up can be guaranteed, as shown on the basis of the parallel implementation of discrete Fourier transformation.
**Keywords.** IEEE1394-1995, cluster computing, parallel systems, isochronous transfers, Linux

## 1 Introduction

Computer networks have to face different, partly contradictory demands. *High bandwidth* seems to be still the measure most people believe in, and most companies use as selling argument for their networks. However, for multimedia (or, more exactly, for continuous data) *guaranteed quality of service* (QoS) at the user-level is more important than raw speed. For parallel applications, user-level *end-to-end latency* is the most important measure. These divergences in the demands lead to a growing heterogeneity in local area network technology. The only common denominator is the requirement for low prices. Currently, we have no standard (or quasi standard) solution that fulfills all requirements for a reasonable price.

We assume that in the coming years a new de facto LAN/SAN standard arises, which is able to provide a good compromise in supporting all the above mentioned requirements. We investigate the IEEE1394-1995 standard (the "FireWire[1]") from this point of view, i.e. its ability to serve as a basis for an "all-round" standard network.

Ethernet is definitely the leading LAN technology, at least regarding the number of installations. There are a number of efforts to raise the bandwidth of Ethernet. 100 Mbps FastEthernet can be regarded as the new standard LAN technology (as being already less expensive than 10 Mbps Ethernet). The Gigabit

---

[1] FireWire is a trademark of Apple Computer, Inc.

P. Zinterhof, M. Vajteršic, A. Uhl (Eds.): ACPC'99, LNCS 1557, pp. 522–532, 1999.
© Springer-Verlag Berlin Heidelberg 1999

Ethernet technology also arises, however, not without contradictions. Latency times seem to get much slower lower than bandwidth is getting higher. The Virtual Interface Architecture (VIA) [4] addresses the problem of low latency and therefore, Ethernet-based systems combined with VIA might have excellent performance in the future. Currently, however, there are no appropriate products yet available.

Some efforts have also be done to provide guaranteed QoS in an Ethernet-based system. In [14] a solution is suggested which requires the replacement of the Ethernet drivers without any change in the hardware. In [15,12] the solution is extended in a way that not even the drivers must be replaced, rather a simple Ethernet switch is used (consisting of a PC with a number of Ethernet cards). These solutions provide satisfactory but by no means optimal results. Summarizing we can say that there is a certain probability that Ethernet will remain the ultimate LAN technology in the future, but this is by no means sure.

Other networks, especially Myrinet [2], provide high bandwidth and good latency times, but provide no guaranteed QoS. Moreover, although Myrinet can be regarded as a cheap solution if it is compared to supercomputers, it is expensive if it is compared to Ethernet. One of the most attractive features of Myrinet is its flexibility, which might be used to add guaranteed QoS to Myrinet.

The IEEE1394-1995 standard [7] addresses first of all the question of multimedia support and QoS. It has, however, some features, that qualify it also as a good basis for parallel computing. In the following we elaborate these features. If IEEE1394-1995 proves as really suitable for parallel computing then it could be an interesting, low-price candidate for future LAN technology.

Previous work covered the design and implementation of a driver for Intel and Alpha based computing nodes running the Linux operating system and optimized for asynchronous data transfers [10]. We use the Adaptec AHA-8940 1394-to-PCI host adapter which supports 200 Mbps data transfer rate at present.

## 2   The IEEE1394-1995 High Performance Serial Bus

### 2.1   Relative Place

We start the investigation with a placement of IEEE1394-1995 among the most important networks used for parallel cluster computing.

Network comparisons regarding pure technical aspects can be found e.g. in [1] and [13]). Table 1 provides a short comparison of essential network technologies. The table reflects beside basic technical characteristics such as network structure, minimal one way latency, maximal access bandwidth and the existence of support for isochronous communication, also aspects of operating system support, manufacturer support and price.

There is obviously no "best" network. Regarding the different operational areas each network has its pros and cons. IEEE1394-1995 belongs to the bests in price, supports isochronous communication, has high bandwidth and low latency. Exactly these are the features that make the further investigation worthwhile.

Table 1. Comparison of networks for parallel cluster computing

|  | Fast-Ethernet | Gigabit-Ethernet | SCI | ATM | Myrinet | IEEE1394-1995 |
|---|---|---|---|---|---|---|
| Network structure[a] | bus | bus | ring | switched | switched | bus |
| Min. one way latency[b] | 20 $\mu$s | 20 $\mu$s[c] | 5 $\mu$s | 120 $\mu$s | 5 $\mu$s | 15 $\mu$s |
| Max. access bandw. | 100 Mbps | 1 Gbps | 4 Gbps | 155 Mbps | 1.2 Gbps | 400 Mbps |
| Isochronous cap. |  |  |  | √ |  | √ |
| Cable len. per link[d] | 200 m | 200 m | 10 m | 100 m | 10 m | 4.8 m |
| Windows-NT supp. | high | high | low | medium | high | medium |
| Solaris support | high | medium | high | medium | high | null |
| Linux support | high | low | low | medium | high | low[e] |
| Manufacturers | a lot | many | few | many | 1 | few |
| Costs per link[f] | $500 | $1500 | $1000 | $3000 | $1800 | $500 |

[a] original network structure, busses can also be switched of course
[b] regarding only the hardware dependent layers
[c] estimated value
[d] when using standard low cost cables
[e] development in progress [11]
[f] host adapter inclusive switching hardware

## 2.2 Features

The IEEE1394-1995 is a low-cost, high-performance ergonomic serial bus [7]. It was designed for operation both in the areas of industrial as in consumer electronics. Its architecture is compatible with other IEEE busses and standards (e.g. [6]) and implements a *memory read/write communication architecture* in contrast to conventional I/O-based communication, so distributed systems with global memory architectures can be mapped without great translation efforts. The shared memory architecture of IEEE1394-1995 is an architecture to provide economic interfaces and low latency. The address model is based on IEEE Std 1212 [6] (CSR Control and Status Register architecture), thus being compatible with SCI (Scalable Coherent Interface). IEEE1394-1995 uses IEEE-1212 "64-bit fixed" addressing, where the first 16 bits are used to represent the *node_ID*, thus allowing up to 64k nodes. Further the *node_ID* is divided into the 10-bit wide *bus_ID* and a 6 bit-wide *physical_ID*. Therefore up to 1023 busses[2], each having up to 63 nodes[3], can be interconnected. The remaining 48 bits are used for addressing the node *memory space*, the *private space* or the *register space*[4]. One or more nodes of the serial bus can be combined to form logical *modules*. An addressing scheme for determining the module number is not provided.

---

[2] bus # 1023 refers to the local bus
[3] node # 63 refers to the broadcast-address
[4] The *private space* is characterized by having 0xFFFFE and the IEEE Std 1212 *register space* by having 0xFFFFF as the leading 20 bits, thus remaining 28 bits for the *private address* resp. the *register address*.

The main feature distinguishing from most other communication technologies is the capability of the *isochronous transfer* of digital data, which is needed when working under *guaranteed timing* or *guaranteed bandwidth* is required, e.g. transfer of multimedia data. Although this feature is also available in the *Asynchronous Transfer Mode (ATM)*, ATM has the disadvantage over IEEE1394-1995 of the requirement of additional expensive switches with high latency.

The serial bus protocols are a set of three stacked layers. The lower two layers can be compared to the ISO/OSI layers 1-2. The highest layer is the so called *transaction layer* and provides *read, write* and *lock transactions*. It provides a path to the *isochronous resource manager* (IRM), which is in fact part of the control and status register (CSR) structure [6]. The middle layer (*link layer*) provides a one-way data transfer with confirmation of request and provides its service to the *transaction layer*. In difference to the *transaction layer* it provides an *isochronous data transfer service* directly to the application using periodic cycles with a cycle time $t_{isoc}$ of 125 $\mu$s. A link-layer-transfer is called a *subaction*. The lowest layer (*physical layer*) translates the logical symbols into electrical symbols, arbitrates the bus and defines the mechanical interfaces for the Serial Bus.

## 2.3 Performance Measurement

Though the Linux IEEE1394-1995 driver is in a preliminary state, we have done some basic performance measurements on asynchronous transactions. The *read quadlet transaction* is used for minimum information transfer to access a single quadlet (four bytes) aligned register of the CSR structure. We measured the performance of consecutive read quadlet transactions, which in turn consist of an acknowledged read request and an acknowledged read response subaction, with the packet length of four resp. five quadlets. The equipment consists of Intel Pentiums 200 MHz running Linux with the 2.0 kernel. Further we use the Adaptec AHA-8940 1394-to-PCI host adapters with the 400 Mbps Adaptec AIC-5800 link controller chip and the 200 Mbps IBM 21S750PFB PHY chip. The tests were done with a data transfer rate of 200 Mbps. A 100000 transaction test showed an average single transaction latency of 129 $\mu$s. The latency consists of the link layer round trip time plus the transaction layer software overhead. For packets with little payload the software and communication overhead is rather high and we expect notably higher throughput with isochronous transfers and asynchronous packets with higher block size.

# 3    Transaction Model for Group Communication

As mentioned in section 10 the highest layer of the Serial Bus protocol stack is the *transaction layer* which provides three different reliable unicast transaction types: the *read, write* and *lock* transactions. For unicast messages services like flow and error control are provided by the underlying *link layer* with *asynchronous subactions*.

The idea of group communication is to let processes communicate with a *group* of other processes *simultaneously*. Examples of group communication are replicated file systems (with coherent caches), replicated program executions (SPMD machines), teleconferencing and financial computing systems. Hardware support influences the efficiency of group communication. The bus architecture, as used at IEEE1394-1995, is an ideal vehicle for one-to-all group communication (broadcasting). Reliable broadcasting, which is related to atomic messages, is an extension to broadcasting and is required for most parallel applications. IEEE1394-1995 provides broadcast write transactions with an unacknowledged request subaction – the standard doesn't define how to complete the transaction to make it reliable. In general it is required to check the needs of the applications for realizing the appropriate broadcast method. For example file-streams can be broadcasted more efficiently using forward error correction mechanisms. In contrast cache coherency protocols have to use broadcasting with causality and total ordering semantics since several processes can transmit "memory-invalid-messages" simultaneously.

Protocols for group communication have been investigated very exhaustively, e.g. [3,5,8]. In this section we want to demonstrate a realization of the BB-method [8], a reliable multicast method, on top of the IEEE1394-1995 link/transaction layer. The CSR architecture can be used very efficiently.

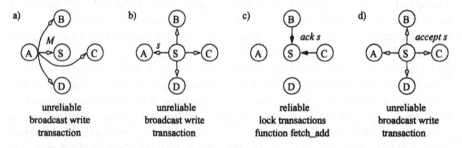

**Fig. 1.** Reliable broadcast method for IEEE1394-1995 using the BB method for five nodes with $r = 2$

The BB method assumes the existence of one processing node with an extra service called *sequencer* S. It is based on unreliable broadcast and reliable unicast messages. The procedure for a single broadcast is demonstrated in figure 1. After node A sends a broadcast of the message $M$ (Fig.1a), S stores $M$ and allocates a sequence number $s$, which is again broadcasted (Fig.1b). To cope with processor failures, $r$ other processors also store the message and have to acknowledge $s$ (Fig.1c). After receiving the $r$ acknowledgements, S broadcasts an *accept s* message and the kernels are allowed to pass $M$ to the application (Fig.1d). To assure that all processing nodes have received the broadcast messages they have to inform the sequencer about their highest received sequence number periodically, so the sequencer and the $r$ other processors are allowed to discard messages from their buffers. In case of loss of a message by single receivers because of transmission errors or because of buffer overflows within the receivers, the lost message can be retransmitted using reliable unicast messages. The correct order of the received broadcast messages is checked by the receiver itself.

The read, write and lock transaction architecture of IEEE1394-1995 is well suited to efficiently implement the BB method. We suggest the following method:

Analogous to the bus manager election process a *sequencer* or *reliable broadcast manager* (RBM) is determined at startup or bus reset. The RBM implements two status registers, which can be read through the CSR address space, one for the actual sequence number and one for the sequence number of the oldest buffered message. These registers can be used to retransmit lost messages. Furthermore each node implements a *broadcast message counter* (BMC) in the RBMs address space, which holds the sequence number of the last successfully received broadcast message. The RBM has to analyze these locally stored BMCs periodically for being able to discard acknowledged messages. Faulty nodes can be also detected by this procedure.

For message broadcasting we use the standardized unacknowledged broadcast transactions. The "fetch_add" lock transaction adds a given value to a memory's value and can be efficiently used to implement the "*ack s*" messages. Every lock transaction decrements a counter, which was initialized with $r$. When the counter reaches zero, the RBM is allowed to complete the reliable broadcast transaction by broadcasting the "*accept s*" message.

## 4  Topologies

The obvious strategy of connecting all nodes to one serial bus is easy to realize, but can have major drawbacks regarding performance. One can separate bus segments by usage of *bus bridges*. Similar to Ethernet bridges, intra-bus traffic is not distributed across bridge ports. Usage of bus bridges means we have to keep the system free of cycles, thus disallowing to build complex communication networks. Therefore we suggest routing on higher layers and connecting nodes to more than one independent bus.

To minimize latency and maximize throughput we have to follow two conflicting targets:

- keep the number of nodes connected to a single bus as small as possible to avoid congestion on the bus
- keep the number of busses small to avoid having messages traveling over many hops which is time-consuming

We suggest building parallel computers by usage of nodes which are connected to more than one bus. Routing occurs by a bridge protocol or by yet a higher layer. We do not use bus bridges because they don't allow cycles and have no advantages over multi-homed nodes. Having the driver bridging seems to be the best solution since we don't need heavy-weight protocol stacks, we can implement our own bridge protocols and policies and can make use of the reliable group communication patterns directly.

For doing routing decisions, performance of standard drivers may be not adequate. In case of the Linux IEEE1394-1995 driver [10], a very low-level cut-through routing implementation, which bypasses the Serial Bus protocol stack is suggested. To achieve real high performance, additional hardware solutions (intelligent PCI-1394 adapters) are required.

The straightforward approach to build an IEEE1394-1995 based "supercomputer" is connecting individual nodes in the form of a grid. In this case vertical and horizontal connections are independent busses. Every host in the grid is therefore connected to two different busses and is capable of routing (figure 2).

Sending a message from node A to node B in a grid of N nodes can be accomplished by the network

- needing one routing decision if B is connected to one of the two busses A is directly connected to ($2\sqrt{N} - 2$ nodes can be reached this way, without any intermediate hop)
- needing two routing decisions (one intermediate hop) if B is connected to any other bus in the network (the rest, or $N - 2\sqrt{N} + 1$ nodes are reached this way)

Routing is done, in the simplest case, only by that one such host which is connected both to the source and target node. This can be extended to arbitrary nodes, demanding a bridge protocol or advanced static policies.

Due to the efficiency of bus-based grids they may be completely satisfying for most networks. You can vary the grid structure by using a different – bigger or smaller – number of busses, thus leaving the "one row/column – one bus" policy.

However, as the node number increases we have to face other topologies, e.g. by connecting each node to three or more busses.

## 5  Parallel Applications Using Isochronous Transfers

One of the great characteristics of IEEE1394-1995 is the *isochronous transfer capability* of digital data. The original thought was to use this sort of transfer for time critical data such as video- and audio streams. The mechanism is to reserve guaranteed communication bandwidth and applications may transfer a limited size of information during time-slots. The idea was to provide guaranteed QoS and to reduce synchronization and buffer management tasks. In this section we show with the help of an easy example (parallel discrete Fourier transform (pDFT)) how isochronous transfers can be used to achieve a *guaranteed speedup* produced by parallel algorithms. Time critical applications (e.g. control systems) have to rely on end-to-end timing guarantees. In contrast to most of todays used networks (see table 1) IEEE1394-1995 is able to provide timely guaranteed delivery. In combination with real time task scheduling end-to-end scheduling can be realized, thus giving us the possibility to calculate the *guaranteed speedup*, which is indeed the lowest bound of the grade of parallelization.

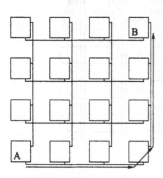

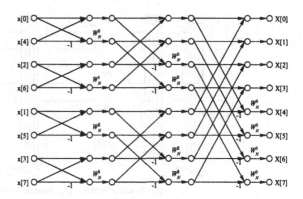

**Fig. 2.** Grid of 16 nodes; Node A sending a message to B

**Fig. 3.** Signal graph of a frequency decimated DFT-algorithm, $N = 8$

## 5.1 DFT Implementation

The discrete Fourier transform (DFT) is applied in the analysis, the design and implementation of time-critical signal processing procedures [9]. The goal is to calculate the DFT of limited sequences with the length $N$ using the formula

$$X[k] = \sum_{n=0}^{N-1} x[n] W_N^{kn} \qquad k = 0, 1, ..., N-1 \tag{1}$$

and $W_N = e^{-j\frac{2\pi}{N}}$. An optimal sequential algorithm, which takes advantage of the symmetrical and periodic characteristic of $W_N^{kn}$ needs $O(N \log N)$ multiplications. The signal graph of an eight point frequency decimated DFT is shown in figure 3. The inputs are fed through a bit reversing operation (butterfly operation). The calculated results are spread to adders with a following multiplicator.

## 5.2 Parallel DFT Implementation with Isochronous Channels

A parallelization can be realized by vertical distribution, so every processing element (PE) has to calculate $N/\#PE$ complex sums and multiplications. For transmitting $N/\#PE$ values at every isochronous cycle the PE has to reserve sizeof($value$) $* N/\#PE$ channel capacity. For calculating the end-to-end processing time we assume a constant processing time $t_p$, which is guaranteed by the operating systems real time scheduler. The overall system's behavior is demonstrated in figure 4 for processing times smaller and larger than the isochronous cycle time $t_{isoc}$. As the result of delayed predecessors we observe that the maximum single idle time for synchronization is $2 * t_{isoc} = 250\mu s$. Although the average aggregate time will be much lower, a guarantee for the maximum aggregate time can be provided by $nt_p + 2(n-1)t_{isoc}$, where $n$ is the number of processing steps. However, in the general case, when $N > \#PE$, only

530    László Böszörményi, Günther Hölzl, and Emanuel Pirker

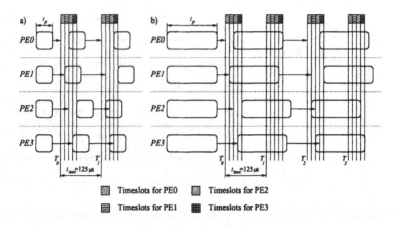

Fig. 4. Parallel implementation of the DFT using isochronous channels and $\#PE = 4$ a) $t_p < t_{isoc}$, b) $t_p > t_{isoc}$

$\log_2 \#PE$ communication steps are needed, the maximum aggregate time will be $t_p \log_2 N + 2t_{isoc}(\log_2 \#PE)$. The *guaranteed speedup* of the parallelization is shown in figure 5 and can be expressed by

$$guaranteed\_speedup_{isoPDFT} = \frac{\#PE \log_2 N}{\log_2 N + 2\frac{t_{isoc}}{t_p} \log_2 \#PE} \qquad (2)$$

# 6   Conclusion and Further Work

We have shown that the IEEE1394-1995 standard (and its followers) might play an important role in the LAN/SAN technology of the next decade. It not only supports the delivery of continuous data with guaranteed quality of service, but it also has the potential of serving as a high-performance, low-latency network, supporting cluster-based parallel computing. Even more, due to the combination of the above features, parallel clusters based on IEEE1394-1995 have the unique feature of *guaranteed speedup*.

We presented some basic figures about the performance of IEEE1394-1995. A next step should be to measure the performance with well-known benchmarks and with some typical applications.

It might be an interesting research to build an *IEEE1394-1995 switch*. Such a switch could be built quite simply with the help of a workstation with a greater number of IEEE1394-1995 adapter cards. With the help of such switches arbitrarily complex networks could be built cost effectively and efficiently.

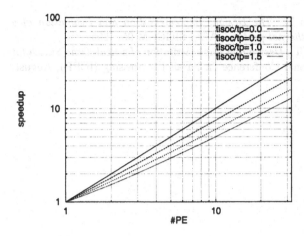

**Fig. 5.** Influence of the communication effort in the guaranteed speedup of the parallel DFT using isochronous channels. Using more PEs for parallelization means relative higher influence of the communication overhead.

# References

1. Bal H.E., Hofman R., and Verstoep K., *A Comparison of Three High Speed Networks for Parallel Cluster Computing*, Workshop on Communication and Architectural Support for Network-based Parallel Computing (CANPC'97), pp. 184-197, San Antonio, Texas, February 1997
2. Boden N.J., Cohen D., Felderman, R.E., Kulawik A.E., Seitz C.L., Seizovic C.L., Su W., *Myrinet: A Gigabit-per-second Local Area Network*, IEEE Micro, 15(1): 29-36, February, 1995
3. Chang J., Maxemchuk N.F., *Reliable Broadcast Protocols*, ACM Transactions on Computer Systems, Vol.2, No.3, August 1984
4. Compaq, Intel, Microsoft, *Virtual Interface Architecture Specification Version 1.0*, http:www/viarch.org, Dec 16, 1997
5. Frank A.J., Wittie L.D., Bernstein A.J., *Multicast Communication on Network Computers*, IEEE Software, May 1985
6. IEEE Computer Society, *IEEE Std 1212, Control and Status Registers (CSR) Architecture for microcomputer buses*, New York, 1994
7. IEEE Computer Society, *IEEE Std 1394-1995, IEEE Standard for a High Performance Serial Bus*, New York, August 1996
8. Kaashoek M.F., Tanenbaum A.S., *Efficient reliable group communication for distributed systems*, Rapport IR-295, Faculteit Wiskunde en Informatica, Vrije Universiteit, July 1992
9. Oppenheim A.V., Schafer R.W., *Discrete-Time Signal Processing*, Prentice Hall, 1989
10. Pirker E., Hölzl G., *The Design, Implementation and Operational Areas of the Linux IEEE-1394 Driver*, Institute of Information Technology, University Klagenfurt, Technical Reports, No 2-98, June 1998
11. Pirker E., *The Linux IEEE1394-1995 Subsystem*, http://www.edu.uni-klu.ac.at/~epirker/ieee1394/
12. Varadarajan S., Chiueh T., *EtheReal: A Host-Transparent Real-Time Fast Ethernet Switch*, State University NY at Stony Brook, TR-45, January 1998
13. Varma A., Raghavendra C.S., *Interconnection Networks for Multiprocessors and Multicomputers: Theory and Practice*, IEEE-press, 1993

14. Venkatramani C., Chiueh T., *The Design, Implementation and Evaluation of a software-based real-time Ethernet protocol*, ACM SIGCOM 95, 1995

15. Venkatramani C., Chiueh T., *Design and Implementation of a Real-Time Switch for Segmented E*, in International Conference on Parallel Processing (ICPP), August, 1998

# Simulating Load Balancing on Heterogeneous Workstation Clusters*

Helmut Hlavacs[1] and Christoph W. Ueberhuber[2]

[1] Institute for Applied Computer Science and Information Systems,
University of Vienna
hlavacs@ani.univie.ac.at

[2] Institute for Applied and Numerical Mathematics, Technical University of Vienna
christof@uranus.tuwien.ac.at

**Abstract.** A new system for simulating dynamic load balancing on heterogeneous workstation clusters is presented. Competing workload can be modeled in various ways, from simple to sophisticated. Instead of running real parallel workload, the program designer builds an application model, which is then run on the simulation system. The simulator provides standard UNIX load averages and can easily be adapted and extended for special purposes.

## 1 Introduction

Heterogeneous workstation clusters are getting increasingly attractive for running parallel software. When running parallel programs on interactively used workstations, program designers have to take care not only of balancing the computational load of their program, but also of reacting to changes of the workstations' workload caused by other user programs. This redistribution of work at runtime requires dynamic load balancing and has been studied extensively in literature (Krommer, Ueberhuber [12], Shivaratri et al. [19], Hac, Jin [9], Burdorf, Marti [4]). In principle, there are various ways of comparing different load balancing strategies with each other. All, though, have to model the occurring workload in one way or the other.

Theoretical approaches include scalability analysis (Kumar, Gramar and Vempaty [13]) and queuing analysis (Wang, Morris [22]). In both techniques, workload is modeled by Poisson arrival processes in most cases.

*Simulation* of load balancing techniques may be performed directly by running benchmark programs on existing workstations (Arpaci, Dusseau, Vahdat [2]) or by using special simulation systems. Simulation systems might either use traces of real user sessions (Zhou [23]) or may be based on Poisson processes as well (Kunz [14]).

There are, however, various ways in modeling workstation workload. Among them are, for instance, Poisson processes (Allen [1]) with constant arrival and departure rates (Kunz [14]), Poisson processes with variable arrival rates during the

* This work was supported by the Austrian Science Fund (*Österreichischer Fonds zur Förderung der wissenschaftlichen Forschung*).

day (Calzarossa, Serazzi [5] [7]), Markovian type models (Haring [10], Calzarossa, Serazzi [6]) and probabilistic context free grammars (Rhagavan, Joseph [17]).

In this paper, a simulator based on MISS-PVM (Kvasnicka, Ueberhuber [15] [16]) is presented, which is designed to simulate different load balancing strategies for parallel programs run on interactively used, heterogeneous workstation clusters. Using this new simulation tool, the workload of workstations can be modeled in various ways, including Poisson processes, user session trace files and user behavior graphs (UBGs). The application of the new simulator is demonstrated by modeling the parallelization of the ion implantation module (Bohmayr, Burenkov, Lorenz, Ryssel, Selberherr [3]) which is part of the technology CAD system VISTA (Strasser, Pichler, Selberherr [20], Grasser et al. [8]).

## 2    The Workstation User Simulator

The *Workstation User Simulator* (WUS) is an add-on to the *Machine Independent Simulation System for PVM3* (MISS-PVM). As with real parallel program runs, the simulated processes are started and use PVM3 (Sunderam et al. [21]) for communication. MISS-PVM, implemented as an independent layer between the user program and PVM, provides a *virtual time* depending on the process CPU usage, the speed of the interconnection network and the size of message packets sent from one process to the other. In principle, the simulated user processes can be started on any single or multi-processor computer.

When sending messages to other processes, MISS-PVM increases the virtual time according to the recently consumed CPU time, information that is provided by all standard UNIX systems. This timing information is then hooked onto the message sent. The receiver uses this time information to increase its own virtual time. The structure of the whole simulation system can be seen in Fig. 1.

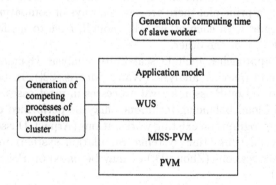

**Fig. 1.** Structure of the simulation system.

When using WUS, the real parallel application has to be replaced by an application model program, creating random CPU requests. These random requests, of course, must follow the observed statistical distribution of the real application. The synthetically generated CPU requests (for instance, one request may refer to $T$ CPU seconds) are then passed to WUS, which puts the request into its run queue and additionally produces competing processes due to the chosen workload model. After the application model has consumed its requested CPU time, it again takes over control and is free to communicate with other programs via MISS-PVM or to perform the chosen load balancing strategy. The currently implemented queuing discipline is *processor sharing*, i.e., all processes receive the same amount of CPU time.

Additionally, load averages similar to the standard UNIX load averages are available, as load balancing algorithms often rely on the exponentially smoothed CPU queue length. Exponential smoothing in essence means trying to estimate the level $Q$ of a stationary process, for which observations $\ldots, X_{t-1}, X_t, X_{t+1}$ at times $\ldots, t-1, t, t+1$ are available (Schlittgen, Streitberg [18]):

$$Q_{t+1} = \alpha Q_t + (1 - \alpha) X_{t+1}.$$

$Q_t$ is the estimate for $Q$ at time $t$. The smoothing constant $\alpha$ defines the weight for the past observations used for the current estimate. If the smoothing is to be done over the last $N$ seconds, then $\alpha = N/(N+1)$. Currently, load averages over the last 5, 30, 60, 300 seconds (5 minutes), and 900 seconds (15 minutes) are available in the simulation tool.

## 3   Using WUS

WUS consists of a set of C++ classes, designed to mimic a normal UNIX computer (Fig. 2). The most important class is called *Computer*. It generates and stores all other classes and interacts with the application model. The class *User* and its derivatives produces competing workload on this virtual machine. Workload is produced by using trace files of real user sessions, Poisson arrival processes with fixed and variable arrival and departure rates, and user behavior graphs.

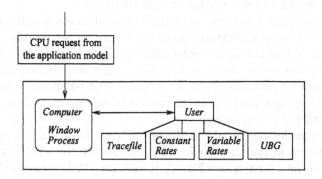

**Fig. 2.** Structure of the Workstation User Simulator (WUS).

Designers of parallel programs wishing to use WUS to test load balancing strategies first have to sample statistical data of the CPU requests of their real parallel application. Using this sampling data, a statistical application model then has to be created. An application model program frame may look like the following example:

```
do communication or initialization
pComputer = new Computer( Workload model );

while( Loop ) {
    runtime = GetRandomRuntime();
    pComputer->RunProcess( runtime );
    loadavg = pComputer->LoadAverage(n);
    do communication or load balancing
}
collect results
```

Initialization is done by creating an instance of the class *Computer*. The type of workload model is passed to the *Computer* class, which in turn creates the necessary *User* classes. Processor requests are passed to *Computer* by calling the member function RunProcess(). Load averages then can be retrieved by calling LoadAverage() and can be used to perform load balancing. Communication to other processes is carried out by normal PVM3 calls (which are replaced by MISS-PVM calls).

## 4    Simulating the VISTA Ion Implantation

VISTA (Strasser, Pichler, Selberherr [20], Grasser et al. [8]), developed at the Institute for Microelectronics at the Technical University of Vienna, is a framework for the design and simulation of process steps involved in semiconductor production. VISTA includes an ion implantation program (Bohmayr et al. [3]), which is based on a Monte Carlo simulator computing the endpoints of ions shot into a substrate. The resulting ion density is needed later on to predict the electrical behavior of the investigated semiconductor.

Monte Carlo ion implantation requires large amounts of computation time and thus has been parallelized to run on the institute's workstation cluster. The workstations of this cluster are not dedicated to ion implantation jobs but are also used by interactive users, as well as by other parallelized programs, like large compilation jobs. Currently, no dynamic load balancing is implemented, causing the implantation program to run out of balance quite frequently. Fig. 3 shows the implantation process.

For the parallel version of the Monte Carlo ion implantation, currently under development at the Institute for Microelectronics, the whole semiconductor region is split into $n$ parts (see Fig. 3). If ions leave their segment they are sent to the owner of the neighboring segment. One important fact limits the parallelization of the ion implantation. In the crystalline case, the penetrating ions

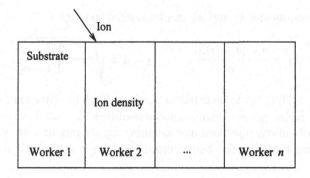

**Fig. 3.** VISTA ion implantation.

change the properties of the substrate. After the implantation of at most 500 ions, the workers must synchronize.

In order to reflect the heterogeneity of the VISTA workstation cluster, each available workstation has been tested using the *Whetstone* floating-point benchmark. Additionally, a user process has logged the CPU and memory demands of all processes run on the whole workstation cluster for two months, resulting in over 500 trace files.

The CPU demands per ion have been modeled by statistical distributions. For the overall ion time, a 3-stage gamma distribution function $G$ has been fitted by using the non-linear minimization capabilities of the statistical package $R$ (Ihaka [11]). The task was to find the set of parameters minimizing

$$SE = \sum_{x_j}(G(x_j) - S_n(x_j))^2,$$

i. e., the squared sum of differences between the observed cumulative distribution function $S_n$ and the model distribution function $G$. Differences are taken at the jump points $x_j$ of the distribution function $S_n$.

Additionally, if ions are sent to a neighbor, the CPU time spent in their segment has been modeled for over 50 different segment sizes. This is important, as parts of segments will be shifted to neighbors in case of load balancing activities. Each of these distributions consist of a *2-stage hyperexponential distribution* with distribution function

$$F(t) = \alpha(1 - e^{-t/\lambda_1}) + (1 - \alpha)(1 - e^{-t/\lambda_2}).$$

Here, fitting has been done by using a mixture of the method of moments (Allen [1]) and non-linear minimization. By computing the first two statistical moments

$$E[X] = \alpha\lambda_1 + (1 - \alpha)\lambda_2 \equiv a$$
$$E[X^2] = 2\alpha\lambda_1^2 + 2(1 - \alpha)\lambda_2^2 \equiv b$$

the distribution means $\lambda_1$ and $\lambda_2$ can be related to $\alpha$ by

$$\lambda_1 = \frac{a - (1 - \alpha)\lambda_1}{\alpha}, \qquad \lambda_2 = a \pm \sqrt{a^2 - \frac{2a^2 - \alpha b}{2(1 - \alpha)}}.$$

Thus, only $\alpha \in [0, 1]$ has to be fitted to the data. Each distribution has been evaluated by the Kolmogorov-Smirnov goodness-of-fit test. Most have been accepted with high probability, rejections at least show significance at a one percent level. Parameters for segments not being considered have been linearly interpolated.

# 5   Simulation Results

Simulation was carried out by starting ten processes, each using a trace file of a real workstation day. From the workstation cluster, two of the fastest machines (320 MWhetstones/second) and eight of the slowest machines (75 MWhetstones/second) were chosen. The data size describing the segment structure was set to 50 MB (anything between a few MBs and 1 GB would have been possible). The simulation start time was set to 4 pm, thus increasing the simulation time accordingly. At 4 pm the master process sends load requests to all processes, then chooses $k$ workstations with the best performance/load ratio.

Fig. 4 shows the simulation results. Load balancing is carried out at each synchronization point. The master receives load information of its workers and starts the load balancing mechanism, in case the load imbalance exceeds some threshold. Load balancing is carried out by shifting the segment limits and sending the according segment data to other workers. The lower limit denotes the case, where no competing processes are started, thus assigning the full computational power to the ion implantation calculations, i. e., the application model under consideration.

# 6   Future Work

The workstation trace files indicate large fluctuations of workload during the day. Especially the faster machines are more likely to get overloaded. Load balancing algorithms to be yet developed must be able to predict the future workload, at least to some degree, to avoid unnecessary work shifts.

Also, the simulator itself will be extended to accept not only application models but real application programs as well. Additional features like other queuing disciplines including the UNIX nice level will be provided.

# 7   Conclusion

In this paper a simulator designed especially for the development of load balancing algorithms on interactively used heterogeneous workstation clusters has been introduced. In this simulator various ways of describing workstation workload

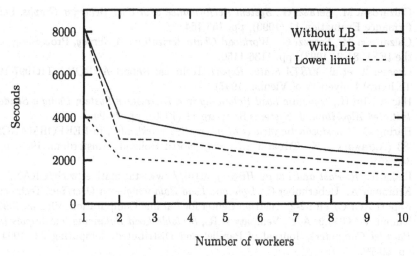

Fig. 4. Effect of load balancing (LB).

can be used, reaching from simple fixed arrival/departure rates up to sophisticated user behavior graphs. The newly developed simulation system will be an important tool for evaluating load balancing algorithms in various applications, amongst them the Monte Carlo ion implantation module of VISTA.

## Acknowledgments

We would like to thank Siegfried Selberherr, Erasmus Langer, Mustafa Radi and Andreas Hössinger (Institute for Microelectronics, Technical University of Vienna) for their cooperation.

Additionally, we would like to acknowledge the financial support of the Austrian Science Fund FWF.

## References

1. Allen A.O., *Probability, Statistics and Queuing Theory*, Academic Press, Orlando, 1990.
2. Arpaci R.H., Dusseau A.C., Vahdat A.M., *The Interaction of Parallel and Sequential Workload on a Network of Workstations*, Performance Evaluation Review 23-1 (1995), pp. 267-278.
3. Bohmayr W., Burenkov A., Lorenz J., Ryssel H., Selberherr S., *Monte Carlo Simulation of Silicon Amorphization During Ion Implantation*, Proceedings SISPAD 96 Conf., (2.-4. September 1996, Tokyo), pp. 17-18.
4. Burdorf C., Marti J., *Load Balancing Strategies for Time Warp on Multi-User Workstations*, The Computer Journal 36-2 (1993), pp. 168-176.
5. Calzarossa M., Serazzi G., *A Characterization of the Variation in Time of Workload Arrival Patterns*, IEEE Transactions on Computers C-34-2 (1985), pp. 156-162.

6. Calzarossa M., Serazzi G., *System Performance with User Behavior Graphs*, Performance Evaluation 11 (1990), pp. 155-164.
7. Calzarossa M., Serazzi G., *Workload Characterization: A Survey*, Proceedings of the IEEE 81-8 (1993), pp. 1136-1150.
8. Grasser T. et al, *VISTA Status Report*, Technical Report AURORA TR1997-16, Technical University of Vienna (1997).
9. Hac A., Jin H., *Dynamic Load Balancing in a Distributed System Using a Sender Initiated Algorithm*, J. Systems Software 11 (1990), pp. 79-94.
10. Haring G., *On stochastic models of interactive workloads"*, in PERFORMANCE '83 (Agrawala A.K., Tripathi S.K. eds), North Holland, Amsterdam, 1983, pp. 345-361.
11. Ihaka R., *R: Past and Future History*, http://www.stat.math.ethz.ch/CRAN/
12. Krommer A., Ueberhuber C., *Dynamic Load Balancing—An Overview*, Technical Report ACPC/TR 92-2, Austrian Center for Parallel Computation, Vienna, 1992.
13. Kumar V., Grama A.Y., Vempaty N.R., *Scalable Load Balancing Techniques for Parallel Computers*, Journal of Parallel and Distributed Computing 22 (1994), pp. 60-79.
14. Kunz T., *The Influence of Different Workload Descriptions on a Heuristic Load Balancing Scheme*, IEEE Transactions on Software Engineering 17-7 (1991), pp. 725-730.
15. Kvasnicka D., Ueberhuber C.W., *Simulating Architecture Adaptive Algorithms with MISS-PVM*, Technical Report AURORA TR1997-16, Technical University of Vienna (1997).
16. Kvasnicka D., Ueberhuber C.W., *Developing Architecture Adaptive Algorithms using Simulation with MISS-PVM for Performance Prediction*, Conference Proceedings of the 1997 ACM/SIGARCH International Conference on Supercomputing, Vienna, Austria, July 7-11, 1997, pp. 333-339.
17. Raghavan S.V., Joseph P.J., *Workload Models for Multiwindow Distributed Environments*, in Quantitative Evaluation of Computing and Communication Systems (Beilner H., Bause F., eds), Springer Heidelberg, 1995.
18. Schlittgen R., Streitberg B., *Zeitreihenanalyse*, R. Oldenbourg Verlag Muenchen, 1995.
19. Shivaratri N.G., Krueger P., Singhal Mukesh, *Load Distributing for Locally Distributed Systems*, Computer 12 (1994), pp. 33-44.
20. Strasser R., Pichler Ch., Selberherr S., *VISTA—A Framework for Technology CAD Purposes*, Proceedings European Simulation Symposium, (19.-22. October 1997, Passau), pp. 445-449.
21. Sunderam V.S., Geist G.A., Dongarra J., Manchek R., *The PVM concurrent computing system: Evolution, experiences and trends*, Parallel Computing 20-4 (1994), pp. 531-545.
22. Wang Yung-Terng, Morris R.J.T, *Load Sharing in Distributed Systems*, IEEE Transactions on Computers C-34-3 (1985), pp. 204-217.
23. Zhou S., *A Trace-Driven Simulation Study of Dynamic Load Balancing*, Technical Report UCB/CSD 87/305 (1986).

# Global Virtual Time Approximation for Split Queue Time Warp

Helge Hagenauer

Institut für Computerwissenschaften und Systemanalyse
Universität Salzburg
Jakob-Haringer-Straße 2
A-5020 Salzburg
Austria
hagenau@cosy.sbg.ac.at

**Abstract.** In the field of distributed discrete event simulation we introduced the split queue time warp algorithm, which is a generalization of the well known time warp algorithm. The main feature is allowing lazy message reception and so the rollback frequency may be reduced. This paper describes a method for global virtual time approximation during a simulation run respecting the specific structure of such a system. This will keep the interference with the underlying simulation at a very low degree.

## 1 Introduction

The *time warp* algorithm for a distributed discrete event simulation system belongs to the so called "optimistic" mechanisms (see [10], [3] and [2] for an introduction). A system under study is viewed as loosely coupled *physical processes*, *PPs*, which are simulated by *logical processes*, *LPs*. The interaction between the *LPs* consists solely of sending and receiving timestamped messages. No central clock and no event list exists like in traditional concepts.

Optimistic behaviour means that a *LP* performs its next step on the base of messages received so far. Possible further input, which should have been considered earlier, is producing causal errors. These are corrected by a special mechanism, which rolls back to former states and cancels unjustified interim output. It is clear that such rollbacks, and especially cascades of them, reduce the benefit of doing work in parallel.

All messages sent to a *LP* are inserted into a single input queue which is maintained in ascending timestamp order. But many real-world processes are "active" in the sense of choosing what kind of information is relevant for their next step, depending on the current state, and what can be delayed. Therefore we proposed to give a *LP* several input queues each of which corresponds to a message type and is read only when necessary ([4], [6]). The two benefits are: First a *LP* has not to store temporarily information that is currently unwanted. Second, the reading of only needed messages does not expand the past of a *LP*

P. Zinterhof, M. Vajteršic, A. Uhl (Eds.): ACPC'99, LNCS 1557, pp. 541–548, 1999.
© Springer-Verlag Berlin Heidelberg 1999

in an superfluous way and consequently the risk of rollbacks is decreased. We call this new algorithm *split queue time warp* or SQTW.

*LP*s are the main building blocks of a SQTW system. But having the focus on an implementation for workstations connected via a network, some additional structural refinements are necessary. In general there are much more *LP*s than available processing nodes (e.g. processors), so several or even many *LP*s have to be allocated to one processing node. We call such a group of *LP*s a submodel, *SM*. It is controlled by a submodel manager, which is responsible for *LP* initialization, *LP* scheduling and gathering results at the end of a simulation. Figure 1 shows the basic structure of such a system.

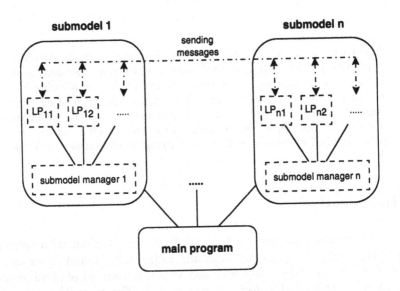

**Fig. 1.** Structure with *LP*s grouped to submodels

Because of not having a central clock a distributed simulation system needs a method of determining the simulation time. This can be done by snapshot algorithms (see [9] and [2]), which are used for computing monotonic functions of the global state. This paper contains a description of an algorithm to compute a lower bound on the simulation time, to which a distributed simulation system has advanced (the so-called *Global Virtual Time* or *GVT*), taking into account the special submodel structure. A main characteristic of a *GVT* value is that a rollback to a state with *LVT* smaller than *GVT* is not possible for any *LP*. Therefore *GVT* values can be used to show the progress of the entire simulation (e.g. displaying intermediate results) and doing fossil collection (freeing memory by deallocation of space used to store states).

The presented algorithm does not depend on SQTW features and thus it is applicable to time warp systems with a similar structure (i.e. *LP*s grouped to submodels). Section 2 describes the general algorithm for $GVT$ approximation. Then, in sections 3 and 4, the tasks of the main program and the submodels are shown together with their collaboration. In section 5 a summing-up and an outlook for future work is done.

## 2    General Algorithm

A distributed discrete event simulation system is composed of a set of $LPs$ which communicate solely via timestamped messages. The simulation time to which $LP_i$ has advanced at real time $t$ is called the *Local Virtual Time*, $LVT_i(t)$. For all $i$ the function $LVT_i(t)$ is not monotonic increasing due to the possibility of rollbacks. If the minimum of all $LVT_i(t)$ at real time $t$ is $LVT_{min}(t)$ and the minimal timestamp of all messages which are in transit at $t$ is $ts_{min}(t)$ then $GVT(t)$ is defined as

$$GVT(t) = \min\left(LVT_{min}(t),\ ts_{min}(t)\right).$$

Following Mattern [9] a general algorithm for $GVT$ approximation is based on so-called "cuts", which divide for each $LP_i$ the process line into two parts – the past and the future. A cut is triggered by cut events which have no direct influence on the underlying simulation (i.e. the central simulation algorithm is not touched by them). In a time diagram a cut can be shown as a zigzag line, the cut line (see figure 2). A cut is consistent if no message is sent in the future of some $LP_i$ and is received in the past of some $LP_j$ ($i = j$ allowed).

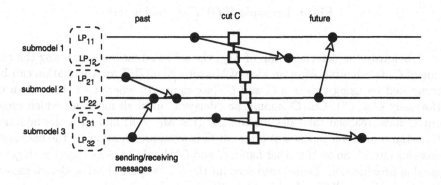

**Fig. 2.** Time diagram with a cut line specific to the submodel structure

Due to this properties a consistent cut $C$ together with $M_C$, the set of all messages crossing the cut line, defines a global state of the system and therefore can be used for $GVT$ approximation.

A general algorithm for determining a consistent cut uses a status information (here the colors white and red) piggybacked onto messages (see [9]):

1. Initially every $LP$ is white and turns red while taking a local snapshot;
2. Every message sent by a white $LP$ is colored white, every message sent by a red $LP$ is colored red;
3. Every $LP$ takes a local snapshot before a red message is possibly received;

Considering the structure in figure 1, the main program is the initiator of a cut algorithm and the submodel managers handle the current color and are responsible for bookkeeping $GVT$ specific data. So for all $LPs$ belonging to one submodel ($LP \in SM$) cut $C$ occurs at the same real time. In figure 2 this is indicated by a vertical cut line within a submodel.

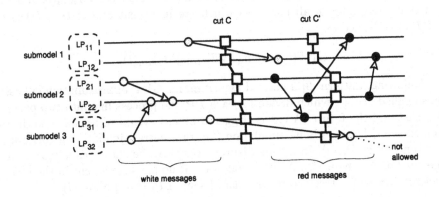

**Fig. 3.** Principle of $GVT$ approximation

Obviously the determination of $M_C$, the set of all messages crossing the cut line of $C$, is essential. After an idea of Mattern [9] a $GVT$ approximation can be computed by using two cuts $C$ and $C'$ (see figure 3), where $C'$ is later than $C$ (formally $C \subseteq C'$). Cut $C'$ should be "delayed" until all messages which cross cut $C$ have reached their destination $LP$ (i.e. $M_C = \emptyset$) and then the minimal timestamp of all messages sent after cut $C$ is an approximation for the messages crossing cut $C'$. So on the other hand, $C$ and $C'$ should be close together to get a good approximation. Taking into account the central part of submodel managers for computing a $GVT$ value the algorithm is characterized by:

– Initially all submodels are colored white (and thus all the $LPs$ too);
– Every submodel counts the white messages sent and received by its $LPs$;
– The first consistent cut $C$ changes the color to red;
– After cut $C$ every submodel remembers the minimum timestamp of all red messages sent by its $LPs$;
– Cut $C$ is consistent with respect to white messages and therefore the number of white messages crossing $C$ can be determined;

- If no (white) messages cross cut $C$ then the minimum over all $LVT_i(C)$ ($LVT$ at cut $C$) is a valid $GVT$ approximation;
- Otherwise a second cut $C'$ is initiated and the minimum over all $LVT_i(C')$ and over all timestamps of sent red messages is determined;
- If all white messages have been received before $C'$ it is a valid $GVT$ approximation, otherwise the result is ignored and a new attempt for $C'$ is started.

# 3   Main Program as $GVT$-Manager

The above described algorithm needs an initiator and a control instance which states if cut $C'$ fulfills the desired criteria or not. This can be a dedicated $LP$ or some other special element. We propose to assign this duty to the main program which serves primarily as the interface to the experimentator and as a controlling instance for starting and closing down the simulation system. So the main program acts also as a "$GVT$-manager".

After telling all submodels to initiate a new $GVT$ approximation (cut $C$) the $GVT$-manager has to collect the relevant data from them. For each submodel those are the difference of sent and received white messages and the minimal timestamp of sent red messages. If the overall difference of white messages is not 0, then a further collection of data is necessary (cut $C'$). After having reached a valid $GVT$ approximation the new $GVT$ value is given to all submodels. The algorithm for the $GVT$ manager is summarized in figure 4, where $LVT_{SM}$ is the actual $LVT$ of a submodel and $md_w$ the submodel difference of white messages (i.e. the difference of white messages sent and received by all $LP_i \in SM$).

to all submodels: start new $GVT$ approximation (corresponds to cut $C$);
**LOOP**
    collect from each $SM$: $LVT_{SM}$ and $md_w$;
    $md = \sum_{\forall SM} md_w$ (overall white message difference);
    **IF** $md = 0$ **THEN**
        exit **LOOP**;
    **ELSE**
        to all submodels: start $GVT$ approximation (corresponds to cut $C'$);
    **END IF**;
**END LOOP**;
$GVT :=$ overall minimum of all $LVT_{SM}$;
to all submodels: report new $GVT$-value;

**Fig. 4.** Algorithm for the $GVT$-manager

It can be seen that the $GVT$-manager is responsible for the start of a new calculation and the number of iterations to reach a valid $GVT$ approximation. It need not handle colors, which is left to the submodels.

# 4  Submodels

As mentioned above, a $GVT$ approximation does not change the central simulation algorithm. But it is clear, that some slow down for the additional computations and communications occurs. To minimize this effect we use our submodel structure (i.e. nearly all work for $GVT$ approximation is done by the submodel managers and the main program).

By doing a step (receiving and sending messages, internal actions, state saving) each $LP$ gets from its submodel manager the current color. Afterwards it reports back the white and red message differences and the minimal timestamp of messages sent with the current color. There is no need for additional attributes to the $LP$.

The submodel manager maintains a color attribute which is changed when a corresponding request from the $GVT$-manager exists (i.e. a new $GVT$ approximation is started). Furthermore counters for white and red message differences over a submodel and the value of the minimal timestamp of sent red messages by all $LP$s of a submodel is updated after each $LP$ step. On request the submodel $LVT$ is computed by finding the minimal $LVT$ value among all $LP$s handled by this submodel. So we have for each submodel

- counters for white and red message differences, $md_w$ and $md_r$ (updated after each $LP$ step);
- the minimal timestamp of messages of the current color sent by all $LP$s of this submodel, $ts_w$ and $ts_r$ (updated after each $LP$ step);
- the submodel $LVT$, $LVT_{SM}$ (determined on request).

On request all this data is given to the $GVT$-manager, which requires only a short interruption of the simulation.

The central algorithm for the submodel manager is summarized in figure 5.

This mechanism needs no direct $LP - LP$ communication for $GVT$ approximation and therefore no auxiliary means like control messages are necessary. Only the data exchange between $GVT$-manager (the main program) and submodel managers requires some additional communication that is not touching $LP$s.

For repeatedly execution of a new $GVT$ approximation the two colors can be used alternately with appropriate meaning. After a new $GVT$ value is computed all the submodel wide message differences of the previous color (i.g. $md_w$ for the second computation) have to be reset to 0 before the next cut.

# 5  Conclusion

The described algorithm for $GVT$ approximation uses the special structure of our SQTW system whereas the main task of a submodel manager is $LP$ scheduling.

IF start $GVT$ approximation is requested **THEN**
   IF new $GVT$ approximation (corresponds to cut $C$) **THEN**
      change color from white to red;
      $ts_r = \infty$;
   **END IF**;
   (for cuts $C$ and $C'$)
   determine $LVT_{SM}$ as the minimal $LVT_i$ over all $LP_i \in SM$;
   $LVT_{SM} = \min(LVT_{SM}, ts_r)$;
   report $LVT_{SM}$ and $md_w$ to $GVT$ manager;
**END IF**;

IF new $GVT$ value available **THEN**
   keep new $GVT$ value;
**END IF**;

**Fig. 5.** Algorithm for submodel manager

So submodel managers are obviously designated for handling $GVT$ specific data. After each $LP$ step the relevant data is updated on the submodel layer. Therefore no control messages sent to $LP$s are needed and thus no direct interruption of the underlying simulation occurs. The cut events are handled by the submodel managers and on request they report the necessary data to the main program, which serves as the $GVT$ manager. The main program has to decide when a $GVT$ approximation is valid and how often it is necessary.

Based on this method we will add a fossil collection to our system. That will enable us to carry out "real" simulations and to compare the results with standard time warp, which can be seen as a special case of SQTW.

## References

1. Chandy, K. N. and J. Misra. 1989. *Parallel Program Design*. Addison-Wesley, Reading, Ma.
2. Ferscha A. 1996. "Parallel and Distributed Simulation of Discrete Event Systems". *Parallel and Distributed Computing Handbook* (A.Y. Zomaya ed.). Mc Graw-Hill.
3. Fujimoto, R. M. 1990. "Parallel Discrete Event Simulation". *Communications ACM* 33(10), pp.31-53.
4. Hagenauer, H. and W. Pohlmann. 1996. "Making Asynchronous Simulation More Asynchronous". Proc. 10th European Simulation Conference, Budapest.
5. Hagenauer H. and W. Pohlmann. 1996. "Prototyping a Parallel Discrete Event Simulation System in Ada". Proc. ACM TRI-ADA96 (S. Carlson ed.). Philadelphia.
6. Hagenauer, H. and W. Pohlmann. 1997. "Using the Ada 95 Distributed Application Framework for a Distributed Simulation System". Proc. 9th European Simulation Symposium ESS97, Passau.
7. Jefferson, D. R. 1985. "Virtual Time". *ACM TOPLAS 7*, pp.404-425.

8. Lin, Yi-Bing and E. Lazowska. 1990. "Determining the Global Virtual Time in a Distributed Simulation". Proc. 1990 International Conference on Parallel Processing.
9. Mattern, F. 1993. "Efficient Algorithms for Distributed Snapshots and Global Virtual Time Approximation". Journal of Parallel and Distributed Computing 18, no 4.
10. Misra, J. 1986. "Distributed Discrete Event Simulation". ACM Computing Surveys 18(1), p.39-65.
11. Pohlmann, W. 1991. "A Fixed Point Approach to Parallel Discrete Event Simulation". Acta Informatica 28, pp.611-629.
12. Samadi, B. 1985. "Distributed Simulation, Algorithms and Performance Analysis". PHD thesis. Computer Science Department, University of California, Los Angeles.

# MPI-parallelized Radiance on SGI CoW and SMP

Roland Koholka, Heinz Mayer, and Alois Goller

Institute for Computer Graphics and Vision (ICG), University of Technology,
Münzgrabenstraße 11, A–8010 Graz, Austria
{koholka,mayer,goller}@icg.tu-graz.ac.at,
http://www.icg-tu.graz.ac.at/~Radiance

**Abstract.** For lighting simulations in architecture there is the need for correct illumination calculation of virtual scenes. The Radiance Synthetic Imaging System delivers an excellent solution to that problem. Unfortunately, simulating complex scenes leads to long computation times even for one frame. This paper proposes a parallelization strategy which is suited for scenes with medium to high complexity to decrease calculation time. For a set of scenes the obtained speedup indicates the good performance of the chosen load balancing method. The use of MPI delivers a platform independent solution for clusters of workstations (CoWs) as well as for shared-memory multiprocessors (SMPs).

## 1  Introduction

Both ray-tracing and radiosity represent a solution for one of the two main global lighting effects, specular and diffuse interreflection. To extend one of the two models to get a complete solution for arbitrary global lighting suffers from long lasting calculations even for scenes with medium complexity.

The Radiance software package offers an accurate solution for that problem which will be described in more detail in section 2. A project about comparison of indoor lighting simulation with real photographs shows that for scenes with high complexity and detailed surface and light source description simulation can take hours or even days [7]. Parallelization seems to be an appropriate method for faster rendering but keeping the level of quality.

Especially in Radiance there exists a gap between rapid prototyping which is possible with the rview program, and the production of film sequences which is best done using queuing systems. Our parallel version of Radiance allows fast generation of images from realistic scenes without any loss of quality within a few minutes.

Since massive parallel systems still are rare and expensive, we primarily focus on parallelizing Radiance for architectures commonly used at universities and in industry, which are CoWs and SMPs. Platform independence and portability are major advantages of Radiance from the systems point of view. These points are met basing our code on the Message Passing Interface (MPI).

P. Zinterhof, M. Vajteršic, A. Uhl (Eds.): ACPC'99, LNCS 1557, pp. 549–558, 1999.

# 2   Structure of Radiance

As any rendering system Radiance tries to offer a solution for the so-called rendering equation [11] for global lighting of a computer generated scene. The basic intention was to support a complete accurate solution. One main difference between Radiance and most other rendering systems is that it can accurately calculate the luminance of the scene which is done by physical definition of light sources and surface properties. For practical issues the founder of the Radiance Synthetic Imaging System implemented a lot of CAD-file translators for easy import of different geometry data. This forms a versatile framework for anyone who has to do physically correct lighting simulations [6]. Technically, Radiance uses a light-backwards ray-tracing algorithm with a special extension to global diffuse interreflections. Global specular lighting is the main contribution of this algorithm and is well situated in computer graphics [1]. The difficulty lies in extending this algorithm to global diffuse illumination which is briefly explained in the next section.

## 2.1   Calculating Ambient Light

Normally a set of rays distributed over the hemisphere is used for calculating diffuse interreflections which is called distributed ray-tracing. One can imagine that in the case of valuable scene complexity calculating multiple reflections of sets of rays is very time consuming. Radiance solves this problem by estimating the gradient in global diffuse illumination for each pixel, then it decides whether the current diffuse illumination can be interpolated from points in the neighborhood or must be calculated from scratch [5]. Exactly calculated points are cached in an octree for fast neighborhood search. Performance is further increased since diffuse interreflections are only calculated at points necessary for a particular view in contrast to other radiosity algorithms. However, all these optimizations appear to be not enough to render a complex scene with adequate lighting conditions in reasonable time on a single workstation. Parallelization seems to be appropriate for further reduction of execution time.

## 2.2   Existing Parallelization: rpiece

Since version 2.3, there is an extra program for parallel execution of Radiance, called rpiece. To keep installation as simple as possible, rpiece only bases on standard UNIX commands like the pipe command. All data are exchanged via files, requiring that NFS runs on all computation nodes. While the presence of NFS usually is no problem, some file lock managers are not designed for fast and frequent change of the writing process. Additionally, remote processes are not initiated automatically but must be invoked by the user.

In his release notes, Ward [4] claimed a speedup of 8.2 on 10 workstations. We can confirm this result only on suitable scenes, and where sequential computation lasts more than one day. However, rpiece performs badly if execution does not last at least several 10 minutes in parallel. Table 1 shows the time rpiece needs on a CoW for smaller scenes. Obviously, there is an enormous overhead in partitioning and splitting the scene as well as in communicating these pieces via NFS. Consequently, rpiece helps rendering very large scenes in shorter time, but does not produce quick results one might be willing to await.

| file | rpict (1 proc.) | rpiece (1 proc.) | rpiece (3 proc.) | rpiece (10 proc.) |
|------|-----------------|------------------|------------------|-------------------|
| A    | 1:49.66         | 19:02.09         | 14:22.93         | 14:48.25          |
| B    | 5:26.79         | 17:38.85         | 12:49.82         | 14:23.04          |
| C    | 17:03.23        | 29:11.56         | 14:15.85         | 15:12.59          |

**Table 1.** Performance (time) of rpiece on a cluster of workstations (CoW).

## 3    Parallelization Strategies

Because of the nature of ray-tracing these algorithms can be easily parallelized by dividing the image space into equal blocks. But since no one can predict the scene complexity per block which depends on the model itself and the current viewpoint, we now discuss strategies which are general and deliver a good speedup for almost all cases.

### 3.1    The Standard Interface: MPI

Using standards is beneficial in many aspects, since they ensure portability and platform-independence is also given. Moreover, standards are well documented and software is well maintained. PVM was a de-facto standard prior to the definition of MPI. Consequently, we use MPI (Message Passing Interface) since now it is the standard for communication on parallel architectures and has widely been accepted.

MPI only defines the interface (API), thus it is subject to the implementation how to handle a specific architecture. Many implementations of MPI exist, the most popular free versions are MPICH from Argonne National Lab. [10] and LAM, which was developed at the Ohio Supercomputer Center and is currently housed with the Laboratory for Scientific Computing at the University of Notre Dame, Indiana [9]. There are also several commercial implementations, as from SGI especially adapted and optimized to run on their PowerChallenge and Origin shared-memory machines. We use MPICH for both the CoW and the SMP, and additionally SGI's own version. MPICH proved to be easy to install, to support many different architectures and communication devices, and to run stable on all tested platforms.

Another possibility would have been to use shared memory communication directly. However, this is only applicable to SMPs. There does not exist any standard for virtual shared memory that is similarly accepted and widespread as MPI. Since Radiance is a software package designed to run on nearly any platform, not using a common means for communication would be a drawback. Thus, we refrained from using shared memory calls, although it might have been beneficial for more efficient distribution of the calculated ambient light values, and for holding the scene description. The upcoming extension to MPI as defined in the MPI-2 standard seems to solve these insufficiencies. We look forward to a stable and publicly available implementation of MPI-2.

## 3.2    Data Decomposition

There are two strategies to implement parallel ray-tracing. One way is to divide the scene into parts, introducing virtual walls [8]. Due to difficulties in load-balancing and the need for very high-bandwidth communication channels virtual walls seem to be not appropriate for the architectures we want to run Radiance.

The second strategy is to divide the frame, implementing the manager/worker paradigm [2]. The manager coordinates the execution of all workers and determines how to split the frame. The workers perform the actual rendering and send the results to the manager, who combines the pieces and writes the output frame to disk. Since a global shared memory cannot be assumed on any architecture, every worker must hold its own version of the scene. This requires much memory and also slows down initialization since every single worker has to read the octree file from disk.

## 3.3    Concurrency Control

Since Radiance calculates a picture line by line, the frame is divided into blocks consisting of successive lines. The manager only has to transmit the start and the end line of the block. Most of the time the manager waits for data. To avoid busy-waiting the manager is suspended for one second each time the test for finished blocks ends negative, as illustrated in figure 1. This way the manager needs less than one percent of CPU-time, and therefore it is possible to run the manager and one worker on the same processor. Suspending the manager for one second causes a nondeterministic behavior that results in slight variations of execution time.

One feature of Radiance is that it can handle diffuse interreflection. These values must be distributed to every worker to avoid needlessly calculations. Broadcasting every new calculated value would cause too much communication ($O(n^2)$), $n$ being the number of workers. To reduce communication traffic, only blocks of 50 values are delivered. This communication must to be non-blocking, because these blocks are only received by a worker when he enters the subroutine for calculating diffuse interreflection.

## 3.4    Load Balancing

The time it takes to calculate a pixel differs for every casted ray, depending on the complexity of the intersections that occur. Therefore load-balancing problems arise when distributing large parts. Radiance uses an advanced ray-tracing

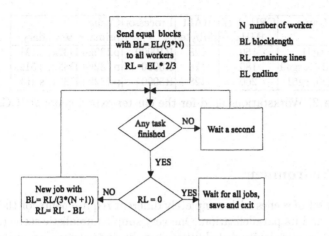

**Fig. 1.** Dynamic load balancing strategy.

algorithm that uses information of neighboring rays to improve speed. Thus, rays are not independent anymore, and scattering the frame in small parts would solve the load-balancing problem, but also decreases the profit from the advanced algorithm.

A compromise that meets these two controversial requirements is to combine dynamic partitioning and dynamic load-balancing as stated in [3]. At the beginning of the calculation relatively large parts (see figure 1) of the frame are sent to the workers to take advantage of the fast algorithm. The distributed portions of the frame become smaller with the progress of the calculation. Near the end the tasks are that small so that they can be executed fast enough to get a good load-balance.

## 4   Computing Platforms

We evaluated performance on computers accessible at our department and at partner institutes. At first, we concentrated on evaluating two different but common architectures.

The first architecture is a cluster of workstations (CoW). It was built at our institute by connecting 11 O2 workstations (see table 2) from SGI via a 10 Mbit Ethernet network. All workstations are equally powerful except machines sg50 and sg51, which are about 10% faster. They are only used for experiments with 10 and 11 computing nodes.

The second platform is a PowerChallenge from SGI containing 20 processors. Each R10000 processor is clocked with 194 MHz and equipped with primary data and instruction cache of 32 Kbytes each, and 2 Mbyte secondary cache (p:I32k+D32k, s:2M). All 20 processors share an 8-way interleaved, global main memory of 2.5 GByte.

| Name | CPU-clock [MHz] | RAM [Mbyte] | Processor | Cache primary, secondary |
|---|---|---|---|---|
| sg41 | 150 | 128 | R10000 | p:I32k+D32k, s:1M |
| sg42 ... sg49 | 180 | 128 | R5000 | p:I32k+D32k, s:512k |
| sg50, sg51 | 200 | 128 | R5000 | p:I32k+D32k, s:1M |

**Table 2.** Workstations used for the cluster experiments at ICG.

## 5 Test Environment

We selected a set of scenes with completely different requirements for the rendering algorithm and its parallelization: One very simple unrealistic scene (spheres) with a bad scene complexity distribution over the image space, an indoor (hall) and an outdoor lighting simulation (townhouse). These 3 scenes come along with the Radiance distribution. The most important test scene is the detailed model of an office at our department with a physical description of light sources (bureau). All scenes and the corresponding parameter settings are listed in table 3, and the corresponding pictures are shown in figure 2.

| name | octree size | image size | options | comment |
|---|---|---|---|---|
| hall0 | 2212157 | -x 2000 -y 1352 | -ab 0 | large hall (indoor simulation) |
| spheres0 | 251 | -x 1280 -y 960 | -ab 0 | 3 spheres, pure ray-tracing |
| spheres2 | 251 | -x 1280 -y 960 | -ab 2 | 3 spheres, ambient lighting |
| bureau0 | 5221329 | -x 512 -y 512 | -ab 0 | bureau room at ICG |
| bureau1 | 5221329 | -x 512 -y 512 | -ab 1 | + ambient lighting |
| bureau2 | 5221329 | -x 512 -y 512 | -ab 2 | + increased ambient lighting |
| townhouse0 | 3327443 | -x 2000 -y 1352 | -ab 0 | townhouse (outdoor simulation) |
| townhouse1 | 3327443 | -x 2000 -y 1352 | -ab 1 | + ambient lighting |

**Table 3.** Used scenes for evaluation of the MPI-parallel version of Radiance.

This test set covers a wide spectrum in size of the octree, complexity of the scene, time required for rendering and visual impression. All scenes were computed with and without diffuse interreflection. Table 4 summarizes the times measured for the sequential version.

Due to the manager/worker paradigm, $x$ processes result in $x - 1$ workers. Consequently, whenever we mention $n$ processors (nodes), we start $n+1$ processes where one of them is the manager. Since the manager is idle nearly all the time, we do not count it for the speedup on the SMP. This is acceptable since we know from tests on the CoW, where the manager and one worker is running on the same workstation, that the result is falsified less than 1%.

On very small scenes no big advantages will be seen. However, an (almost) linear speedup should be noticed for larger scenes if no diffuse interreflections

A: Hall (clipped).

B: 3 spheres.

C: ICG room.

D: Townhouse.

Fig. 2. The 4 scenes used for evaluation.

| name | time on sg48 | time on SMP |
|------|-------------|-------------|
| hall0 | 1:19:13.48 | 33:50.34 |
| spheres0 | 40.89 | 19.27 |
| spheres2 | 8:27:00 | 3:10.11 |
| bureau0 | 2:32:55.84 | 1:14:42.73 |
| bureau1 | 3:56:03.65 | 1:54:23.19 |
| bureau2 | 5:13:30.72 | 2:29:16.35 |
| townhouse0 | 5:17.98 | 2:40.64 |
| townhouse1 | 9:21.82 | 3:55.07 |

Table 4. Execution times of the sequential Radiance for the test scenes.

are to be calculated (-ab 0). Due to heavy all-to-all communication to distribute newly calculated ambient lighting values, performance will decrease on such runs (-ab 2). This effect will mainly influence the performance on the CoW, since network bandwidth is very limited. It should show nearly no effects on the SMP.

## 6    Results

The times for sequential execution $T(1)$ are already shown in table 4. For good comparison of the platforms and to visualize the performance, we use the speedup $S(n) := \frac{T(1)}{T(n)}$ and the efficiency $\varepsilon(n) := \frac{S(n)}{n}$, being $n$ the number of processors.

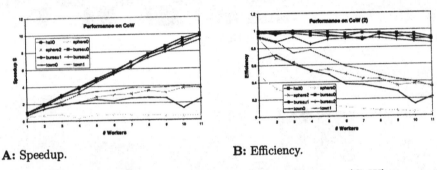

A: Speedup.                    B: Efficiency.

**Fig. 3.** Performance on the cluster of workstations (CoW).

Although one node of the cluster is only half as fast as one SMP-processor, all larger scenes show a good parallel performance. Utilizing 11 processors, time for rendering the bureau scene without ambient lighting decreases from 2:32:55.84 to 15:14.35 giving a speedup of 10. Even more processors could be utilized since the efficiency always exceeds 85%, as shown in figure 3.

The four smaller scenes illustrate that speedup cannot be increased if the time for parallel execution falls below about one minute. This is mainly caused by the remote shells MPICH has to open in combination with the (slow) 10-Mbit Ethernet network. However, MPICH proved to run stable during all test. Furthermore, the chosen load-balancing strategy appears to be well suited even for the CoW, since no communication bottlenecks occurred during the all-to-all communication when calculating diffuse interreflection.

When comparing figures 3 and 4 one can see no big difference. The four smaller scenes are rendered a little bit more efficiently on the SMP, but above 13 workers, effects of the all-to-all communication decrease performance. Changing the communication pattern to $\log n$-broadcast instead of sequentially communicating to every other node would reduce this effect. Improving this communication pattern will be a near future goal.

Most surprisingly, execution time could not be moved well below two minutes, even if there would have been enough processors. As we experienced, sequential

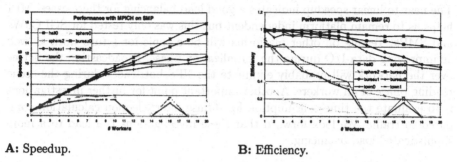

A: Speedup.                                          B: Efficiency.

**Fig. 4.** Performance on the PowerChallenge using MPICH.

overhead starting the parallel tasks and minor load imbalances prohibit scaling execution time down to seconds even on a shared-memory machine. As on the cluster, MPICH also worked well on this architecture.

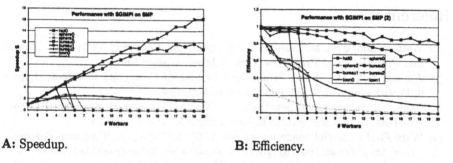

A: Speedup.                                          B: Efficiency.

**Fig. 5.** Performance on the PowerChallenge using native MPI.

Initially we expected the native MPI version of SGI to be much faster than MPICH. However, as figure 5 illustrates, only a few scenes could be rendered using an arbitrary number of processors. The number of open communication channels seems to be the problem. Moreover, we could find no advantage regarding performance when using native MPI.

## 7   Conclusion and Future Work

Our main goal was to find a parallelized version of the Radiance Synthetic Imaging System for scenes with medium to high complexity to close the gap between the rview and rpict programs. Another prerequisite was the platform independence of the overall system. In section 6 we show that our implementation works well for most of the selected scenes running on an SMP and also on a CoW.

The near to linear speedup indicates a good load balancing for these cases. MPI helps us to remain platform independent but this was only tested for SGI CoWs and SMPs. Testing on other platforms will be a topic for future work as well as improving data I/O during the initialization phase. Broadcasting the octree over the network will probably eliminate the NFS bottleneck during the scene reading phase of all workers. Another important point for further investigations is to evaluate the time consumption for diffuse interreflection calculation compared to traffic. We look forward that these investigations will lead to a more sophisticated load balancing.

## Acknowledgments

This work is financed by the Austrian Research Funds "Fonds zur Förderung der wissenschaftlichen Forschung (FWF)", Project 7001, Task 1.4 "Mathematical and Algorithmic Tools for Digital Image Processing". We also thank the Research Institute for Software Technology (RIST++), University of Salzburg, for allowing us to run our code on their 20-node PowerChallenge.

## References

[1] Watt A. and Watt M. *Advanced Animation and Rendering Techniques, Theory and Practice*. ACM Press New York, Addison-Wesley, 1992.

[2] Zomaya A. Y. H., editor. *Parallel and Distributed Computing Handbook*, chapter 9, Partitioning and Scheduling. McGraw-Hill, 1996.

[3] Pandzic I.-S. and Magnenat-Thalmann N. Parallel raytracing on the IBM SP2 and T3D. In *Supercomputing Review*, volume 7, November 1995.

[4] Ward G. J. Parallel rendering on the ICSD SPARC-10's. Radiance Reference Notes, http://radsite.lbl.gov/radiance/refer/Notes/parallel.html.

[5] Ward G. J. A ray tracing solution for diffuse interreflexion. In *Computer Graphics (SIGGRAPH '88 Proceedings)*, volume 22, pages 85–92, August 1988.

[6] Ward G. J. The radiance lighting simulation and rendering system. In *Computer Graphics (SIGGRAPH '94 Proceedings)*, volume 28, pages 459–472, July 1994.

[7] Karner K. *Assessing the Realism of Local and Global Illumination Models*. PhD thesis, Computer Graphics and Vision (ICG), Graz University of Technology, 1996.

[8] Menzel K. Parallel Rendering Techniques for Multiprocessor Systems. In *Proceedings of the Spring School on Computer Graphics (SSCG '94)*, pages 91–103. Comenius University Press, 1994.

[9] LAM/MPI parallel computing. Home page of LAM — Local Area Multicomputer: http://www.lsc.nd.edu/lam/.

[10] MPICH — A portable implementation of MPI. Home page of MPICH: http://www.mcs.anl.gov/mpi/mpich/.

[11] Kajiya J. T. The rendering equation. In *Computer Graphics (SIGGRAPH '86 Proceedings)*, volume 20, pages 143–150, August 1986.

# Parallel Sub-collection Join Query Algorithms for a High Performance Object-Oriented Database Architecture

David Taniar[1] and J. Wenny Rahayu[2]

[1] Monash University - GSCIT, Churchill, Vic 3842, Australia
[2] La Trobe University, Dept. of Computer Sc. & Comp. Eng., Australia

**Abstract.** One of the differences between relational and object-oriented databases (OODB) is attributes in OODB can be of a collection type (e.g. sets, lists, arrays, bags) as well as a simple type (e.g. integer, string). Consequently, explicit join queries in OODB may be based on collection attributes. One form of collection join queries in OODB is *sub-collection join queries*, where the joins are based on collection attributes and the queries check for whether one collection is a sub-collection of the other. The sub-collection predicate can be in a form of subset, sublist, proper subset, or proper sublist. We propose two algorithms for parallel processing of sub-collection join queries. The first one is based on sort-merge-nested-loop, whereas the second is based on sort-hash.

## 1 Introduction

In *Object-Oriented Databases* (OODB), although path expression between classes may exist, it is sometimes necessary to perform an explicit join between two or more classes due to the absence of pointer connections or the need for value matching between objects. Furthermore, since objects are not in a normal form, an attribute of a class may have a collection as a domain. ODMG (Object Data Management Group) has formulated four kinds of collection types, namely *set* (unordered, no duplicate), *list/array* (ordered, allow duplicate), and *bag* (unordered, allow duplicate) [1]. Consequently, object-oriented join queries may also be based on attributes of any collection type. Such join queries are called *collection join queries* [7]. Our previous work reported in Taniar and Rahayu [7] classify three different types of collection join queries, namely: *collection-equi join*, *collection-intersect join*, and *sub-collection join*. In this paper, we would like to focus on sub-collection join queries. We are particularly interested in formulating parallel algorithms for processing such queries. The algorithms are non-trivial to *parallel object-oriented database systems*, since most conventional join algorithms (e.g. hybrid hash join, sort-merge join) [4] deal with single-valued attributes and hence most of the time they are not capable to handle collection join queries.

An interest in *parallel OODB* among database community has been growing rapidly, following the popularity of multiprocessor servers and the maturity

P. Zinterhof, M. Vajteršic, A. Uhl (Eds.): ACPC'99, LNCS 1557, pp. 559–569, 1999.
© Springer-Verlag Berlin Heidelberg 1999

of OODB. The emerging between parallel technology and OODB has shown promising results [2], [5], [10]. However, most research done in this area concentrated on path expression queries with pointer chasing. Explicit join processing exploiting collection attributes has not been given much attention.

The work presented in this paper is actually part of a larger project on parallelization of collection join queries. Parallelization of sub-collection join is the third and final stage of the project. The first two stages dealt with parallelization of collection-equi and collection-intersect join. We have reported some of the results obtained from the first two stages at IDEAS'98 (parallelization of collection-equi join [6]) and EuroPar'98 (parallelization of collection-intersect join [9]).

The main difference between parallelization of sub-collection join and parallelization of collection-equi join is shown not only in the data partitioning step, in which parallelization of sub-collection join employs a non-disjoint partitioning whereas parallelization of collection-equi join adopts a disjoint partitioning, but also in the local join step due to the nature of the two different join predicates. The local join in the parallelization of sub-collection join is also much more complex than that in the parallelization of collection-intersect join. However, parallelization of sub-collection join has less option in the data partitioning method to be used, as it relies only on the divide and broadcast partitioning and its variant, whereas parallelization of collection-intersect join, as well as using the same data partitioning in parallelization of sub-collection join, it may also adopt an alternative data partitioning based on a simple replication. As a summary, there are some degrees of differences among the three parallelization models for collection join queries.

To provide enough background for the proposed algorithms, we first define sub-collection join queries. *Sub-collection Join Queries* contain a join predicate where one collection attribute of the join predicate is compared against another collection attribute and the predicate checks for a subset, sublist, proper subset, or proper sublist. The difference between *proper* and *non-proper* is that the proper predicates require both join operands to be properly sub-collection. That means that if both operands are the same, they do not satisfy the predicate, and hence return a false. The difference between subset and sublist originated from the basic differences between sets and lists [1]. In other words, subset predicates are applied to sets/bags, whereas sublists are applied to lists/arrays.

As a running example, consider the data shown in Figure 1. Suppose class $A$ and class $B$ are *Journal* and *Proceedings*, respectively. Both classes contain a few objects shown by their OIDs (e.g., objects $a$ to $i$ are Journal objects and objects $p$ to $w$ are Proceedings objects). The join attributes are *editor-in-chief* of Journal and *program-chair* of Proceedings; and are of type collection of *Person*. The OID of each person in these attributes are shown in the brackets. For example $a(250,75)$ denotes a Journal object with OID $a$ and the editors of this journal are Persons with OIDs 250 and 75.

A *subset* predicate can be written by applying an intersection between the two sets and comparing the intersection result with the smaller set. It is normally

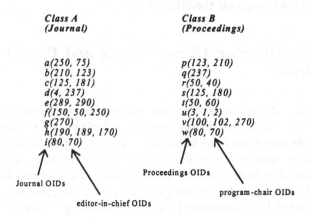

Fig. 1. Sample data

in a form of (attr1 intersect attr2) = attr1. Attributes attr1 and attr2 are of type set. If one or both of them are of type bag, they must be converted to sets. Suppose the attribute *editor-in-chief* of class *Journal* and the attribute *program-chair* of class *Proceedings* are of type sets of *Person*. An example of a sub-collection join is to retrieve pairs of Journal and Proceedings, where the program-chairs of a conference are a subset of the editors-in-chief of a journal. The query expressed in OQL (Object Query Language) [1] can be written as follows:

```
Select A, B
From A in Journal, B in Proceedings
Where (A.editor-in-chief intersect B.program-chair)
                                      = B.program-chair
```

If the join predicate is a *proper subset*, the where clause of the above query must be ANDed with A.editor-in-chief != B.program-chair. This is necessary to enforce that both operands are not identical.

The *sublist* predicate checks whether the first list is a sublist of the second. Two identical lists are regarded as one list being a sublist of the other. If the predicate is a *proper sublist* instead, identical lists are not allowed. The difference between sublist and subset predicate is determined by the type of the operand.

The rest of this paper is organized as follows. Section 2 describes a data partitioning method for parallel sub-collection join query processing, namely "Divide and Partial Broadcast". Section 3 presents a proposed parallel join algorithm for sub-collection join queries based on sort-merge and nested-loop techniques. This algorithm is called "Parallel Sort-Merge Nested-Loop" algorithm. Section 4 introduces another algorithm called "Parallel Sort-Hash" algorithm, which is based on a combination of sort and hash techniques. Finally, section 5 draws the conclusions and explains the future work.

## 2  Data Partitioning Using "Divide and Partial Broadcast"

Parallel join algorithms normally are decomposed into two steps: *data partitioning* and *local join*. Data partitioning creates parallelism, as it divides the data to multiple processors, so that the join can then be performed locally in each processor without interfering others. For sub-collection join queries, it is not possible to have non-overlap partitions, due to the nature of collections which may be overlapped. Hence, some data needs to be replicated. We propose a partitioning strategy for parallel sub-collection join query processing, which is based on the *Divide and Broadcast* technique [3], called "Divide and Partial Broadcast".

The Divide and Partial Broadcast algorithm, shown in Figure 2, proceeds in two steps. The first step is a *divide* step, and the second step is a *partial broadcast* step. We divide class $B$ and partially broadcast class $A$. The *divide* step is explained as follows. Divide class $B$ into $n$ number of partitions. Each partition of class $B$ is placed in a separate processor (e.g. partition $B1$ to processor 1, partition $B2$ to processor 2, etc). Partitions are created based on the largest element of each collection. For example, object $p(123, 210)$; the first object in class $B$, is partitioned based on element 210, as element 210 is the largest element in the collection. Then, object $p$ is placed on a certain partition, depending on the partition range. For example, if the first partition is ranging from the largest element 0 to 99, the second partition is ranging from 100 to 199, and the third partition is ranging from 200 to 299, then object $p$ is placed in partition $B3$, and subsequently in processor 3. This is repeated for all objects of class $B$.

```
Procedure DividePartialBroadcast
Begin
   Step 1 (divide):
      1. Divide class B based on the largest element in each collection.
      2. For each partition of B (i = 1, 2, ..., n)
            Place partition Bi to processor i
         End For

   Step 2 (partial broadcast):
      3. Divide class A based on the smallest element in each collection.
      4. For each partition of A (i = 1, 2, ..., n)
            Broadcast partition Ai to processor i to n
         End For
End Procedure
```

Fig. 2. Divide and Partial Broadcast Algorithm

The *partial broadcast* step can be described as follows. First, partition class A based on the smallest element of each collection. Clearly, this partitioning method is exactly the opposite of that in the divide step. Then for each partition $Ai$ where $i=1$ to $n$, broadcast partition $Ai$ to processors $i$ to $n$. This broadcasting technique is said to be partial, since the broadcasting goes down as the partition number goes up. For example, partition $A1$ is basically replicated to all processors, partition $A2$ is broadcast to processor 2 to $n$ only, and so on. In regard to the load of each processor, the load of the last processor may be the heaviest, as it receives a full copy of class A and a portion of class B. The load goes down as class A is divided into smaller size (e.g., processor 1). Load balanced can be achieved by applying the same algorithm to each partition but with a reverse role of A and B; that is, *divide A* and *partial broadcast B*. It is beyond the scope of this paper to evaluate the Divide and Partial Broadcast partitioning method. This has been reserved for future work. Some preliminary results have been reported in [8].

# 3    Parallel SORT-MERGE-NESTED-LOOP Join Algorithm for Sub-collection Join Query Processing

After data partitioning is completed, each processor has its own data. The join operation can then be done independently. The local join is made of a simple sort-merge and a nested-loop structure. A sort operator is applied to each collection, and then a nested-loop construct is used in join-merging the collections. There are two important things to mention regarding the sorting: one is that sorting is applied to the *is_subset* predicate only, and two is that to avoid a repeated sorting especially in the inner loop of the nested loop, sorting of the second class is taken out from the nested loop. The algorithm uses a nested-loop structure, because of not only its simplicity but also the need for all-round comparisons among all objects.

In the merging phase, the *is_subcollection* function is invoked, in comparing each pair of collections from the two classes. In the case of a subset predicate, after converting the sets to lists, the *is_subcollection* function is executed. The result of this function call becomes the final result of the subset predicate. If the predicate is a sublist, the *is_subcollection* function is directly invoked, without the necessity to convert the collections into lists, since the operands are already lists.

The *is_subcollection* function receives two parameters: the two collections to compare. The function first finds a match of the first element of the smallest list in the bigger list. If a match is found, subsequent element comparisons of both lists are carried out. Whenever the subsequent element comparison fails, the process has to start finding another match for the first element of the smallest list again (in case of duplicate items exist). Figure 3 shows the complete parallel sort-merge-nested-loop join algorithm for sub-collection join queries.

```
Parallel-Sort-Merge-Nested-Loop-Subcollection:
Begin
  // step 1 (Data Partitioning):
     Call DividePartialBroadcast

  // step 2 (Local Join in each processor)
     // a) sort class B (is_subset predicate only)
          For each object of class B
             Sort collection b of class B
          End For
     // b) nested loop and sort class A
          For each collection a of class A
             For each collection b of class B
                If the predicate is subset or proper subset Then
                   Convert a and b to list type
                End If
                // c) merging
                If is_subcollection(a,b) Then
                   Concatenate the two objects into the query result
                End If
             End For
          End For
End Program

Function is_subcollection (L1, L2: list) : Boolean
Begin
  Set i to 0
  For j=0 to len(L2)
    If L1[i] = L2[j]
       Set Flag to TRUE
       If i = len(L1)-1 Then Break     // end of L1
       Else i++ End If                 // find the next match
    Else
       Set Flag to FALSE // reset the flag
       Reset i to 0
    End If
  End For
  If Flag = TRUE and i = len(L1)-1 Then
     Return TRUE
     (for the is_proper predicate: If len(L1) != len(L2) Return TRUE
                                   Else Return FALSE End If)

  Else
     Return FALSE
  End If
End Function
```

**Fig. 3.** Parallel Sort-Merge-Nested-Loop Sub-collection Join Algorithm

Using the sample data in Figure 1, the result of a subset join is $(g,v),(i,w)$, and $(b,p)$. The last two pairs will not be included in the results, if the join predicate is a proper subset, since the two collections in each pair are equal.

# 4  Parallel SORT-HASH Join Algorithm for Sub-collection Join Queries

Parallel Sort-Hash Sub-collection join algorithm consists of two main steps: *data partitioning* and *local join*. In the data partitioning step, the *divide and partial broadcast* algorithm is employed. The local join is based on a combination of *sort* and *hash* methods. In the local join step, if the collection attributes, on where the join is based, are sets or bags, they are sorted first. The next step would be hash and probe. The following sections describe the hashing and probing mechanisms.

## 4.1  Hashing

In this section we introduce a parallel join algorithm based on a combination of *sort* and *hash* methods for sub-collection join queries. In the hashing part, hash tables are used. Hash tables for collection joins are different from those for relational hash join queries. For sub-collection join queries, *multiple hash tables* are used.

Each hash table contains all elements of the same position of all collections. For example, entries in hash table 1 contain all first elements in the collections. The number of hash tables is determined by the largest collection among objects of the class to be hashed. If the collection is a *list/array*, the position of the element is as the original element composition in each collection. If the collection is a *set/bag*, the smallest element within each collection will be hashed into the first hash table, the second smallest element is hashed to the second hash table, and so on. Set/bag hashing will be enhanced if the set/bag is *preprocessed* by means of *sorting*, so that the hashing process will not have to search for the order of the elements within the set/bag. Figure 4 shows an example where three objects are hashed into multiple hash tables. Case 1 is where the objects are arrays, and case 2 is where the objects are sets.

In the case of collision, it will be resolved as per normal collision handling. However, it must be noted that the same element value and position in another collection (especially set) may not necessary result in a collision. For example, a new collection $h(150,50,25)$ is to be hashed. If collection $h$ is a set, the elements 150, 50, and 25 will be hashed to the hash tables 3, 2, and 1, respectively. There will be no collision occurs between $h(150,50,25)$ and collection $f(150,50,250)$. Collision will occur however if collection $h$ is a list. The element $150(h)$ will be hashed to hash table 1 and will collide with $150(f)$. Subsequently, the element $150(h)$ will go to the next available entry in hash table 1, as the result of the collision.

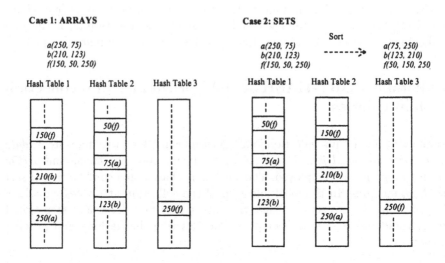

**Fig. 4.** Multiple Hash Tables

## 4.2 Probing

In the hashing part, the *super* collection is hashed. Once the multiple hash tables are built, the probing process begins. In the probing part, each *sub* collection is probed one by one. If the join predicate is an *is_proper* predicate, it has to make sure that the two matched collections are not equal. This can be implemented in two separate checking using an XOR operator. It checks either the first matched element is *not* from the first hash table, or the collection of the first class has not been reached. If either condition (not both) is satisfied, the matched collections are put into the query result.

If the join predicate is a normal subset/sublist (i.e. non-proper), apart from the probing process, no other checking is necessary. Figure 5 gives the pseudo-code for the complete parallel sort-hash sub-collection join algorithm.

The central part of the join processing is basically the probing process. The main probing function for sub-collection join queries is called *function some*. It recursively checks for a match for the first element in the collection. Once a match is found, another function called *function universal* is called. This function checks recursively whether a collection exists in the multiple hash table and the elements belong to the same collection. Figure 6 shows the pseudocode for the two functions for the sort-hash version of parallel sub-collection join algorithm.

## 5   Conclusions and Future Work

The need for join algorithms especially designed for sub-collection join queries is clear, as the conventional parallel join algorithms were not designed for collection types. In this paper, we present two algorithms especially design for sub-collection join queries, namely *Parallel Sort-Merge-Nested-Loop* and *Parallel Sort-Hash* algorithms. Data partitioning method, which creates parallelism, is based on *Divide and Partial Broadcast*. Once the data partitioning method is applied, local join is carried out by either a sort-merge operator (in the case of parallel sort-merge-nested-loop) or a hash function (in the case of parallel sort-hash).

```
Parallel-Sort-Hash-Subcollection-Join:
Begin
  // step 1 (data partitioning):
    Call DivideAndPartialBroadcast partitioning

  // step 2 (local joining): In each processor
    // a. preprocessing (sorting) (set/bags only)
        For each collection of class A and class B
          Sort each collection
        End For

    // b. hash
        For each object of class B
          Hash the object into multiple hash table
        End For

    // c. probe
        For each object of class A
          Case is_proper predicate:
            If some(1,1) Then      // element 1, hash table 1
              If first match is not from the first hash table XOR
              not end of collection of the first class Then
                Put the matching pairs into the result
              End If
            End If
          Case is_non_proper predicate:
            If some(1,1) Then
              Put the matching pair into the result
            End If
          End Case
        End For
End Program
```

**Fig. 5.** Parallel Sort-Hash Sub-collection Join Algorithm

```
1) Function some (element i, hash table j) Return Boolean
   Begin
      Hash and Probe element i to hash table j
      If matched Then        // match the element and the object
        Increment i and j
        If end of collection is reached Then
          // check for end of collection of the probing class
          Return TRUE
        End If
        If hash table j exists Then    // check for the hash table
          result = universal (i, j)
        Else
          Return FALSE
        End If
      Else
        increment j
        result = some (i, j)
        // continue searching to the next hash table (recursive).
      End If
      Return result
   End Function

2) Function universal (element i, hash table j) Return Boolean
   Begin
      Hash and Probe element i to hash table j
      If matched Then        // match the element and the object
        Increment i and j
        If end of collection is reached Then
          // check for end of collection of the probing class
          Return TRUE
        End If
        If hash table j exists Then   // check for the hash table
          result = universal (i, j)
        Else
          Return FALSE
        End If
      Else
        Return FALSE
      End If
      Return result
   End Function
```

**Fig. 6.** Probing Functions

Our future plan includes investigating the possibility to use a *hash-hash* method, instead of a sort-hash method like in the Parallel Sort-Hash algorithm. Due to the limited space, we have not yet included the performance evaluation of the proposed algorithms in this paper. This is also reserved for a future paper.

# References

1. Cattell, R.G.G. (ed.), *The Object Database Standard: ODMG-93*, Release 1.1, Morgan Kaufmann, 1994.
2. Kim, K-C., "Parallelism in Object-Oriented Query Processing", *Proceedings of the Sixth International Conference on Data Engineering*, pp. 209-217, 1990.
3. Leung, C.H.C. and Ghogomu, H.T., "A High-Performance Parallel Database Architecture", *Proceedings of the Seventh ACM International Conference on Supercomputing*, pp. 377-386, Tokyo, 1993.
4. Mishra, P. and Eich, M.H., "Join Processing in Relational Databases", *ACM Computing Surveys*, **24**(1), pp. 63-113, March 1992.
5. Taniar, D., and Rahayu, W., "Parallelization and Object-Orientation: A Database Processing Point of View", *Proceedings of the Twenty-Fourth International Conference on Technology of Object-Oriented Languages and Systems TOOLS ASIA '97*, Beijing, China, pp. 301-310, 1997.
6. Taniar, D. and Rahayu, J.W., "Parallel Collection-Equi Join Algorithms for Object-Oriented Databases", *Proceedings of International Database Engineering and Applications Symposium IDEAS'98*, IEEE Computer Society Press, Cardiff, UK, July 1998.
7. Taniar, D. and Rahayu, J.W., "A Taxonomy for Object-Oriented Queries", a book chapter in *Current Trends in Database Technology*, Idea Group Publishing, in press (1998).
8. Taniar, D. and Rahayu, J.W., "Divide and Partial Broadcast Method for Parallel Collection Join Queries", *High-Performance Computing and Networking*, P.Sloot et al (eds.), Springer-Verlag LNCS 1401, pp. 937-939, 1998.
9. Taniar, D. and Rahayu, J.W., "Collection-Intersect Join Algorithms for Parallel Object-Oriented Database Systems", *EuroPar'98 Parallel Processing*, Springer-Verlag LNCS 1470, D.Pritchard and J.Reeve (eds.), 1998.
10. Thakore, A.K. and Su, S.Y.W., "Performance Analysis of Parallel Object-Oriented Query Processing Algorithms", *Distributed and Parallel Databases* 2, pp. 59-100, 1994.

# An Evaluation of Parallel Computing in PC Clusters with Fast Ethernet

Manuel Acacio, Oscar Cánovas, José M. García, and Pedro E. López-de-Teruel

Dpto. de Ingeniería y Tecnología de Computadores
Universidad de Murcia
Campus de Espinardo, s/n, 30080 Murcia (Spain)
{meacacio,jmgarcia,pedroe}@ditec.um.es
ocanovas@fcu.um.es

**Abstract.** Currently, clusters of Personal Computers (PCs) with sharing Fast Ethernet may offer a significant price/performance ratio in parallel processing. This paper presents a study of the promising possibilities of parallelism in this environment, using both a classical numerical method to solve a differential equation and an innovative algorithm to solve the statistical problem of density estimation. In addition, and to compare, some comparisons with an IBM SP2 are carried out.

## 1 Motivations

The evolution of CPU technology has brought high-end Personal Computers (PCs) to performance levels in the range of workstations at a very competitive cost. So, PCs are already a very cost effective alternative to workstations as processing nodes in NOW platforms [2,3]. The major limitation of PC clusters is the high communication overhead when exchanging messages between nodes. Unlike a multiprocessor that has a custom low latency/high bandwidth network, nodes in a PC cluster are loosely connected by a LAN with a lower bandwidth.

Our work is situated in the context of an ambitious research project, called ELDORADO, whose aim is to show how using a high-performance comunication channel as the underlying network, clusters of powerful PCs can even overtake traditional parallel machines in a wide range of applications. We show the performance of two real problems implemented with MPI: a numerical algorithm (the parallel Jacobi method for solving differential equations) and a new parallel algorithm used for non-parametric density estimation from uncertain samples. Both algorithms have been successfully executed in a cluster of PCs with Fast Ethernet, and the results we have achieved are very good. A comparison with an IBM SP2 parallel computer has been performed for the first algorithm.

## 2 Preliminary Results, Conclusions, and Future Work

Jacobi relaxation method is an algorithm for solving a differential equation called Laplace's equation. We have used it in the iterative algorithm that calculates

P. Zinterhof, M. Vajteršic, A. Uhl (Eds.): ACPC'99, LNCS 1557, pp. 570–571, 1999.

the temperature of each particle in a body. The EDR algorithm [1], on the other side, is an extension of the Parzen method. It is used for nonparametric density estimation when working with uncertain input samples. We have developed and measured MPI parallel versions of these algorithms. Our tests have been carried out on a cluster of Intel Pentium 200 MHz processor (32 MB RAM) interconnected by a Fast Ethernet 3Com 905-network. The operating system was Linux 2.0.32 and the MPI implemetation MPICH v.1.0.13. Programs were compiled with the best optimization options. Figure 1 shows some illustrative results.

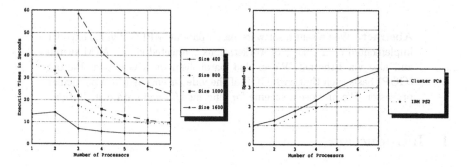

**Fig. 1.** Execution times in seconds for several sizes of the problem in the EDR algorithm, and speed-up comparison with IBM SP2 for the Jacobi method

As a conclusion, we can say that parallel computing on clusters of workstations and PCs has very high potential, since it takes advantage of existing hardware and software. Performance tests of the implementations show that they are superior to much existing parallel programming environments for some application problems.

Our research group is currently working in practical applications in different fields, such as real-time computer vision, and computationally expensive statistical and machine learning problems.

This work has been partially supported by the Spanish CICYT under grant TIC97-0897-C04-02.

# References

1. López-de-Teruel, P.E., Acacio, M.E., Cánovas, O., García, J.M.: EDR: An Algorithm for Parallel Implementation of Parzen Density Estimation from Uncertain Observations. Submitted to the 2nd Merged Symposium IPPS/SPDP (1999)
2. Nupairoj, N., Ni, L.: Performance Evaluation of Some MPI Implementations on Workstations Clusters. Procceedings of the 1994 Scalable Parallel Libraries Conference, (1994) 98–105
3. Wang, X. Blum, E.K.: Parallel Execution of Iterative Computations on Workstations Clusters. Journal of Parallel and Distributed Computing, vol. 34 (1996) 218–226

# Parallel MPEG-2 Encoder on ATM and Ethernet-Connected Workstations

Shahriar M. Akramullah[1], Ishfaq Ahmad[2], and Ming L. Liou[1]

[1] Department of EEE,
[2] Department of Computer Science
Hong Kong University of Science & Tech., Clear Water Bay, Kowloon, Hong Kong

**Abstract.** We present a software-based parallel MPEG-2 video encoder implemented on a cluster of workstations connected via an ATM switch and also via Ethernet. We exploit parallelism on a Group of Picture (GOP) basis such that each GOP of the video is encoded by a particular processor. We propose a scheme for efficient I/O and data distribution.

## 1 Introduction

Video compression can be hardware-based [1] or software-based [2,3]. However, a software solution is more flexible, and thus allows algorithmic improvements. Nonetheless, very high computation requirements of video applications can often overwhelm a single-processor sequential computer [2]. Therefore, it is natural to exploit the enormous computing power offered by parallel computing systems. With their high performance-cost ratio and efficient communication hardware/software, clusters of workstations are the most suitable candidates for computation-intensive applications like video compression.

In this work, we explore the use of general-purpose workstations enabling other jobs to run while compressing video data. The environment is off-line, but the aim is to achieve maximum possible encoding rate beyond the real-time speed. We compare the performance of the encoder when the interconnection is via Ethernet with that when it is via an ATM switch. We propose a scheme for efficient I/O and data distribution. This scheme provides fast data retrieval as well as efficient scheduling and matching of I/O and compression rates such that the entire system operates in a highly balanced fashion without any bottlenecks.

## 2 The Parallel Encoder

In order to achieve maximum possible encoding rate, we keep all of the processors busy in reading the data, performing the encoding, and writing the coded bitstream by scheduling the disks, I/O processors, and computing processors. In other words, the processor waiting times are minimized.

In distributing data, an I/O processor reads $m$ batches of uncompressed video frames and sends one batch to each compute processor in the group. While the

P. Zinterhof, M. Vajteršic, A. Uhl (Eds.): ACPC'99, LNCS 1557, pp. 572–574, 1999.
© Springer-Verlag Berlin Heidelberg 1999

$m$ compute processors are encoding, the I/O processor reads the next $m$ batches and waits. As soon as a compute processor finishes the encoding of a batch it sends a request to the I/O processor, which sends the next batch to the requester. After serving all the requests, the I/O processor reads the next $m$ batches and waits for the requests. The compute processors save the compressed data into a buffer, and write it to the disk when the buffer is full.

## 3  Experimental Results

We use a video test sequence named *football* of CCIR-601 format ($720 \times 480$) and SIF format ($360 \times 240$) as input to our parallel encoder. For message-passing among the workstations in the cluster, which consists of 20 Sun Ultra-1 workstations, we use Message Passing Interface (MPI) environment.

**Table 1.** The frame encoding rate for the *football* sequence.

| No. of Processors | Group Size | Frame/sec. using ATM switch | | | Frame/sec. using Ethernet | | |
|---|---|---|---|---|---|---|---|
| | | SIF | | CCIR-601 | SIF | | CCIR-601 |
| | | $B=3$ | $B=6$ | $B=1$ | $B=3$ | $B=6$ | $B=1$ |
| 4 | 3 | 5.32 | 5.41 | 1.05 | 5.19 | 5.28 | 0.90 |
| 8 | 7 | 12.55 | 12.78 | 2.13 | 6.92 | 7.31 | 1.31 |
| 12 | 11 | 19.38 | 20.15 | 4.20 | 6.98 | 7.64 | 1.84 |
| 16 | 15 | 23.81 | 24.15 | 4.85 | 7.17 | 7.72 | 2.42 |
| 20 | 19 | 24.20 | 25.66 | 5.11 | 7.33 | 7.57 | 2.77 |

There is only one I/O processor which reads the uncompressed data from only one disk and sends to 3, 7, 11, 15 and 19 compute processors respectively. For SIF, a batch size ($B$) of 6 frames results in less overall waiting time and therefore yields better results than a smaller batch size. However, for CCIR-601, due to limitation of memory, only a batch of one frame is used. The cluster of workstations performs significantly better while connected via the ATM switch due to its higher bandwidth (155 Mb/s) compared to the Ethernet (10 Mb/s).

## 4  Conclusion

We have discussed a scheme for combined scheduling of I/O, disks and processors. We have achieved an encoding rate of 25.66 and 5.11 frames/sec. for SIF and CCIR-601 respectively, using only 20 workstations. However, for optimal performance, a careful balance is necessary in determining the batch and group sizes (depending on the I/O processor's memory), for which work is currently underway. This work was partially supported by HKTIIT grant 92/93.001.

574    Shahriar M. Akramullah, Ishfaq Ahmad, and Ming L. Liou

## References

1. T. Akiyama et al., "MPEG2 Video Codec using Image Compression DSP", *IEEE Tran. on Consumer Electronics*, VOL. 40, NO. 3, Aug. 1994, pp. 466-472.
2. S. M. Akramullah, I. Ahmad and M. L. Liou, "Performance of a Software-Based MPEG-2 Video Encoder on Parallel and Distributed Systems", *IEEE Tran. on Circuits and Systems for Video Technology*, VOL. 7, NO. 4, Aug. 1997, pp. 687-695.
3. K. Shen, L. A. Rowe, E. J. Delp, "A Spatial-Temporal Parallel Approach for Real-time MPEG Video Compression", *Proc. of 1996 International Conf. on Parallel Processing*, VOL. 2, 1996, pp. 100-107.

# Block and Partitioned Neville Elimination

Pedro Alonso[1] and Juan M. Peña[2]

[1] Universidad de Oviedo, E.U.I.T. Informática de Gijón,
Campus de Viesques s/n, 33203 Gijón-Spain
[2] Universidad de Zaragoza, Edificio de Matemáticas,
50009 Zaragoza-Spain

**Abstract.** In this paper we analyze block and partitioned Neville elimination, necessary steps in order to adapt Neville elimination to many high performance computer architectures.

## 1   Block Neville Elimination

Scalar Neville elimination is an elimination method which creates zeros in a column of a matrix by adding to each row an appropriate multiple of the previous one. This procedure, alternative to Gaussian elimination, has proved to be useful when dealing with some special classes of matrices such as totally positive matrices. In [1] one can find references and proofs of the results of this paper.

Let $k, l, n \in N$, $k \leq n$ and let $Q_{k,n}$ denote the set of strictly increasing sequences of $k$ natural numbers less than or equal to $n$. Let $A = (A_{ij})_{1 \leq i,j \leq p}$ be an $n \times n$ matrix, $A_{ij}$ a $q \times q$ matrix $(n = pq)$ for each $i, j$ and let $\alpha = (\alpha_1, \ldots, \alpha_k) \in Q_{k,n}$, $\beta = (\beta_1, \ldots, \beta_l) \in Q_{l,n}$. Then $A[\alpha|\beta] := (A_{\alpha_i, \beta_j})_{1 \leq i \leq k, 1 \leq j \leq l}$ and $A[\alpha] := A[\alpha|\alpha]$. For simplicity, 0 and $I$ will denote a zero and an identity matrix, respectively.

Block Neville elimination consists of $p - 1$ successive major steps, resulting in a sequence of matrices $A^{(t)} = (A_{ij}^{(t)})_{1 \leq i,j \leq p}$ $(1 \leq t \leq p)$ where $A^{(p)}$ is a block upper triangular matrix. The matrix $A^{(t)}$ $(1 \leq t \leq p)$ has zero blocks below its main diagonal in the first $t - 1$ columns.

The next result provides a sufficient condition to guarantee that we can perform block Neville elimination to an $n \times n$ matrix $A = (A_{ij})_{1 \leq i,j \leq p}$.

**Theorem 1.** *Let $A = (A_{ij})_{1 \leq i,j \leq p}$ be an $n \times n$ matrix, $A_{ij}$ a $q \times q$ submatrix of $A$ for $1 \leq i, j \leq p$. If all submatrices of the form $A[k, k+1, \ldots, k+l-1|1, \ldots, l]$ are nonsingular for all $l \in \{1, 2, \ldots, p\}$ and $k \in \{1, 2, \ldots, p-l+1\}$ then we can perform block Neville elimination on $A$ without row exchanges.*

## 2   Partitioned Neville Elimination

In this section we analyze how to perform Neville elimination (by columns) when we have partitioned the data. We shall assume that we have a nonsingular matrix $A$ such that (scalar)Neville elimination can be performed without row exchanges.

P. Zinterhof, M. Vajteršic, A. Uhl (Eds.): ACPC'99, LNCS 1557, pp. 575–576, 1999.
© Springer-Verlag Berlin Heidelberg 1999

For an $n \times n$ matrix $A$ and a given block size $q$, write $A = (A_{ij})_{1 \leq i,j \leq p}$ ($pq = n$). We want to obtain a partitioned $LU$-factorization of $A$.

It is a scalar elimination and for $p = n$ the process can be matricially described as $L^{n-1}L^{n-2} \cdots L^1 A = U$. We shall present how to perform partitioned Neville elimination, where the matrices $L^k = (L_{ij}^k)_{1 \leq i,j \leq p}$ ($k = 1, 2, \ldots n - 1$) are block bidiagonal matrices containing the multipliers of Neville elimination.

For each $L^k$, $k = 1, 2, \ldots, n - 1$, $L_{ij}^k$ is a $q \times q$ matrix ($pq = n$). The matrices $L_{ii}^k$, $i = 1, 2, \ldots, p$, are bidiagonal lower triangular matrices with unit diagonal . When $(m - 1)q + 1 \leq k \leq mq$, with $m = 1, 2, \ldots, p$ and $k = 1, 2, \ldots, n - 1$, one has that $L_{i-1,i-1}^k = I$, and $L_{i,i-1}^k = 0$, for $i = 2, \ldots, m$.

Let us now see how to calculate the elements of $L^k$ and $A^{(k+1)}$ from those of $A^{(k)}$ ($L^k A^{(k)} = A^{(k+1)}$), with $A^{(k+1)} = (A_{ij}^{(k+1)})_{1 \leq i,j \leq p}$ . We shall focus our study on the block matrices $L^k[m, m + 1, \ldots, p]$ and $A^{(k+1)}[m, m + 1, \ldots, p]$.

At step $i + 1$ ($0 \leq i \leq p - m$) and for $j = m, m + 1, \ldots, p$, we have

$$(L_{m+i,m+i-1}^k \times A_{m+i-1,j}^{(k)}) + (L_{m+i,m+i}^k \times A_{m+i,j}^{(k)}) = A_{m+i,j}^{(k+1)}. \tag{1}$$

Let us first calculate the only nonzero element of $L_{m+i,m+i-1}^k$ for $1 \leq i \leq p - m$. Using (1) for $j = m$ we can derive $L_{m+i,m+i-1}^k$. In a similar way, we can calculate the elements below the main diagonal of $L_{m+i,m+i}^k$ ($i \in \{1, 2, \ldots, p - m\}$). In the case $i = 0$, let us recall that $L_{m,m-1}^k = 0$.

When we have obtained all the elements of the matrices $L_{m+i,m+i-1}^k$ and $L_{m+i,m+i}^k$ ($i \in \{0, 1, \ldots, p - m\}$), we can derive from (1) the elements of $A_{m+i,j}^{(k+1)}$.

The advantage of the algorithms just described is that they introduce matrix operations, which is of great interest in high performance computation. Moreover, the data partition is also very useful in parallel computation, since independent tasks can be done simultaneously in different processors. For instance, if we want to calculate $A_{m+i,j}^{(k+1)}$ with $j = m, m + 1, \ldots, p$, in a MIMD multiprocessor system, and taking into account that each processor contains the elements of some $A_{m+i,j}^{(k)}$, we can send the last row of $A_{m+i-1,j}^{(k)}$ and the adequate multipliers to corresponding processor, so that each of them gets $A_{m+i,j}^{(k+1)}$, simultaneously producing the computations (with the corresponding reduction of time).

Furthermore, an algorithm's efficiency also depends on the time devoted to communications. By Neville elimination's nature, communications are performed between neighbour processors, what reduces communications time.

Finally, data partitioning also allows us a more efficient use of the given computer memory, which is designed taking into account the inverse relationship between size and access speed.

# References

1. Alonso, P., Peña, J.M.: Development of block and partitioned Neville elimination. Preprint.

# An Object-Oriented DataBase for Movies-on-Demand: Two Approaches[§]

Francesco Amalfitano[1], Antonio d'Acierno[1], Ivana Marra[1], Lucio Sansone[2]

[1] IRSIP – CNR, Via P. Castellino 111 - 80131 Naples (Italy)
{amalfitano.f, dacierno.a, marra.i}@irsip.na.cnr.it
[2] DIS - Università degli Studi di Napoli "Federico II"
Via Claudio 21 - 80125 Naples (Italy)
sansone@unina.it

## 1. Introduction

Interactive multimedia technology has seen a great deal of recent attention since it will allow its users with the flexibility of selecting and receiving specific information; many telecommunication industries and research groups are thus employing a lot of their resources to study the feasibility of new video services, nowadays known as *Video on Demand* (VoD). Among these services, **T-VoD** (True-VOD) services for movie on demand require the use of a bi-directional signal so that the user should be able to *play* a film with full (virtual) VCR capabilities.

In a typical T-VOD system there should be some content archives connected via an high-speed network to many local (regional) distribution sites; this data distribution increases availability and reliability, and is easier to manage since each local system is responsible for its own billing and accounting. Concerning with the DataBase Management Systems (DBMSs), the research is oriented on Object-Oriented DBMS (OODBMS). In this paper we briefly discuss the organization of video-data on the regional server by describing the work performed to implement an application to manage digital movies; our application is based on the ObjectStore OODBMS.

## 2. The Movie DataBase

Our database schema consists of a core entity, the *Film*, logically stored in a set of *Archives*, which provide more modularity and flexibility for the stream; each archive is then segmented into a set of logical units, the *Scene*. To each scene are associated, at store time, two integer values that measure its violence and sex content; at play time, if an user-defined threshold is exceeded, such a critical scene is discarded.

Each scene is then divided into a set of *Sequences*, in accordance with the physical structure of the MPEG-encoded video-stream, each sequence being a set of *GOPs*.

---

[§] This research has been supported by EC FESR Project "Sviluppo di tecnologie Digitali di Grafica Avanzata per Applicazioni Industriali e Commericiali ".

P. Zinterhof, M. Vajteršic, A. Uhl (Eds.): ACPC'99, LNCS 1557, pp. 577-578, 1999.

Last, each GOP is an ordered set of *Frames*, each of which can be an I frame, a B frame as well as a P frame. Some classes supporting additional information contents of the movie database are also provided: there is the possible *Genre* of the film that an *User* can play, the *Session* an user starts to play a film, information about the *Artist* interpreting a character in the film or directing the selected film. Moreover, for a film we can store one or more *Reviews*.

In order to "navigate" the database, we are implementing the CORBA *View Interface*. Through this object, developed by following the DSM-CC specifications, the user can exploit its methods to query the database by means of SQL-like statements, so obtaining a "relational view" of the object-oriented database.

Using our data model, "to play" a film simply means to play each element of an array of archives and so forth until each frame is actually played. Trivially, the *fast forward* method simply means "play a film by skipping B frames" while *rewind* means "play a film in reverse order by skipping B and P frames"(*reverse order* of course means that arrays have to be traversed from last position to the first one). Moreover, since each GOP, according to the MPEG standard, contains its own start time, we can easily define start and end time of each sequence, the start and end time of each scene and so on. This allows fast and easy implementation of methods, referring to the time parameter, like the *SeekToSecond* method, allowing to seek a frame related to a given second.

As concerning the actual implementation there are two possibilities. First we can have a *full-database* approach in which each frame has an attribute (an array of char) that contains the data to be played. Another possibility (*half-database*) assumes that each frame has three type of information, one representing the position of the actual frame in a standard file, one its size, and the last one the frame type (I, P, or B).

To test our prototypes we also implemented a *"no-database"* approach, in which, to each MPEG data stream, are associated two files, one representing all the data coded into the MPEG protocol, the other one being a set of records (one for each frame) storing the offset of each frame.

The tests have been performed on a SUN UltraE work-station, using a stripped file-system; (4 controllers and 8Kbyte of strip-size). As a preliminary test-case we used 16 minutes of a film (*Never Say Never Again*) coded as an Elementary Video Stream (without audio) at 4.5 Mb/secs, so deriving a stream size of about 0.5 Gbyte. By assuming that just a consumer process reads data (i.e. no visualization is performed) we obtained 30 MB/secs for the *no-database* approach, 26 MB/secs for the *half-database* approach and 3.5 MB/secs for the full-database approach.

The obtained results show that the *half-database* approach allows to obtain a performance that is reasonably high compared with the *no-database* approach (the fastest one), while offers many of the advantages of a typical DBMS (data integrity, transaction processing, data access control, concurrency control and so on). The *full-database* approach, on the other hand, offers the DBMS advantages also on the stream data. Roughly speaking, the *half-database* approach sustains, with the tested stream and system, about 45 clients.

The key problem we are currently dealing with is the generalization of the *half-database* approach to stream composed both of audio and video.

# Parallel Tree Algorithms for N-body Simulations

Vincenzo Antonuccio-Delogu[1], Ugo Becciani[1], Mario Gambera[2], and
Antonio Pagliaro[3]

[1] Astrophysical Observatory of Catania, Città Universitaria,
Viale A. Doria 6, I-95125 Catania, ITALY
van@sunct.ct.astro.it,ube@sunct.ct.astro.it
[2] Istituto di Astronomia, Università di Catania, Città Universitaria,
Viale A. Doria 6, I-95125 Catania, ITALY
mga@sunct.ct.astro.it
[3] SRON - Sorbonnelaan 2, 3584 CA Utrecht, THE NETHERLANDS
A.Pagliaro@sron.nl

**Abstract.** We describe a new parallel implementation of the octal-
hierarchical tree N-body algorithm on SHared Memory systems (SHM)
we have recently developed. Pursuing an effort to optimize as much as
possible the code on a generic SHM we have modified the original algorithm introduced by Barnes and Hut, introducing a new scheme of
"grouping" of particle interactions. We present speedup and efficiency
results. ...

## The Physical Problem.

Systems of particles interacting through long-range interactions occur often in
physics. Probably the best known example of such an interaction is the gravitational one, and indeed the simulation of astrophysical systems like globular
clusters, galaxies and clusters of galaxies still represents a challenging task for
numerical analysis and computational science. The main reason for this lies precisely in the *truly long-range* character of the gravitational interaction, which
makes impossible to neglect the effect of relatively "far" particles when compared to "nearby" ones. This is a generic feature of forces decaying as $\approx r^{-(2+n)}$
with $n \leq 0$.
There exist two type of algorithms to deal with this problem, namely mesh-
and tree-based algorithms. The *octal hierarchical tree* algorithm introduced by
Barnes and Hut ([3], hereafter BH) does not make use of any FFT for the calculation of the force, and it is the one we started from. We refer the reader to more
specialized papers for a more detailed description of this algorithm ([3], [6], [7])

## Parallel Work- and Shared-Memory Treecodes.

During the past four years we have developed and studied different parallel versions of the BH algorithm, starting with a message-passing version based on the
*locally essential tree* version by Salmon ([6]) (Antonuccio-Delogu et al., [1]). Here

P. Zinterhof, M. Vajteršic, A. Uhl (Eds.): ACPC'99, LNCS 1557, pp. 579–580, 1999.
© Springer-Verlag Berlin Heidelberg 1999

we show results for our most recent direct data- and work-shared approaches, where there is a single, physically distributed tree mapped over all the available processors (Becciani et al., [4], [5]). In the Table below we show the *efficiency*, measured as: $e = 1/(t_{step} * N_{PE})$, where $t_{step}$ is the execution time (in seconds) of the gravitational acceleration calculation and $N_{PE}$ is the number of processors used. All the runs were performed on a Cray T3E-128 at Cineca, Bologna, ITALY. In the case of ideal scaling one should expect that the efficiency should be constant with $N_{PE}$. Apart for the value for the run with 2 million partciles and 128 processors, where due to the small number of particles the communication overhead dominates, all the values are constant within $10^{-2}$ of the average, demonstrating the correct scalability of the grouping algorithm applied to treecodes. More details and results will be made available in a forthcoming paper ([2]).

**Table 1.** Efficiency for different sizes of the running job $N_{bodies}$. We run simulations for 2 million and 16 million particles, starting with clustered, realistic initial conditions.

| $N_{PE}$ | $e_{2ml}$ | $e_{16ml}$ |
|---|---|---|
| 16 | 1.69e-4 | - |
| 32 | 1.72e-4 | 1.83e-5 |
| 64 | 1.68e-4 | 1.79e-5 |
| 128 | 1.13e-4 | 1.78e-5 |

# References

1. Antonuccio-Delogu, V., Becciani, U.: A Parallel Tree N-Body Code for Heterogeneous Clusters, in: Dongarra, J. and Wasniewsky, J. eds.,*Parallel Sciebtific Computing - PARA '94* (Springer Verlag: 1994), 17
2. Antonuccio-Delogu, V., Becciani, U., Pagliaro, A. and Gambera, M: J. Comp. Phys., in preparation
3. Barnes, J., Hut, P.: Nature **324** (1986) 446-448
4. Becciani, U., Antonuccio-Delogu, V., Pagliaro, A.: A work- and data-sharing parallel tree N-body code, Comp. Phys. Comms., **99** (1996) 9-17
5. Becciani, U., Ansaloni, R., Antonuccio-Delogu, V., Erbacci, G., Gambera, M. and Pagliaro, A.: WDSH-PTc parallel code for large Nbody simulation. Dynamic Load Balance and data distribution on CRAY T3D system, Comp. Phys. Comms., **103** (1997) 19-25
6. Salmon, J.K.: Parallel hierarchical N-body methods, Ph. D.. Thesis, *unpublished* (California Institute of Technology: 1991)
7. Salmon, J.K., Warren, M.S.: Skeletons from the Treecode closet, J. Comp. Phys., **111** (1995) 136

# Parallel Numerical Algorithms for Distributed Memory Machines

Pierre Bassomo[1], Ibrahima Sakho[2], and Annie Corbel[1]

[1] Equipe RIM (SIMADE) Ecole des mines, Saint-Etienne, France
[2] LRIM, Université de Metz, Metz, France

Distributed memory machines consisting of multiple autonomous processors connected by a network are becoming commonplace. Unlike specialized machines like *systolic arrays*, such systems of autonomous processors provide virtual parallelism through standard *message passing* libraries (*PVM*[GBD+94], *MPI* [GLS94]). Their very high level of flexibility, scalability and performance has allowed many parallel algorithms to be designed.

Most of those algorithms arise from scientific computing and their parallelism is based on *data and nested loops* parallelism. Such parallel algorithms are generally suited for problems dealing with regular computations and data structures. The evaluation of their correctness and performance is generally achieved through the simulation of moderate scale problems. So, we definitely need a methodology to design parallel algorithms systematically because not only is a methodology a tool which helps with design, but it also avoids the difficult problem of solving correctness and performance issues, since only the method need to be proven.

Our methodology aims at reusing the *systolic model* principles for the display of potential parallelism inside nested loops, and justifying the aggregation of iteration of loops so as to reduce communication overhead while exploiting coarse-grained parallelism. Each aggregate is a *block* of fine-grained computations located in different hyperplanes of a given space. It also defines an atomic unit of computation i.e no synchronization or communication is necessary during the execution of the fine-grained computations inside a block. Thus all necessary data must be available before such atomic executions. This imposes the constraint that splitting the set of fine-grained computations does not result in deadlocks.

The present methodology is based on *systolic algorithmic* ideas as described by *Miranker*[MW84], *Moldovan*[Mol83], *Moreno*[ML88] and *Quinton*[Qui83]. It proceeds in six steps.

First, we use *uniform recurrence equations* to describe the sequential algorithm and model it into a directed graph, namely $G$. In this graph, $D$ is the set of vertices and $E$ the set of arcs. Each point $u \in D$ represents a fine-grained computation. A pair $(u, v)$ in $E$ means that the fine-grained computation at point $v$ requires data from $u$. Note that in the graph $G$, vertices are locally connected according to the uniform recurrence equations model.

P. Zinterhof, M. Vajteršic, A. Uhl (Eds.): ACPC'99, LNCS 1557, pp. 581–583, 1999.

Second, we design an optimal scheduling of the directed graph. To do this, one needs an optimal time function. A time function is optimal if its maximum value is equal to the number of vertices of the longest path in $G$. Hereafter, we shall be mainly interested in optimal time algorithms. The sequence of computations located on a path of $G$ must obviously be done in a sequential manner. Therefore, the minimal time required to achieve the parallel computation, say $T_{sys}$, is equal to the number of vertices of the longest path in $G$.

Third, we study the set of fine-grained computations executed at the same time $t$, say $D(t)$, and derive the complexity in terms of processors. Up to this point, we simply apply systolic arrays design methodology.

Fourth, we describe and justify the block aggregation strategy. Given $P_{sys}$ and $T_{sys}$, as previously defined, our aim is to "split" $G$ into different blocks of nodes. Each of them will corespond to a single node in a new graph, say $G'$. Let $u$ and $v$ be two nodes in $G'$ and $A(u, v)$ be the set of arcs linking the two blocks of nodes in $G$ corresponding to $u$ and $v$. If $A(u, v) \neq \emptyset$ then a pair $(u, v)$ defines an arc in $G'$. The graph $G'$ is built in order to be computed in a time bounded above by the execution time required to compute $G$ with $P$ processors. We observe that the optimal time to compute $G$ with $P$ processors is bounded above by $T_{sys} \frac{P_{sys}}{P}$. So the bloc aggregation strategy deals with: -"splitting" $G$ into a "similar" one, say $G'$, -choosing a block size that allows to compute $G'$ when using $P$ processors with a time bounded above by $T_{sys} \frac{P_{sys}}{P}$. Note that, if the unit time is chosen as being the maximum of the execution time of the block, then the time to compute $G'$ is estimated using the second step of our methodology.

Fifth, we propose a logic of allocating computations and data to available processors. According to uniform recurrence, a final result, say $Y$, is obtained after successive modifications of an input piece of data, say $Y(u_0)$. Let $(Y(u_k))_{k \geq 0}$ be the sequence of these successive modifications. Note that $u_k$ is a point of $G'$ and $Y(u_k)$ is a value produced by the node $u_k$ and required by $u_{k+1}$ for its computations. We note that, for many numerical algorithms, piece of data $Y$ is reused to store the terms $Y(u_k)$ during one or more iterations. Besides, in several cases we also notice that $\forall k \geq k_0$ terms $Y(u_k)$ may be localized along the same line. Then the idea of allocation strategy is: -to consider the intersection of $D$ with a set of parallel hyperplanes containing such lines, -to allocate such lines to available processors under the constraint of the optimal time function.

Finally, we describe and analyze coarse-grained algorithms. Since each vertex of $G'$ is associated with the modification of locally reused data, one associates to such data a data structure, say a *block*, which contains the set of data $Y$, the dates at which they are reused and the processors required for the data exchange at these dates. Let $P(u)$ be the program associated to the vertex $u$. The basic principle followed by the program of a given processor is: for each locally reused data, $P(u)$ is executed according to its successive modifications, and as soon as possible.

# References

[GBD+94] Al Geist, Adam Beguelin, Jack Dongarra, Weicheng Jiang, Robert Manchek, and Vaidy Sunderam. *PVM: Parallel Virtual Machine A Users' Guide and Tutorial for Networked Parallel Computing.* MIT Press, year 1994.

[GLS94] William Gropp, Ewing Lusk, and Anthony Skjellum. *Using MPI: Portable Parallel Programming with the Message-Passing Interface.* MIT Press, year 1994.

[ML88] J.H Moreno and T. Lang. Graph-based partitioning for matrix algorithms for systolic arrays: Application to transitive closure. *ICPP*, 1:28–31, 1988.

[Mol83] D.I Moldovan. On the design of algorithms for vlsi systolic arrays. *Proc.IEEE*, 71:113–120, 1983.

[MW84] W. L. Miranker and A. Winkler. Space-time representation of computational structures. *Computing*, 32:93–114, 1984.

[Qui83] P. Quinton. The systematic design of systolic arrays. Technical Report 193, IRISA, 1983.

# Dynamic Scheduling on a Network Heterogeneous Computer System

Janez Brest, Viljem Žumer, and Milan Ojsteršek

University of Maribor
Faculty of Electrical Engineering and Computer Science
Smetanova 17, 2000 Maribor, Slovenia
janez.brest@uni-mb.si

**Abstract.** Dynamic scheduling on a heterogeneous system was implemented to minimize the application program execution time. Our method decomposes the program workload into computationally homogeneous subtasks, which may be of different size, depending on the current load of each machine in the heterogeneous computer system.

## 1 Introduction

Research in the field of heterogeneous computing began in the mid 1980s; since then it has grown tremendously [2]. From the scientific community to the federal government, heterogeneous computing has become an important area of research and interest. In general, the goal of heterogeneous computing is to assign each subtask to one of the machines in the system so that the total execution time (computation time and inter-machine communication time) of the application program is minimized. Mapping can be specified statically or determined at runtime by load balancing algorithms [4].

## 2 Task Scheduling and Load Balancing

The problem of load partitioning and scheduling in a multiple-processor system has been an area of active and sustained research over the past two decades [2]. The most critical aspect of a task-scheduling algorithm is the strategy used to allocate problems to slaves.

The arbitrarily divisible load model can be used in applications where the load consists of a large number of small elements with identical processing requirements. Examples can be found in applications for image and signal processing, and also in iterative algorithms.

We have modeled the computing capacity of each computer with a single parameter: its response time needed for the execution of a task. The new task size is a function of two previous of response times and task sizes values:

$$s(t+2) = f(s(t+1), \tau(t+1), s(t), \tau(t)), \quad t = 0, 1, 2, \ldots \tag{1}$$

P. Zinterhof, M. Vajteršic, A. Uhl (Eds.): ACPC'99, LNCS 1557, pp. 584–585, 1999.
© Springer-Verlag Berlin Heidelberg 1999

This single parameter includes several aspects of heterogeneity of each computer during the given operating conditions.

We have implemented the algorithm for dynamic scheduling on a heterogeneous computer system using the master/slave scheme [3]. The master task performs the dynamic scheduling algorithm so that it also distributes the tasks into the process unit on the heterogeneous computer system. The function (1) is used by the master task to calculate the new task size.

## 3 Results

We have tested our dynamic scheduling strategy on two classes of optimization problems. We used LAM/MPI [1].

The first one was the problem of continuous speech recognition, where we implemented an algorithm for bigram word clustering [5] In the second case we chose the traveling salesman problem [6] to test the presented dynamic scheduling on a heterogeneous system.

Utilizations (the ratio between the work time and the total execution time) of slaves were about 0.9. Our method has found better final solutions of both problems.

## 4 Conclusion

A method for dynamic scheduling on a heterogeneous computer system is presented in this paper. Our method is simple, but the results obtained are comparable to other approaches described in the literature. In our future research we intend to extend the power of the method, and to apply it to a wider range of practical problems.

## References

1. Gregory D. Burns, Raja B. Daoud, and James R. Vaigl. Lam: An Open Cluster Environment for MPI. In *Supercomputing Symposium '94*, Toronto, Canada, June 1994.
2. Mary. M. Eshagian, editor. *Heterogeneous Computing*. Artech House, Inc., Norwood, MA 02062, ISBN 0-89006-552-7, 1996.
3. Ian Foster. *Designing and Building Parallel Programs*. Addison-Wesley, ISBN 0-201-57594-9, 1995.
4. Emile Haddan. Load Balancing and Scheduling in Network Heterogeneous Computing. In Mary. M. Eshagian, editor, *Heterogeneous Computing*, pages 224–276, Norwood, MA 02062, ISBN 0-89006-552-7, 1996. Artech House, Inc.
5. Sven Martin, Jörg Liermann, and Hermann Ney. Algorithms for bigram and trigram word clustering. *EUROSPEECH'95. 4th European Conference on Speech Communication and Technology*, pages 203–213, September 1995.
6. Gerhard Reinelt. TSPLIB - a traveling salesman problem library. *European Journal of Operations Research*, 52(1):125, 1991.

# Interaction between PVM Parameters and Communication Performances on ATM Networks

Maurizio Giordano, Mario Mango Furnari, and Francesco Vitobello

Istituto di Cibernetica del CNR, Via Toiano, 6
I-80072 Arco Felice (NA), Italy
{M.Giordano, mf, F.Vitobello}@cib.na.cnr.it

**Abstract.** This short paper presents preliminary results of an investigation to improve PVM communication on *workstation clusters* over *ATM*. The experiments show that a better exploitation of ATM bandwidth could be gained by tuning parameters like PVM *packet fragment size* and TCP *socket buffer size*.

## 1 Introduction

PVM [4] is the *de facto* standard for parallel computing on *workstation clusters* interconnected by a LAN. Since communication between processors is a critical point for such systems, the use of high–speed switched networks, such as ATM [2], are expected to significantly increase communication performances [3].

PVM is based on the *BSD Socket* programming interface, and uses *TCP* and *UDP* as transport protocols. Some works have been done in implementing PVM directly on top of *ATM Adaptation Layers* protocols [5], but the maximum achieved throughput is still far below the full potential of high-speed networks.

We claim that it is possible to achieve improved PVM communication performances on ATM networks by tuning some PVM parameters, like *PVM message fragment size* and *TCP socket buffer size*, respectively at the PVM interface and implementation level. With this approach we do not need to re-implement PVM directly on ATM protocols (avoiding TCP/IP layers).

## 2 Preliminary Results & Considerations

The hardware setup was composed of two SUN SPARCstations 20 (with Solaris 2.5), an Ethernet adapter connected to a 10 Mbit/s Fore Systems ES3810 Etherswitch, and a 155 Mbit/s Fore Systems SBA-200E ATM adapter connected to a Fore Systems LE155 ATM switch (switching capability of 2.5 Gbit/s).

We measured *latency* as half of *round–trip time* (*RTT*), i.e. the time needed by a packet to go from source to destination host and return back. *Throughput* is the number of bytes sent divided by half of RTT. Measurements were based on a "ping–pong" test performed by two programs, exchanging one message of

P. Zinterhof, M. Vajteršic, A. Uhl (Eds.): ACPC'99, LNCS 1557, pp. 586–587, 1999.
© Springer-Verlag Berlin Heidelberg 1999

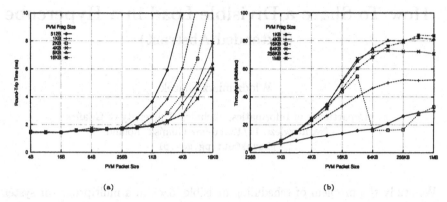

(a)                                              (b)

**Fig. 1.** (a) RTTs and (b) throughputs on ATM varying packet fragment and size.

given size. We run the test program by changing both PVM packet size and fragmentation with four TCP socket buffer sizes: 16, 32, 64 and 128 Kbytes.

Preliminary experiments showed that Ethernet latency and throughput are not affected by these parameters. This is not true for ATM: TCP socket buffer size significantly affects measurements for packet size larger than 4 Kbytes and the maximum throughput increases by 20% changing the size from 16 to 64 Kbytes and then it saturates. PVM packet fragmentation affects both ATM latency and throughput (see figures 1(a) and 1(b)): larger are packet fragments, longer is the range of packet sizes in which latency is kept low. Also throughput increases with fragment size and saturates for fragments larger than 256 Kbytes.

These results show that communication performances could be improved by means of PVM parameter tuning without the need of re-implementing PVM directly on top of the ATM protocols. Some works in this directions [1,6] show that the achieved throughput is still far below the full potential of ATM.

## References

1. Chang, S.L., Du, D.H.C., Hsieh, J., Lin, M., Tsang, R.P.: Enhanced PVM Communication over a High-Speed Local Area Network. In: Proceeding of the 1st Intern. Workshop on High-Speed Network Computing. Santa Barbara, California (1995)
2. de Prycker, M. (eds.): Asynchronous Transfer Mode solution for broadband ISDN. Ellis Horwood series in Computer Communications and Networking, (1991)
3. Dongarra, J., T. Dunigan, T.: Message-Passing Performance of Various Computers. Technical Report, Oak Ridge National Laboratory (1997)
4. Geist, A., Beguelin, A., Dongarra, J., Jiang, W., Manchek, R., Sunderam, V. (eds.): PVM: Parallel Virtual Machine - A Users Guide and Tutorial for Network Parallel Computing. MIT Press, (1994)
5. Lin, M., Hsieh, J., Du, D.H.C., Thomas, J.P., MacDonald, J.A.: Distributed Network Computing over Local ATM Networks. IEEE Journal on Selected Areas in Communication, 13(4) (1995) 733–748
6. Zhou, H., Geist, A.: Faster Message Passing in PVM. Technical Report, Oak Ridge National Laboratory (1994)

# How To Share a Divisible Load in a Hypercube
## (Extended Abstract)

Włodzimierz Głazek

Department of Informatics, Technical University of Gdańsk
Narutowicza 11/12, 80-952 Gdańsk, Poland
glazek@eti.pg.gda.pl

We study the problem of scheduling divisible loads in a multiprocessor system with hypercube interconnection topology and linear communication delays. A *divisible* load consists of large number of fine grain data elements, each requiring the same type of processing [1]. In the absence of precedence constraints such a load may be partitioned among processors in the system in every suitable way so that each part may be processed separately of and simultaneously with all other parts of the load. Initially, the whole load resides in the local memory of a selected node, called originator. The goal is to partition the load into fractions and communicate them to other processors in such way so that the entire load is transmitted and processed in the shortest possible time. Earlier solutions for this optimization problem were obtained under the assumption that each processor in the system can use all its communication links at the same time [2,3,4]. While this is not a problem in small hypercubes, it can present some difficulties in very large machines (e.g. machines with thousands of processors). We present a scheduling algorithm that restricts each processor to use only one communication link at a time. We model the process of data dissemination and computation with a set of recursive equations. By solving it we obtain a closed-form expression for the schedule length as a function of the load size and key parameters of the system such as hypercube dimension, processing speed of one node and communication bandwidth of one link.

Our algorithm, called NEAREST LAYER FIRST, is based upon the Johnson and Ho spanning binomial tree algorithm SBT($d$) for one-to-all single item scattering in $d$-dimensional hypercube $H_d$ [5]. In step 1 of algorithm SBT($d$) the originator sends $2^{d-1}$ items to a neighbor over dimension 1 link. Subsequently, in step $i, 2 \leq i \leq d$, each processor which already received data sends $2^{d-i}$ items to a neighbor over dimension $i$ link. We define a *layer* to be a set of processors which receive data in the same step of scattering.

Note that in one-to-all single item scatter problem each processor in a hypercube must receive a *dedicated* data item of unit size as fast as possible. In our problem data is *nameless*, i.e. a particular data item can by processed by any processor in the system. For simplicity of the presentation we assume that results of load processing are not returned back to the originator. Let $a_i, 0 \leq i \leq d$, be the fraction of the load assigned for computing to a processor of layer $i$ and $v_i$ be the part of the load communicated to a processor of layer $i$ from its predecessor.

P. Zinterhof, M. Vajteršic, A. Uhl (Eds.): ACPC'99, LNCS 1557, pp. 588–589, 1999.
© Springer-Verlag Berlin Heidelberg 1999

Algorithm NEAREST LAYER FIRST($d$)
step 1: originator starts to process fraction $a_0$ of load and
sends fraction $v_1$ of load over 1-st link;
step $i = 2, 3, \ldots, d$: each processor which received data in step
$i - 1$ starts to process fraction $a_{i-1}$ of load; each processor
which received data in or before step $i - 1$ sends fraction $v_i$
of load over $i$-th link;
step $d+1$: each processor which received data in step $d$ (say,
fraction $a_d$ of load) starts to process it.

It can be observed that in the optimal schedule without returning of results
all processors must finish computation at the same moment of time. Due to
communication delays equal partition of the load among processors does not
satisfy this rule. Instead, the share of the load to be processed by each processor
in layer $i$ can be determined from a recursive relation of processing times in layer
$i$ and layer $i + 1$ with respect to the common finish time constraint. Namely, for
each $i, 0 \le i \le d - 1$, computation in layer $i$ must take as long as the time of
communication to layer $i+1$ plus the time of computation in layer $i+1$. Moreover,
part $v_i$ communicated to a processor in layer $i$ must accommodate data for all
its descendants and be equal to $v_i = a_i + a_{i+1} + 2a_{i+2} + \ldots + 2^{d-i-1}a_d$. In this
way we obtain a set of $d$ non-homogeneous recursive equations for $d+1$ variables
$a_i$'s. The last $(d + 1)$-th equation constrains the size of load shares assigned to
particular processors so that they add up to the load volume.

The model may have no feasible solution for some values of $d$. This situation
happens if the load volume is small and communication speed of the network is
low. Then, the front layers finish their work assignments before data can arrive
to the most distant layers. In that case a smaller hypercube should be considered. The maximum size of the usable hypercube can be found from analysis
of the closed-form solutions of the model. Further analysis with respect to the
attained schedule length demonstrates that processing a load of fixed volume
in a medium sized hypercube can be often almost as effective as in a big one.
Another conclusion is that speed of interconnection is more important for good
time performance than speed of individual processors.

# References

1. Bharadwaj V., Ghose D., Mani V., Robertazzi T.G.: Scheduling Divisible Loads
   in Parallel and Distributed Systems. IEEE Computer Society Press, Los Alamitos
   (1996)
2. Blanc J.-Y., Trystram D.: Implementation of Parallel Numerical Routines Using
   Broadcast Communication Schemes, Proc. CONPAR'90, LNCS 457 (1990) 469–478
3. Błażewicz J., Drozdowski M.: Scheduling Divisible Jobs on Hypercubes. Parallel
   Computing 21 (1995) 1945–1956
4. Drozdowski M.: Selected Problems of Scheduling Tasks in Multiprocessor Computer
   Systems. Poznań University of Technology Press (1997)
5. Johnson S.L., Ho C.-T.: Optimum Broadcasting and Personalized Communication
   in Hypercubes. IEEE Transactions on Computers 35, 9 (1989) 1249–1268

# Overlapped Four-Step FFT Computation*

Herbert Karner and Christoph W. Ueberhuber

Institute for Applied and Numerical Mathematics
Technical University of Vienna
Wiedner Hauptstrasse 8–10/115, A–1040 Vienna, Austria
karner@titania.tuwien.ac.at christof@uranus.tuwien.ac.at

**Abstract.** In this paper a new approach is presented in order to overlap all communication intensive steps appearing in the four-step FFT algorithm—initial data distribution, matrix transpose, and final data collection—with computation. The presented method is based on a Kronecker product factorization of the four-step FFT algorithm.

One of the primary objectives in the programming of parallel algorithms is to reduce the effects of the overhead introduced when a given problem is parallelized. A key contributor to overhead is communication time. One way to reduce the communication overhead is to hide communication by overlapping it with computation.

Let $N = n_1 n_2$ with $n_1 = m_1 p$, $n_2 = m_2 p$, and $N > p^2$, where $p$ denotes the number of processors. Let the data vector $x \in \mathbb{C}^N$ be arranged into a matrix $x_{n_1 \times n_2} \in \mathbb{C}^{n_1 \times n_2}$ in column major order. The nodes are arranged as a logical $p \times 1$ mesh and the matrix $x_{n_1 \times n_2}$ is distributed as (CYCLIC, *), where $X_{j,0} \in \mathbb{C}^{m_1 \times n_2}$, $j := 0 : p-1$, is assigned to processor $P_{j,0}$. With this distributed data structure each processor $P_{j,0}$ has to compute (see Karner, Ueberhuber [3]):

$$
\begin{align}
&(1) & X_{j,0} &:= X_{j,0}(F_p \otimes I_{m_2})^\top \\
&(2) & X_{j,0} &:= X_{j,0}(L_{m_2}^{n_2})^\top \\
&(3) & X_{j,0} &:= X_{j,0}(T_p^{n_2})^\top \\
&(4) & x_{loc} &:= (I_{m_2} \otimes L_{m_1}^{m_1 p}) x_{loc} \\
&(5) & x_{loc} &:= (F_{m_2} \otimes I_{m_1} \otimes I_p) x_{loc} \\
&(6) & \text{Commun}&\text{ication for } x := (I_{m_1 m_2} \otimes L_p^{p^2}) x \\
&(7) & X_{j,0} &:= F_N^\top (j : n_1 - 1 : p, 0 : n_2 - 1) * X_{j,0} \\
&(8) & x_{loc} &:= (I_{m_2} \otimes F_{m_1} \otimes I_p) x_{loc} \\
&(9) & x_{loc} &:= (L_{m_1 p}^{m_1 p m_2}) x_{loc} \\
&(10) & X_{j,0} &:= X_{j,0}(L_p^{n_1})^\top \\
&(11) & X_{j,0} &:= X_{j,0}(T_{m_1}^{n_1})^\top \\
&(12) & X_{j,0} &:= X_{j,0}(F_p \otimes I_{m_1})^\top
\end{align}
$$

where $x_{loc} \in \mathbb{C}^{m_1 p m_2} = \mathrm{vec}(X_{j,0})$.
(Background information about 4-step FFT algorithms can be found in Bailey [1], Van Loan [4], and Karner, Ueberhuber [2]).

---

* The work was supported by the Special Research Program SFB F011 "AURORA" of the Austrian Science Fund FWF.

P. Zinterhof, M. Vajteršic, A. Uhl (Eds.): ACPC'99, LNCS 1557, pp. 590–591, 1999.
© Springer-Verlag Berlin Heidelberg 1999

The overlapping is divided into three phases:

**Phase I:** Overlapping of steps 1, 2, and 3 with the initial distribution of the matrices $X_{0,0}, \cdots, X_{p-1,0}$.
**Phase II:** Overlapping of steps 5, 7, and 8 with communication step 6.
**Phase III:** Overlapping of steps 10, 11, and 12 with the final collection of the data.

**Phase I.** It is assumed that processor $P_{0,0}$ contains $x_{n_1 \times n_2} \in \mathbb{C}^{n_1 \times n_2}$ and has to distribute $X_{1,0}, \cdots, X_{p-1,0}$ row-by-row to the remaining $p-1$ node processors. This is done by a nonblocking *scatter* operation. In each processor $P_{j,0}$, $j := 1 :$ $p-1$, the locally performed computational steps (1) – (3) on row $X_{j,0}(i-1,:)$ are overlapped with a nonblocking receive of row $X_{j,0}(i,:)$.

**Phase II.** The computation of $x := (I_{m_1 m_2} \otimes L_p^{p^2})x$ involves a *complete exchange* operation (all-to-all personalized communication). In this paper the *pairwise direct exchange* algorithm is assumed. On each processor $X_{j,0} \in \mathbb{C}^{m_1 \times p m_2}$ is partitioned as $X_{j,0} = (A_0 \mid A_1 \mid \cdots \mid A_{p-1})$, where $A_j \in \mathbb{C}^{m_1 \times m_2}$, $j := 0 : p-1$. Each processor $P_{j,0}$, $j := 0 : p-1$, has to exchange block $A_{j \oplus i}$ with processor $P_{j \oplus i,0}$, $i := 0 : p-1$ ($\oplus$ denotes the bitwise XOR operation). The direct exchange algorithm needs $p$ communication steps. During the communication steps the multi-row FFTs, step (5), on the block to be sent in the following communication step, block $A_{j \oplus (i+1)}$, and the scaling operation and multi-column FFTs, steps (7) and (8), on the block received in the previous communication step, block $A_{j \oplus (i-1)}$, are overlapped with the communication of the block sent in the current communication step, block $A_{j \oplus i}$.

**Phase III.** Each processor $P_{j,0}$, $j := 1 : p-1$, has to send $X_{j,0}$ row-by-row to processor $P_{0,0}$. The locally performed computational steps (10) – (12) on row $X_{j,0}(i,:)$ are overlapped with a nonblocking send of row $X_{j,0}(i-1,:)$.

# References

1. Bailey, D. H.: FFTs in External or Hierarchical Memory. J. Supercomputing **4** (1990) 23–35
2. Karner, H., Ueberhuber, C. W.: Architecture Adaptive FFT Algorithms. In: Bukhres, O., El-Rewini, H. (eds.): Proceedings of the Second IASTED International Conference on European Parallel and Distributed Systems (Euro-PDS'98). IASTED/ACTA Press, Anaheim Calgary Zürich (1998) 331–334
3. Karner, H., Ueberhuber, C. W.: Parallel FFT Algorithms with Reduced Communication Overhead. AURORA Tech. Report TR1998-14, Institute for Applied and Numerical Mathematics, Technical University of Vienna (1998)
4. Van Loan, C. F.: Computational Frameworks for the Fast Fourier Transform. SIAM Press, Philadelphia (1992)

# Design of Parallel Processing System for Facial Image Retrieval

Hyung Lee[1], Kyung-Ae Moon[2], and Jong-Won Park[1]

[1]Chungnam National University, Taejeon, Korea
[2]Electronic and Telecommunications Research Institute, Taejeon, Korea
hyung@crow.chungnam.ac.kr

**Abstract.** Method of enhancing the processing speed of automatic recognition and retrieval of human faces was investigated. Here, we proposed a parallel program based on mesh method and a parallel processing system that is similar to an SIMD architecture. The simulation result yielded 61.9 times faster speed than the serial processing.

## 1 Algorithm for Facial Image Retrieval

Many studies have been done on building systems for the automatic recognition and retrieval of human faces[1]. However, many difficulties still exist. One of them is the lower speed of processing. In this paper, we propose a parallel processor system with MAMS(Multi-Access Memory System)[2] and an adaptable algorithm to this system. Images were allowed to have some limitations. Those remained in gray level without a background, those looking forward without glasses, and the part of facial components was not to be covered by the hair. In the first step, hair on the face scanned was eliminated and the region containing eyes was detected using the histogram of the gray intensity of an image after sobel edge operation. Based on the length between eyes, the region containing eyes, nose, and mouth was extracted. In order to extract the feature region with binary image, the former region was equalized and binarized using a combined technique of P_tile global binarization with local window based method. The feature region yielded the first feature pairing the length and the width. We then normalized the feature region to compare the same facial images with different sizes. The normalized region yielded 3 feature matrices as features and this was became the fourth feature. Each element of matrix, which is the second feature, was calculated for each 3x3 mesh of the normalized region. The third feature was calculated using 6x6 mesh.

## 2 Parallel Processing System

The design of parallel processor architecture is important for improving processing speed. We designed the parallel processing system, as depicted in Fig. 1-(a). The system was made up of a host computer, 144 processing elements, and MAMS[2].

P. Zinterhof, M. Vajteršic, A. Uhl (Eds.): ACPC'99, LNCS 1557, pp. 592-593, 1999.
© Springer-Verlag Berlin Heidelberg 1999

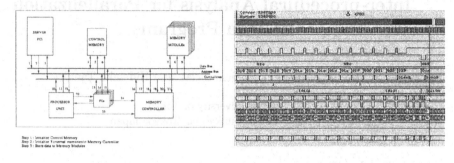

(a) block diagram                    (b) waveform

Fig. 1 Block diagram of parallel processing system and waveform generated by simulation

It was programmed in Verilog HDL, simulated by CADENCE Verilog-XL(a hardware simulation package), and synthesized to ALTERA FPGA devices.

# 3 Experimental Results and Conclusions

One hundred image scans were randomly acquired from an alumni album. For the test, 53 out of 100 images were randomly selected and their sizes were to four-ninth. The results were shown in Table 1 and the waveform of the system simulated is shown in Fig 1-(b).

Table 1. The results of facial retrieval

|  | The first order | The second order | Failure | Total |
|---|---|---|---|---|
| No. of images | 47 | 5 | 1 | 53 |
| retrieval rate(%) | 88.6 | 9.6 | 1.8 | 100 |

On the result of simulation, we compared serial processing with parallel one as following :

$$Tss = Tsr + (Tss1 + Tss2 + Tss3 + Tss4) \times 53 = T + 85899750 \ ns$$
$$Tps = T + Tps1 + Tps2 + Tps3 + Tps4 \times 5 + Tpci \times 2 = T + 1385720 \ ns.$$

The proposed parallel processing is 61.9 times faster than the serial processing.

# References

1. Ashok Samal and Prasana A. Iyenger : Automatic recognition and analysis of human faces and Facial expression : A survey, Pattern Recognition, Vol. 25, No. 1,(1992)65-77
2. J.W.Park : An Efficient memory system for image processing, IEEE Trans. Computers, Vol. C-35, No. 7, NewYork(1986)169-174

# Inter-procedural Analysis for Parallelization of Java Programs

Antonio Magnaghi, Shuichi Sakai, and Hidehiko Tanaka

The University of Tokyo
7-3-1 Hongo, Bunkyo-ku, Tokyo, Japan

## 1 Introduction

Parallelization of Java programs is a complex task due to inheritance, dynamic method dispatching and aliases. Our research [2] aims to perform static analysis of Java programs in order to identify implicit parallelism. In this paper, we discuss first the inter-procedural analysis technique we are studying and implementing to characterize data-dependency. And then we enhance this framework with type-based alias analysis.

## 2 Inter-procedural Analysis

Some of the fundamental characteristics of Object Oriented Programming (OOP), such as information hiding, lead us to stress the role of inter-procedural analysis for Java programs because access to objects relies heavily on method dispatching. Through the example below we demonstrate how implicit parallelism can be identified in source programs. Let us focus on instructions S1 and S2 of method p in class Y (using a notation borrowed from C++:

```
class Y {
private X myX;
void p() {
S1:myX.m();
S2:myX.n(3);}}
class X {
private char c;
private int i;
void m() {
S3:String s="X"+c;}
void n(int e) {
S4:i=e; }}
```

Y::p()). The member field myX of class Y is a reference object comprising two primitive member fields: c and i. In S1 and S2, the actions performed by methods m and n on the receiver myX can alter the state of its constituent fields. Method X::m() performs an assignment that uses as input the value of member field c. Hence we can conclude that: 1.) no alteration is produced on the state of the receiver by X::m(); 2.) the execution of X::m() requires member field c as input information. We model this situation by associating two sets to method X::m(), a set IN(X::m())={this.c} containing objects externally visible to the method and used as input to the task performed by the method.

And similarly, OUT(X::m())=∅ is the set of objects which are modified. The following sets are produced for X::n(int): IN(X::n(int))={e}, OUT(X::n(int))= {this.i}. Because it is not possible for the two member fields of myX (c and i) to be alias of one another, we conclude that S1 and S2 in Y::p do not interfere with each other even if they are invoked on the same object. Therefore S1 and S2 can be issued simultaneously in a multi-threaded manner.

P. Zinterhof, M. Vajteršic, A. Uhl (Eds.): ACPC'99, LNCS 1557, pp. 594–595, 1999.

# 3 Type-Based Pointer Analysis

In more general contexts than the example above, alias problems arise [1]. We can formalize our type-based approach for aliases as follows. Let $Class$ and $Object$ respectively be the sets containing all program classes, and all program objects. Let $Type$ be a function that returns the class type of an object. Let $Comp$ be a function from $Class$ to $2^{Class}$. It maps every class to the set of reference types encountered by recursively traversing the data structure of the input class. If $cls \in Class$, and $S$ is the set of nodes of the sub-tree rooted in $cls$ in the program taxonomy, then we designate $Comp^*(cls) = \bigcup_{i \in S} Comp(i)$. If we consider two objects in $Object$, $obj_1$ and $obj_2$, we assume that an alias may take place iff the following condition holds: $Comp^*(Type(obj_1)) \cap Romp^*(Type(obj_2)) \neq \emptyset$

# 4 Analyzer Structure

Current implementation produces dependency information based on the algorithms expressed above, conveniently addressing inheritance and method (constructor) overloading. The structure of the analyzer can be outlined as follows: *1.) First Pass:* the input program is parsed and information about call sides is collected. *2.) Second Pass:* it gathers additional information that becomes available only after parsing the whole input program. The following activities are carried out: call graph construction; topological sorting of methods (constructors) based on call graph analysis (the concept of $Transfer$ set is introduced as a generalization to OOP of the $Extension$ in [3]); evaluation of $IN$ and $OUT$ sets for every call site and method (constructor) by type-based alias analysis.

# 5 Conclusions

Preliminary evaluation has been carried out. On simple benchmarks like Linpack and Primes, we identified parallelism associated to immutable objects or stateless methods. Current research activity is improving the alias analysis algorithms by refining the type information associated with every program object.

# References

1. A. Diwan, K. McKinley, E. Moss. Type-based Alias Analysis. In *ACM SIGPLAN98 Conference on Programming Language Design and Implementation*, pp. 106-117, 1998.
2. A. Magnaghi, S. Sakai, H. Tanaka. An Inter-procedural Approach for Optimizations of Java Programs. In *Proceedings of the Information Processing Society of Japan*, vol. 57, pp. 299-300, 1998.
3. M. Rinard, P. Diniz. Commutativity Analysis: a New Analysis Technique for Parallelizing Compilers. *ACM Transactions on Programming Languages and Systems*, vol. 19, no. 6, pp. 942-991, 1997.
4. H. Zima, B. Chapman. Supercompilers for Parallel and Vector Computers. *ACM Press*, 1992.

# Fast Recursive Computation of Local Axial Moments by Using Primitive Kernel Functions

Roman M. Palenichka

Institute of Physics and Mechanics, Lviv, UKRAINE,
pal@vision.ipm.lviv.ua

**Abstract.** Computation of local axial moments from images in a running mode (per pixel) is computationally very extensive procedure, especially, when the local moments must be computed for many orientation angles and for a large window size at the same time. This paper considers efficient computation of local axial moments by using the concept of primitive kernel functions. It allows to reduce the computation time from $O(L^2)$ to $O(1)$ per pixel, where LxL is the window size.

The computation of many image local properties can be done by using a linear filtering, where the kernel of the filter defines a local property, e.g. the computation of local moments of different orders can be reduced to the linear filtering combined with other operations. In order to define some interesting cases of the filtering kernel function, the 1D linear filtering is considered. If the kernel is separable, then the generalization to 2D case is straightforward.

*Definition 1.* The discrete linear filtering of a function $g(i)$ with the kernel function $h(i)$ of the length $L$ (window size) is computed by the equation:

$$f(i) = \sum_{k=0}^{L-1} h(k) \cdot g(i - k),\qquad(1)$$

where the function $h(i)$ is sampled at equal spacing from the continuous kernel function $h(x)$, and $h(x)$ is a bounded function, i.e. $|h(x)| < \infty$.

*Definition 2.* A differentiable function with all bounded derivatives, the $p$-th derivative of which $h^{(p)}(x)$ is equal to a constant, $h^{(p)}(x) = const$, or to a constant multiplied by this same function, $h^{(p)}(x) = const \cdot h(x)$, is called a *primitive kernel function* of $p$-th order.

The first example of such a function is the function $x^p$. Another example of the primitive kernel functions of second order are the trigonometric functions $cos(x)$ and $sin(x)$.

*Theorem 1.* The computational complexity of the linear filtering algorithm by using the primitive kernel function of order $p$ does not depend on the kernel size $L$ (usually, $p << L$) and has the order of computation no more than $O(p^2)$ per point.

The proof is based on the Taylor series expansion in a current point $i$ and the definition of primitive kernel function, which provides recursive expressions

P. Zinterhof, M. Vajteršic, A. Uhl (Eds.): ACPC'99, LNCS 1557, pp. 596–597, 1999.
© Springer-Verlag Berlin Heidelberg 1999

for partial results. These recursive equations are reduced to the so-called basic primitive recursion in $O(p^2)$ steps. The basic primitive recursion, in turn, allows fast recursive averaging in a sliding window of L points in $O(1)$ operations per point.

*Definition 3.* The function of local axial moments $M_\alpha(i,j)$ relatively to the axis $\alpha$ is the linear filter of order $L \times L$, where $L = 2r + 1$ and $r$ is odd,

$$M_\alpha(i,j) = \sum_{k=0}^{L-1} \sum_{l=0}^{L-1} h_\alpha(k,l) \cdot g(i-k, j-l), \qquad (2)$$

where $h_\alpha(k,l)$ is the kernel function for the axial moment which depends on a chosen distance function $d(i)$ to the axis with the slope $\alpha$ and each value of $h_\alpha(k,l)$ denotes the distance of related point $(k.l)$ to the axis.

The computation of local axial moments per pixel according to Eq. (2) requires $L^2$ operations of multiplication and $L^2 - 1$ additions, i.e. its computational complexity is $O(L^2)$. However, it can be substantially reduced by using the results of application of the recursion principle and assuming a specific distance function.

*Theorem 2.* The computational complexity of the local axial moment function $M_\alpha(i,j)$ of size $L \times L$ points relatively to four main orientations (where $\alpha = 0, 1, 2, 3$) can be reduced to $O(p^2)$ operations per pixel for the four main axes of symmetry if the distance function $d(i)$ is a primitive kernel function of order $p$, where $p << L$.

Fast computation of the function of local axial moments for an arbitrary axis of symmetry is based on the so-called *rotational recursion* principle using the already calculated per point values of the main four axial moments.

*Theorem 3.* The computational complexity of the function $M_\alpha(i,j)$ of local axial moments for an arbitrary axis of symmetry with angle $\alpha$ has the order $O(1)$ per pixel if the number of pixels included between the current axis and the horizontal (or vertical) symmetry axis has the order $O(1)$ not including $L$ pixels lying on the horizontal (or vertical) axis, where $L \times L$ is the window size.

The fast recursive computation of axial moments is independent of the kernel size $L$ in Eq.(2) and yields significant saving of the run time, however it is not sufficient for many applications due to the large image size and large number of orientation angles. Two ways of parallel implementation have been investigated which are readily realisable in the framework of a SIMD/MIMD architecture of parallel computer systems: angle parallelism (different angles of orientation) and segment parallelism (different image segments). The angle parallelism should be used at small number of available processors, namely four processors is the minimal suitable number of processors. The segment parallelism is suitable at large number of available processors and large image size as well. Parallel systolic implementation is also applicable for the main orientations similar to the known systolic structure of central moment computation in a running mode. Its time complexity is $O(1)$ per point and space complexity equals $O(p^2)$, where $p$ is the order of primitive kernel function used as the distance function in Eq. (2).

# Speed Up Estimation for a Parallel Method for Systems of Linear Ordinary Differential Equations

Miron Pavluš

Deapartment of Mathematics, Technical University,
Vysokoškolská 4, 042 00 Košice, Slovakia
pavlus@tuke.sk
http://svfsun.tuke.sk/svf/pracoviska/km/mate/pavlus/index.html

**Abstract.** We determine the speed up of a recently developed parallel algorithm of solving of systems of linear ODEs on large parallel MIMD computers. The used numerical method for solving systems of linear ODEs is the Runge-Kutta method. An optimal number of subintervals (or processors) and an optimal number of equidistant points for an individual processor are assessed if a total interval is subdivided into N equal parts. It can be proven that the speed up is proportional to $N^{1/2}$.

## 1 Introduction

Recently Podlubny [1] suggested a parallel algorithm for solving linear ODE and their systems with non-constant coefficients; the method is based on the solutions of the so-called local problems. A modification of Podlubny's method for linear systems of ODEs with constant coefficients was given by Török [2]. One of the most important characteristic of each parallel algorithm – the speed up – was not investigated properly yet. This paper shows some results for the speed up of Podlubny's algorithm.

## 2 The Studied Method

Let us briefly remind the parallel algorithm for solving systems for the non-homogeneous linear ODEs which has been described in detail in [1]. We consider the following Cauchy problem:

$$Y'(x) = C(x)Y(x) + F(x) \tag{1}$$
$$Y(a) = Y_0 \tag{2}$$

where $C(x)$ is a given matrix which elements are $c_{ij}(x)$, $i, j = 1, 2, \ldots, m$; $F(x)$ and $Y_0$ are given vectors with elements $f_i(x), y_i$, respectively and $Y(x)$ is a wanted vector which elements are $y_i(x)$, $i = 1, 2, \ldots, m$. $Y(x)$, $F(x)$, $C(x)$ are continuous in the closed interval $[a, b]$. Let us divide $[a, b]$ in $nq$ subintervals:

$$a = x_0 < x_1 < x_2 < \ldots < x_{nq-1} < x_{nq} = b$$

P. Zinterhof, M. Vajteršic, A. Uhl (Eds.): ACPC'99, LNCS 1557, pp. 598–599, 1999.
© Springer-Verlag Berlin Heidelberg 1999

and look for the solution of the problem (1)-(2) for the interval $[a, b]$ in the following form:

$$Y(x) = U_i(x)A_i + V_i(x), x \in I_i, I_i = [x_{q(i-1)}, x_{qi}], (i = \overline{1, n}) \qquad (3)$$

where $U_{ik}(x)$ (the columns of $U_i(x)$) and $V_i(x)$ are the solutions of the following problems:

$$U'_{ik}(x) = C(x)U_{ik}(x), x \in I_i, U_{ik}(x) = e_k, (k = \overline{1, m}) \qquad (4)$$

$$V'_i(x) = C(x)V_i(x) + F(x), x \in I_i, V_i(x_{i-1}) = 0 \qquad (5)$$

$U_{ik}(x)$ and $V_i(x)$ are equal to the zero vector outside their subinterval $I_i$. We call $U_{ik}(x)$ and $V_i(x)$ the "local solutions". In (4) we denoted by $e_k$ a vector which all components are zeros but $k$-th one is equal to 1. The constant vector $A_i$ must be determined by satisfying the initial condition (2) and the condition of continuity of $Y(x)$ in $[a, b]$. The last condition reduces to satisfy continuity of $Y(x)$ for $x = x_i, (i = \overline{1, n})$.

## 3   Speed Up Estimation

Let us denote

$$A(m) = [1 + 12m + 4(2m^2 + \sum_{i=1}^{m} r_i + \sum_{i=1}^{m}\sum_{j=1}^{m} r_{ij})]/(2m^2) \qquad (6)$$

$$B(m) = [1 + 13m + 18m^2 + 8m^3 + 4\sum_{i=1}^{m} r_i + 4(m+1)\sum_{i=1}^{m}\sum_{j=1}^{m} r_{ij}]/(2m^2) \qquad (7)$$

where $r_i$ and $r_{ij}$ are the numbers of arithmetical operations needed for a single value of the functions $f_i(x)$ and $c_{ij}(x)$, respectively. It can be proven

**Theorem 1.** *If the interval $[a, b]$ is divided into $N = nq$ equal parts then for parallel performance of the problem (1)-(2) by RK method an optimal number of intervals $I_i = [x_{(i-1)q}, x_{iq}], i = \overline{1, n}$ (or processors) is equal to $n^* = (NB)^{1/2}$ and an optimal number of RK steps within each interval is equal to $q^* = (N/B)^{1/2}$. For large $N$ the speed up function value in the optimal point $(m, n^*, q^*)$ is approximately equal to*

$$\mathcal{F}(m, n^*, q^*) \approx \frac{A(m)}{2(B(m))^{1/2}} N^{1/2}$$

*where $A(m), B(m)$ are given by relations (6)-(7).*

## References

1. Podlubny, I.: Parallel Algorithms for Initial and Value problems for Linear Ordinary Differential Equations and Their Systems, Kybernetika, Prague, **32** 3 (1996) 251-260

2. Török, C.: On the Parallel Algorithms of Initial Value Problems for Systems of Linear ODEs, Preprints of the 113th Pannonian Applied Mathematical Meeting, Bardejovské Kúpele, (1996) 231-237

# Efficient Parallel Algorithms for Dense Cholesky Factorization

Eunice E. Santos and Pei-Yue Pauline Chu

Department of Electrical Engineering and Computer Science,
Lehigh University, Bethlehem, PA 18015, USA.

## 1 Introduction

Dense Cholesky factorization arises in area such as linear programming, boundary element methods and radiosity computations. In this paper we focus on designing efficient parallel algorithms for Cholesky factorization.

In order to take portability into account, we utilize LogP [1] in our performance analysis. LogP is a general parallel model utilizing point-to-point communication. The model captures important characteristics of parallel machines without taking into account specifics such as the interconnection network in order to deal with portability issues. The four parameters of LogP are: (**L**): latency; (**o**): overhead; (**g**): gap; and (**P**): number of processor/memory pairs. For ease of discussion, we assume that $L = g$ and $o = 0$.

## 2 Assumptions and Results

Given a dense symmetric positive definite matrix $A$ of size $n$ by $n$, there are 3! ways of executing Cholesky factorization algorithm by rearranging the order of the triple outer **for** loops. However, because of the inherent serial dependency of each Cholesky factor $L_{ij}$ on the component of rows less than $i$ and of columns less than $j$, [3] suggested that row processing incurs too many communications to be considered for parallelization. We concentrate on two algorithms using column processing which [2,3] had termed column-Cholesky (fan-in) and submatrix-Cholesky (fan-out) These algorithms use $CMOD$ and $CDIV$ primitives to modify the column and to do the column division respectively.

**Parallel Column-Cholesky "fan-in"** A processor, when processing data for each column $j$ of a matrix of size $n$ by $n$, computes aggregates to send to the owner of column $j$. Aggregates for column $j$ are the sum of multiplications within $CMOD$ operations. If the processor was allotted column $j$ then it waits to receive the aggregates for column $j$ from the other processors and also one from its own. For each aggregate array the processor receives, it is subtracted from column $j$. Finally, $CDIV$ operations are done to obtain the $L_{ij}$ for column $j$.

P. Zinterhof, M. Vajteršic, A. Uhl (Eds.): ACPC'99, LNCS 1557, pp. 600–602, 1999.

**Parallel Submatrix-Cholesky "fan-out"** When processing data for each column $j$ of a dense matrix, if the processor is allocated the column $j$, it performs $CDIV$ to get the $L[j]$ array and then sends it to all other processors. It also uses the result from $CDIV$ (i.e. $L[j]$) to perform $CMOD$ on those columns it owns which are greater than $j$. On the other hand, if it does not own column $j$, it then waits to receive the $L[j]$ from the processor which owns column $j$ and then perform $CMOD$ on those columns that the processor owns which are greater than column $j$.

**Parallel Issues** When designing and analyzing a parallel algorithm three important items must be taken into account: (i) local computation tasks, (ii) data layout, and (iii) the communication schedule.

**Local Computation** For both fan-in and fan-out, an inherent parallel computation lower bound is $\Omega(\frac{n^3}{P})$. For brevity, we have omitted an in-depth analysis which provided us with a precise computation lower bound.

**Data Layout** We define column-wrapped data layout as: Column $j$ is assigned to processor $P_i$ if $j = kP + i$ where $0 \le k \le \frac{n}{P} - 1$. We assume our data layout is column-wrapped and that there exists an integer $w$ where $n = wP$.

**Communication** For studying the communication lower bound, we computed the minimum number of messages sent and received respectively for both fan-in and fan-out. From our detailed analysis of the problem, we saw that the number of messages sent for both fan-in and fan-out is $\Omega(\frac{n^2}{P})$. Furthermore, we have devised a communication scheme which overlaps communication and computation. However, for brevity, this discussion has been omitted.

**Lower Bound** Clearly, the lower bound for both parallel Cholesky factorization is the maximum of the lower bounds between communication and local computation giving us $\Omega(\frac{n^3}{P} + \frac{n^2}{P}g)$.

**Parallel Running Time** We were able to design an algorithm with parallel running time of $O(\frac{n^3}{P} + g\frac{n^2}{P})$. For brevity, we have omitted the analysis of the precise running time as well as a discussion of our algorithm design.

# 3 Conclusion

Some existing research have found that using aggregates in **fan-in** has resulted in a better performance than **fan-out**. However, in our study we have found that the difference is insignificant due to the communication scheme. We plan to study how performance may be improved by exploring issues such as other data distribution methods such panel and block distributions.

# References

[1] D. Culler, R. Karp, D. Patterson, A. Sahay, E. Santos, K.Schauser, R. Subramonian and T. von Eicken. A Practical Model of Parallel Machine. Communications of ACM. November, 1996.
[2] J.W. Demmel, et al. Parallel Numerical Linear Algebra. ACTA Numerica, 1992.
[3] J.M.Ortega. Intro to Parallel and Vector solutions of Linear System. Plenum Press, 1988.

# Author Index

604     Author Index

# Springer
# and the
# environment

At Springer we firmly believe that an international science publisher has a special obligation to the environment, and our corporate policies consistently reflect this conviction.

We also expect our business partners – paper mills, printers, packaging manufacturers, etc. – to commit themselves to using materials and production processes that do not harm the environment. The paper in this book is made from low- or no-chlorine pulp and is acid free, in conformance with international standards for paper permanency.

Springer

# Lecture Notes in Computer Science

For information about Vols. 1–1473
please contact your bookseller or Springer-Verlag

Vol. 1510: J.M. Zytkow, M. Quafafou (Eds.), Principles of Data Mining and Knowledge Discovery. Proceedings, 1998. XI, 482 pages. 1998. (Subseries LNAI).

Vol. 1511: D. O'Hallaron (Ed.), Languages, Compilers, and Run-Time Systems for Scalable Computers. Proceedings, 1998. IX, 412 pages. 1998.

Vol. 1512: E. Giménez, C. Paulin-Mohring (Eds.), Types for Proofs and Programs. Proceedings, 1996. VIII, 373 pages. 1998.

Vol. 1513: C. Nikolaou, C. Stephanidis (Eds.), Research and Advanced Technology for Digital Libraries. Proceedings, 1998. XV, 912 pages. 1998.

Vol. 1514: K. Ohta, D. Pei (Eds.), Advances in Cryptology – ASIACRYPT'98. Proceedings, 1998. XII, 436 pages. 1998.

Vol. 1515: F. Moreira de Oliveira (Ed.), Advances in Artificial Intelligence. Proceedings, 1998. X, 259 pages. 1998. (Subseries LNAI).

Vol. 1516: W. Ehrenberger (Ed.), Computer Safety, Reliability and Security. Proceedings, 1998. XVI, 392 pages. 1998.

Vol. 1517: J. Hromkovič, O. Sýkora (Eds.), Graph-Theoretic Concepts in Computer Science. Proceedings, 1998. X, 385 pages. 1998.

Vol. 1518: M. Luby, J. Rolim, M. Serna (Eds.), Randomization and Approximation Techniques in Computer Science. Proceedings, 1998. IX, 385 pages. 1998.

1519: T. Ishida (Ed.), Community Computing and Support Systems. VIII, 393 pages. 1998.

Vol. 1520: M. Maher, J.-F. Puget (Eds.), Principles and Practice of Constraint Programming - CP98. Proceedings, 1998. XI, 482 pages. 1998.

Vol. 1521: B. Rovan (Ed.), SOFSEM'98: Theory and Practice of Informatics. Proceedings, 1998. XI, 453 pages. 1998.

Vol. 1522: G. Gopalakrishnan, P. Windley (Eds.), Formal Methods in Computer-Aided Design. Proceedings, 1998. IX, 529 pages. 1998.

Vol. 1524: G.B. Orr, K.-R. Müller (Eds.), Neural Networks: Tricks of the Trade. VI, 432 pages. 1998.

Vol. 1525: D. Aucsmith (Ed.), Information Hiding. Proceedings, 1998. IX, 369 pages. 1998.

Vol. 1526: M. Broy, B. Rumpe (Eds.), Requirements Targeting Software and Systems Engineering. Proceedings, 1997. VIII, 357 pages. 1998.

Vol. 1527: P. Baumgartner, Theory Reasoning in Connection Calculi. IX, 283. 1999. (Subseries LNAI).

Vol. 1528: B. Preneel, V. Rijmen (Eds.), State of the Art in Applied Cryptography. Revised Lectures, 1997. VIII, 395 pages. 1998.

Vol. 1529: D. Farwell, L. Gerber, E. Hovy (Eds.), Machine Translation and the Information Soup. Proceedings, 1998. XIX, 532 pages. 1998. (Subseries LNAI).

Vol. 1530: V. Arvind, R. Ramanujam (Eds.), Foundations of Software Technology and Theoretical Computer Science. XII, 369 pages. 1998.

Vol. 1531: H.-Y. Lee, H. Motoda (Eds.), PRICAI'98: Topics in Artificial Intelligence. XIX, 646 pages. 1998. (Subseries LNAI).

Vol. 1096: T. Schael, Workflow Management Systems for Process Organisations. Second Edition. XII, 229 pages. 1998.

Vol. 1532: S. Arikawa, H. Motoda (Eds.), Discovery Science. Proceedings, 1998. XI, 456 pages. 1998. (Subseries LNAI).

Vol. 1533: K.-Y. Chwa, O.H. Ibarra (Eds.), Algorithms and Computation. Proceedings, 1998. XIII, 478 pages. 1998.

Vol. 1534: J.S. Sichman, R. Conte, N. Gilbert (Eds.), Multi-Agent Systems and Agent-Based Simulation. Proceedings, 1998. VIII, 237 pages. 1998. (Subseries LNAI).

Vol. 1535: S. Ossowski, Co-ordination in Artificial Agent Societies. XV; 221 pages. 1999. (Subseries LNAI).

Vol. 1536: W.-P. de Roever, H. Langmaack, A. Pnueli (Eds.), Compositionality: The Significant Difference. Proceedings, 1997. VIII, 647 pages. 1998.

Vol. 1538: J. Hsiang, A. Ohori (Eds.), Advances in Computing Science – ASIAN'98. Proceedings, 1998. X, 305 pages. 1998.

Vol. 1539: O. Rüthing, Interacting Code Motion Transformations: Their Impact and Their Complexity. XXI,225 pages. 1998.

Vol. 1540: C. Beeri, P. Buneman (Eds.), Database Theory – ICDT'99. Proceedings, 1999. XI, 489 pages. 1999.

Vol. 1541: B. Kågström, J. Dongarra, E. Elmroth, J. Waśniewski (Eds.), Applied Parallel Computing. Proceedings, 1998. XIV, 586 pages. 1998.

Vol. 1542: H.I. Christensen (Ed.), Computer Vision Systems. Proceedings, 1999. XI, 554 pages. 1999.

Vol. 1543: S. Demeyer, J. Bosch (Eds.), Object-Oriented Technology ECOOP'98 Workshop Reader. 1998. XXII, 573 pages. 1998.

Vol. 1544: C. Zhang, D. Lukose (Eds.), Multi-Agent Systems. Proceedings, 1998. VII, 195 pages. 1998. (Subseries LNAI).

Vol. 1545: A. Birk, J. Demiris (Eds.), Learning Robots. Proceedings, 1996. IX, 188 pages. 1998. (Subseries LNAI).

Vol. 1546: B. Möller, J.V. Tucker (Eds.), Prospects for Hardware Foundations. Survey Chapters, 1998. X, 468 pages. 1998.

Vol. 1547: S.H. Whitesides (Ed.), Graph Drawing. Proceedings 1998. XII, 468 pages. 1998.

Vol. 1548: A.M. Haeberer (Ed.), Algebraic Methodology and Software Technology. Proceedings, 1999. XI, 531 pages. 1999.

Vol. 1551: G. Gupta (Ed.), Practical Aspects of Declarative Languages. Proceedings, 1999. VIII, 367 pgages. 1999.

Vol. 1553: S.T. Andler, J. Hansson (Eds.), Active, Real-Time, and Temporal Database Systems. Proceedings, 1997. VIII, 245 pages. 1998.

Vol. 1557: P. Zinterhof, M. Vajteršic, A. Uhl (Eds.), Parallel Computation. Proceedings, 1999. XV, 604 pages. 1999.

Vol. 1567: P. Antsaklis, M. Lemmon, A. Nerode, S. Sastry (Eds.), Hybrid Systems V. X, 445 pages. 1999.